High Dynamic Range Video
From Acquisition to Display and Applications

High Dynamic Range Video
From Acquisition to Display and Applications

Edited by

Frédéric Dufaux
CNRS, Télécom ParisTech

Patrick Le Callet
University of Nantes

Rafał K. Mantiuk
University of Cambridge

Marta Mrak
British Broadcasting Corporation

AMSTERDAM • BOSTON • HEIDELBERG • LONDON
NEW YORK • OXFORD • PARIS • SAN DIEGO
SAN FRANCISCO • SINGAPORE • SYDNEY • TOKYO
Academic Press is an imprint of Elsevier

Academic Press is an imprint of Elsevier
125 London Wall, London, EC2Y 5AS, UK
The Boulevard, Langford Lane, Kidlington, Oxford OX5 1GB, UK

Notices

Knowledge and best practice in this field are constantly changing. As new research and experience broaden our understanding, changes in research methods, professional practices, or medical treatment may become necessary.

Practitioners and researchers must always rely on their own experience and knowledge in evaluating and using any information, methods, compounds, or experiments described herein. In using such information or methods they should be mindful of their own safety and the safety of others, including parties for whom they have a professional responsibility.

To the fullest extent of the law, neither the Publisher nor the authors, contributors, or editors, assume any liability for any injury and/or damage to persons or property as a matter of products liability, negligence or otherwise, or from any use or operation of any methods, products, instructions, or ideas contained in the material herein.

Library of Congress Cataloging-in-Publication Data
A catalog record for this book is available from the Library of Congress

British Library Cataloguing-in-Publication Data
A catalogue record for this book is available from the British Library

ISBN: 978-0-08-100412-8

For information on all Academic Press publications
visit our website at https://www.store.elsevier.com/

Working together
to grow libraries in
developing countries

www.elsevier.com • www.bookaid.org

Contents

PART I   CONTENT ACQUISITION AND PRODUCTION

CHAPTER 2   Unified Reconstruction of Raw HDR Video Data 63

J. Unger, S. Hajisharif, J. Kronander

PART IV   DISPLAY

PART V   PERCEPTION AND QUALITY OF EXPERIENCE

Contributors

M.A. Abebe
Technicolor, Cesson-Sévigné, France

N. Adami
University of Brescia, Brescia, Italy

D. Agrafiotis
University of Bristol, Bristol, United Kingdom

L. Albani
Barco FIMI, Saronno (VA), Italy

M.A. Ali
University of Toronto, Toronto, ON, Canada

P. Andrivon
Technicolor, Cesson-Sevigne, France

A. Artusi
University of Girona, Girona, Spain

A. Badano
US Food and Drug Administration, Silver Spring, MD, United States

R. Boitard
IRISA, Rennes; Technicolor, Cesson-Sevigne, France

S. Bonfiglio
Barco FIMI, Saronno (VA), Italy

P. Bordes
Technicolor, Cesson-Sevigne, France

K. Bouatouch
IRISA, Rennes, France

R. Brémond
Paris Est University, Marne-la-Vallée, France

D.R. Bull
University of Bristol, Bristol, United Kingdom

A. Chalmers
University of Warwick, Coventry, United Kingdom

R. Cozot
IRISA, Rennes, France

M.P. Da Silva
University of Nantes, Nantes, France

S. Daly
Dolby Laboratories, Inc., San Francisco, CA, United States

N.-T. Dang
Paris Est University, Marne-la-Vallée, France

F. De Simone
Telecom ParisTech, CNRS LTCI, Paris, France

K. Debattista
University of Warwick, Coventry, United Kingdom

F. Dufaux
Telecom ParisTech, CNRS LTCI, Paris, France

G. Eilertsen
Linköping University, Norrköping, Sweden

S. Forchhammer
Technical University of Denmark, Bristol, Denmark

E. François
Technicolor, Cesson-Sevigne, France

J. Froehlich
Dolby Laboratories, Inc., San Francisco, CA, United States

O. Gallo
NVIDIA Research, Santa Clara, CA, United States

G. Guarnieri
University of Trieste, Trieste, Italy

F. Guerrini
University of Brescia, Brescia, Italy

S. Hajisharif
Linköping University, Norrköping, Sweden

B. Karr
Kennedy Space Center, FL, United States; University of Warwick, Coventry, United Kingdom

J. Korhonen
Technical University of Denmark, Bristol, Denmark

J. Kronander
Linköping University, Norrköping, Sweden

T. Kunkel
Dolby Laboratories, Inc., San Francisco, CA, United States

M.-C. Larabi
University of Poitiers, Poitiers, France

S. Lasserre
Technicolor, Cesson-Sevigne, France

P. Le Callet
University of Nantes, Nantes, France

F. Le Léannec
Technicolor, Cesson-Sevigne, France

C. Lèbre
Binocle, Paris, France

R. Leonardi
University of Brescia, Brescia, Italy

O. Letz
Binocle, Paris, France

C. Loscos
University of Reims Champagne-Ardenne, Reims, France

S. Mann
University of Toronto; Rotman School of Management CDL, Toronto, ON, Canada; Meta, Redwood City, CA, United States

C. Mantel
Technical University of Denmark, Bristol, Denmark

R.K. Mantiuk
University of Cambridge, Cambridge; Bangor University, Bangor, United Kingdom

I. Martin
University of Girona, Girona, Spain

S. Miller
Dolby Laboratories, Inc., San Francisco, CA, United States

M. Narwaria
University of Nantes, Nantes, France

M. Okuda
University of Kitakyushu, Kitakyushu, Japan

R.R. Orozco
University of Girona, Girona, Spain

T. Pouli
Technicolor, Cesson-Sévigné, France

Y. Pupulin
Binocle, Paris, France

G. Ramponi
University of Trieste, Trieste, Italy

E. Reinhard
Technicolor, Cesson-Sévigné, France

T. Richter
University of Stuttgart, Stuttgart, Germany

T. Ritschel
MPI Informatik; Saarland University, Saarbrücken, Germany

U. Seger
Robert Bosch GmbH, Leonberg, Germany

P. Sen
University of California, Santa Barbara, CA, United States

X. Shu
McMaster University, Hamilton, ON, Canada

J. Unger
Linköping University, Norrköping, Sweden

G. Valenzise
Telecom ParisTech, CNRS LTCI, Paris, France

C. Villa
Paris Est University, Marne-la-Vallée, France

X. Wu
McMaster University, Hamilton, ON, Canada

Y. Zhang
University of Bristol, Bristol, United Kingdom

Editor Biographies

Frédéric Dufaux is a Centre National de la Recherche Scientifique (CNRS) research director at Telecom ParisTech. He is also the Editor-in-Chief of *Signal Processing: Image Communication*.

Frédéric received his M.Sc. in physics and his Ph.D. in electrical engineering from the Swiss Federal Institute of Technology in Lausanne in 1990 and 1994, respectively. He has more than 20 years of experience in research, previously holding positions at the Swiss Federal Institute of Technology in Lausanne, Emitall Surveillance, Genimedia, Compaq, Digital Equipment, MIT, and Bell Labs. He has been involved in the standardization of digital video and imaging technologies, being a member of the MPEG and JPEG committees. He is the recipient of two ISO awards for his contributions. Frédéric is a fellow of IEEE. He was Vice General Chair of the 2014 International Conference on Image Processing. He is an elected member of the IEEE Image, Video, and Multidimensional Signal Processing Technical Committee and the IEEE Multimedia Signal Processing Technical Committee. He is also Chair of the European Association for Signal Processing (EURASIP) Special Area Team on Visual Information Processing.

His research interests include image and video coding, distributed video coding, 3D video, high dynamic range imaging, visual quality assessment, video surveillance, privacy protection, image and video analysis, multimedia content search and retrieval, and video transmission over wireless networks. He is the author or coauthor of more than 120 research publications and holds 17 patents issued or pending.

Patrick Le Callet is Professor at the University of Nantes/Polytech Nantes and leads research at the Institut de Recherche en Communications et Cybernétique de Nantes (IRCCyN)/Centre National de la Recherche Scientifique (CNRS) laboratory.

He received both an M.Sc. and a Ph.D. in image processing from École Polytechnique de l'Université de Nantes. He was also a student at École Normale Superieure de Cachan, where he sat the *aggrégation* (competitive examination) in electronics of the French national education system. He worked as an assistant professor from 1997 to 1999 and as a full-time lecturer from 1999 to 2003 in the Department of Electrical Engineering of the Technical Institute of the University of Nantes. Since 2003 he has taught at École Polytechnique de l'Université de Nantes (Engineering School) in the Electrical Engineering Department and the Computer Science Department, where is now a full professor. Since 2006, he has been the head of the Image and Video Communication Laboratory at CNRS IRCCyN, a group of more than 35 researchers. He is mostly engaged in research dealing with the application of human vision modeling in image and video processing. His interests currently center on 3D image and video quality assessment, watermarking techniques, and visual attention modeling and applications. He is the coauthor of more than 200 publications and communications and the co-inventor named on 13 international patents on these topics. He also cochairs within the Video Quality Expert Group, the HDR Group, and 3DTV activities. He is currently serving as an associate editor of *IEEE Transactions on Image Processing*, *IEEE Transactions on Circuit System and Video Technology*, *EURASIP Journal on Image and Video Processing*, and *SPIE Electronic Imaging*.

Rafał K. Mantiuk is a senior lecturer in the Computer Laboratory, University of Cambridge (UK). He received his Ph.D. from the Max Planck Institute for Computer Science (Germany, 2006), was a postdoctoral researcher at the University of British Columbia (Canada) and a lecturer at Bangor University (UK). He has published numerous journal articles and conference papers presented at ACM SIGGRAPH, Eurographics, IEEE Computer Vision and Pattern Recognition, and SPIE Human Vision and Electronic Imaging conferences, has been awarded several patents, and was recognized with the Heinz Billing Award (2006). He has made contributions to the field of high dynamic range imaging in the areas of video compression, tone mapping, and quality assessment. In his work he investigates how knowledge of the human visual system and perception can be incorporated within computer graphics and imaging algorithms. His recent interests focus on the design of imaging algorithms that adapt to human visual performance and viewing conditions in order to deliver the best images given limited resources, such as bandwidth, computation time, and display contrast.

Marta Mrak received Dipl. Ing. and M.Sc. in electronic engineering from the University of Zagreb, Croatia, and a Ph.D. from Queen Mary University of London, UK. Before joining the Research and Development Department at the BBC in 2010 to work on video compression research and the H.265/HEVC standardization, she was a postdoctoral researcher at the University of Surrey and Queen Mary University of London. In 2002, she was awarded a German Academic Exchange Service (DAAD) scholarship for video compression research at the Heinrich Hertz Institute, Germany. She has coauthored more than 100 articles, book chapters, and standardization contributions, and coedited the book *High-Quality Visual Experience* (Springer, 2010). She has been involved in several projects funded by European and UK research councils in roles ranging from researcher to scientific coordinator. She is a member of the Multimedia Signal Processing Technical Committee of the IEEE, a senior member of the IEEE, and an area editor of *Signal Processing: Image Communication*, and has been a guest editor of several special issues in relevant journals in the field.

Preface

Producing truly realistic video is widely seen as the holy grail toward further improving the quality of experience for end users of multimedia services. Additionally, numerous professional applications can be made possible only by the use of video signals with properties beyond the widespread, but very limited, parameter space, such as standard dynamic range. Several new approaches have been proposed and often successfully implemented in the history of the applications that use video data. Currently investigated directions include high spatial resolution, high frame rates, wide color gamut and high-bit-depth rendering.

The human visual system is able to perceive a wide range of colors and luminous intensities, as present in outdoor scenes in everyday life, ranging from bright sunshine to dark shadows. However, current traditional imaging technologies cannot capture nor reproduce such a broad range of luminance. The objective of high dynamic range (HDR) imaging is to overcome these limitations, hence leading to more realistic video content and a greatly enhanced user experience.

HDR applied to still images has been an active field of research and development for many years, especially for photography. However, its extension to video content has been considered only recently. Thanks to rapid technological advancements, commercial HDR video cameras and HDR monitors are becoming available. Nevertheless, the effective deployment of HDR video technologies involves redefining common interfaces for end-to-end content delivery, which in turn entails many technical and scientific challenges for both academia and industry.

This book provides an overview of recently researched technological directions for enabling HDR video in key application areas. More specifically, it covers the state of the art, discusses the effectiveness of various techniques, reviews some of the standardization efforts during the time of writing, and explores new research directions.

By providing a broad coverage, including general and advanced topics, the book should be of interest to a large readership with different backgrounds and expectations. Target readers include researchers, students, engineers, practitioners, and managers. This book has been edited with the aim for it to be the first reference on many topics that encompass HDR video, and to be a relevant handbook for those involved or interested in this field.

The aim is to cover numerous aspects required to establish HDR video systems, ranging from content acquisition to display technologies. Although several applications that use HDR video technologies are discussed, the range of topics covered provides a solid base for the establishment of new HDR application areas.

This book is composed of six parts. After an introduction to the background and fundamentals of HDR imaging, Part I addresses the problem of HDR *content acquisition and production*. Different approaches are described to capture HDR video, including the use of multiple synchronized sensors, spatial multiplexing of the sensor response, merging of multiple images captured with different exposure times, and capturing multiview HDR video. The impact of HDR on cinematographic shooting is also discussed.

Part II considers several aspects of *processing* for HDR video. In particular, video tone mapping operators are introduced and evaluated. Next, the way visual anomalies can be used to improve realism is studied. Finally, solutions for color management in the context of HDR imaging are described.

Efficient *representation and coding* of HDR video is another important topic, and is addressed in Part III. An overview of compression techniques for HDR video is given first. Initial standardization activities related to HDR and wide color gamut are then described. Lastly, the JPEG XT standard for still images, which is backward compatible with the JPEG standard, is presented.

To complete the chain of components in HDR video systems, the issue of *display* is discussed in Part IV. HDR displays are first characterized and modeled. Next, a dual modulation technique for LED-backlit HDR displays is presented.

As stated before, HDR imaging is the key element for further improvement of the end user experience. For this purpose, we dedicate a full part (Part V) to these aspects, discussing topics related to *perception and quality of experience*. The perceptual behavior of the human visual system in the context of HDR is discussed first, together with how this knowledge can be exploited to design a perceptually accurate imaging pipeline. The concept of quality of experience in the context of HDR is then carefully introduced, with practical methods and use cases to measure it also being provided. Finally, objective measures to predict HDR image and video quality are described as part of an important component of the quality of experience.

To conclude the book, we illustrate how HDR video technology has already impacted some application fields. Part VI provides such an overview, presenting several *applications* of HDR imaging in the automotive industry, medical imaging, spacecraft imaging, driving simulation, and watermarking.

The Editors
Frédéric Dufaux
CNRS, Télécom ParisTech
Patrick Le Callet
University of Nantes
Rafał K. Mantiuk
University of Cambridge
Marta Mrak
British Broadcasting Corporation

Acknowledgments

The editors gratefully acknowledge all the authors for their invaluable contributions. This book would not have been possible without their commitment and efforts.

Moreover, the editors express their deepest appreciation to the publisher, Elsevier, and in particular to Tim Pitts for helping to get this project off the ground and Charlie Kent for diligently following and supporting us throughout the preparation of this book.

The Editors
Frédéric Dufaux
Patrick Le Callet
Rafał K. Mantiuk
Marta Mrak

THE FUNDAMENTAL BASIS OF HDR: COMPARAMETRIC EQUATIONS

1

S. Mann*,†,‡, M.A. Ali*

University of Toronto, Toronto, ON, Canada
Rotman School of Management CDL, Toronto, ON, Canada†
Meta, Redwood City, CA, United States‡

CHAPTER OUTLINE

High Dynamic Range Video. http://dx.doi.org/10.1016/B978-0-08-100412-8.00001-2

1.1   INTRODUCTION TO HIGH DYNAMIC RANGE IMAGING

High dynamic range (HDR) imaging originated in the 1970s through digital eyeglass (DEG) and wearable computing as a seeing aid and, more generally, wearable technology as a sensory aid, which includes also HDR metasensing (the sensing of sensors and the sensing of their capacity to sense). HDR video as a way of seeing included other senses as well, such as HDR radar for the blind (Mann, 2001), and as a way of seeing radio waves with augmented reality overlays (Mann, 1992, 2001) (http://wearcam.org/swim/). Wearable computing gives rise to a rich sensory landscape that includes video, audio, and radar, plus physiological signals such as electrocardiogram and electroencephalogram, all of which made use of HDR signal capture and processing (Mann, 2001). This work is part of the general field of research and practice known as "sousveillance" (undersight), defined as "wearing and implanting various sensors, effectors, and multimedia computation in order to redefine personal space and modify sensory perception computationally" (Mann, 2004; Shakhakarmi, 2014).

1.1.1   THE FUNDAMENTAL CONCEPT OF HDR SENSING AND METASENSING

We begin by introducing the fundamental concept of HDR sensing, which allows the *dynamic range* of a sensor to be increased toward its *"dynamage range"* — that is, for a sensor to capture the full

FIGURE 1.1

Left: World's earliest known photograph, taken in 1826 on a plate coated in bitumen (petroleum tar, the asphalt commonly used as roofing material). Image from Wikimedia Commons. Center: A typical rooftop with modified bitumen shingles. Bitumen is commonly covered in stones or granules to protect it from damage by sunlight. Nevertheless, the south-facing side of the roof shows extensive damage due to exposure to sunlight. Right: Closeup showing sunlight-damaged southern exposure.

range of signals all the way to its limit, as defined as the limit to what it can sense without permanent damage to it.

Photography has a long and interesting history. For many years, it has been known that objects exposed to bright sunlight fade. Of particular note is the fact that bitumen (petroleum tar, asphalt, pitch, or the like), commonly used on rooftops, is easily damaged and becomes hard, brittle, and fragile with prolonged exposure to light (Fig. 1.1, center and right). The world's first known photograph was captured, in 1826, on a plate coated in bitumen (Fig. 1.1, left). Subsequently, various improvements to photography were made to make materials more sensitive to light, but the dynamic range of photographs was typically less than that of human vision. Similarly, the invention of motion pictures and of television gave rise to video capture, but also with similar problems regarding dynamic range.

Early video cameras (Fig. 1.2, left) used image pickup tubes that could be easily damaged by overexposure. As a result, the user must be careful with the f-stop on the camera lens, not to open it too far, and many cameras used to have an iris control button to protect the camera from damage:

> **Iris control button**: A feature that closes down the iris or aperture of the lens to protect the sensitive video camera tube when the camera is not in operation. Camera tubes that are exposed to overly bright light or sun develop a 'burn' that may become permanent" (Jack and Tsatsulin, 2002).

Also, "some lenses even have a 'C' setting after the highest f-stop which means the lens is completely closed, letting no light through at all" (Inman and Smith, 2009).

Because early video cameras were easily damaged by excessive light exposure, the user had to be careful with the f-stop on the camera lens, and open it up only enough to get a proper exposure.

HDR video was made possible by the invention of sensors (eg, camera image sensors) that can be overexposed without permanent damage. Unlike old vidicon cameras, for which there was less difference between the dynamic range and dynamage range, modern cameras have a much greater dynamage range than their dynamic range. This allows them to produce images that are massively

FIGURE 1.2

Left: Early video cameras (television cameras), such as this one at the 1936 Summer Olympics, used camera tubes that could be permanently damaged by overexposure. Image from Wikimedia Commons. Center and right: Early television camera lenses typically had an iris with a "C" setting, which means that the iris is completely closed. This was to protect the sensor from being damaged by light. When the camera was not in use, or was passing into a region of bright light, the lens could be closed to protect the sensor from damage. Generally, the camera operator had to be very careful to open up the iris only enough for a proper exposure, but not so far as to permanently damage the camera sensor.

overexposed, thus making it possible to capture images at an extreme range of exposures, and it is these extreme exposures that allow us to see extreme shadow detail (due to the massively overexposed images) and extreme highlight detail (due to having one or more massively underexposed images as well).

HDR allows us to increase the dynamic range, ideally, all the way up to being equal to the dynamage range.

1.1.2 THE FUNDAMENTAL PRINCIPLE OF HDR: DYNAMIC RANGE AND DYNAMAGE RANGE

HDR works whenever the *dynamage range* of a sensor exceeds its *dynamic range*.

Dynamic range and dynamage range are defined as follows:

Dynamic range is the ratio between the largest and smallest nonnegative quantity, such as magnitude, amplitude, energy, or the like, of sound, light, or the like, for which a small incremental difference in the quantity can still be sensed (ie, the range over which changes in the quantity remain discernible) (Mann et al., 2011, 2012).

Dynamage range is the ratio between the largest quantity that will not damage a sensor or device or receiver, and the smallest nonnegative quantity for which changes in the quantity remain discernible (Mann et al., 2011, 2012).

1.1.3 **HDR IMAGING TECHNIQUES**

HDR imaging is the set of techniques that computationally extend the usual or standard dynamic range of a signal. HDR imaging has arisen in multiple fields, such as computational photography, computer graphics, and animation. HDR signals may be produced in several ways: by the combining of multiple lower dynamic range signals for HDR reconstruction; synthetically, by simulation or raytracing; or by use of HDR sensors for data acquisition.

1.1.4 **HDR FROM MULTIPLE EXPOSURES**

According to Robertson et al. (2003), "the first report of digitally combining multiple pictures of the same scene to improve dynamic range appears to be (Mann, 1993)."

HDR imaging by reconstruction from multiple exposures is defined as follows:

> **Definition of HDR reconstruction:** The estimation of at least one photoquantity from a plurality of differently exposed images of the same scene or subject matter (Mann, 1993, 2000, 2001; Mann and Picard, 1995a; Ali and Mann, 2012; Robertson et al., 2003; Reinhard et al., 2005).

Specifically, HDR reconstruction returns an estimate of a photoquantity (or sequence of estimates in the case of video), $q(x, y)$ (any possibly spatially or temporally varying q or $q(x)$, $q(x, y, z)$, $q(x, y, t)$, or $q(x, y, z, t)$), on the basis of a plurality of exposures $f_i = f(k_i q(x, y))$, at exposure settings k_i, where there is also noise in each of these exposures f_i, through a camera response function, f, which is often unknown (although it may also be known, or it may be linear, or it may be the identity map). The exposure settings k_i may also be unknown.

A separate optional step of tone mapping the photoquantigraph, q, may be taken, if desired — for example, to produce an output image that can be printed or displayed on low dynamic range (LDR) output media. In situations where there is no need for a human-viewable HDR image (eg, HDR-based input to a computer vision system such as the wearable face-recognizer Mann, 1996b), the photoquantigraph may have direct use without the need to convert it to an LDR image.

A typical approach to generate q from f_i is to transform each of the input images f_i to estimates of that photoquantity, and then to combine the results with use of a weighted sum (Mann, 1993, 2000, 2001; Mann and Picard, 1995a; Debevec and Malik, 1997; Robertson et al., 2003). Other approaches are probabilistic in nature, and typically use nonlinear optimization (Ali and Mann, 2012; Pal et al., 2004).

1.2 **HISTORICAL MOTIVATION FOR HDR IMAGING**

HDR reconstruction from multiple exposures originated with author S. Mann (described as "the father of the wearable computer," IEEE International Solid-State Circuits Conference, February 7, 2000), through a childhood fascination with sensing and metasensing that led to the invention of the DEG (Fig. 1.3). This includes the use of wearable sensors to process and mediate daily life, from wearable technologies as a photographic art form (Mann, 1985; Ryals, 1995), to gesture-based augmented/augmediated reality (AR) (Mann, 1997b) for the capture, enhancement, and rendering of

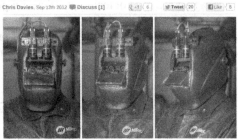

Traditional welding helmets use a sheet of smoked glass for the eyepiece, cutting down on the dangerous glare from the welding process itself, but also reducing overall visibility. The HDRrchitecture system, instead, processes images coming from one or more cameras, rendering a Full HD, 30fps stream with the brighter elements stripped out but the core details retained, all in real-time.

FIGURE 1.3

Author S. Mann with DEG and a DEG-based HDR welding helmet. Left: Original DEG, in comparison with a more recent commercial product. Right: Early DEG-based welding helmet prototype. The EyeTap principle was used to capture eyeward-bound light and process it with HDR video overlays, along with augmediated reality: augmented in dark areas with 3D computer-generated overlays that adapt to appear where they are least distracting, while the vision is deliberately diminished in bright areas of the electric welding arc.

everyday experience, a generalized form of self-sensing known as "sousveillance" ("undersight"), in contrast to the more familiar and common practice of "surveillance" ("oversight"). Sousveillance, as a field of research, has recently expanded greatly, and been given a variety of new names — for example, "lifelogging," "quantified self," "self-quantifying,' "self-quantification," "personal imaging," "personal informatics," "personal sensing," "self-tracking," "self-analytics," "autoveillance," "self-(sur/sous)veillance," "body hacking," "personal media analytics," and "personal informatics." Sousveillance also includes metaveillance. Metaveillance is the seeing of sight, visualization of vision, sensing of sensors, and sensing their capacity to sense (Fig. 1.4). According to Nicholas Negroponte, Founder, Director, and Chairman of the MIT Media Lab, "Steve Mann is the perfect example of someone... who persisted in his vision and ended up founding a new discipline." From Bangor Daily News — Sep. 26, 1997; later appears in Toronto Star — Jul. 8, 2001.

1.3 THEORY OF HDR IMAGING

The theory and practice of quantigraphic image processing, with comparametric equations, arose out of the field of sousveillance (wearable computing, quantimetric self-sensing, etc.) within the context of mediated reality (Mann, 1997a) and personal imaging (Mann, 1997b). However, it has potentially much more widespread applications in image processing than just the wearable photographic and videographic vision systems for which it was developed. Accordingly, a general formulation that does not necessarily involve a wearable camera system will be given. This section follows very closely, if not identically, that given in Mann (2000) and the textbook *Intelligent Image Processing* (Mann, 2001).

 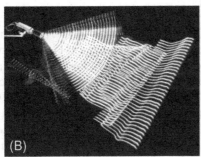 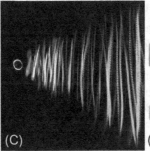 

(A)   (B)   (C)   (D)

FIGURE 1.4

Sousveillance: (A) Sequential Wave Imprinting Machine developed by author S. Mann in the 1970s and early 1980s for making invisible fields such as sound waves and radio waves, visible, and also to sense sensing itself (Mann, 1992). In this example, a television receiver ("rabbit ears" antenna) picked up the video signal from a wireless surveillance camera and amplified the signal with enough strength to directly drive an array of 35 electric light bulbs. (B) Because of video feedback, the lamps illuminated when they were within the field of view of the surveillance camera. The effect was an AR display rendering the camera's sightfield (Mann, 2014) visible to anyone watching the light wand being waved back and forth (no need for special eyewear). A problem with metasensing is the massive dynamic range present in the sightfield (not just the camera's field/angle of view, but also the wide variation in visual acuity, so rendered: more light meant more sight, but the light saturated and made it difficult to see differences in the amount of sight present at each point in space). (C) To overcome this limitation, pseudocolor metasensing was used. Here multiple-exposure photographs were taken through various colored filters while a metasensory device was moved through the space around the sensor (camera or human eye). Recent work on veillametrics (Janzen and Mann, 2014) also uses pseudocolor to render HDR metasensory images from a large collection of highly accurate scientific measurements of camera or human sight. (D) Author S. Mann, in 1996, with video (electrovisuogram), electrocardiogram, electroencephalogram, respiration, skin conductivity, and various other sensors that provide a highly complete capture, processing, recording, and transmission of physiological body function plus surrounding environmental information.

From Mann, S., 2001. Intelligent Image Processing. John Wiley and Sons, New York, p. 384.

1.3.1 THE WYCKOFF PRINCIPLE AND THE RANGE OF LIGHT

The quantity of light falling on an image sensor array, or the like, is a real-valued function $q(x, y)$ of two real variables x and y. An image is typically a degraded measurement of this function, where degradations may be divided into two categories, those that act on the domain (x, y) and those that act on the range q. Sampling, aliasing, and blurring act on the domain, while noise (including quantization noise) and the nonlinear response function of the camera act on the range q.

Registering and combining multiple pictures of the same subject matter will often result in an improved image of greater definition. There are four classes of such improvement:

1. increased spatial resolution (domain resolution),
2. increased spatial extent (domain extent),
3. increased tonal fidelity (range resolution), and
4. increased dynamic range (range extent).

1.3.2 WHAT'S GOOD FOR THE DOMAIN IS GOOD FOR THE RANGE

The notion of producing a better picture by combining multiple input pictures has been well studied with regard to the *domain* (x, y) of these pictures. Horn and Schunk (1981), for example, provided a means of determining optical flow, and many researchers have used this result to spatially *register* multiple images in order to provide a single image of increased spatial resolution and increased spatial extent. Subpixel registration methods such as those proposed by Irani and Peleg (1991) and Mann and Picard (1994) attempt to increase *domain resolution*. These methods depend on a slight (subpixel) shift from one image to the next. Image compositing (mosaicing) methods such as those proposed by Mann (1993), Mann and Picard (1995c), and Szeliski (1996) attempt to increase *domain extent*. These methods depend on large shifts from one image to the next.

Methods that are aimed at increasing *domain resolution* and *domain extent* tend to also improve tonal fidelity, to a limited extent, by virtue of a signal-averaging and noise-reducing effect. However, we will see in what follows a generalization of the concept of signal averaging called "quantigraphic signal averaging." This generalized signal averaging allows images of different exposure to be combined to further improve on tonal fidelity (*range resolution*), beyond improvements possible by traditional signal averaging. Moreover, the proposed method drastically increases dynamic range (*range extent*). Just as spatial shifts in the domain (x, y) improve the image, we will also see how exposure shifts (shifts in the range, q) can, with the proposed method, result in even greater improvements to the image.

1.3.3 EXTENDING DYNAMIC RANGE AND IMPROVEMENT OF RANGE RESOLUTION BY COMBINING DIFFERENTLY EXPOSED PICTURES OF THE SAME SUBJECT MATTER

The principles of quantigraphic image processing and the notion of the use of differently exposed pictures of the same subject matter to make a picture composite of extended dynamic range were inspired by the pioneering work of Wyckoff (1962, 1961), who invented so-called "extended response film."

Most everyday scenes have a far greater dynamic range than can be recorded on a photographic film or electronic imaging apparatus. However, a set of pictures that are identical except for their exposure collectively show us much more dynamic range than any single picture from that set, and also allow the camera's response function to be estimated, to within a single constant scalar unknown (Mann, 1993, 1996a; Mann and Picard, 1995b).

A set of functions

$$f_i(\mathbf{x}) = f(k_i q(\mathbf{x})), \tag{1.1}$$

where k_i are scalar constants, is known as a Wyckoff set (Mann, 1993, 1996a). A Wyckoff set of functions $f_i(\mathbf{x})$ describes a set of images differing only in exposure when $\mathbf{x} = (x, y)$ is the continuous spatial coordinate of the focal plane of an electronic imaging array (or piece of film), q is the quantity of light falling on the array (or film), and f is the unknown nonlinearity of the camera's (or combined film's and scanner's) response function. Generally, f is assumed to be a pointwise function (eg, invariant to $\mathbf{x}$).

1.3.4 **THE PHOTOQUANTITY, q**

The quantity, q, in Eq. (1.1), is called the "*photoquantigraphic* quantity (Mann, 1998)," or just the "photoquantity" (or "photoq") for short. This quantity is neither radiometric (eg, neither *radiance* nor *irradiance*) nor photometric (eg, neither *luminance* nor *illuminance*). Most notably, because the camera will not necessarily have the same spectral response as the human eye, or, in particular, that of the photopic spectral luminous efficiency function as determined by the CIE and standardized in 1924, q is not brightness, lightness, luminance, or illuminance. Instead, quantigraphic imaging measures the quantity of light integrated over the spectral response of the particular camera system,

$$q = \int_0^\infty q_s(\lambda)s(\lambda)\,d\lambda, \tag{1.2}$$

where $q_s(\lambda)$ is the quantity of light falling on the image sensor and s is the spectral sensitivity of an element of the sensor array. It is assumed that the spectral sensitivity does not vary across the sensor array.

1.3.5 **THE CAMERA AS AN ARRAY OF LIGHT METERS**

The quantity q reads in units that are quantifiable (eg, linearized or logarithmic), in much the same way that a photographic light meter measures in quantifiable (linear or logarithmic) units. However, just as the photographic light meter imparts to the measurement its own spectral response (eg, a light meter using a selenium cell will impart the spectral response of selenium cells to the measurement), quantigraphic imaging accepts that there will be a particular spectral response of the camera, which will define the quantigraphic unit q. Each camera will typically have its own quantigraphic unit. In this way, the camera may be regarded as an array of light meters, each being responsive to the quantigral:

$$q(x, y) = \int_0^\infty q_{ss}(x, y, \lambda)s(\lambda)\,d\lambda, \tag{1.3}$$

where q_{ss} is the spatially varying spectral distribution of light falling on the image sensor.

Thus, varying numbers of photons of lesser or greater energy (frequency times Planck's constant) are absorbed by a given element of the sensor array, and, over the temporal quantigration time of a single frame in the video sequence (or the exposure time of a still image) result in the photoquantity given by Eq. (1.3).

In the case of a color camera, or other color processes, $q(x, y)$ is simply a vector quantity. Color images may arise from as few as two channels, as in the old bichromatic (orange and blue) motion pictures, but more typically arise from three channels, or sometimes more as in the four-color offset printing, or even the high-quality Hexachrome printing process. A typical color camera might, for example, include three channels — for example, $[q_r(x, y), q_g(x, y), q_b(x, y)]$ — where each component is derived from a separate spectral sensitivity function. Alternatively, another space such as YIQ, YUV, or the like may be used, in which, for example, the Y (luminance) channel has full resolution and the U and V channels have reduced (eg, half in each linear dimension giving rise to one quarter the number of pixels) spatial resolution and reduced quantizational definition. Part III of this book covers representation and coding of HDR video in general.

In this chapter, the theory will be developed and explained for grayscale images, where it is understood that most images are color images, for which the procedures are applied either to the separate color channels or by way of a multichannel quantigrahic analysis. Thus, in both cases (grayscale and color) the continuous spectral information $q_s(\lambda)$ is lost through conversion to a single number q or to typically three numbers, q_r, q_g, q_b. Although it is easiest to apply the theory in this chapter to color systems having distinct spectral bands, there is no reason why it cannot also be applied to more complicated polychromatic, possibly tensor, quantigrals.

Ordinarily, cameras give rise to noise — for example, there is noise from the sensor elements and further noise within the camera (or equivalently noise due to film grain and subsequent scanning of a film, etc.). Thus, a goal of quantigraphic imaging is to attempt to estimate the photoquantity q in the presence of noise. Because $q_s(\lambda)$ is destroyed, the best we can do is to estimate q. Thus, q is the fundamental or "atomic" unit of quantigraphic image processing.

1.3.6 THE ACCIDENTALLY DISCOVERED COMPANDER

In general, cameras do not provide an output that varies linearly with light input. Instead, most cameras contain a dynamic range compressor, as illustrated in Fig. 1.5. Historically, the dynamic range compressor in video cameras arose because it was found that televisions did not produce a linear response to the video signal. In particular, it was found that early cathode ray screens provided a light output approximately equal to voltage raised to the exponent of 2.5. Rather than build a circuit into every television to compensate for this nonlinearity, a partial compensation (exponent of

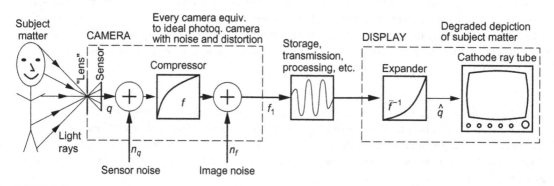

FIGURE 1.5

Typical camera and display. Light from subject matter passes through a lens (typically approximated with simple algebraic projective geometry, eg, an idealized "pinhole") and is quantified in units "q" by a sensor array where noise n_q is also added, to produce an output which is compressed in dynamic range by a typically unknown function f. Further noise n_f is introduced by the camera electronics, including quantization noise if the camera is a digital camera and compression noise if the camera produces a compressed output such as a JPEG image, giving rise to an output image $f_1(x, y)$. The apparatus that converts light rays into $f_1(x, y)$ is labeled "CAMERA." The image f_1 is transmitted or recorded and played back into a display system (labeled "DISPLAY"), where the dynamic range is expanded again. Most cathode ray tubes exhibit a nonlinear response to voltage, and this nonlinear response is the expander. The block labeled "expander" is generally a side effect of the display, and is not usually a separate device. It is depicted as a separate device simply for clarity. Typical print media also exhibit a nonlinear response that embodies an implicit "expander."

1/2.22) was introduced into the television camera at much lesser total cost because there were far more televisions than television cameras in those days before widespread deployment of video surveillance cameras and the like. Indeed, the original model of television is suggested by the names of some of the early players: American Broadcasting Corporation (ABC), National Broadcasting Corporation (NBC), etc. Names such as these suggest that they envisioned a national infrastructure in which there would be one or two television cameras and millions of television receivers.

Through a very fortunate and amazing coincidence, the logarithmic response of human visual perception is approximately the same as the inverse of the response of a television tube (eg, human visual response is approximately the same as the response of the television camera) (Poynton, 1996). For this reason, processing done on typical video signals will be on a perceptually relevant tone scale. Moreover, any quantization on such a video signal (eg, quantization into 8 bits) will be close to ideal in the sense that each step of the quantizer will have associated with it a roughly equal perceptual change in perceptual units.

Fig. 1.6 shows plots of the compressor (and expander) used in video systems together with the corresponding logarithm $\log(q + 1)$ and antilogarithm $\exp(q) - 1$ plots of the human visual system and its inverse. (The plots have been normalized so that the scales match.)

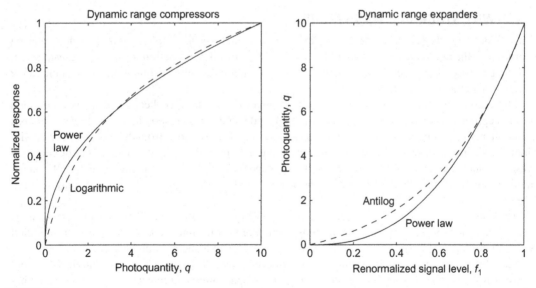

FIGURE 1.6

The power law dynamic range compression implemented inside most cameras has approximately the same shape of curve as the logarithmic function, over the range of signals typically used in video and still photography. Similarly, the power law response of typical cathode ray tubes, as well as that of typical print media, is quite similar to the antilogarithmic function. Therefore, the act of doing conventional linear filtering operations on images obtained from typical video cameras, or from still cameras taking pictures intended for typical print media, is, in effect, homomorphic filtering with an approximately logarithmic nonlinearity.

With images in print media, there is a similarly expansive effect in which the ink from the dots bleeds and spreads out on the printed paper, such that the midtones darken in the print. For this reason, printed matter has a nonlinear response curve similar in shape to that of a cathode ray tube (eg, the nonlinearity expands the dynamic range of the printed image). Thus, cameras designed to capture images for display on video screens have approximately the same kind of built-in dynamic range compression suitable for print media as well.

It is interesting to compare this naturally occurring (and somewhat accidental) development in video and print media with the deliberate introduction of companders (compressors and expanders) in the audio field. Both the accidentally occurring compression and expansion of picture signals and the deliberate use of logarithmic (or mu-law) compression and expansion of audio signals serve to allow 8 bits to be used to often encode these signals in a satisfactory manner. (Without dynamic range compression, 12–16 bits would be needed to obtain satisfactory reproduction.)

Most still cameras also have dynamic range compression built into the camera. For example, the Kodak DCS-420 and DCS-460 cameras capture images internally in 12 bits (per pixel per color) and then apply dynamic range compression, and finally output the range-compressed images in 8 bits (per pixel per color). Recently, as digital cameras have become dominant, range compression of images is still performed; however, modern cameras typically either emulate photographic film or perform computational tone mapping.

1.3.7 WHY STOCKHAM WAS WRONG

When video signals are processed, with linear filters, there is an implicit homomorphic filtering operation on the photoquantity. As should be evident from Fig. 1.5, operations of storage, transmission, and image processing occur between approximately reciprocal nonlinear functions of dynamic range compression and dynamic range expansion.

Many users of image-processing methods are unaware of this fact, because there is a common misconception that cameras produce a linear output, and that displays respond linearly. In fact there is a common misconception that nonlinearities in cameras and displays arise from defects and poor-quality circuits, when in fact these nonlinearities are fortuitously present in display media and deliberately present in most cameras. While CMOS and CCD response to light (electron counts) is usually linear, nonlinearities are introduced because of the need to reduce bit depth or produce display-referred images.

Thus, the effect of processing signals such as f_1 in Fig. 1.5 with linear filtering is, whether one is aware of it or not, homomorphic filtering; most computer vision cameras or RAW images from digital SLR cameras are linear, so in this case the assumption of linear camera output is correct.

Stockham advocated a kind of homomorphic filtering operation in which the logarithm of the input image was taken, followed by linear filtering (eg, linear space invariant filters), followed by the taking of the antilogarithm (Stockham, 1972).

In essence, what Stockham did not appear to realize is that such homomorphic filtering is already manifest in the application of ordinary linear filtering on ordinary picture signals (whether from video, film, or otherwise). In particular, the compressor gives an image $f_1 = f(q) = q^{1/2.22} = q^{0.45}$ (ignoring noise n_q and n_f) which has the approximate effect of $f_1 = f(q) = \log(q+1)$ (eg, roughly the same shape of curve, and roughly the same effect, eg, to brighten the midtones of the image before processing), as shown in Fig. 1.6. Similarly, a typical video display has the effect of undoing (approximately) this compression — for example, darkening the midtones of the image after processing with $\hat{q} = f^{-1}(f_1) = f_1^{2.5}$.

Thus, in some sense what Stockham did, without really realizing it, was to apply dynamic range compression to already range-compressed images, then do linear filtering, then apply dynamic range expansion to images being fed to already expansive display media.

1.3.8 THE VALUE OF DOING THE EXACT OPPOSITE OF WHAT STOCKHAM ADVOCATED

There exist certain kinds of image processing for which it is preferable to operate linearly on the photoquantity q. Such operations include sharpening of an image to undo the effect of the point spread function blur of a lens. It is interesting to note that many textbooks and articles that describe image restoration (eg, deblurring an image) fail to take into account the inherent nonlinearity deliberately built into most cameras.

What is needed to do this deblurring and other kinds of quantigraphic image processing is an *antihomomorphic filter*. The manner in which an antihomomorphic filter is inserted into the image-processing path is shown in Fig. 1.7.

Consider an image acquired through an imperfect lens that imparts a blurring to the image. The lens blurs the actual spatiospectral (spatially varying and spectrally varying) quantity of light $q_{ss}(x, y, \lambda)$, which is the quantity of light falling on the sensor array just before the light is *measured* by the sensor array:

$$\tilde{q}_{ss}(x, y, \lambda) = \int \int B(x - u, y - v)q_{ss}(u, v, \lambda)\, du\, dv. \tag{1.4}$$

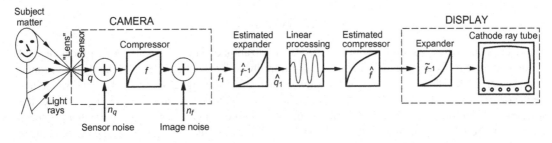

FIGURE 1.7

The antihomomorphic filter. Two new elements $\hat{f}^{-1}$ and $\hat{f}$ have been inserted as compared with Fig. 1.5. These are *estimates* of the inverse and forward nonlinear response functions of the camera. Estimates are required because the exact nonlinear response of a camera is generally not part of the camera specifications. (Many camera vendors do not even disclose this information if asked.) Because of noise in the signal f_1, and also because of noise in the estimate of the camera nonlinearity f, what we have at the output of $\hat{f}^{-1}$ is not q but, rather, an estimate, $\tilde{q}$. This signal is processed with linear filtering, and then the processed result is passed through the estimated camera response function, $\hat{f}$, which returns it to a compressed tone scale suitable for viewing on a typical television, computer, or the like, or for further processing.

This blurred spatiospectral quantity of light $\tilde{q}_{ss}(x, y, \lambda)$ is then photoquantified by the sensor array:

$$
\begin{aligned}
q(x, y) &= \int_0^\infty \tilde{q}_{ss}(x, y, \lambda)s(\lambda)\, d\lambda \\
&= \int_0^\infty \int_{-\infty}^\infty \int_{-\infty}^\infty B(x - u, y - v)q_{ss}(u, v, \lambda)s(\lambda)\, du\, dv\, d\lambda \\
&= \int_{-\infty}^\infty \int_{-\infty}^\infty B(x - u, y - v)\left(\int_0^\infty q_{ss}(u, v, \lambda)s(\lambda)\, d\lambda \right) du\, dv \\
&= \int_{-\infty}^\infty \int_{-\infty}^\infty B(x - u, y - v)q(u, v)\, du\, dv,
\end{aligned}
\tag{1.5}
$$

which is just the blurred photoquantity q.

Thus the antihomomorphic filter of Fig. 1.7 can be used to undo the effect of lens blur better than traditional linear filtering, which simply applies linear operations to the signal f_1 and therefore operates homomorphically rather than linearly on the photoquantity q.

Thus, we see that in many practical situations there is an articulable basis for doing exactly the opposite of what Stockham advocated (eg, expanding the dynamic range of the image before processing and compressing it afterward as opposed to what Stockham advocated, which was to compress the dynamic range before processing and expand it afterward).

1.3.9 USING DIFFERENTLY EXPOSED PICTURES OF THE SAME SUBJECT MATTER TO GET A BETTER ESTIMATE OF q

Because of the effects of noise (quantization noise, sensor noise, etc.), in practical imaging situations, the Wyckoff set that describes a plurality of pictures that differ only in exposure (1.1) should be rewritten as follows:

$$
f_i(\mathbf{x}) = f(k_i q(\mathbf{x}) + n_{q_i}) + n_{f_i},
\tag{1.6}
$$

where each image has, associated with it, a separate realization of a quantigraphic noise process n_q and an image noise process n_f, which includes noise introduced by the electronics of the dynamic range compressor f and other electronics in the camera that affect the signal *after* its dynamic range has been compressed. In the case of a digital camera, n_f also includes quantization noise (applied after the image has undergone dynamic range compression). Furthermore, in the case of a camera that produces a data-compressed output, such as the Kodak DC260, which produces JPEG images, n_f also includes data-compression noise (JPEG artifacts, etc., which are also applied to the signal after it has undergone dynamic range compression). Refer again to Fig. 1.5.

If it were not for noise, we could obtain the photoquantity q from any one of a plurality of differently exposed pictures of the same subject matter — for example as

$$
q = \frac{1}{k_i} f^{-1}(f_i),
\tag{1.7}
$$

where the existence of an inverse for f follows from the semimonotonicity assumption. Semimonotonicity follows from the fact that we expect pixel values to either increase or stay the same with increasing quantity of light falling on the image sensor.[1] However, because of noise, we obtain an advantage by capturing multiple pictures that differ only in exposure. The dark ("underexposed") pictures show us highlight details of the scene that would have been overcome by noise (eg, washed out) had the picture been "properly exposed." Similarly, the light pictures show us some shadow detail that would not have appeared above the noise threshold had the picture been "properly exposed."

Each image thus provides us with an estimate of the actual photoquantity q:

$$q = \frac{1}{k_i}\left(f^{-1}(f_i - n_{f_i}) - n_{q_i}\right), \tag{1.8}$$

where n_{q_i} is the quantigraphic noise associated with image i, and n_{f_i} is the image noise for image i. This estimate of q, $\hat{q}$, may be written as

$$\hat{q}_i = \frac{1}{\hat{k}_i}\hat{f}^{-1}(f_i), \tag{1.9}$$

where $\hat{q}_i$ is the estimate of q based on our considering image i, and $\hat{k}_i$ is the estimate of the exposure of image i based on our considering a plurality of differently exposed images. The estimated $\hat{q}_i$ is also typically based on an estimate of the camera response function f, which is also based on our considering a plurality of differently exposed images. Although we could just assume a generic function $f(q) = q^{0.45}$, in practice, f varies from camera to camera. We can, however, make certain assumptions about f that are reasonable for most cameras, such as the fact that f does not decrease when q is increased (that f is semimonotonic), and that it is usually smooth, and that $f(0) = 0$. In what follows, it will be shown how k and f are estimated from multiple differently exposed pictures. For the time being, let us suppose that they have been successfully estimated, so that we can calculate $\hat{q}_i$ from each of the input images i. Such calculations, for each input image i, give rise to a plurality of estimates of q, which in theory would be identical, were it not for noise. However, in practice, because of noise, the estimates $\hat{q}_i$ are each corrupted *in different ways*. Therefore, it has been suggested that multiple differently exposed images may be combined to provide a single estimate of q which can then be turned into an image of greater dynamic range, greater tonal resolution, and lesser noise (Mann, 1993, 1996a). In particular, the criteria under which collective processing of multiple differently exposed images of the same subject matter will give rise to an output image which is acceptable at every point (x, y) in the output image, are summarized as follows:

The Wyckoff signal/noise criteria: $\forall (x_0, y_0) \in (x, y), \quad \exists\, k_i q(x_0, y_0)$ such that

1. $k_i q(x_0, y_0) \gg n_{q_i}$ and
2. $c_i(q(x_0, y_0)) \gg c_i\left(\frac{1}{k_i}f^{-1}(n_{f_i})\right)$.

The first criterion indicates that for every pixel in the output image, at least one of the input images provides sufficient exposure at that pixel location to overcome sensor noise, n_{q_i}. The second criterion states that at least one input image provides an exposure that falls favorably (eg, is neither overexposed nor underexposed) on the response curve of the camera, so as not to be overcome by camera noise n_{f_i}.

The manner in which differently exposed images of the same subject matter are combined is illustrated, by way of an example involving three input images, in Fig. 1.8.

---

[1] Except in rare instances where the illumination is so intense as to damage the imaging apparatus — for example, when the sun burns through photographic negative film and appears black in the final print or scan.

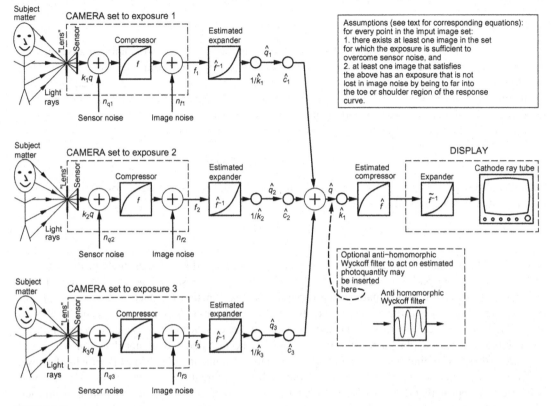

FIGURE 1.8

The Wyckoff principle. Multiple differently exposed images of the same subject matter are captured by a single camera. In this example there are three different exposures. The first exposure (CAMERA set to exposure 1), gives rise to an exposure k_1q, the second exposure gives rise to an exposure k_2q, and the third exposure gives rise to an exposure k_3q. Each exposure has a different realization of the same noise process associated with it, and the three noisy pictures that the camera provides are denoted f_1, f_2, and f_3. These three differently exposed pictures constitute a noisy Wyckoff set. To combine them into a single estimate, the effect of f is undone with an estimate $\hat{f}$ that represents our best guess of what the function f is. While many video cameras use something close to the standard $f = kq^{0.45}$ function, it is preferable to attempt to estimate f for the specific camera in use. Generally, this estimate is made together with an estimate of the exposures k_i. After the dynamic ranges have been reexpanded with $\hat{f}^{-1}$, the inverse of the estimated exposures $1/\hat{k}_i$ is applied. In this way, the darker images are made lighter and the lighter images are made darker, so that they all (theoretically) match. At this point the images will all appear as if they were taken with identical exposure, except that the pictures that were brighter to start with will be noisy in lighter areas of the image and those that were darker to start with will be noisy in dark areas of the image. Thus, rather than application of ordinary *signal averaging*, a weighted average is taken. The weights are the spatially varying *certainty functions*, $c_i(x, y)$. These certainty functions are the derivative of the camera response function shifted up or down by an amount k_i. In practice, because f is an estimate, so is c_i, so it is denoted $\hat{c}_i$ in this figure. The weighted sum is $\hat{q}(x, y)$, the estimate of the photoquantity $q(x, y)$. To view this quantity on a video display, it is first adjusted in exposure, and may be adjusted to an exposure level different from any of the exposure levels used in the taking of the input images. In this figure, for illustrative purposes, it is set to the estimated exposure of the first image, $\hat{k}_1$. The result is then range-compressed with $\hat{f}$ for display on an expansive medium (DISPLAY).

Moreover, it has been shown (Mann and Picard, 1995b) that the constants k_i as well as the unknown nonlinear response function of the camera can be determined, up to a single unknown scalar constant, given nothing more than two or more pictures of the same subject matter, in which the pictures differ only in exposure. Thus, the reciprocal exposures used to tonally register (tonally align) the multiple input images are estimates, $1/\hat{k}_i$, in Fig. 1.8. These exposure estimates are generally made by application of an estimation algorithm to the input images, either while f is simultaneously estimated or as a separate estimation process (because f has to be estimated only once for each camera but the exposure k_i is estimated for every picture i that is taken).

Owing to the large dynamic range that some Wyckoff sets can cover, small errors in f tend to have adverse effects on the overall estimate $\hat{q}$. Thus, it may be preferable to estimate f as a separate process (eg, by the taking of hundreds of exposures with the camera under computer program control). Once f is known (previously measured), k_i can be estimated for a particular set of images.

The final estimate for q, depicted in Fig. 1.8, is given by

$$\hat{q}(x,y) = \frac{\sum_i \hat{c}_i \hat{q}_i}{\sum_i \hat{c}_i} = \frac{\sum_i \frac{\hat{c}_i(\hat{q}(x,y))}{\hat{k}_i} \hat{f}^{-1}(f_i(x,y))}{\sum_i \hat{c}_i(\hat{q}(x,y))}, \tag{1.10}$$

where $\hat{c}_i$ is given by

$$\hat{c}_i(\log(q(x,y))) = \frac{df_i(x,y)}{d\log\hat{q}(x,y)} = \frac{d\hat{f}(\hat{k}_i\hat{q}(x,y))}{d\log\hat{q}(x,y)}, \tag{1.11}$$

from which we can see that $c_i(\log q)$ are just shifted versions of $c(\log q)$ — for example, dilated versions of $c(q)$. While this analysis is useful for insight into the process, the certainty and uncertainty functions in this form ignore other sources of noise (eg, photon noise, readout noise), which are dominant for modern cameras. See Granados et al. (2010) for details on handling camera noise.

The intuitive significance of the certainty function is that it captures the slope of the response function, which indicates how quickly the output (pixel value or the like) of the camera varies for given input. In the case of a noisy camera, especially a digital camera, where quantization noise is involved, generally the output of the camera will be most reliable where it is most sensitive to a fixed change in input light level. This point where the camera is most responsive to changes in input is at the peak of the certainty function, c. The peak in c tends to be near the middle of the camera's exposure range. On the other hand, where the camera exposure input is extremely large or small (eg, the sensor is very much overexposed or very much underexposed), the change in output for a given input is much less. Thus, the output is not very responsive to the input, and the change in output can be easily overcome by noise. Thus, c tends to fall off toward zero on either side of its peak value.

The certainty functions are functions of q. We may also write the uncertainty functions, which are functions of pixel value in the image (eg, functions of gray value in f_i), as

$$U(x,y) = \frac{dF(f_i(x,y))}{df_i(x,y)}, \tag{1.12}$$

and its reciprocal is the certainty function C in the domain of the image (eg, the certainty function in *pixel coordinates*):

$$C(x,y) = \frac{df_i(x,y)}{dF(f_i(x,y))}, \tag{1.13}$$

where $F^{-1} = \log f$. Note that C is the same for all images (eg, for all values of image index i), whereas c_i was defined separately for each image. For any i, the function c_i is a shifted (dilated) version of any other certainty function, c_j, where the shift (dilation) depends on the log exposure, K_i (the exposure k_i).

The final estimate of q (1.10) is simply a weighted sum of the estimates from q obtained from each of the input images, where each input image is weighted by the certainties in that image.

1.3.10 EXPOSURE INTERPOLATION AND EXTRAPOLATION

The architecture of this process is shown in Fig. 1.9, which depicts an image acquisition section (in this illustration, of three images), followed by an analysis section (to estimate q), followed by a resynthesis section to generate an image again at the output (in this case four different possible output images are shown).

The output image can look like any of the input images, but with improved signal-to-noise ratio, better tonal range, better color fidelity, etc. Moreover, an output image can be an interpolated or extrapolated version in which it is lighter or darker than any of the input images. This process of interpolation or extrapolation provides a new way of adjusting the tonal range of an image. The process is illustrated in Fig. 1.9. The image synthesis portion may also include various kinds of deblurring operations, as well as other kinds of image sharpening and lateral inhibition filters to reduce the dynamic range of the output image without loss of fine details, so that it can be printed on paper or presented to an electronic display in such a way as to have optimal tonal definition.

1.4 COMPARAMETRIC IMAGE PROCESSING: COMPARING DIFFERENTLY EXPOSED IMAGES OF THE SAME SUBJECT MATTER

As previously mentioned, comparison of two or more differently exposed images may be done to determine q, or simply to tonally register the images without determining q. Also, as previously mentioned, tonal registration is more numerically stable than estimation of q, so there are some advantages to comparametric analysis and comparametric image processing in which one of the images is selected as a reference image and others are expressed in terms of this reference image, rather than in terms of q. Typically the dark images are lightened and/or the light images are darkened so that all the images match the selected reference image. In such lightening and darkening operations, full precision is retained for further comparametric processing. Thus, all but the reference image will be stored as an array of floating point numbers.

1.4.1 MISCONCEPTIONS ABOUT GAMMA CORRECTION

So-called *gamma correction* (raising the pixel values in an image to an exponent) is often used to lighten or darken images. While gamma correction does have important uses, such as lightening or darkening images to compensate for incorrect display settings, it will now be shown that when one uses gamma correction to lighten or darken an image to compensate for incorrect exposure that, whether one is aware of it or not, one is making an unrealistic assumption about the camera response function.

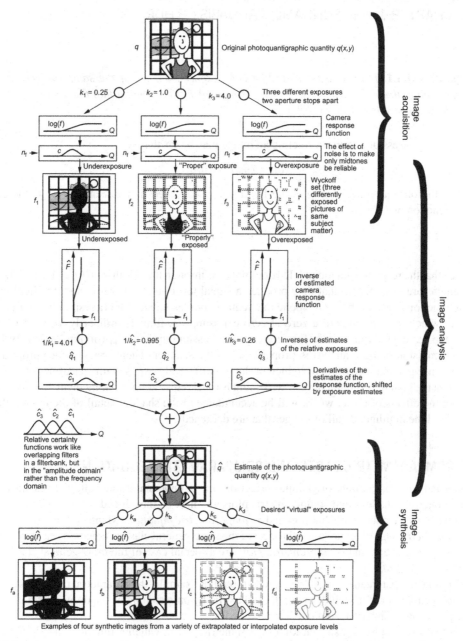

FIGURE 1.9

Quantigraphic exposure adjustment on a Wyckoff set. Multiple (in this example, three) differently exposed images are acquired. Estimates of q from each image are obtained and combined by weighted sum. Weights are estimates of the certainty function shifted along the exposure axis by an amount given by the estimated exposure for each image. From the estimated photoquantity $\hat{q}$, one or more output images may be generated by multiplication by the desired synthetic exposure and the passing of the result through the estimated camera nonlinearity. In this example, four synthetic pictures are generated. These are extrapolated and interpolated versions of the input exposures. The result is a "virtual camera" (Mann, 1999) where a picture can be generated as if the user were free to select the original exposure settings that had been used on the camera originally taking the input images.

Proposition 1.4.1. *Tonally registering differently exposed images of the same subject matter by gamma correcting them with exponent* $\gamma = k^\Gamma$ *is equivalent to assuming that the nonlinear response function of the camera is* $f(q) = \exp(q^\Gamma)$.

Proof. The process of gamma correcting an image may be written as

$$g(q) = f(kq) = (f(q))^\gamma, \tag{1.14}$$

where f is the original image and g is the lightened or darkened image. Solving for f, the camera response function, we obtain

$$f(q) = \exp q^\Gamma. \tag{1.15}$$

$\square$

We see that the response function (1.15) does not pass through the origin — $f(0) = 1$, not 0. Because most cameras are designed so that they produce a signal level output of zero when the light input is zero, the function $f(q)$ does not correspond to a realistic or reasonable camera response function. Even a medium which does not record a zero at zero exposure (eg, film) is ordinarily scanned in such a way that the scanned output is zero for zero exposure, assuming that d_{min} (minimum density for the particular emulsion being scanned) is properly set in the scanner. Therefore, it is inappropriate and incorrect to use gamma correction to lighten or darken differently exposed images of the same subject matter when the goal of this lightening or darkening is tonal registration (making them look the "same," apart from the effects of noise, which will be accentuated in the shadow detail of the images that are lightened and the highlight detail of images that are darkened).

1.4.2 COMPARAMETRIC PLOTS AND COMPARAMETRIC EQUATIONS

To understand the shortcomings of gamma correction, and to understand some alternatives, the concept of comparametric equations and comparametric plots will now be introduced.

Eq. (1.14) is an example of what is called a *comparametric equation* (Mann, 1999).

Comparametric equations are a special case of the more general class of equations called *functional equations* (Aczél, 1966), and *comparametric plots* are a special case of the more general class of plots called *parametric plots*.

The notion of a parametric plot is well understood. For example, the parametric plot $(r \cos q, r \sin q)$ is a plot of a circle of radius r. It does not depend explicitly on q so long as the domain of q includes at least all points on the interval from 0 to 2π, modulo 2π.

A comparametric plot is a special kind of parametric plot in which a function f is plotted against itself, and in which the parameterization of the ordinate is a linearly scaled parameterization of the abscissa.

More precisely, the comparametric plot is defined as follows:

Definition 1.4.1. A plot along coordinates $(f(q), f(kq))$ is called a *comparametric plot* (Mann, 1999) of the function $f(q)$.

A function $f(q)$ has a family of comparametric plots, one for each value of the constant k, which is called the *comparametric ratio*.

Proposition 1.4.2. *When a function f(q) is monotonic, the comparametric plot (f(q), f(kq)) can be expressed as a monotonic function g(f) not involving q.*

Thus, the plot in Definition 1.4.1 may be rewritten as a plot $(f, g(f))$ not involving q. In this form, the function g is called the *comparametric function*, and expresses the range of the function $f(kq)$ as a function of the range of the function $f(q)$ independently of the domain, q, of the function f.

The plot g defines what is called a *comparametric equation*:

Definition 1.4.2. Equations of the form $g(f(q)) = f(kq)$ are called comparametric equations (Mann, 1999).

A better understanding of comparametric equations may be had if we refer to the following diagram:

$$
\begin{array}{ccc}
 & k & \\
q & \longrightarrow & kq \\
f \downarrow & & \downarrow f \\
f(q) & \longrightarrow & f(kq) \\
 & g &
\end{array}
\tag{1.16}
$$

wherein it is evident that there are two equivalent paths to follow from q to $f(kq)$,

$$g \circ f = f \circ k. \tag{1.17}$$

Eq. (1.17) may be rewritten as

$$g = f \circ k \circ f^{-1}, \tag{1.18}$$

which provides an alternative definition of a *comparametric equation* to that given in Definition 1.4.2.

Eq. (1.14) is an example of a comparametric equation, and Eq. (1.15) is a solution of Eq. (1.14).

It is often preferable that comparametric equations be on the interval from 0 to 1 in the range of f. Equivalently stated, we desire comparametric equations to be on the interval from 0 to 1 in the domain of g and the range of g. In this case, the corresponding plots and equations are said to be *unicomparametric*. (Actual images typically range from 0 to 255 and must thus be rescaled so that they range from 0 to 1, for unicomparametric image processing.)

Often we also impose a further constraint that $f(0) = 0$, and the constraint that $g(0) = 0$ and $g(1) = 1$.

Solving a comparametric equation is equivalent to determining the unknown camera response function from a pair of images that differ only in exposure when the comparametric equation represents the relationship between gray values in the two pictures and the comparametric ratio, k, represents the ratio of exposures (eg, if one picture was taken with twice the exposure of the other, then $k = 2$).

1.4.3 ZETA CORRECTION OF IMAGES

An alternative to *gamma correction* is proposed. This alternative, called *zeta correction*, will also serve as another example of a comparametric equation.

For zeta correction, we simply adjust the exponential solution (1.15) of the comparametric equation given by traditional gamma correction so that the solution passes through the origin: $f(q) = \exp(q - 1)$. (For simplicity, and without loss of generality, γ has been set to k, the comparametric exposure ratio.)

Thus, $g = (bf + 1)^\gamma = (bf + 1)^k$. Preferably (to be unicomparametric) we would like to have $g(1) = 1$, so we use the response function

$$f(q) = \frac{e^q - 1}{\sqrt[k]{2} - 1}, \tag{1.19}$$

which is a solution to the corresponding comparametric equation:

$$g = ((\sqrt[k]{2} - 1)f + 1)^k - 1, \quad \forall k \neq 0. \tag{1.20}$$

The comparametric equation (1.20) forms the basis for zeta correction of images:

$$g = \begin{cases} \left((2^\zeta - 1)f + 1\right)^{\frac{1}{\zeta}} - 1, & \forall \, \zeta \neq 0. \\ 2^f - 1, & \text{for } \zeta = 0. \end{cases} \tag{1.21}$$

Implicit in zeta correction of images is the assumption of an exponential camera response function, which, although not realistic (given that the exponential function expands dynamic range, and most cameras have compressive response functions rather than expansive response functions), is preferable to gamma correction because of the implicit notion of a response function for which $f(0) = 0$.

With standard IEEE arithmetic, values of ζ can range from approximately -50 to $+1000$.

1.4.4 THE AFFINE COMPARAMETRIC EQUATION AND AFFINE CORRECTION OF IMAGES

In this section, one of the two most useful (in the authors' opinion) comparametric equations is introduced, giving rise to *affine correction* of images. Affine correction is an improvement over zeta correction (which itself was an improvement over gamma correction).

First consider the classic model

$$f(q) = \alpha + \beta q^\gamma \tag{1.22}$$

used by photographers to characterize the response of a variety of photographic emulsions, including so-called *extended response* film (Wyckoff, 1962). It is well known that Eq. (1.22) becomes the equation of a straight line when expressed in logarithmic coordinates if we subtract α (as many scanners such as Photo CD scanners attempt to do by prompting the user to scan a piece of blank film from the film trailer before scanning the rest of the roll of film):

$$\log(f(q) - \alpha) = \gamma \log q + \beta. \tag{1.23}$$

It is an interesting coincidence that the comparametric plot of this function (1.22) is also a straight line:

Proposition 1.4.3. *The comparametric plot corresponding to the standard photographic response function (1.22) is a straight line. The slope is k^γ, and the intercept is $\alpha(1 - k^\gamma)$.*

Proof. $g(f(kq)) = f(kq) = \alpha + \beta(kq)^\gamma$. Rearranging the equation to eliminate q gives $g = k^\gamma(\alpha + \beta q^\gamma) + \alpha(1 - k^\gamma)$, so

$$g = k^\gamma f + \alpha(1 - k^\gamma). \qquad (1.24)$$

$\square$

The constant β does not appear in the comparametric equation and thus we cannot determine β from the comparametric equation. The physical (intuitive) interpretation is that we can determine the nonlinear response function of a camera only up to a single unknown scalar constant.

Eq. (1.14) looks quite similar in form to Eq. (1.22), and in fact is identical if we set $\alpha = 0$ and $\beta = 1$. However, one must recall that Eq. (1.14) is a comparametric equation and Eq. (1.22) is a solution to a (different) comparametric equation; thus, we must be careful not to confuse the two. The first corresponds to gamma correction of an image, while the second corresponds to the camera response function that is implicit in the application of Eq. (1.24) to lighten or darken the image. To make this distinction clear, application of Eq. (1.24) to lighten or darken an image will be called *affine correction* (eg, correction by the modeling of the comparametric function with a straight line). The special case of *affine correction* when the intercept is equal to zero will be called *linear correction*.

Preferably affine correction of an image also includes a step of clipping values greater than 1 to 1, and values less than 0 to 0, in the output image:

$$g = \min(\max(k^\gamma f + \alpha(1 - k^\gamma), 0), 1). \qquad (1.25)$$

If the intercept is 0 and the slope is greater than 1, the effect, disregarding noise, of Eq. (1.25) is to lighten the image in a natural manner that properly simulates the effect of the image having been exposed with greater exposure. In this case, the effect is theoretically identical to that which would have been obtained by the use of a greater exposure on the camera, assuming the response function of the camera follows the power law $f = q^\gamma$, as is the case for many cameras. Thus, it has been shown that the correct way to lighten an image is to apply linear correction, not gamma correction (apart from correction of an image to match an incorrectly adjusted display device or the like, where gamma correction is still the correct operation to apply).

Here we have worked forward, starting with the solution (1.22) and deriving the comparametric equation (1.24), of which Eq. (1.22) is a solution. It is much easier to generate comparametric equations from their solutions than it is to solve comparametric equations.

The earlier comparametric equation is both useful and simple. The simplicity is in the ease with which it is solved, and in the fact that the solution happens to be the most commonly used camera response model in photography. As we will later see, when images are being processed, the comparametric function can be estimated by the fitting of a straight line through data points describing the comparametric relation between images. However, there are two shortcomings to affine correction:

1. It is not inherently unicomparametric, so it must be clipped to 1 when it exceeds 1 and to 0 when it falls below 0, as shown in Eq. (1.25).
2. Its solution, $f(q)$ describes the response of cameras only within their normal operating regime. Because the art of quantigraphic image processing involves a great deal of image processing done on images that have been deliberately and grossly overexposed or underexposed, there is a need for a comparametric model that captures the essence of cameras at both extremes of exposure (eg, both overexposure and underexposure).

1.4.5 THE PREFERRED CORRECTION OF IMAGES

Although affine correction was an improvement over zeta correction, which itself was an improvement over gamma correction, affine correction still has the two shortcomings listed earlier. Therefore, another form of image exposure correction is proposed, and it will be called the *preferred correction*. This new exposure correction is unicomparametric (bounded in normalized units between 0 and 1) and also has a parameter to control the softness of the transition into the *toe* and *shoulder* regions of the response function, rather than the hard clipping introduced by Eq. (1.25).

As with affine correction, the preferred correction will be introduced first by its solution, from which the comparametric equation will be derived. The solution is

$$f(q) = \left(\frac{e^b q^a}{e^b q^a + 1} \right)^c,$$

(1.26)

which has only three parameters. Thus, no extra unnecessary degrees of freedom (which might otherwise capture or model noise) have been added over and above the number of degrees of freedom in the previous model (1.22).

An intuitive understanding of Eq. (1.26) can be better had if we rewrite it:

$$f = \begin{cases} \exp\left[\dfrac{1}{\left(1 + e^{-(a \log q + b)}\right)^c} \right]^c, & \forall\, q \neq 0, \\ 0, & \text{for } q = 0, \end{cases}$$

(1.27)

where the soft transition into the toe (region of underexposure) and shoulder (region of overexposure) is evident by the shape of this curve on a logarithmic exposure scale.

This model may, at first, seem like only a slight improvement over (1.22), given our common intuition that most exposure information is ordinarily captured in the central portion, which is linear on the logarithmic exposure plot. However, it is important that we unlearn what we have been taught in traditional photography, where incorrectly exposed images are ordinarily thrown away rather than used to enhance the other images! It must be emphasized that comparametric image processing differs from traditional image processing in the sense that in comparametric image processing (following the Wyckoff principle, as illustrated in Fig. 1.8) the images typically include some that are *deliberately* underexposed and overexposed. In fact this overexposure of some images and underexposure of other images is often deliberately taken to extremes. Therefore, the additional sophistication of the model (1.26) is of great value in capturing the essence of a set of images where some extend to great extremes in the *toe* or *shoulder* regions of the response function.

Proposition 1.4.4. *The comparametric equation of which the proposed photographic response function (1.26) is a solution is given by*

$$g(f) = \frac{f k^{ac}}{\left(\sqrt[c]{f}(k^a - 1) + 1 \right)^c},$$

(1.28)

where K = log(k).

Function (1.28) gives rise to *the preferred correction* of images (eg, the preferred recipe for lightening or darkening an image). Again, $g(f)$ does not depend on b, which is consistent with our knowledge that the comparametric equation captures the information of $f(q)$ up to a single unknown scalar proportionality constant.

1.4.6 SOME SOLUTIONS TO SOME COMPARAMETRIC EQUATIONS THAT ARE PARTICULARLY ILLUSTRATIVE OR USEFUL

Some examples of comparametric equations and their solutions are summarized in Table 1.1.

1.4.7 PROPERTIES OF COMPARAMETRIC EQUATIONS

As stated previously, the comparametric equation provides information about the actual photoquantity only up to a single unknown scalar quantity — for example, if $f(q)$ is a solution of comparametric equation g then so is $f(\beta q)$. In general we can think of this as a coordinate transformation from q to

Table 1.1 Illustrative or Useful Examples of Comparametric Equations and Their Solutions

Comparametric Equation $g(f(q)) = f(kq)$	Solution (Camera Response Function) $f(q)$
$g = f^{\gamma}$	$f = \exp q^{\Gamma}, \quad \gamma = k^{\Gamma}$
$g = k^{\gamma} f$	$f = q^{\gamma}$
$g = af + b, \; \forall a \neq 1, \text{or} \quad b = 0$	$f = \alpha + \beta q^{\gamma}, \quad a = k^{\gamma}, b = \alpha(1 - k^{\gamma})$
$g = f + b$	$f = B \log(\beta q), \quad b = B \log k$
$g = \left(\sqrt[\gamma]{f} + \log k \right)^{\gamma}$	$f = \log^{\gamma} q$
$g = \begin{cases} ((2^{\zeta} - 1)f + 1)^{\frac{1}{\zeta}} - 1, & \forall \zeta \neq 0 \\ 2^f - 1, & \text{for } \zeta = 0 \end{cases}$	$\dfrac{e^q - 1}{2^{\zeta} - 1}$
$g = e^b f^a = e^{\alpha(1 - k^{\gamma})} f^{(k^{\gamma})}$	$\log f = \alpha + \beta q^{\gamma}$
$g = \exp((\log f)^{(k^b)})$	$f = \exp(a^{(q^b)})$
$g = \exp(\log^k f)$	$f = \exp(a^{bq})$
$g = \frac{2}{\pi} \arctan \left(k \tan(\frac{\pi}{2} f) \right)$	$f = \frac{2}{\pi} \arctan q$
$g = \frac{1}{\pi} \arctan(b\pi \log k + \tan((f - 1/2)\pi)) + \frac{1}{2}$	$f = \begin{cases} \frac{1}{\pi} \arctan(b\pi \log q) + \frac{1}{2}, & \forall q \neq 0 \\ 0, & \text{for } q = 0 \end{cases}$
$g = \left(\dfrac{\sqrt[c]{f} \; k^a}{\sqrt[c]{f}(k^a - 1) + 1} \right)^c$	$f = \left(\dfrac{e^b q^a}{e^b q^a + 1} \right)^c = \begin{cases} \left(\frac{1}{1 + e^{-(a \log q + b)}} \right)^c, & \forall q \neq 0 \\ 0, & \text{for } q = 0 \end{cases}$
$g = \exp \left(\left(\dfrac{\sqrt[c]{\log f} \; k^a}{\sqrt[c]{\log f}(k^a - 1) + 1} \right)^c \right)$	$f = \exp \left(\left(\dfrac{e^b q^a}{e^b q^a + 1} \right)^c \right) = \begin{cases} \exp \left(\left(\frac{1}{1 + e^{-(a \log q + b)}} \right)^c \right), & \forall q \neq 0 \\ 0, & \text{for } q = 0 \end{cases}$

Notes: The third from the top and second from the bottom were found to describe a large variety of cameras and have been used in a wide variety of quantigraphic image-processing applications. The second one from the bottom is the one that is most commonly used by the authors.

βq, in the domain of f. More generally, the comparametric plot $(f(q), g(f(q)))$ has the same shape as the comparametric plot $(f(h(q)), g(f(h(q))))$, for all bijective h. From this fact we can construct a property of comparametric equations in general:

Proposition 1.4.5. *A comparametric equation* $\breve{g}(f(q)) = g(f(h(q))) = g(\breve{f}(q))$ *has solution* $\breve{f}(q) = f(h(q))$ *for any bijective function h.*

Likewise, we can also consider coordinate transformations in the range of comparametric equations, and their effects on the solutions:

Proposition 1.4.6. *A comparametric equation* $g(f) = h(\breve{g})$ *has solution* $f = h(\breve{f})$, *where* $\breve{g}(q) = \breve{f}(kq)$ *is a comparametric equation with solution* $\breve{f}(q)$

Properties of comparametric equations related to their coordinate transformations are presented in Table 1.2.

Some simple but illustrative examples of the use of coordinate transformation properties to solve comparametric equations are now provided:

Example 1. Let $\breve{g} = a\breve{f} + b$, which we know has solution $\breve{f} = \alpha + \beta q^\gamma$, with $a = k^\gamma$ and $b = \alpha(1 - a)$. Let $h() = () + n_f$ (eg, h is a transformation which consists in simply adding noise.) Thus, $g(f) = h(\breve{g}) = \breve{g} + n_f = a\breve{f} + b + n_f$, so $g = a(f - n_f) + b + n_f$ has solution $f = h(\breve{f}) = \alpha + \beta q^\gamma + n_f$.

Example 2. From Table 1.1, observe that the comparametric equation $\breve{g} = a\breve{f} + b$ has solution $\breve{f} = \alpha + \beta q^\gamma$. Let $h() = \exp()$. We can thus solve $g(f) = h(\breve{g}) = \exp(a\breve{f} + b) = \exp(a \log(f) + b) = e^b f^a$ by noting that $f = h(\breve{f}) = \exp(\breve{f}) = \exp(\alpha + \beta q^\gamma)$.

This solution also appears in Table 1.1. We may also use this solution to seed the solution of the comparametric equation second from the bottom in Table 1.1 by using $h(x) = x/(x + 1)$. The equation second from the bottom in Table 1.1 may then be further coordinate-transformed into the equation at the bottom of Table 1.1 by use of $h(x) = \exp(x)$. Thus, properties of comparametric equations, such as those summarized in Table 1.2, can be used to help solve comparametric equations, such as those listed in Table 1.1.

Table 1.2 Some Properties of Comparametric Equations

Comparametric Equations $g(f(q)) = f(kq)$	Solutions (Camera Response Functions) $f(q)$
$\breve{g}(f) = g(\breve{f})$, where $\breve{g}(f(q)) = g(f(h(q)))$	$\breve{f}(q) = f(h(q))$, $\forall$ bijective h
$g(f) = \breve{g}(f)$, where $\breve{g}(f(q)) = \breve{f}(\beta q)$	$f(q) = \breve{f}(\beta q)$
$g(f) = g(h(\breve{f})) = h(\breve{g})$, where $\breve{g}(q) = \breve{f}(kq)$	$f = h(\breve{f})$
$h^{-1}(g) = \breve{g}(\breve{f})$	$f = h(\breve{f})$

Notes: This table could be extended over several pages, much like an extensive table listing properties of Laplace transforms, or a table of properties of Fourier transforms, or the like.

1.5 PRACTICAL IMPLEMENTATIONS

This section pertains to the practical implementation of the theory presented in previous sections.

1.5.1 COMPARING TWO IMAGES THAT DIFFER ONLY IN EXPOSURE

Without loss of generality, consider two differently exposed pictures of the same subject matter, f_1 and f_2, and recognize that in the absence of noise the relationship between the two images would be

$$\frac{1}{k_1}f^{-1}(f_1) = q = \frac{1}{k_2}f^{-1}(f_2), \tag{1.29}$$

so

$$f_2 = f(kf^{-1}(f_1)) = F^{-1}(F(f_1) + K), \tag{1.30}$$

where $k = k_2/k_1$, $K = \log(k_2) - \log(k_1)$, and $F = \log(f^{-1})$. It is evident that Eq. (1.30) is a comparametric equation.

This process (1.30) of "registering" the second image with the first differs from the image registration procedure commonly used in much of machine vision (Horn, 1986; Horn and Weldon, 1988; Faugeras and Lustman, 1988; Laveau and Faugeras, 1994) and image resolution enhancement (Irani and Peleg, 1991; Szeliski, 1996; Mann and Picard, 1995c) because it operates on the *range*, $f(q(\mathbf{x}))$ (tonal range), of the image $f_i(\mathbf{x})$ as opposed to its *domain* (spatial coordinates) $\mathbf{x} = (x, y)$.

1.5.2 JOINT HISTOGRAMS AND COMPARAGRAMS

The joint histogram between two images is a square matrix of size $M \times N$, where M is the number of gray levels in the first image and N is the number of gray levels in the second image (Mann and Picard, 1995b; Mann, 1996a). When the two images have the same number of gray levels, $M = N$, the joint histogram is a square $N \times N$ array. When the two images are (possibly) differently exposed images of the same subject matter, the joint histogram of these images is called a *comparagram* (Mann, 2000). When the two images are identical, the comparagram is zero except along the diagonal, and the diagonal values are those of the histogram of the input image.

It can be seen from Eq. (1.30) that the general problem of solving Eq. (1.30) can be done directly on the comparagram instead of the original pair of images. This comparametric approach has the added advantage of breaking the problem down into two separate simpler steps:

1. *Comparametric regression:* Finding a smooth semimonotonic function, g, that passes through most of the highest bins in the comparagram.
2. *Solving the comparametric equation:* Unrolling this function, $g(f(q)) = f(kq)$ into $f(q/q_0)$ by regarding it as an *iterative map* onto itself (see Fig. 1.10.) The iterative map (*logistic map*) is most familiar in chaos theory (Woon, 1995; Berger, 1989), but here, because the map is monotonic, the result is a deterministic function.

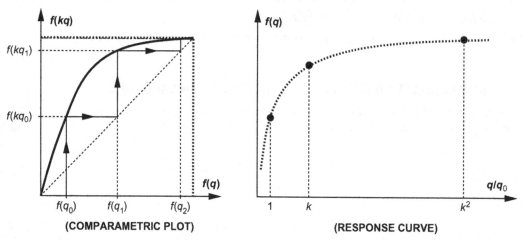

FIGURE 1.10

Comparametric procedure for finding the pointwise nonlinearity of an image sensor from two pictures differing only in their exposures. Left: Plot of pixel values in one image against corresponding pixel values in the other. Right: Points on the response curve, found from only the two pictures, without any knowledge of the characteristics of the image sensor. These discrete points are only for illustrative purposes. If a logarithmic exposure scale is used (as is done by most photographers), then the points fall uniformly on the $Q = \log(q/q_0)$ axis.

Separating this estimation process into two stages also allows us to have a more direct route to "registering" the image domains if, for example, we do not need to know f but require only g, which is the recipe for expressing the range of $f(kq)$ in the units of $f(q)$. In particular, we can lighten or darken images to match one another without ever having to solve for q. The first part of the earlier two-step process allows us to determine the relationship between two pictures that differ only in exposure, so that we can directly perform operations such as image exposure interpolation and extrapolation as in Fig. 1.9, but skip the intermediate step of computing q. Not all image-processing applications require q to be determined, so there is great value in simply understanding the relationship between differently exposed pictures of the same subject matter.

1.5.3 COMPARAMETRIC REGRESSION AND THE JOINT HISTOGRAM

In situations where the image data are extremely noisy, and/or where a closed-form solution for $f(q)$ is desired, a parameterized form of the comparametric function is used, in which a function $g(f)$ corresponding to a suitably parameterized response function $f(q)$ is selected. The method amounts to a *curve fitting* problem in which the parameters of g are selected so that g best fits one or more comparagrams (joint histograms constructed from two or more differently exposed images under analysis).

1.5.4 COMPARAMETRIC REGRESSION TO A STRAIGHT LINE

Eq. (1.24) suggests that we can determine $f(q)$ from two differently exposed images by applying linear regression to the joint histogram of the images, J, treating each entry as a data point and weighting it by the number of bin counts $J(m, n)$ at each point. One often does this by combining the input images, weighting them with $(J(m, n))^\lambda$. For example, $\lambda = 0$ (assuming empty bins are not counted) provides the classic linear regression problem in which all nonempty bins are weighted equally and the slope and intercept of the best-fit line through nonempty bins are found. Generally, λ is chosen somewhere between 0.25 and 2.

A simple example is presented, that of reverse engineering the standard Kodak Photo CD scanner issued to most major photographic processing and scanning houses. In most situations, a human operator runs the machine and decides, by visual inspection, what "brightness" level to scan the image at (there is also an automatic exposure feature which allows the operator to preview the scanned image and decide whether or not the chosen "brightness" level needs to be overridden). By scanning the same image at different "brightness" settings, one obtains a Wyckoff set. This allows the scanner to capture nearly the entire dynamic range of the film, which is of great utility because typical photographic negative film captures far greater dynamic range than is possible with the scanner as it is ordinarily used. A photographic negative taken from a scene of extremely high contrast (a sculpture on exhibit at the List Visual Arts Center, in a completely darkened room, illuminated with a bare flashlamp from one side only) was selected because of its great dynamic range that could not be captured in a single scan. A Wyckoff set was constructed by the scanning of the same negative at five different "brightness" settings (Fig. 1.11). The settings were controlled by a slider that was calibrated in arbitrary units from −99 to +99 while Kodak's proprietary scanning software was being run. Kodak provides no information about what these units mean. Accordingly, the goal of the experiment was to find a closed-form mathematical equation describing the effects of the "brightness" slider on the scans, and to recover the unknown nonlinearity of the scanner. To make the problem a little more challenging and, more

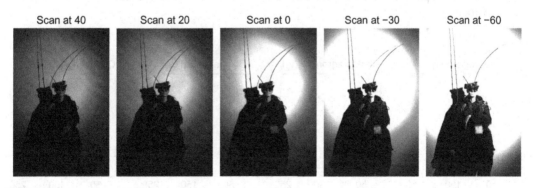

FIGURE 1.11

These scans from a photographic negative differ only in the choice of the "brightness" setting selected with the slider provided on the X-Windows screen by the proprietary Kodak Photo CD scanning software. The slider is calibrated in arbitrary units from −99 to +99. Five scans were done, and the setting of the slider is noted above each scan.

importantly, to better illustrate the principles of comparametric image processing, the *dmin* procedure of scanning a blank film at the beginning of the roll was overridden.

Joint histograms J_{01}, J_{12}, J_{23}, and J_{34} were computed from the five images (v_0 through v_4) in Fig. 1.11, and are displayed as density plots (eg, they are treated as images of dimension 256×256 pixels, where the darkness of the image is proportional to the number of counts — darkness rather than lightness to make it easier to see the pattern) in Fig. 1.12. Linear regression was applied to the data, and the best-fit straight line is shown passing through the data points. Because the *dmin* procedure was overridden, notice that the plots do not pass through the origin. The two leftmost plots had nearly identical slopes and intercepts, and likewise for the two rightmost plots, which indicates that the arbitrary Kodak units of "brightness" are self-consistent (eg, J_{01}, which describes the relationship between a scan at a "brightness" of 40 units and a scan at a "brightness" of 20 units, is essentially the same as J_{12}, which describes the relationship between a scan at a "brightness" of 20 units and a scan at a "brightness" of 0 units). Because there are three parameters in Eq. (1.22), k, α, and γ, which describe only two degrees of freedom (slope and intercept), γ may be chosen so that $k = \sqrt[\gamma]{a}$ is linearly proportional to arbitrary Kodak units. Thus, setting $(\sqrt[\gamma]{a_{\text{left}}})/(\sqrt[\gamma]{a_{\text{right}}}) = 20/30$ (where a_{left} is the average slope of the two leftmost plots in Fig. 1.12 and a_{right} is the average slope of the two rightmost plots in Fig. 1.12) results in $\gamma = 0.2254$. From this we obtain $\alpha = b/(1 - a) = 23.88$. Thus, we have

$$f(k_i q) = 23.88 + (kq)^{0.2254}, \tag{1.31}$$

where k_i is in arbitrary Kodak units (eg, $k_0 = 40$ for the leftmost image, $k_1 = 20$ for the next image, $k_2 = 0$ for the next image, $k_3 = -30$ for the next image, and $k_4 = -60$ for the rightmost image in Fig. 1.11). Thus, Eq. (1.31) gives us a closed-form solution that describes the response curve associated with each of the five exposures $f(k_i q), i \in \mathbb{Z}, 0 <= i <= 4$. The curves $f(k_i q)$ may be differentiated, and if these derivatives are evaluated at $q_i = \frac{1}{k_i} \sqrt[\alpha]{f_i(x, y)} - \alpha$, the so-called *certainty images*, shown in Fig. 1.13, are obtained.

In the next example, the use of the certainty functions to construct an optimal estimate, $\hat{q}(x, y)$ will be demonstrated.

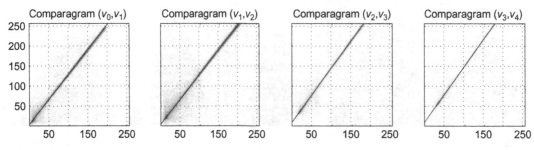

FIGURE 1.12

Pairwise joint histograms of the images in Fig. 1.12. It is evident that the data are well fitted by a straight line, which suggests that Kodak must have used the standard nonlinear response function $f(q) = \alpha + \beta q^\gamma$ in the design of its Photo CD scanner.

c_0 c_1 c_2 c_3 c_4

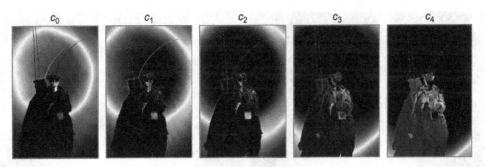

FIGURE 1.13

The *certainty functions* express the rate of change of $f(q(x, y))$ with $Q(x, y)$. The certainty functions may be used to compute the *certainty images*, $f(c_i)$. White areas in one of the certainty images indicate that pixel values $f(q)$ change fastest with a corresponding change in the photoquantity, Q. When the camera is used as a light meter (eg, a quantigraphic instrument to estimate q), it will be most sensitive where the certainty images are white. White areas of these certainty images correspond to mid-gray values (*midtones*) of the corresponding original images in Fig. 1.11, while dark areas correspond to extreme pixel values (either highlights or shadows) of the original images in Fig. 1.11. Black areas of the certainty image indicate that Q changes drastically with small changes in pixel value, and thus an estimate of Q in these areas will be overcome by image noise n_{f_i}.

1.5.5 COMPARAMETRIC REGRESSION TO THE PREFERRED MODEL

For this second example, the comparametric model proposed in Eq. (1.28) will be used.

In many practical situations, real-world images are very noisy. Accordingly, an example of noisy images that constitute a Wyckoff set (Fig. 1.14), in which an extremely poor scan was deliberately used to scan images from a publication (Mann and Picard, 1995b), is now considered.

That the images in Fig. 1.14 are of very poor quality is evidenced by their comparagrams (Fig. 1.15A). With use of regression of Eq. (1.28) to the comparagram combined with the knowledge (from the publication from which the images were obtained Mann and Picard, 1995b) that $K = 2$, it was found that $a = 0.0017$ and $c = -3.01$. These data provide a closed-form solution for the response function. The two effective response functions, which are shifted versions of this single response function, where the relative shift is K, are plotted in Fig. 1.16, together with their derivatives. (Recall that the derivatives of the response functions are the certainty functions.) Because a closed-form solution has been obtained, it may be easily differentiated without the further increase in noise that usually accompanies differentiation. Otherwise, when one is determining the certainty functions from poor estimates of f, the certainty functions would be even noisier than the poor estimate of f itself. The resulting certainty images, denoted by $c(f_i)$, are shown in Fig. 1.17. Each of the images, $f_i(x, y)$, gives rise to an actual estimate of the quantity of light arriving at the image sensor (1.9). These estimates were combined by way of Eq. (1.10), resulting in the composite image shown in Fig. 1.18.

The resulting image $\hat{I}_1$ looks very similar to f_1, except that it is a floating-point image array, of much greater tonal range and image quality.

Furthermore, given a Wyckoff set, a composite image may be rendered at any in-between exposure from the set (exposure interpolation), as well as somewhat beyond the exposures given (exposure

FIGURE 1.14

Noisy images badly scanned from a publication. These images are identical except for exposure and a good deal of quantization noise, additive noise, scanning noise, etc. Left: The darker image shows clearly the eight people standing outside the doorway, but shows little of the architectural details of the dimly lit interior. Right: The lighter image shows the architecture of the interior, but it is not possible to determine how many people are standing outside, let alone to recognize any of them.

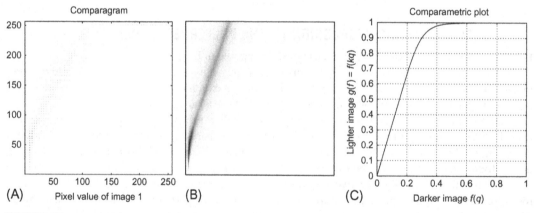

FIGURE 1.15

Comparametric regression. (A) Comparagrams. Because the images were extremely noisy, the comparagram is spread out over a fat ridge. Note also the gaps in the comparagram owing to the poor quality of the scanning process. (B) Even the comparagram of the images before the deliberately poor scan of them is itself quite spread out, indicating the images were quite noisy to begin with. (C) Comparametric regression is used to solve for the parameters of the comparametric function. The resulting comparametric plot is a noise-removed version of the comparagram (eg, provides a smoothly constrained comparametric relationship between the two differently exposed images).

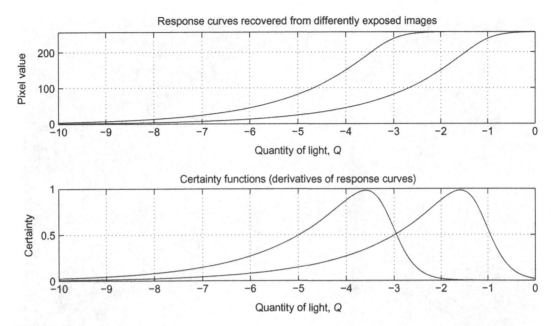

FIGURE 1.16

Relative response functions $F^{-1}(K_i Q)$ recovered from the images in Fig. 1.14, plotted together with their derivatives. The derivatives of these response functions suggest a degree of confidence in the estimate $\hat{Q}_i = F(f_i) - K_i$ derived from each input image.

FIGURE 1.17

Certainty images which will be used as weights when the weighted sum of estimates of the actual quantity of light is computed. Bright areas correspond to large degrees of certainty.

FIGURE 1.18

Composite image made by simultaneous estimation of the unknown nonlinearity of the camera as well as the true quantity of light incident on the camera's sensor array, given the two input images from Fig. 1.14. The combined optimal estimate of $\hat{q}$ is expressed here, in the coordinates of the lighter (rightmost) image in Fig. 1.14. Although nothing has been done to appreciably enhance this image (eg, the procedure of estimating q and then just converting it back into a picture again may seem pointless), while the image appears much like the rightmost input image, the clipping of the highlight details has been softened somewhat.

From © Steve Mann, 1993.

extrapolation). This result suggests the "virtual camera" (Mann, 1999), which allows images to be rendered at any desired exposure, once q has been computed.

This capability is somewhat similar to QuickTime VR and other image-based rendering systems, except that it operates in the range of the images $\hat{f}_i$ rather than their domain.

1.6 TONE MAPPING IN HDR SYSTEMS

Ordinarily, most print and display media have limited dynamic range. Thus, one might be tempted to argue against the utility of the Wyckoff principle based on this fact — for example, one might ask why, because televisions and print media cannot display more than a very limited dynamic range, we should bother building a Wyckoff camera that can capture such dynamic ranges. Why should we bother capturing the photoquantity q with more accuracy than is needed for display?

Some possible answers to this question are as follows:

1. Estimates of q are still useful for machine vision and for other applications that do not involve direct viewing of a final picture. An example is the wearable face-recognizer (Mann, 1996b) which determines the identity of an individual from a plurality of differently exposed pictures of that person, and then presents the identity in the form of a text label (virtual name tag) on the retina of an eye of the wearer of the eyeglass-based apparatus. Because $\hat{q}$ need not be displayed, the problem of output dynamic range, etc., of the display (eg, number of distinct intensity levels of the laser beam shining into the lens of the eye of the wearer) is of no consequence.

2. Even though the ordinary dynamic range and the range resolution (typically 8 bits) is sufficient for print media (given the deliberately introduced nonlinearities that best use the limited range resolution), when operations such as deblurring are performed, noise artifacts become more evident. In general, sharpening involves high-pass filtering, and thus sharpening will often tend to noise artifacts being uncovered that would normally exist below the perceptual threshold when viewed through ordinary display media. In particular, sharpening often uncovers noise in the shadow areas, making dark areas of the image appear noisy in the final print or display. Thus, in addition to the benefits of performing sharpening quantigraphically by applying an antihomomorphic filter as in Fig. 1.7 to undo the blur of Eq. (1.5), there is also further benefit from doing the generalized antihomomorphic filtering operation at the point $\hat{q}$ in Fig. 1.8, rather than just that depicted in Fig. 1.7.

3. A third benefit of capturing more information than can be displayed is that it defers the choice of which information to get rid of. For example, a camera could be constructed in such a way that it had no exposure adjustments: neither automatic nor manual settings. Instead, the camera would be more like an array of light meters that would capture an array of light measurements. Decisions as to what subject matter is of importance could then be made at the time of viewing or the time of printing. Such a camera has been incorporated into eyeglasses (Mann, 1998), allowing the wearer to completely forget about the camera, with no need to worry about settings or adjustments. In this way the wearer can capture once-in-a-lifetime moments such as a baby's first steps, and worry about adjusting the exposure settings later. Exposure can then be adjusted in the peaceful quiet of the living room, long after the picture has been captured and the confusing excitement of the moment has passed. In this way exposure can be adjusted carefully in a quiet setting away from the busy and distracting action of everyday life. Because these decisions are made later, they can also be changed, as there is no need to commit to one particular exposure setting. Moreover, deferring exposure decisions may have forensic value. For example, ordinary everyday subject matter and scenes may later become crime scenes, such that pictures previously taken in those spaces may help solve a crime. A family photographing their child's first trip to a grocery store may inadvertently capture an image of a fire exit illegally chained shut in the background. A fatal fire at some time in the future might call for evidence against the owner of the shop, where deferred choices of exposure may assist in the production of a picture exposed optimally for the fire exit in the background rather than the child in the foreground. Because the wearable apparatus transmits images to the World Wide Web, various viewers can each adjust the image interactively to suit their own display and perceptual capabilities, as well as their own preferences. These capabilities are a major reason that digital SLR cameras provide RAW data formats suitable for postcapture processing.

4. A fourth benefit from capturing a true and accurate measurement of the photoquantity, even if all that is desired is a nice looking picture (eg, even if what is desired is not necessarily a true or accurate depiction of reality), is that additional processing may be done to produce a picture in which the limited dynamic range of the display or print medium shows a much greater dynamic range of input signal, through the use of further image processing on the photoquantity before display or printing.

It is this fourth benefit that will be further described, as well as illustrated through a very compelling example, in this section. Pursuit of this benefit, of spatially varying tonal adjustment, has created the field commonly referred to as "tone mapping," which is a substantial field in the HDR area. Also see Part II of this book, where tone mapping and simulation of perceptual phenomena are discussed.

Ordinarily, humans cannot directly perceive the "signal" we process numerically, but, rather, we perceive the effects of the "signal" on perceptible media such as television screens or the like. In particular, to display $\hat{q}(x, y)$, it is typically converted into an image $f(\hat{q}(x, y))$ and displayed, for example, on a television screen.

Fig. 1.18 is an attempt to display, on the printed page, a signal which contains much greater dynamic range than can be directly represented on the page. To do this, the estimate $\hat{q}$ was converted into an image by evaluation of $\hat{f}(\hat{k}_2 \hat{q})$. Even though we see some slight benefit over f_2 (one of the input images), the benefit has not been made fully visible in this print.

1.6.1 AN EXTREME EXAMPLE WITH SPATIOTONAL PROCESSING OF PHOTOQUANTITIES

To fully appreciate the benefits of quantigraphic image processing, let us consider a seemingly impossible scene to photograph reasonably (in a natural way without bringing in lighting equipment of any kind).

Fig. 1.19 depicts a scene in which there is a dynamic range in excess of a 10^6:1. In this case, two pictures were captured with several orders of magnitude difference between the two exposures. Thus, the quantigraphic estimate $\hat{q}$ has far greater dynamic range than can be directly viewed on a television or on the printed page. Display of $\hat{f}(\hat{k}_1 q)$ would fail to show the shadow details, while display of $\hat{f}(\hat{k}_2 q)$ would fail to show the highlight details.

In this case, even if we use the virtual camera architecture depicted in Fig. 1.9, there is no single value of display exposure k_d for which a display image $f_d = \hat{f}(k_d \, \hat{q})$ will capture both the inside of the abandoned fortress and the details looking outside through the open doorway.

Therefore, a strong high-pass (sharpening) filter, S, is applied to $\hat{q}$ to sharpen the photoquantity $\hat{q}$, as well as to provide lateral inhibition similarly to the way in which the human eye functions. Then the filtered result, $\hat{f}\left(k_d S \hat{q} \left(\frac{\hat{A}_2 \mathbf{x} + \hat{\mathbf{b}}_2}{\hat{\mathbf{c}}_2 \mathbf{x} + \hat{d}_2}\right)\right)$, is displayed on the printed page (Fig. 1.19C) in the projective coordinates of the second (Fig. 1.19B) image, $i = 2$. Note the introduction of spatial coordinates $\mathbf{A}, \mathbf{b}, \mathbf{c}$, and d. These compensate for projection (eg, if the camera moves slightly between pictures), as described in Mann (1996a, 1998). In particular, the parameters of a projective coordinate transformation are typically estimated together with the nonlinear camera response function and the exposure ratio between pictures (Mann, 1996a, 1998); see also Chapter 3.

As a result of the filtering operation, notice that there is no longer a monotonic relationship between input photoquantity q and output level on the printed page. Notice, for example, that the sail is as dark

FIGURE 1.19

Extreme example to illustrate nonmonotonic processing. (A) An underexposed picture shows details such as the horizon and the sail of a boat, as seen through an open doorway, even though the sail is backlit with extremely bright light. (B) The picture is taken from inside an abandoned fortress with no interior lights. Light coming in from the open door is largely lost in the vastness of the dark interior, so a much longer exposure is needed to show any detail of the inside of the fortress. (C) Sharpened (spatiotonally filtered) quantigraphic estimate $\hat{f}\left(k_d S\hat{q}\left(\frac{\hat{A}_2\mathbf{x}+\hat{\mathbf{b}}_2}{\hat{\mathbf{c}}_2\mathbf{x}} + \hat{d}_2\right)\right)$ expressed in the projective coordinates of the second image in the image sequence (B). A dynamic range in excess of 10^6:1 was captured in $\hat{q}$, and the estimate was then quantigraphically sharpened, with filter S, resulting in a lateral inhibition effect so that the output is no longer monotonically related to the input. Notice, for example, that the sail is as dark as some shadow areas inside the fortress. Because of this filtering, a tremendous dynamic range has been captured and reduced to that of printed media.

From © Steve Mann, 1991.

as some shadow areas inside the fortress. Because of this filtering, the dynamic range of the image may be reduced to that of printed media, while still revealing details of the scene. This example answers the question of why we should capture more dynamic range than we can display.

1.7 ANALYTICAL SOLUTION OF COMPARAMETRIC EQUATIONS

1.7.1 OVERVIEW

In this section, we demonstrate the connection between comparametric equations and the scaling operator arising in quantum field theory, and provide a general method of solution to the comparametric class of functional equations.

The development of the scaling operator here follows the standard formulation for operators based on generators of infinitesimal transformations. This theory can be found in any standard graduate-level introductory modern quantum mechanics text covering quantum field theory, such as Chapter 1 of Sakurai (1994) or Chapter 3 of Ballentine (1998).

1.7.2 FORMAL SOLUTION BY SCALING OPERATOR

In quantum field theory we have a scaling operator S_k (which is also referred to as the dilation operator D_k). We will derive an explicit expression for this operator. First consider an infinitesimally small scaling of the function f. Let ϵ be a very small number. We have, up to first order in ϵ,

$$f((1 + \epsilon)q) \approx f(q) + \epsilon\, q\frac{\partial}{\partial q}f(q). \tag{1.32}$$

By repeated application of the infinitesimal scaling of q, we can scale q by any amount e^Λ

$$e^\Lambda q = \lim_{N \to \infty} \left(1 + \frac{\Lambda}{N}\right)^N q. \tag{1.33}$$

Application of Eq. (1.32) N times with $\epsilon = \frac{\Lambda}{N}$ for N large gives

$$f(e^\Lambda q) = \lim_{N \to \infty} \left(1 + \frac{\Lambda}{N}q\frac{\partial}{\partial q}\right)^N f(q), \tag{1.34}$$

$$= \exp\left(\Lambda q\frac{\partial}{\partial q}\right) f(q). \tag{1.35}$$

Choosing $\Lambda = \log(k)$, we scale q by k and thus define the scaling operator S_k as

$$S_k = \exp\left(\log k \times q\frac{\partial}{\partial q}\right). \tag{1.36}$$

The exponential here is defined as the formal series

$$\exp\left(\log k \times q\frac{\partial}{\partial q}\right) = \sum_{n=0}^{\infty} \frac{(\log k)^n}{n!} \left(q\frac{\partial}{\partial q}\right)^n. \tag{1.37}$$

Each differential operator in this series acts on every term to the right of it. The inverse of the scaling operator is then

$$(S_k)^{-1} = S_{1/k} = \exp\left(-\log k \times q\frac{\partial}{\partial q}\right). \tag{1.38}$$

Now, given $g(k,f)$, we can write $f(q)$ formally as

$$f(q) = S_{1/k}g(k,f(q)). \tag{1.39}$$

Although this is not a convenient formulation for explicit computation of $f(q)$, it opens the possibility for further analysis of the general comparametric problem using the machinery of the well-known operators arising in quantum field theory. Because Eq. (1.39) holds for any value of k, we can take k close to 1 (ie, $k = 1 + \epsilon$ for $\epsilon \approx 0$) and we see that

$$f(q) = \exp\left(-\log(1+\epsilon)q\frac{\partial}{\partial q}\right)g(1+\epsilon,f). \tag{1.40}$$

If we expand this in ϵ up to first order, noting that higher-order terms vanish, we find

$$f(q) = \left(1 - \epsilon q\frac{\partial}{\partial q}\right)\left(g(1,f) + \epsilon\frac{\partial g(k,f)}{\partial k}\right)\bigg|_{k=1}. \tag{1.41}$$

Using the identity $f = g(1,f)$, we end up with an ordinary differential equation,

$$\frac{df}{dq} = \frac{1}{q}\frac{\partial g(k,f)}{\partial k}\bigg|_{k=1}, \tag{1.42}$$

from which we can obtain $f(q)$. This equation is always separable, and the solution family always contains at least one valid solution when $g(f)$ is monotonic and smooth. This analytical form provides the benefit that any arbitrary camera response function can be solved exactly, and that the behavior of noise terms can be modeled as shown in Section 1.8.

1.7.3 SOLUTION BY ORDINARY DIFFERENTIAL EQUATION

The machinery of the previous section allows us to proceed directly, merely using the result as a recipe to solve any given analytical comparametric equations. To begin, we examine the comparametric equation given by

$$f(kq) = g(k,f(q)). \tag{1.43}$$

Consider two cases. In the first case, the function f is known. Then g is easily found. For example, consider the classical model

$$f(q) = \alpha + \beta q^\gamma. \tag{1.44}$$

Then,

$$f(kq) = \alpha + \beta k^\gamma q^\gamma. \tag{1.45}$$

From Eq. (1.44) it follows that

$$q = \left(\frac{f - \alpha}{\beta}\right)^{1/\gamma}. \tag{1.46}$$

Substituting this in Eq. (1.45), we find

$$g(k,f) = \alpha + \beta\, k^\gamma \left(\frac{f - \alpha}{\beta}\right). \tag{1.47}$$

In the second, more difficult case, $g(k,f)$ is given and we have to find $f(q)$. This is actually solving the comparametric equation. The way to do this is as follows. By partial differentiation of Eq. (1.47) by k, and substituting $k = 1$, we find

$$\left.\frac{\partial f(kq)}{\partial k}\right|_{k=1} = qf'(kq)\big|_{k=1} = qf'(q) \tag{1.48}$$

$$= \left.\frac{\partial g(k,f)}{\partial k}\right|_{k=1},$$

where f' is the derivative of f. So f satisfies the following ordinary differential equation:

$$\frac{df}{dq} = \frac{1}{q}\left.\frac{\partial g(k,f)}{\partial k}\right|_{k=1}, \tag{1.49}$$

which we derived in the previous section using the scaling operator. This differential equation is easily solved because it is always separable. For example, take

$$g(k,f) = \alpha\,(1 - k^\gamma) + k^\gamma f. \tag{1.50}$$

We have

$$\left.\frac{\partial g(k,f)}{\partial k}\right|_{k=1} = \left.((f - \alpha)\,\gamma\, k^{\gamma-1})\right|_{k=1} \tag{1.51}$$

$$= (f - \alpha)\,\gamma.$$

Now f satisfies

$$\frac{df(q)}{dq} = \frac{1}{q}(f - \alpha)\,\gamma. \tag{1.52}$$

By separating the variables, integrating, and taking the exponential of both sides, we obtain

$$f(q) = \beta q^\gamma + \alpha, \tag{1.53}$$

where β appears as a constant of integration.

1.8 COMPOSITING AS BAYESIAN JOINT ESTIMATION

Up to this point in the discussion, we have not characterized the noise terms inherent in the imaging process, nor have we used them to improve the estimation of the light quantity present in the original scene. In this section we develop a technique to incorporate a noise model into the estimation process.

Our approach for creating an HDR image from N input LDR images begins with the construction of a notional N-dimensional inverse camera response function that incorporates the different exposure and weighting values between the input images. Then we could use this to estimate the photoquantity $\hat{q}$ at each point by writing $\hat{q}(\mathbf{x}) = f^{-1}(f_1, f_2, \ldots, f_N)/k_1$. In this case f^{-1} is a joint estimator that could potentially be implemented as an N-dimensional lookup table (LUT). Recognizing the impracticality of this for large N, we consider pairwise recursive estimation for larger N values in the next section. The joint estimator $f^{-1}(f_1, f_2, \ldots, f_N)$ may be referred to more precisely as a *comparametric inverse camera response function* because it always has the domain of a comparagram and the range of the inverse of the response function of the camera under consideration.

Pairwise estimation

Assume we have N LDR images that are a constant change in exposure value apart, so that $\Delta\text{EV} = \log_2 k_{i+1} - \log_2 k_i$ is a positive constant $\forall\, i \in \{1, \ldots, N-1\}$, where k_i is the exposure of the ith image. Now consider specializing to the case $N = 2$ so we have two exposures, one at $k_1 = 1$ (without loss of generality, because exposures have meaning only in proportion to one another) and the other at $k_2 = k$. Our estimate of the photoquantity may then be written as $\hat{q}(\mathbf{x}) = f^{-1}_{\Delta\text{EV}}(f_1, f_2)$, where $\Delta\text{EV} = \log_2 k$.

To apply this pairwise estimator to three input LDR images, each with a constant difference in exposure between them, we can proceed by writing

$$f(\hat{q}) = f(f^{-1}_{\Delta\text{EV}}(f(f^{-1}_{\Delta\text{EV}}(f_1, f_2)), f(f^{-1}_{\Delta\text{EV}}(f_2, f_3)))). \qquad (1.54)$$

In this expression, we first estimate the photoquantity based on images 1 and 2, and then the photoquantity based on images 2 and 3, then we combine these estimates using the same joint estimator, by first putting each of the earlier round (or "level") of estimates through a virtual camera f, which is the camera response function.

This process may be expanded to any number N of input LDR images, by use of the recursive relation

$$f^{(j+1)}_i = f(f^{-1}_{\Delta\text{EV}}(f^{(j)}_i, f^{(j)}_{i+1})),$$

where $j = 1, \ldots, N-1$, $i = 1, \ldots, N-j$, and $f^{(N)}_1$ is the final output image, and in the base case, $f^{(1)}_i$ is the ith input image. This recursive process may be understood graphically as in Fig. 1.20. This process forms a graph with estimates of photoquantities as the nodes, and comparametric mappings between the nodes as the edges. A single estimation step using a CCRF is illustrated in Fig. 1.21.

For efficient implementation, rather than computing at runtime or storing values of $f^{-1}(f_1, f_2)$, we can store $f(f^{-1}(f_1, f_2))$. We call this the *comparametric camera response function* (CCRF). It is the comparametric inverse camera response function evaluated via (or "imaged" through, because we are in effect using a virtual camera) the camera response function f. This means at runtime we require $N(N-1)/2$ recursive lookups, and we can perform all pairwise comparisons at each level in parallel, where a level is a row in Fig. 1.20.

The reason we can use the same CCRF throughout is because each virtual comparametric camera $f \circ f^{-1}$ returns an exposure that is at the same exposure point as the less exposed of the two input images (recall that we set $k_1 = 1$), so ΔEV between images remains constant at each subsequent level.

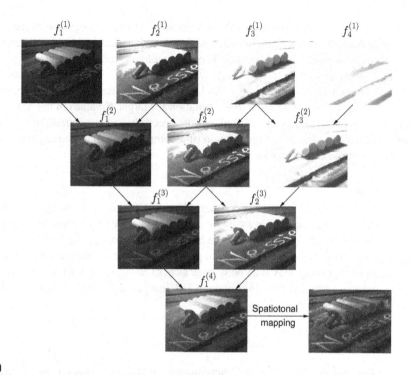

FIGURE 1.20

Graph structure of pairwise comparametric image compositing. The HDR image $f_1^{(4)}$ is composited from the LDR source camera images $f_{1...4}^{(1)}$. Nodes $f_i^{(j>1)}$ are rendered here by the scaling and rounding of the output from the *comparametric camera response function* (CCRF). To illustrate the details captured in the highlights and lowlights in the LDR medium of this paper, we include a spatiotonally mapped LDR rendering of $f_1^{(4)}$.

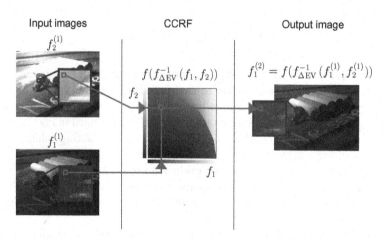

FIGURE 1.21

CCRF-based compositing of a single pixel. The floating-point tonal values f_1 and f_2 are the arguments to the CCRF $f \circ f_{\Delta EV}^{-1}$, which returns a refined estimate of an ideal camera response to the scene being photographed. This virtual camera's exposure setting is equal to the exposure of the lower-exposure image f_1.

In comparison with Fig. 3 in Mann and Mann (2001), wherein the objective is to recover the camera response function and its inverse, in this case we are using a similar hierarchical structure to instead combine information from multiple source images to create a single composite image.

The memory required to store the entire pyramid, including the source images, is $N(N + 1)$ times the amount of memory needed to store a single uncompressed source image with floating-point pixels. Multichannel estimation (eg, for color images) can be done by use of separate response functions for each channel, at a cost in compute operations and memory storage that is proportional to the number of channels.

Alternative graph topology

Other connection topologies are possible — for example, we can trade memory usage for speed by compositing using the following form for the case $N = 4$:

$$f(\hat{q}) = f(f_{2\Delta EV}^{-1}(f(f_{\Delta EV}^{-1}(f_1, f_2)), f(f_{\Delta EV}^{-1}(f_3, f_4)))),$$

in which case we perform only three lookups at runtime, instead of six with the previous structure. However, we must store twice as much lookup information in memory: for $f \circ f_{\Delta EV}^{-1}$ as before, and for $f \circ f_{2\Delta EV}^{-1}$, because the results of the inner expressions are no longer ΔEV apart, but instead are twice as far apart in exposure value, $2\Delta EV$, as shown in Fig. 1.22. As a recursive relation for $N = 2^n, n \in \mathbb{N}$ we have

$$f_i^{(j+1)} = f(f_{j\Delta EV}^{-1}(f_{2i-1}^{(j)}, f_{2i}^{(j)})),$$

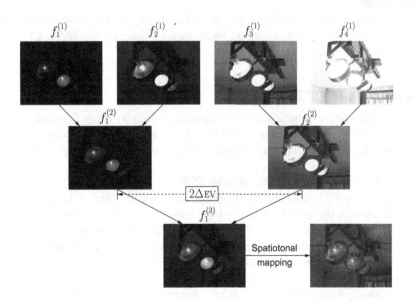

FIGURE 1.22

Example of an alternative graph structure for pairwise comparametric image compositing.

where $j = 1, \ldots, \log_2 N$ and $i = 1, \ldots, N/2^{j-1}$. The final output image is $f_1^{(\log_2 N+1)}$, and $f_i^{(1)}$ is the ith input image. This form requires $N - 1$ lookups. In general, by combining this approach with the previous graph structure, we can see that comparametric image composition can always be done in $O(N)$ lookups $\forall N \in \mathbb{N}$.

The alternative form described so far is only a single option of many possible configurations. For example, use of multiple LUTs provides less locality of reference, causing cache misses in the memory hierarchy, and uses more memory. In a memory-constrained environment, or one in which memory access is slow, we could use a single LUT while still using this alternate topology, as in

$$\hat{q}(\mathbf{x}) = f_{\Delta EV}^{-1}(f(f_{\Delta EV}^{-1}(f_1, f_2)/2), f(f_{\Delta EV}^{-1}(f_3, f_4)/2)). \tag{1.55}$$

This comes at the expense of our performing more arithmetic operations per comparametric lookup.

Constructing the CCRF

To create a CCRF $f \circ f^{-1}(f_1, f_2, \ldots, f_N)$, the ingredients required are a camera response function $f(q)$ and an algorithm for creating an estimate $\hat{q}$ of photoquantity by combining multiple measurements. Once these have been selected, $f \circ f^{-1}$ is the camera response evaluated at the output of the joint estimator, and is a function of two or more tonal inputs f_i.

To create a LUT means sampling through the possible tonal values, so, for example, to create a 1024×1024 LUT we could execute our $\hat{q}$ estimation algorithm for all combinations of $f_1, f_2 \in \{0, \frac{1}{1023}, \frac{2}{1023}, \ldots, 1\}$ and store the result of $f(\hat{q})$ in a matrix indexed by $[1023f_1, 1023f_2]$, assuming zero-based array indexing. Intermediate values may be estimated by linear or other interpolation.

Incremental updates

In the common situation that there is a single camera capturing images in sequence, we can easily perform updates of the final composited image incrementally, using partial updates, by only updating the buffers dependent on the new input.

1.8.1 EXAMPLE JOINT ESTIMATOR

In this section we describe a simple joint photoquantity estimator, using nonlinear optimization to compute a CCRF. This method executes in real time for HDR video, using pairwise comparametric image compositing. Examples of the results of this estimator can be seen in Figs. 1.20 and 1.22.

1.8.1.1 Bayesian probabilistic model for the CCRF

In this section we propose a simple method for estimating a CCRF. First, we select a comparametric model, which determines the analytical form of the camera response function. As an example, we illustrate our compositing approach using the "preferred saturation" camera model (Mann, 2001), for which an analytical is known and can be verified by use of the approach in the previous section.

The next step is to determine the model parameters, as in Fig. 1.23; however, any camera model with good empirical fit may be used with this method.

Let scalars f_1 and f_2 form a Wyckoff set from a camera with zero-mean Gaussian noise, and let random variables $X_i = f_i - f(k_i q), i \in \{1, 2\}$ be the difference between the observation and the model, with $k_1 = 1$ and $k_2 = k$.

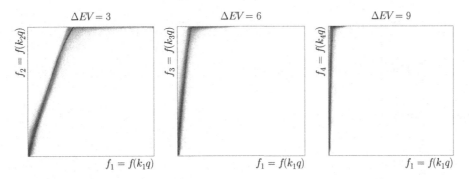

FIGURE 1.23

Comparametric model fitting. Preferred saturation model parameters were found via nonlinear optimization, by the method of least squares with the Levenberg-Marquardt algorithm. The optimal comparametric model function, determined per color channel, is plotted directly on empirical comparasums to verify a good fit. Comparasums are sums of comparagrams from the same sensor with the same difference in exposure value ΔEV. They are shown range compressed with the log function, and color inverted, to show finer variation. The best results for comparametric compositing are found when the camera response function model parameters are optimized against a range of k values. Here $k_1 = 1$ and $k_2 = 8$, $k_3 = 64$, and $k_4 = 512$, which implies that for comparametric image compositing we would use $\Delta EV = 3$.

We can estimate the variances of X_i can be estimated between exposures by calculating the *interquartile range* along each column (for X_1) and row (for X_2) of the comparagram with the ΔEV of interest (ie, using the "fatness" of the comparagram). A robust statistical formula, based on the quartiles of the normal distribution, gives $\hat{\sigma} \approx$ interquartile range$/1.349$, which can be stored in two one-dimensional vectors.

Discontinuities in $\hat{\sigma}_{X_i}$ with respect to f_i can be mitigated by Gaussian blurring of the sample statistics, as shown in Fig. 1.24. Using interpolation between samples of the standard deviation, and extrapolation beyond the first and last samples, we can estimate for any value of f_1 or f_2 the corresponding constant σ_{X_1} or σ_{X_2}.

The probability of $\hat{q}$, given f_1 and f_2, is

$$
\begin{aligned}
P(q = \hat{q} | f_1, f_2) &= \frac{P(q)P(f_1 | q, f_2)P(f_2 | q)}{P(f_1, f_2)} \\
&= \frac{P(q)P(f_1 | q)P(f_2 | q)}{P(f_1 | f_2)} \\
&= \frac{P(q)P(f_1 | q)P(f_2 | q)}{\int_0^\infty P(f_1 | q)P(f_2 | q) \, dq} \\
&\propto P(q = \hat{q})P(f_1 | q)P(f_2 | q).
\end{aligned}
$$

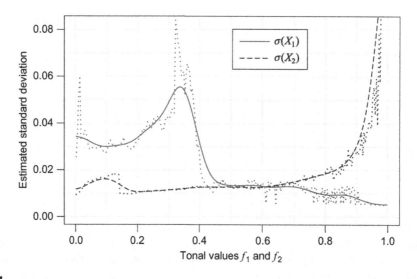

FIGURE 1.24

Trace plot of estimated standard deviations from a comparagram. Each estimate is proportional to the IQR, calculated from each column f_1 and row f_2 of a comparagram; here with $\Delta EV = 3$ as given in Fig. 1.23. Gaussian smoothing is applied to reduce discontinuities due to edge effects, quantization, and other noise.

For simplicity, we choose a uniform prior, which gives us $P_{\text{prior}}(q = \hat{q}) = \text{constant}$. Using X_i, we have

$$P_{\text{model}}(f_i|q) = \text{normal}(\mu_{X_i} = 0, \sigma_{X_i}^2)$$

$$= \frac{1}{\sqrt{2\pi}\sigma_{X_i}} \exp\left(-\frac{(f_i - f(k_iq))^2}{2\sigma_{X_i}^2}\right).$$

To maximize $P(q = \hat{q}|f_1, f_2)$ with respect to q, we remove constant factors and equivalently minimize $-\log(P)$. Then the optimal value of q, given f_1 and f_2, is

$$q = \underset{q}{\text{argmin}} \left(\frac{(f_1 - f(q))^2}{\sigma_{X_1}^2} + \frac{(f_2 - f(kq))^2}{\sigma_{X_2}^2}\right).$$

In practice, good estimates of optimal q values can be found with use of, for example, the Levenberg-Marquardt algorithm.

1.8.2 DISCUSSION REGARDING COMPOSITING VIA THE CCRF

Use of direct computation for nonlinear iterative methods as in Pal et al. (2004) is not feasible for real-time HDR video, because the time required to converge to a solution on a per-pixel basis is too long.

Table 1.3 Performance of Pairwise Composition Versus Direct Calculation of a Composite HDR Image on Four Input LDR Images

	Method			
	Direct Calculation	CCRF, Full Update	CCRF, Incremental	
Platform	Speed (output frames per second)			Speedup
CPU (serial)	0.0154	51	78	5065 times
CPU (threaded)	0.103	191	265	2573 times
GPU	–	272	398	–

For our simplistic probabilistic model given in Section 1.8.1, it takes more than 1 min (approximately 65 s) to compute each output frame with a single processor. With the method proposed in Section 1.8, the multicore speedup is more than 2500 times for CPU-based computation, and 3800 times for graphics processing unit (GPU)-based computation (versus CPU), as shown in Table 1.3.

The selection of the size of the LUT depends on the range of exposures for which it is used. It was found empirically that 1024×1024 samples of a CCRF is enough for the practical dynamic range of our typical setups. Further increases in the size of the LUT resulted in no noticeable improvement in output video quality.

Because GPUs implement floating-point texture lookup with linear interpolation in hardware, and can execute highly parallelized code, the GPU execution would seem to be a natural application of general-purpose GPU computation. However, for this application much of the time is spent waiting for data transfer between the host and the GPU — the pairwise partial update is useful in this context because we can reuse partial results from the previous estimate, and transfer only the new data.

1.8.3 REVIEW OF ANALYTICAL COMPARAMETRIC EQUATIONS

In this section have we developed the general solution to any comparametric problem. This solution converts the comparametric system from a functional equation to a separable ordinary differential equation. The solution to this differential equation can then be used to perform image compositing based on photoquantity estimation.

We have further illustrated comparametric compositing as a novel computational method that uses multidimensional LUTs, recursively if necessary, to estimate HDR output from LDR inputs. The runtime cost is fixed irrespective of the algorithm implemented if it can be expressed as a comparametric lookup. Pairwise estimation decouples the specific compositing algorithm from runtime, enabling a flexible architecture for real-time applications, such as HDR video, that require fast computation. Our experiments show that we approach a data transfer barrier rather than a compute-time limit. We demonstrated a speedup of three orders of magnitude for nonlinear optimization–based photoquantity estimation.

1.9 EFFICIENT IMPLEMENTATION OF HDR RECONSTRUCTION VIA CCRF COMPRESSION

High-quality HDR typically requires large computational and memory requirements. We seek to provide a system for efficient real-time computation of HDR reconstruction from multiple LDR samples, using very limited memory. The result translates into hardware such as field-programmable gate array (FPGA) chips that can fit within eyeglass frames or other miniature devices.

The CCRF enables an HDR compositing method that performs a pairwise estimate by taking two pixel values p_i and p_j at an exposure difference of ΔEV and outputs $\hat{q}$, the photoquantigraphic quantity (Ali and Mann, 2012). The CCRF results can be stored in a LUT. The LUT contains $N \times N$ elements of precomputed CCRF results on pairs of discretized p_i and p_j values. The benefit of the use of a CCRF is that the final estimate of photoquantity can incorporate multiple samples at different exposures to determine a single estimate of the photoquantity at each pixel.

Quadtree representation

The CCRF LUT can be represented in a tree structure. For $N \times N$ elements, we can generate a quadtree (a parent node in a such tree contains four child nodes) to fully represent the CCRF LUT. Such a quadtree is a complete tree with $\log_4 N^2$ or $\log_2 N$ levels.

One method of generating such a tree is to recursively divide a unit square into four quadrants (four smaller but equally sized squares). We can visualize the center of a divided unit square as the parent node of the four quadrants. The center of each quadrant is considered a child node. Such a process is performed recursively in each quadrant until the root unit square is divided into $N \times N$ equally sized squares. The bottom nodes of the quadtree are the leaves of the tree, each of which stores the CCRF lookup value of the corresponding pixel pair (p_i, p_j).

Reducing the quadtree

Storing the leaves of the complete quadtree representation of the CCRF costs as much space as the CCRF LUT itself. To reduce the number of elements needed for storage, we proceed to interpolate the CCRF value $\hat{p}_{i,j}$ of a specific pair (p_i, p_j) on the basis of its neighbor CCRF lookups. The value $\hat{p}_{i,j}$ interpolated on the basis of its neighbors is compared against the actual CCRF lookup value $p_{i,j}$, which gives $e_{i,j}$, the error per lookup entry:

$$e_{i,j} = |\hat{p}_{i,j} - p_{i,j}|. \tag{1.56}$$

We accept the approximated result if $e_{i,j}$ is within a fraction of $\frac{1}{2^D}$ as the error threshold e_{th}:

$$e_{\text{th}} = \alpha \cdot \frac{1}{2^D}, \tag{1.57}$$

where α is the fraction constant and D is the bit depth of the pixel value. We define the neighbor CCRF points as the four corner CCRF points of the square. We interpolate all CCRF (p_i, p_j) within the same square on the basis of these four corner values. We denote $I\left(e_{i,j}, e_{\text{th}}\right)$ as the indicator function of unsatisfied error condition:

$$I\left(e_{i,j}, e_{\text{th}}\right) = \begin{cases} 1, & \text{if } e_{i,j} \geq e_{\text{th}}, \\ 0, & \text{otherwise}, \end{cases} \tag{1.58}$$

We divide the square into four quadrants if

$$\sum I\left(e_{i,j}, e_{\text{th}}\right) > 0, \ \forall \, p_{i,j} \in \text{square}. \tag{1.59}$$

The purpose of the division is to obtain new corner values that are closer to the point. The closer corner values to the point for interpolation may yield lower $e_{i,j}$ as the CCRF LUT generally varies smoothly over a continuous and wide range of p_i and p_j values. Therefore, we expect that the density of the divisions corresponds to the local gradient of the CCRF LUT: the higher the local gradient, the more recursive divisions are required to bring corners closer to the point, whereas the points formed of a large and smooth region of the CCRF with low local gradient share the same corners.

Error weighting and tree depth criteria

The error of the interpolation against the original lookup value of the input pair (p_i, p_j) should be within e_{th}. This error is affected by the interpolation method. Empirically, we find that bilinear interpolation works better than quartic interpolation in terms of minimizing the number of lookup points while satisfying the error constraint.

Statistically, we observe that the most frequently accessed CCRF values lie along the comparagram, as shown in Fig. 1.25. Therefore, higher precision on interpolation may not be necessary for CCRF lookup points that are distant from the comparagram. This suggests that the error constraint for the pair (p_i, p_j) should vary depending on its likelihood of occurrence. To further compress the CCRF LUT, we can scale $e_{i,j}$ by the number of observed occurrences of the pair (p_i, p_j). This information is obtained through the construction of the comparagram. For each entry of the CCRF lookup, we weight the interpolation error by directly multiplying it by the occurrence count observed on the comparagram:

$$e_{i,j} \cdot (B_{i,j} + 1), \tag{1.60}$$

where $B_{i,j}$ is the count of the number of occurrences on the comparagram entry of (p_i, p_j). The result with the weighted errors also favors bilinear interpolation over quartic interpolation in terms of minimizing the number of entries of the CCRF for storage, as shown in Fig. 1.26. Reconstruction of the original CCRF from this quadtree is shown in Fig. 1.27, with the corresponding error shown in Fig. 1.28.

The CCRF LUT we use has 1024 entries for p_i and 1024 entries for p_j. Therefore, there is no point in constructing a tree that has a depth of more than $\log_2 1024 = 10$. We may constrain the depth of the tree to fewer than 10 levels as long as the error constraint is met. This affects the resulting number of entries in the CCRF quadtree, as well as the number of iterations required to find a leaf node.

Corner value access

Each node of the quadtree is the center point of the square that contains it. To access the corner values of a leaf node, we can perform a recursive comparison of pair (p_i, p_j) with (p_x, p_y) of the nonleaf nodes in the tree until a leaf node has been reached, seen in Algorithm 1. The leaf nodes contain memory addresses of the corresponding corner values. The corner values of a leaf node are stored in the memory for retrieval.

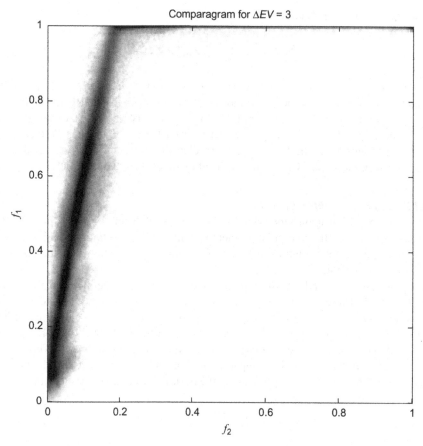

FIGURE 1.25

A comparagram is a joint histogram of pixel values from images of the same scene taken with different exposures. For any given sensor, the comparagram is directly related to the camera response function. Areas that are dark in this plot correspond to joint values that are expected to occur in practice.

1.9.1 IMPLEMENTATION

The algorithm can be implemented efficiently on a medium-sized FPGA. Given a finalized quadtree data structure, a system can be generated with software. The four corner values are stored in ROMs implemented with on-chip block RAM, and then selected by multiplexer chains based on the inputs f_1 and f_2. An arithmetic circuit that follows can then calculate the result on the basis of bilinear interpolation. Thus, as shown in Fig. 1.29, the system consists of two major parts: an addressing circuit and an interpolation circuit.

Because the size of the quadtree can grow to 10 levels, a C program is written to generate the implementation of the two circuits in Verilog hardware description language. Given a compressed

(0,1)   (1,1)

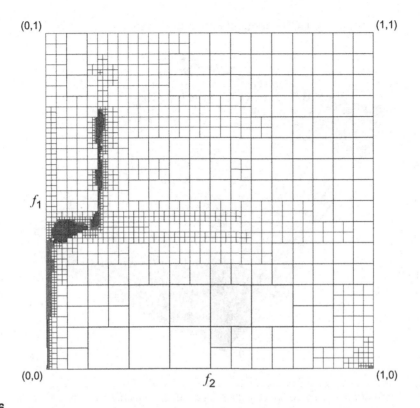

f_1

(0,0) f_2 (1,0)

FIGURE 1.26

Quadtree-based representation of the CCRF based on weighting from the comparagram shown in Fig. 1.25. Areas of rapid change or high use are more finely subdivided for greater accuracy. Inside each square the CCRF value is approximated by bilinear interpolation based on the corner values.

quadtree data structure, this program generates four ROM initialization files and a circuit that retrieves corner values stored in the ROMs based on the inputs.

1.9.1.1 Addressing circuit

Each of the quadtree leaves needs a unique address. This address is used to retrieve the corresponding corner values from ROMs. As shown in Fig. 1.30, the circuit outputs the address by comparing f_1 and f_2 with constant boundary values, in the same way as we traverse the quadtree. The main function of the boundary comparator is to send the controlling signal to traverse the multiplexer tree, on the basis of the given input pair. At each level, it compares the input values with the prestored center coordinate values, and determines which branch (if exits) it should take next. Otherwise, the current node is a quadtree leaf and a valid address will be selected.

The algorithm generates the circuit by traversing the quadtree. Because a new unique address is needed for every leaf being visited, a global counter is used to determine the addresses. The width of the circuit data path is then determined with the last address generated (ie, the maximum address).

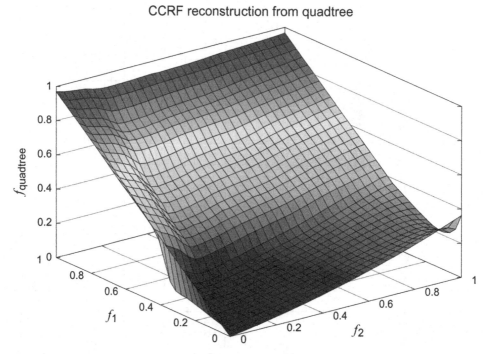

FIGURE 1.27

Reconstruction of the CCRF LUT based on a compressed quadtree with an error constraint of within one pixel value (α set to 1.0). The error bound is weighted by the expected usage, so values used more often have a smaller error bound.

Algorithm 1 Recursive quadtree search

procedure GET CORNERS(p_i, p_j, Node)
    if Node is not a leaf then
        if $p_i <$ Node$\rightarrow p_x$ then
            if $p_j <$ Node$\rightarrow p_y$ then
                get corners(p_i, p_j, Node$\rightarrow$Child(left, down))
            else
                get corners(p_i, p_j, Node$\rightarrow$Child(left, up))
        else
            if $p_j <$ Node$\rightarrow p_y$ then
                get corners(p_i, p_j, Node$\rightarrow$Child(right, down))
            else
                get corners(p_i, p_j, Node$\rightarrow$Child(right, up))
    else
        retrieve corner values
end procedure

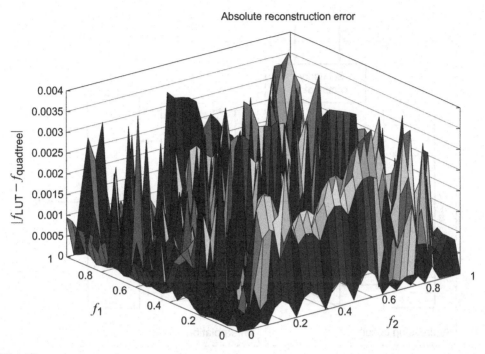

FIGURE 1.28

The absolute difference between the original and the reconstructed CCRF LUTs. The allowed error is greater in areas of the table that correspond to highly unlikely situations.

1.9.1.2 Interpolation circuit

The circuit takes the address and uses it to look up values that are prestored in the block RAM. These values can be used to perform an arithmetic operation (as shown in Fig. 1.31) for bilinear interpolation. To maintain high throughput, the intermediate stages are pipelined with use of registers.

1.9.2 COMPRESSION PERFORMANCE

The compressed CCRF LUT requires storage of all corner values per pair (p_i, p_j) for interpolation. Without any compression, this method would require as many as four times the storage space because of redundancies of identical CCRF values shared between adjacent lookups. However, the compression is able to reduce the number of lookup entries by a factor greater than four times. This compression factor depends on the selection of α.

We wrote the compression in the C programming language to output the compressed CCRF LUT. In Table 1.4, we list the minimum compression factor and the expected error over the entire CCRF range, for four selections of α.

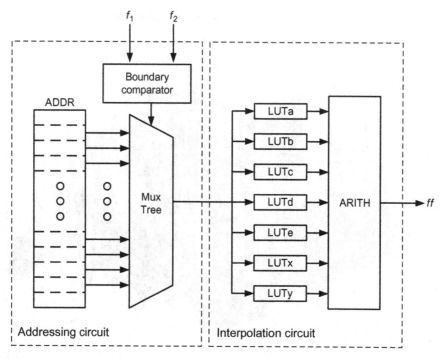

FIGURE 1.29

The top-level system consists of two main parts: the addressing circuit and the interpolation circuit. Input pixel values are normalized first to 16-bit fixed-point representation (f_1 and f_2) before entering the boundary block. According to the boundary conditions of the two values, the circuit generates controlling signals to the multiplexer tree, which selects the correct address that corresponds to the original quadtree node. The address is then used in the interpolation circuit to retrieve the stored values that are necessary for bilinear interpolation. After arithmetic circuit operation, the output *ff* can be further combined with *ff* from another instance of the same circuit.

The minimum compression factor summarizes the amount of compression achieved by taking the ratio of the total number of entries in the CCRF LUT to the number in the compressed one:

$$\text{Minimum compression factor} = \frac{N^2}{\text{number of leaf nodes} \times 4}. \qquad (1.61)$$

The constant 4 is the upper bound of the maximum redundancy of the corner value storage that overlaps with adjacent CCRF lookup points.

1.9.2.1 Hardware resource usage

For resource estimation without consideration of optimization effort performed by the CAD tool, we have

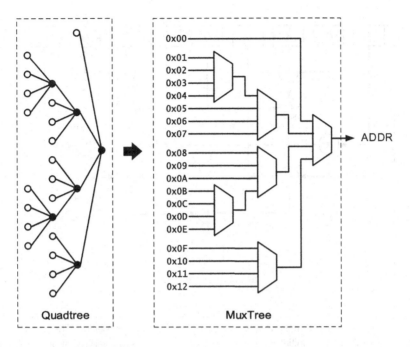

FIGURE 1.30

The relation between the original quadtree and its multiplexer implementation makes it very easy to generate the Verilog using the same quadtree data structure in software (ie, generate software that generates the Verilog code that describes the hardware design). Efficient use of four-to-one multiplexers in the six-LUT FPGA architecture can significantly reduce resource usage (ie, each multiplexer is mapped to one logic slice, instead of three slices if a two-to-one multiplexer is used) and code generation algorithm.

Table 1.4 Quadtree Properties Depending on the Choice of α				
Maximum Depth = 8	$\alpha = 1$	$\alpha = 1/2$	$\alpha = 1/4$	$\alpha = 1/16$
Number of entries	3315	4828	6508	10,807
Compression factor	79.1	54.2	161.1	97.0
Mean depth	5.8	6.2	6.6	7.4
Expected depth	7.7	7.8	7.8	7.9
Error constraint	0.0039	0.0019	0.00098	0.00024
Expected error	0.00042	0.00038	0.00036	0.00035

Notes: The compression factor is calculated on the basis of the number of CCRF lookup entries required divided by the number of the entries after compression. The CCRF without compression contains 10^6 floating-point entries.

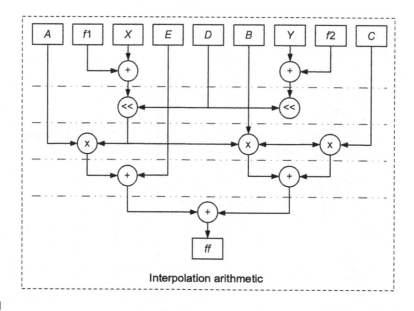

FIGURE 1.31

This circuit performs pipelined arithmetic that is necessary for bilinear interpolation. The inputs to this circuit are load from block RAM, which is initialized according to the compressed quadtree structure. The dotted red (dark gray in print versions) lines indicate the stage after which the data can be pipelined in order to have higher throughput.

$$Number\ of\ multiplexers = number\ of\ nonleaf\ nodes,$$
$$Number\ of\ comparators = number\ nonleaf\ nodes \times 2,$$
$$Number\ of\ addresses = number\ of\ leaves.$$

The actual number of resources needed to implement the system is much less (almost halved) than what we expected. The detailed data gathered from both expectation (using counters) and the CAD outputs are summarized in Table 1.5.

1.9.3 CONCLUSION

This section presented an architecture for accurate HDR imaging that is amenable for implementation on highly memory constrained low-power FPGA platforms. Construction of the compositing function is performed by nonlinear optimization of a Bayesian formulation of the compositing problem, where the selected prior creates an accurate estimator that is smooth for robustness and enhanced compression. The estimator is solved over a regular grid in the unit square, forming a two-dimensional LUT. Implementation of this solution on an FPGA then relies on the compression of the LUT into a quadtree form that allows random access, and uses bilinear interpolation to approximate values for intermediate points. This form allows selective control over error bounds, depending on the expected use of the

Table 1.5 The Resource Usage of the Implementation on a Kintex 7 (Xilinx XC7K325T) FPGA

	Depth = 4	Depth = 6	Depth = 8	Depth = 10
Expected slice usage	63	633	3315	21,207
Portion used (%)	0.0309	0.31	1.6	10
Actual slice usage	2	119	737	11,101
Portion used (%)	9.8×10^{-4}	0.058	0.36	5.5

Notes: The total number of logic slices used in this FPGA is 203,800. The difference between the expected and actual usage is due to the optimization present in the synthesis process, enabling more efficient use of resources.

table, which is easily obtained for a particular sensor. This results in compression of more than 60 times relative to the original LUT, with visually indistinguishable results.

ACKNOWLEDGMENTS

The authors acknowledge assistance or contributions to this project from Antonin Kimla of York Radio and TV, Kopin, Kodak, Digital Equipment, Compaq, Xybernaut, WaveRider, CITO (Communications and Information Technology Ontario), and NSERC (Natural Sciences and Engineering Research Council of Canada).

REFERENCES

Aczél, J., 1966. Mathematics in science and engineering. A series of monographs and textbooks edited by Richard Bellman, University of Southern California. In: Lectures on Functional Equations and Their Applications, vol. 19. Academic Press, New York/London, p. 510.

Ali, M.A., Mann, S., 2012. Comparametric image compositing: computationally efficient high dynamic range imaging. In: IEEE International Conference on Acoustics, Speech and Signal Processing (ICASSP), 2012, pp. 913–916.

Ballentine, L., 1998. Quantum Mechanics — A Modern Development. World Scientific Publishing Co. Pte. Ltd., Singapore.

Berger, M.A., 1989. Random affine iterated function systems: curve generation and wavelets. SIAM Rev. 31 (4), 614–627.

Debevec, P.E., Malik, J., 1997. Recovering high dynamic range radiance maps from photographs. In: SIGGRAPH, pp. 369–378.

Faugeras, O.D., Lustman, F., 1988. Motion and structure from motion in a piecewise planar environment. Int. J. Pattern Recognit. Artif. Intell. 2 (3), 485–508.

Granados, M., Ajdin, B., Wand, M., Theobalt, C., Seidel, H.P., Lensch, H.P.A., 2010. Optimal HDR reconstruction with linear digital cameras. In: IEEE Conference on Computer Vision and Pattern Recognition (CVPR), 2010, pp. 215–222.

Horn, B.K.P., 1986. Robot vision. In: MIT Electrical Engineering and Computer Science Series. McGraw-Hill Book Company, Cambridge, MA.

Horn, B., Schunk, B., 1981. Determining optical flow. Artif. Intell. 17, 185–203.

Horn, B.K.P., Weldon Jr., E.W., 1988. Direct methods for recovering motion. Int. J. Comput. Vis. 2 (1).

Inman, R., Smith, G., 2009. Television Production Handbook. http://www.tv-handbook.com/Camera.html.

Irani, M., Peleg, S., 1991. Improving resolution by image registration. CVGIP 53, 231–239.

Jack, K., Tsatsulin, V., 2002. Dictionary of Video and Television Technology. Gulf Professional Publishing, Houston.

Janzen, R., Mann, S., 2014. An information-bearing extramissive formulation of sensing, to measure surveillance and sousveillance. In: IEEE 27th Canadian Conference on Electrical and Computer Engineering (CCECE), 2014, pp. 1–10.

Laveau, S., Faugeras, O., 1994. 3-D scene representation as a collection of images. In: Proc. IEEE Conference on Computer Vision and Pattern Recognition, Seattle, Washington.

Mann, S., 1992. Wavelets and Chirplets: time-frequency perspectives, with applications. In: Archibald, P. (Ed.), Advances in Machine Vision, Strategies and Applications. World Scientific, Singapore/New Jersey/London/Hong Kong, pp. 99–128.

Mann, S., 1993. Compositing multiple pictures of the same scene. In: Proc. Annual IS&T Conference, Cambridge, MA, pp. 50–52.

Mann, S., 1996a. Joint parameter estimation in both domain and range of functions in same orbit of the projective-Wyckoff group. In: IEEE International Conference on Image Processing (ICIP-96), Lausanne, Switzerland, pp. III-193–III-196.

Mann, S., 1996b, Wearable, tetherless computer-mediated reality: WearCam as a wearable face-recognizer, and other applications for the disabled. TR 361, M.I.T. Media Lab Perceptual Computing Section. MIT, Cambridge, MA. Also appears in AAAI Fall Symposium on Developing Assistive Technology for People with Disabilities, Nov. 9–11, 1996. http://wearcam.org/vmp.htm.

Mann, S., 1997a. An historical account of the 'WearComp' and 'WearCam' projects developed for 'personal imaging'. In: International Symposium on Wearable Computing. IEEE, Cambridge, MA, pp. 66–73.

Mann, S., 1997b. Wearable computing: a first step toward personal imaging. IEEE Comput. 30 (2), 25–32.

Mann, S., 1998. Humanistic intelligence/humanistic computing: 'wearcomp' as a new framework for intelligent signal processing. Proc. IEEE 86 (11), 2123–2151.

Mann, S., 1999. Personal imaging and lookpainting as tools for personal documentary and investigative photojournalism. ACM Mob. Netw. 4 (1), 23–36.

Mann, S., 2000. Comparametric equations with practical applications in quantigraphic image processing. IEEE Trans. Image Proc. 9 (8), 1389–1406. ISSN 1057-7149.

Mann, S., 2001. Intelligent Image Processing. John Wiley and Sons, New York, p. 384 .

Mann, S., 2004. Sousveillance: inverse surveillance in multimedia imaging. In: Proceedings of the 12th Annual ACM International Conference on Multimedia. ACM, New York, pp. 620–627.

Mann, S., 1985. Light painting. Impulse 12 (2).

Mann, S., 2014. The Sightfield: visualizing computer vision, and seeing its capacity to "see". In: IEEE Conference on Computer Vision and Pattern Recognition Workshops (CVPRW), 2014, pp. 618–623.

Mann, S., Mann, R. 2001. Quantigraphic imaging: estimating the camera response and exposures from differently exposed images. In: CVPR, pp. 842–849.

Mann, S., Picard, R., 1995a. Being 'undigital' with digital cameras: extending dynamic range by combining differently exposed pictures. In: Proc. IS&T's 48th Annual Conference, Washington, DC, May 7–11, 1995. Also appears in M.I.T. M.L. T.R. 323, 1994, http://wearcam.org/ist95.htm.

Mann, S., Picard, R., 1995b, Being 'undigital' with digital cameras: extending dynamic range by combining differently exposed pictures. Tech. Rep. 323, M.I.T. Media Lab Perceptual Computing Section. MIT, Cambridge, MA. Also appears in Proc. IS&T's 48th Annual Conference, Washington, DC, May 7–11, 1995, pp. 422–428.

Mann, S., Picard, R.W., 1994. Virtual bellows: constructing high-quality images from video. In: Proceedings of the IEEE First International Conference on Image Processing, Austin, Texas, Nov. 13–16, 1994, pp. 363–367.

Mann, S., Picard, R.W., 1995c. Video orbits of the projective group; a simple approach to featureless estimation of parameters. TR 338. MIT, Cambridge, MA. http://hi.eecg.toronto.edu/tip.html. Also appears in IEEE Trans. Image Proc. 1997, 6 (9), 1281–1295.

Mann, S., Janzen, R., Hobson, T., 2011. Multisensor broadband high dynamic range sensing. In: Proc. Tangible and Embedded Interaction (TEI 2011), pp. 21–24.

Mann, S., Lo, R.C.H., Ovtcharov, K., Gu, S., Dai, D., Ngan, C., Ai, T., 2012. Realtime HDR (high dynamic range) video for eyetap wearable computers, FPGA-based seeing aids, and glasseyes (EyeTaps). In: 25th IEEE Canadian Conference on Electrical & Computer Engineering (CCECE), 2012, pp. 1–6.

Pal, C., Szeliski, R., Uyttendaele, M., Jojic, N., 2004. Probability models for high dynamic range imaging. In: Proceedings of the 2004 IEEE Computer Society Conference on Computer Vision and Pattern Recognition, 2004. CVPR 2004, vol. 2, pp. II-173–II-180.

Poynton, C., 1996. A Technical Introduction to Digital Video. John Wiley & Sons, New York.

Reinhard, E., Ward, G., Pattanaik, S., Debevec, P., 2005. High Dynamic Range Imaging: Acquisition, Display, and Image-Based Lighting (The Morgan Kaufmann Series in Computer Graphics). Morgan Kaufmann Publishers Inc., San Francisco, CA.

Robertson, M.A., Borman, S., Stevenson, R.L., 2003. Estimation-theoretic approach to dynamic range enhancement using multiple exposures. J. Electron. Imaging 12, 219.

Ryals, C., 1995. Lightspace: a new language of imaging. PHOTO Electron. Imaging 38 (2), 14–16.

Sakurai, J., 1994. Modern Quantum Mechanics, Revised Edition. Addison-Wesley Publishing Co. Inc., Reading, MA.

Shakhakarmi, N., 2014. Next generation wearable devices: smart health monitoring device and smart sousveillance hat using D2D communications in LTE assisted networks. Int. J. Interdiscip. Telecommun. Netw. 6 (2), 25–51.

Szeliski, R., 1996. Video mosaics for virtual environments. Comput. Graph. Appl. 16 (2), 22–30.

Stockham Jr., T.G., 1972. Image processing in the context of a visual model. Proc. IEEE 60 (7), 828–842.

Woon, S., 1995. Period-harmonic-tupling jumps to chaos and fractal-scaling in a class of series. Chaos, Solitons Fractals 5 (1), 125–130.

Wyckoff, C.W., 1962. An experimental extended response film. S.P.I.E. Newsletter, June–July, 16–20.

Wyckoff, C.W., 1961. An experimental extended response film. Tech. Rep. NO. B-321. Edgerton, Germeshausen & Grier, Inc., Boston, MA.

PART

I

CONTENT ACQUISITION AND PRODUCTION

UNIFIED RECONSTRUCTION OF RAW HDR VIDEO DATA

J. Unger, S. Hajisharif, J. Kronander
Linköping University, Norrköping, Sweden

CHAPTER OUTLINE

2.1 INTRODUCTION

Most real scenes exhibit a dynamic range on the order of 1,000,000:1 or more. This is far more than what a typical digital camera can capture in a single exposure. Current commercial high-end cameras with CMOS and CCD imaging sensors typically capture images with a dynamic range on the order of 10,000:1, quantized into 12–14 bits per pixel. There are a few specialized image sensors designed to have a logarithmic response that achieves a higher dynamic range than standard CMOS and CCD sensors. These are, however, in most applications still not accurate enough because of problems with

High Dynamic Range Video. http://dx.doi.org/10.1016/B978-0-08-100412-8.00002-4

low image resolution, because of excessive image noise in darker regions, the entire dynamic range is usually quantized into a 10–12 bit output. In recent years, we have seen a significant increase in the computational power onboard cameras as well as a higher degree of control over the actual sensors. This produces much greater freedom in the way we capture and reconstruct images, even with consumer cameras. Currently there are three major approaches to capturing high dynamic range (HDR) video with off-the-shelf image sensor technology:

(1) The traditional approach is based on temporal multiplexing, where a sequence of images are captured with varying exposure times and then merged to form an HDR image. However, this approach works best for static scenes. For dynamic scenes, nonrigid registration of the individual exposures is necessary, making the problem much less robust for HDR video capture and in general requires computationally demanding registration algorithms to be run, limiting real-time capture. An overview of the problems and solutions for systems of this type is given in Chapter 3.

(2) The second approach uses optical elements (eg, beamsplitters) to project the incident optical image onto multiple image sensors. To achieve varying exposures of the scene on the different sensors, optical filters can be inserted in front of the individual sensors or different ISO/gain settings can be used. These optical designs also make it possible to vary the percentage of incident light that is projected onto each sensor. A major advantage of these systems is that use of a common exposure time for all sensors enables robust capture of a dynamic scene.

(3) The third approach is to use spatial multiplexing. Here, a single image sensor is used where the response to incident light varies over the sensor. The spatially varying response is often achieved by the placing of an optical filter array in front of the sensor, but there are also approaches in which, for example, the ISO/gain response is varied over the sensor. This design avoids the need for more than one sensor, and allows for robust HDR video capture in dynamic scenes. Its most familiar application is color imaging via a color filter array (CFA) (commonly a Bayer pattern).

In this chapter we describe a unified framework for capture and reconstruction of HDR video and images using the second and third approaches — that is, systems splitting the incident light onto multiple sensors and those that use a spatially varying sensor response. A common criterion for the systems we discuss in this chapter is that they do not vary the exposure time, but instead use optical setups, neutral density (ND) filters, or ISO variations so that the different sensors or pixels capture different exposures. The main advantage of these systems is that they do not suffer from problems with temporal registration and motion blur artifacts often encountered when images captured with varying exposure times are fused.

To reconstruct HDR video frames from the output of these optical designs, it is necessary to perform several steps, such as demosaicing (reconstruction of full resolution color from CFA data), realignment (correction for possible geometric misalignments between multiple sensors), HDR assembly to fuse the input low dynamic range (LDR) images into the output HDR image, and denoising to reduce excessive image and color noise.

Most existing HDR reconstruction techniques (eg, Debevec and Malik, 1997; Ajdin et al., 2008; Granados et al., 2010; Tocci et al., 2011) treat these steps in a pipeline fashion and perform demosaicing (and realignment) either before or after HDR assembly; see Fig. 2.1. This pipeline approach introduces several problems. Demosaicing before HDR assembly as in Debevec and Malik (1997) causes problems with bad or missing data around saturated pixels. This is especially problematic as color channels usually saturate at different levels. Demosaicing after HDR assembly as in Ajdin et al. (2008) and

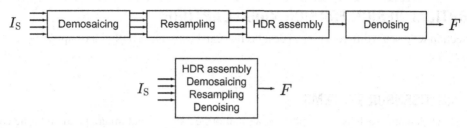

FIGURE 2.1

The vast majority of previous HDR reconstruction methods have considered *demosaicing, resampling, HDR assembly*, and *denoising* as separate problems. We describe a unified approach in which all these steps are performed in a single filtering step.

Tocci et al. (2011) causes problems with blur and ghosting unless the sensors are perfectly aligned. For multisensor systems using high-resolution sensors, it is problematic, costly, and sometimes even impossible to match the sensors so that the CFA patterns align correctly. Tocci et al. (2011) reported misalignment errors around the size of a pixel despite considerable alignment efforts. This means that overlapping pixels from different sensors might not have the same spectral response. Also, treating demosaicing, resampling, and HDR assembly in separate steps makes it difficult to analyze and incorporate sensor noise in the HDR reconstruction in a coherent way.

The reconstruction framework described in this chapter unifies these steps into a single processing operation. Based on a general image formation model that includes both multisensor and spatially multiplexed capturing system, the reconstruction takes into account the entire set of nonsaturated low dynamic range samples available around the output HDR pixel, and performs demosaicing, resampling, HDR assembly, and denoising in a single operation. An essential foundation of our approach is the inclusion of a statistical noise model describing the heterogenous sensor noise. The noise model enables us to express the reconstruction as a statistical estimation problem. To recover the true irradiance reaching the sensor, we fit local polynomial approximations (LPA) to measured pixel values in a local region of the reconstruction point using a local maximum likelihood approach (Kronander et al., 2013). To reconstruct sharp edges we use adaptive anisotropic filtering supports, taking the correlation between color channels into account (Kronander et al., 2014).

2.1.1 OUTLINE

We start by discussing different optical setups for robust HDR video capture using multiple sensors and spatially varying sensor responses in Section 2.2. We then model these systems using a general image formation model in Section 2.3, detailing how the incident radiant power is transformed into noisy pixel responses. On the basis of the image formation model, we then show how the measured pixel responses can be used to reconstruct high-quality, high SNR HDR images in Section 2.4. To validate our approach, in Section 2.5 we show two example applications using our reconstruction framework: reconstructions from multisensor data and reconstructions from input data captured with use of spatially varying pixel responses. The spatial variations were captured with a Canon 5D consumer camera where the per-pixel gain (ISO) varies over the sensor.

2.2 OPTICAL DESIGN FOR HDR VIDEO CAPTURE

In this section we discuss different optical designs that can be used for HDR video capture with a single exposure time.

2.2.1 MULTISENSOR SYSTEMS

Using optical elements such as beamsplitters, we can project the optical image incident to the camera onto several sensors. We can achieve different exposures for the sensors by placing ND filters in front of the sensors, using different ISO/gain settings for the sensors, or splitting the light unevenly onto the sensors. To eliminate registration and motion blur problems, the exposure time of the sensors is often synchronized.

By placing the beamsplitters in front of the lens, we can direct the light directly into several standard cameras each with a separate lens. Froehlich et al. (2014) used such a setup with two commercial Arri Alexa cameras to capture a dynamic range of up to 18 f-stops. This procedure can also be applied recursively to construct so-called optical splitting trees where the light is projected onto multiple cameras. McGuire and Hughes (2007) presented a framework for optimally selecting components for such optical splitting trees given specific target goals, such as dynamic range, spectral sensitivity, and cost budgets. However, these systems are limited as the separate lenses must be perfectly matched, and zoom and focus settings can be difficult to maintain between them. In addition, putting the beamsplitter in front of the camera lens often places limitations on the field of view. Setups of this type tend to be quite bulky, often prohibiting the design of a single handheld unit.

Alternatively, the beamsplitters can be placed behind the camera lens, inside the camera housing. Aggarwal and Ahuja (2004) presented one of the earliest multisensor systems for HDR video capture using a mirror pyramid to split the incoming light and project it through ND filters before it reached separate sensors. A modern multisensor camera developed by Spheron VR and Linköping University is shown in Fig. 2.2. The camera uses four sensors with different ND filters introduced in front, and can capture a dynamic range of up to 24 f-stops. While traditional setups often waste a large part of the incident light because of the ND filters, Tocci et al. (2011) presented a light-efficient compact multisensor HDR video system utilizing up to 99.96% of the incident light. Instead of the system

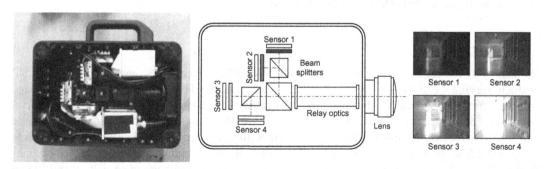

FIGURE 2.2

A standard multisensor camera system.

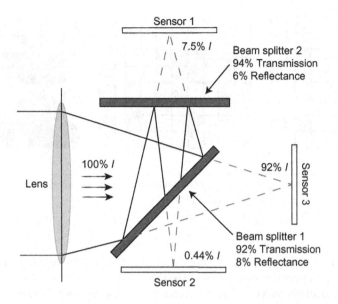

FIGURE 2.3

Example of a light-efficient multisensor system, proposed by Tocci et al. (2011).

splitting the light equally and relying on ND filters to achieve different exposures, the optical system is designed so that the light is split unevenly between the sensors. This is achieved by recursive splitting of the light. One can design a compact system by directing some of the light passing through the first beamsplitter back through the beamsplitter a second time, thus reusing the optical elements; see Fig. 2.3.

The use of traditional beamsplitters often results in custom-built setups with specialized camera bodies. To enable HDR capture with off-the-shelf cameras, Manakov et al. (2013) recently proposed a small optical element that can be inserted between the lens and body of a standard commercial DSLR camera. This element splits the incident light into four identical images, optically filters each image with a separate ND filter, and finally projects each image to a quarter of the same sensor. This setup thus introduces a trade-off between spatial resolution on the sensor and the number of subimages/exposures used, but can be used directly with standard cameras.

2.2.2 SPATIALLY VARYING SENSOR EXPOSURE

One of the simplest optical designs to achieve single-shot HDR imaging would introduce an optical mask with spatially varying transmittance over the image sensor, allowing the amount of light reaching the pixels to change over the sensor. The mask can be introduced just in front of the camera sensor, similarly to a traditional Bayer pattern used to capture color, or in the lens element. This introduces a trade-off between the spatial resolution, noise level, and dynamic range that can be captured. This approached was first proposed by Nayar and Mitsunaga (2000), who introduced an ND filter mask with

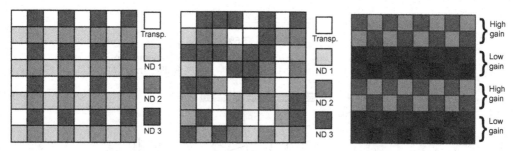

FIGURE 2.4

Spatially varying filter mask (left), spatially varying filter mask with random filter positions (middle), and varying ISO/gain in combination with a Bayer filter mask (right).

four different optical densities (transparent, highly transmissive, moderately transmissive, and almost opaque) in a regular grid over the image sensor; see Fig. 2.4.

When designing spatially varying exposure systems, one has to consider two main design criteria. The first is the number of optical filters that are used in the mask (ie, how many different exposures of the scene are being captured). A higher number of filters enables a wider dynamic range to be captured but leads to an overall lower spatial resolution, which may introduce visual artifacts in the reconstructed HDR images. Secondly, the spatial distribution of the different filters will affect the acquired data. Both regular patterns and stochastic (random) patterns can be used. The choice of pattern is important as pixels with highly transmissive filters will most likely be saturated in high-intensity regions, and interpolation will be required to reconstruct these values. If the sampling patterns are regular, aliasing artifacts may appear in the interpolation. In contrast, if a random or pseudorandom pattern is used, aliasing can be avoided or significantly suppressed (Schoberl et al., 2012; Aguerrebere et al., 2014). However, these patterns can be harder to use for practical purposes.

For HDR color capture, the spatially varying exposure approach can be combined with Bayer filter designs by use of a spatially varying exposure and a CFA (Narasimhan and Nayar, 2005) instead of a traditional Bayer filter. A further generalization was presented by Yasuma et al. (2010), who optimized the pixel filter patterns given the optical resolution limit of the imaging optics in both the spatial domain and the spectral domain.

One disadvantage with approaches that introduce a filter array over the sensor is that some of the incident light is blocked by the filters and never reaches the image sensor. This can be a major concern in darker scenes, and in general leads to increased image noise. An alternative solution that uses all the incident light is based on recent (commercial) imaging sensors where the pixel gain can be varied over the sensor (Unger et al., 2004; Unger and Gustavson, 2007; alex, 2013; Hajisharif et al., 2014). The analog pixel gain is proportional to the ISO setting found on most cameras. A low gain setting leads to a high saturation threshold but a lower signal-to-noise ratio in dark regions compared with a high gain setting. This approach can also be combined with traditional Bayer patterns, resulting in a multigain RAW sensor image where color is captured with a CFA (eg, a Bayer pattern). Varying sensor gain can be achieved in existing commercial digital cameras with a customized firmware update (eg, the Magic Lantern firmware for Canon cameras). This firmware allows the sensor to use two ISO settings simultaneously in a single frame, with varying gain for every second row. The Bayer pattern that the dual-ISO module creates is depicted in Fig. 2.4. Since the ISO values are changing for every second row, it covers the RGGB Bayer pattern for that exposure setting.

2.3 IMAGE FORMATION MODEL

This section describes how HDR video systems based on spatial multiplexing and multisensor setups can be described with a single unified image formation model. This model then forms the foundation for the statistically motivated reconstruction framework presented in Section 2.4.

We view each pixel in the input images as a measurement of the incident radiant power at the pixel location, scaled by an unknown factor due to the quantum efficiency of the sensor. For convenience, we will express the incident radiant power reaching the image as the number of photoinduced electrons collected per unit time. Each sensor image, I_s, samples the incident radiant power, f, at a set of discrete pixel locations. Using a linear index i for pixels in each sensor image, we define the measured digital sample value at a pixel i in image I_s as $y_{s,i}$. The samples, $y_{s,i}$, contain measurement noise that is dependent on the incident radiant power, $f_{s,i}$, as well as the sensor characteristics. During capture, we assume that the sensors are synchronized or that the scene is static, so that the motion blur characteristics are the same for all sensors. The parameters that are allowed to vary between the sensors are the sensor gain/ISO setting, g_s, and the exposure scaling coefficient, n_s, which depends on the fraction of the incident light reaching the sensor (eg, because of ND filters or uneven splitting of the incident light onto the different sensors in a multisensor system).

The (noisy) digital pixel values $y_{s,i}$ in the input images, I_s, are first converted to radiant power estimates $\hat{f}_{s,i}$. This is done by subtraction of a bias frame and our taking into account the exposure settings (exposure time, exposure scaling, global gain) and per-pixel gain variations. The variance of the radiant power, $\sigma^2_{f_{s,i}}$, can then be estimated by a calibrated radiometric camera noise model. Similarly to Granados et al. (2010), we assume the noise follows a Gaussian distribution. This assumption is described in detail in Section 2.3.1, where we give an in-depth overview of our radiometric noise model adapted for HDR video. Our framework can use any radiometric noise model that assumes an approximately Gaussian noise distribution.

2.3.1 SENSOR NOISE MODEL

Each observed digital pixel value, $y_{s,i}$, is obtained with an exposure time, t_s, a gain, $g_{s,i}$, and an exposure scaling, $n_{s,i}$. The accumulated number of photoelectrons, $e_{s,i}$, collected at the pixel during the exposure time, t_s, follows a Poisson distribution with expected value

$$E[e_{s,i}] = t_s \left(a_{s,i} n_s f_{s,i} \right) \tag{2.1}$$

and variance $\text{Var}\left[e_{s,i}\right] = E[e_{s,i}]$, where $a_{s,i}$ is a per pixel factor due to a nonuniform photoresponse. The recorded digital value, $y_{s,i}$, is also affected by the sensor gain and signal independent readout noise,

$$y_{s,i} = g_s(e_{s,i}) + r_{s,i}(g), \tag{2.2}$$

where g_s is the analog amplifier/sensor gain (proportional to the native ISO settings on modern cameras) and $r_{s,i}(g)$ is the readout noise, which generally depends on the gain setting of the sensor. As we focus on video applications with frame rates of 25 frames per second or more, the dark current noise is ignored. For most modern camera sensors, dark current noise has a negligible effect for exposures of less than 1 s (Hasinoff et al., 2010). In contrast to previous work (Hasinoff et al., 2010; Kronander et al., 2013), we do not assume a simple parametric model for the readout noise dependence on the gain, as we have found that such models do not generally provide a satisfactory fit to measured readout noise at different gain settings for modern camera sensors. To handle sensors with different gain settings, we instead calibrate the readout noise, $r_{s,i}(g)$, for each separate gain/ISO setting considered.

Before reconstruction of the HDR output frame, each digital pixel value, $y_{s,i}$, is independently transformed to an estimate of the number of photoelectrons reaching the pixel per unit time, $\hat{f}_{s,i}$. To compensate for the readout noise bias (blacklevel), $E[d_{s,i}(g, t)]$, we subtract a bias frame, $b_{s,i}$, from each observation. The bias frame is computed as the average of a large set of black images captured with the same camera settings as the observations but with the lens covered, so that no photons reach the sensor. The radiant power $f_{s,i}$ can then be estimated as

$$\hat{f}_{s,i} = \frac{y_{s,i} - b_{s,i}}{g_s t_s a_{s,i} n_s}. \tag{2.3}$$

2.3.2 VARIANCE ESTIMATE

We assume that $\hat{f}_{s,i}$ follows a normal distribution with mean $f_{s,i}$ and standard deviation $\sigma_{\hat{f}}$. For bright image regions, the Poisson distributed shot noise is well approximated by a normal distribution. For low light levels, the photon shot noise is generally dominated by signal-independent readout noise, which can also be approximated to follow a normal distribution. Assuming there are no saturated pixel values, the variance of $\hat{f}_{s,i}$ is given by

$$\text{Var}\left[\hat{f}_{s,i}\right] = \sigma_{\hat{f}}^2 = \frac{g_s^2 \left(\text{Var}\left[e_{s,i}\right]\right) + \text{Var}\left[r_{s,i}(g, t)\right]}{g_s^2 t_s^2 a_{s,i}^2 n_s^2}. \tag{2.4}$$

Because of the photon shot noise, to estimate, $\text{Var}\left[e_{s,i}\right]$ we should ideally know the true value of $f_{s,i}$; however, in practice we are instead forced to use the (noisy) estimate, $\hat{f}_{s,i}$, obtained from Eq. (2.3). Thus, we form an approximate variance estimate, $\hat{\sigma}_{\hat{f}_{s,i}}^2$, as

$$\hat{\sigma}_{\hat{f}_{s,i}}^2 = \frac{g_s^2 t_s a_{s,i} n_s \hat{f} + \text{Var}\left[r_{s,i}(g, t)\right]}{g_s^2 t_s^2 a_{s,i}^2 n_s^2}. \tag{2.5}$$

Assuming that the effect of pixel cross talk is negligible, we make the approximation that the variances, $\sigma_{\hat{f}_{s,i}}^2$, are independent of each other.

The analysis above assumes there are nonsaturated pixels. In practice, some pixels (eg, direct light sources or the Sun) may, of course, be saturated. To avoid the spreading of unreliable, saturated, or dead pixels, each digital pixel value, $y_{s,i}$, is compared with a *saturation frame* describing at which value each pixel saturates. Dead pixels are set to zero in the saturation frame to effectively remove any possible error introduced. More complicated models of sensor noise also include the effects of saturation on the digital value variance, $\text{Var}\left[y_{s,i}\right]$ (see, eg, Foi, 2009). However such models assume $y_{s,i}$ follows a clipped normal distribution, which requires an impractical and computationally costly nonlinear estimation when one is computing the coefficients for the LPAs discussed in Section 2.4.

2.3.2.1 Parameter calibration

The exposure scaling, n_s, and per-pixel nonuniformity, $a_{s,i}$, can, in practice, be estimated with a flat field image computed as the average over a large sequence of images, running the sensors at different exposure times to ensure well-exposed pixels for all sensors. If we choose one sensor as a reference, the exposure scaling, n_s, can be estimated as the spatial mean of the ratio, computed pixelwise between the radiant power estimates, Eq. (2.3), of the sensors. The nonuniformity, $a_{s,i}$, can then be computed as the per-pixel deviation from the exposure scaling, n_s.

The variance of the readout noise, $\text{Var}\left[r_{s,i}(g,t)\right]$, can be estimated, similarly to the bias frame, from a set of black images captured with the same camera settings, (g,t), as the observations but with the lens covered so that no photons reach the sensor.

The sensor gain, g_s, can be calibrated with the relation,

$$\frac{\text{Var}\left[y_{s,i}\right] - \text{Var}\left[b_{s,i}\right]}{E[y_{s,i}] - E[b_{s,i}]} = \frac{g_s^2 \text{Var}\left[e_{s,i}\right]}{g_s E[e_{s,i}]} = g_s, \tag{2.6}$$

where the second equality follows from $e_{s,i}$ being Poisson distributed shot noise with $E[e_{s,i}] = \text{Var}\left[e_{s,i}\right]$. $E[y_{s,i}]$ and $E[b_{s,i}]$ can be estimated by averaged flat fields and the bias frame, respectively, and $\text{Var}\left[b_{s,i}\right]$ as described above.

2.4 HDR RECONSTRUCTION

The HDR reconstruction is performed in a virtual reference space, corresponding to a virtual sensor placed somewhere in the focal plane. The virtual sensor dimensions are chosen to reflect the desired output frame, unconstrained by the resolution of the input frames. For single sensor setups using spatial multiplexing, the virtual sensor is usually chosen to coincide with the real sensor. For multisensor systems, it is often necessary to register the sensors. Assuming that the sensors, s, lie in the same virtual focal plane in the split optical path, the relations between the sensors are described by affine transformations, T_s, mapping sensor pixel coordinates, $x_{s,i}$, to the coordinates of the reference output coordinate system, $X_{s,i} = T_s(x_{s,i})$. The transformations T_s can be estimated by a simple calibration procedure (eg, imaging a known target with features visible in all sensors). If the sensors, s, keep their configuration between frames, the transformed coordinates, $X_{s,i}$, can be precomputed once. In general, the transformed input pixel coordinates, $X_{s,i}$, are not integer valued in the output image coordinates, and for general affine transformations the sample locations $X_{s,i}$ will become irregularly positioned. An example for a multisensor system using three separate sensors and a single sensor using a spatially varying ND filter is shown in Fig. 2.5.

To reconstruct the HDR output frame, we estimate the relative radiant power at each pixel location in the virtual sensor separately in a nonparametric fashion using the transformed samples $\hat{f}_{s,i}(X_{s,i})$. For each output pixel j with integer-valued, regularly spaced coordinates X_j, a local polynomial model is

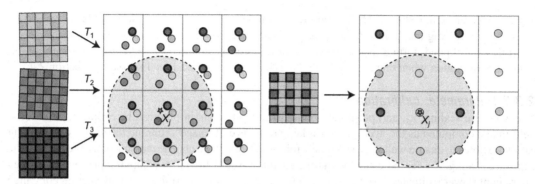

FIGURE 2.5

HDR reconstruction is performed for a virtual reference system with arbitrary resolution and mapping. The sensor samples are mapped to the reconstruction grid with use of the transformations T_s. The pixel at position X_j in the HDR image is estimated by an LPA of nearby samples (within the shaded circle). Saturated and defective pixels are discarded, and samples near the black level, local outliers, and samples farther from the reconstruction point contribute less to the radiance estimate.

fitted to observed samples $\hat{f}_{s,i}(X_{s,i})$ in a local neighborhood; see Fig. 2.5. We first discuss how a single color channel, $c = \{R\|G\|B\}$, can be reconstructed independently of the other channels with an LPA with isotropic filtering support. In Section 2.4.4, we then show how the LPA estimate can be improved by use of adaptive anisotropic filtering support. Use of adaptive filtering support also enables us to use cross-correlation between color channels to improve the estimate. This is discussed in Section 2.4.5.

2.4.1 LOCAL POLYNOMIAL MODEL

To estimate the radiant power, $f(x)$, at an output pixel, we use a generic local polynomial expansion of the radiant power around the output pixel location $X_j = [x_1, x_2]^T$. Assuming that the radiant power $f(x)$ is a smooth function in a local neighborhood around the output location X_j, an Mth order Taylor series expansion is used to predict the radiant power at a point X_i close to X_j as

$$\tilde{f}(X_i) = C_0 + C_1(X_i - X_j)$$
$$+ C_2 \text{tril}\{(X_i - X_j)(X_i - X_j)^T\} + \cdots, \tag{2.7}$$

where tril lexicographically vectorizes the lower triangular part of a symmetric matrix and where

$$C_0 = f(X_j), \tag{2.8}$$

$$C_1 = \nabla f(X_j) = \left[\frac{\partial f(X_j)}{\partial x_1}, \frac{\partial f(X_j)}{\partial x_2} \right], \tag{2.9}$$

$$C_2 = \frac{1}{2}\left[\frac{\partial^2 f(X_j)}{\partial x_1^2}, 2\frac{\partial^2 f(X_j)}{\partial x_1 \partial x_2}, \frac{\partial^2 f(X_j)}{\partial x_2^2} \right]. \tag{2.10}$$

Given the fitted polynomial coefficients, $C_{1:M}$, we can thus predict the radiant power at the output location X_j by $C_0 = f(X_j)$, and the first-order gradients by C_1. In this work, we consider only local expansions of order $M \leq 2$.

2.4.2 MAXIMUM LOCALIZED LIKELIHOOD FITTING

To estimate the coefficients, C, of the local polynomial model, we maximize a localized likelihood function (Tibshirani and Hastie, 1987) defined by a smoothing window centered around X_j:

$$\mathcal{W}_H(X) = \frac{1}{\det(H)} \mathcal{W}(H^{-1}X), \tag{2.11}$$

where H is a 2×2 smoothing matrix that determines the shape and size of the window. In Section 2.4.3.1, we discuss how the shape and size of the window function and smoothing can be selected. To lessen the notational burden, the observed samples in the neighborhood, $\left\{ \hat{f}_{s,i}(X_{s,i}) : X_{s,i} \in \text{supp}(\mathcal{W}_H(X)) \right\}$, are denoted below by f_k with a linear index $k = 1 \ldots K$. Under the assumption of normally distributed radiant power estimates f_k, see Section 2.3.1, the log of the localized likelihood for the polynomial expansion centered at X_j is given by

$$L(X_j, C) = \sum_k \log \left(N \left(f_k | \tilde{f}(X_k), \hat{\sigma}_{f_k}^2 \right) \mathcal{W}_H(X_k) \right)$$

$$= \sum_k \left[f_k - C_0 - C_1 (X_i - X_j) - C_2^T \text{tril} \left\{ (X_i - X_j)(X_i - X_j)^T \right\} - \cdots \right]^2 \frac{\mathcal{W}_H(X_k)}{\hat{\sigma}_{f_k}} + R,$$

where R represents terms independent of C. The polynomial coefficient, $\tilde{C}$, maximizing the localized likelihood function is found by the weighted least squares estimate:

$$\tilde{C} = \arg\max_{C \in \mathcal{R}^M} (L(X_j, C))$$

$$= (\Phi^T W \Phi)^{-1} \Phi^T W \tilde{f}, \tag{2.12}$$

where

$$\tilde{f} = [f_1, f_2, \ldots f_K]^T,$$

$$W = \text{diag} \left[\frac{\mathcal{W}_H(X_1)}{\hat{\sigma}_{f_1}^2}, \frac{\mathcal{W}_H(X_2)}{\hat{\sigma}_{f_2}^2}, \ldots, \frac{\mathcal{W}_H(X_K)}{\hat{\sigma}_{f_k}^2} \right],$$

$$\Phi = \begin{bmatrix} 1 & (X_1 - X_j) & \text{tril}^T \left\{ (X_1 - X_j)(X_1 - X_j)^T \right\} & \cdots \\ 1 & (X_2 - X_j) & \text{tril}^T \left\{ (X_2 - X_j)(X_2 - X_j)^T \right\} & \cdots \\ \vdots & \vdots & \vdots & \vdots \\ 1 & (X_K - X_j) & \text{tril}^T \left\{ (X_K - X_j)(X_K - X_j)^T \right\} & \cdots \end{bmatrix}.$$

Note that similarly to the kernel regression framework (Takeda et al., 2007) using equivalent kernels, we can also formulate the solution to the weighted least squares problem as a locally adaptive linear filtering.

2.4.3 PARAMETERS

The expected mean square error of the reconstructed image depends on a trade-off between bias and variance of the estimate. This trade-off is determined by the order of the polynomial basis M, the window function $\mathcal{W}$, and the smoothing matrix H.

For a piecewise constant polynomial, $M = 0$, the estimator corresponds to an ordinary locally weighted average of neighboring sensor observations:

$$\hat{f}(X_j) = \frac{\sum_k \frac{\mathcal{W}_H(X_k)}{\hat{\sigma}_{f_k}^2} f_k}{\sum_k \frac{\mathcal{W}_H(X_k)}{\hat{\sigma}_{f_k}^2}}. \tag{2.13}$$

However, locally weighted averages can exhibit severe bias at image boundaries and in regions around sensor saturation, due to the asymmetry of the number of available sensor measurements in these locations. By instead fitting a linear polynomial (a plane, $M = 1$), we can reduce the bias significantly; see Fig. 2.6. Introducing higher-order polynomials is possible, but may lead to increased variance in the estimates.

2.4.3.1 Window and scale selection

Here, we show only results obtained with a Gaussian window function because of its simplicity and widespread use:

$$\mathcal{W}_H(X_k) = \frac{1}{2\pi \det(H)} \exp\left\{-(X_k - X_j)^T H^{-1}(X_k - X_j)\right\}. \tag{2.14}$$

There are also other possible choices of symmetric smooth window functions (eg, Epanechnikov or Tricube windows Astola et al., 2006).

The 2×2 smoothing matrix, H, affects the shape of the Gaussian. It is important that the support of the smoothing window is extended to incorporate enough samples into the estimate in Eq. (2.12) so

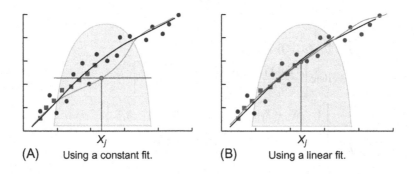

X_j	X_j
(A) Using a constant fit.	(B) Using a linear fit.

FIGURE 2.6

One-dimensional example showing the effect of the polynomial order, M, in areas affected by sensor saturation. The squares represent a sensor saturating at lower values than the sensor represented by the circles. (A) Use of a constant fit, $M = 0$, introduces significant bias as more unsaturated measurements are available to the left of the reconstruction point. (B) With use of a linear fit, $M = 1$, the bias is reduced drastically.

that ($\Phi^T W \Phi$) is invertible (full rank). The simplest choice of the smoothing matrix is $H = hI$, where h is a global scale parameter and I is the identity matrix. This corresponds to an isotropic filter support. The choice of h is dependent on the sensor characteristics (noise) and the scene. We therefore treat h as a user parameter. To reconstruct sharp images in low-noise conditions, we generally choose the smallest possible h such that ($\Phi^T W \Phi$) is invertible. If the captured images exhibit severe noise, a larger scale parameter may be beneficial. For Bayer CFA sensors, we use different scale parameters for the green and red/blue color channels, $h_G = \frac{h_{R,B}}{\sqrt{2}}$, as there are more green samples per unit area. For the remainder of this chapter, we will refer to the reconstruction method using a global smoothing matrix, $H = hI$, as the *local polynomial approximation (LPA)*.

2.4.4 ADAPTIVE LPA RECONSTRUCTION

The truncated local polynomial expansion assumes that the radiant power, $f(x)$, is a smooth function (up to order M) in a neighborhood of the reconstruction point, X_j. In many image regions these assumptions are not valid. For example, if an output pixel is located near a sharp edge, it cannot be represented accurately with a finite polynomial expansion. It is therefore desirable to adapt the window support so that it includes only observations on the same side of the edge.

To adapt the support of the local polynomial model to the data, we use a two-step approach. First, we use the LPA with $M \geq 1$ and isotropic window support to compute an initial estimation of the local gradients, $\left[\frac{\partial f(X_i)}{\partial x_1}, \frac{\partial f(X_i)}{\partial x_2} \right]$. Using the estimated gradients, we then locally adapt the smoothing matrix, H_j, to the local structure of the signal at each output pixel X_j.

We define an oriented, anisotropic smoothing matrix as $H_j^s = hC_j$, where C_j represents the local gradient covariance of the signal estimated from the data. To estimate C_j, we use a parametric approach similar to that of Takeda et al. (2007). Specifically, we consider a decomposition of the covariance matrix as

$$C_j = \gamma_j U_{\theta_j} \Lambda_j U_{\theta_j}, \tag{2.15}$$

$$U_{\theta_j} = \begin{bmatrix} \cos(\theta_j) & \sin(\theta_j) \\ -\sin(\theta_j) & \cos(\theta_j) \end{bmatrix}, \tag{2.16}$$

$$\Delta_j = \begin{bmatrix} \sigma_j & 0 \\ 0 & \frac{1}{\sigma_j} \end{bmatrix}. \tag{2.17}$$

This corresponds to describing the covariance matrix as a function of three parameters: σ_j, describing an elongation along the principal directions; θ_i, describing a rotation angle; and γ_i, describing an overall scaling. The elongation, rotation, and scaling parameters are estimated from a truncated singular value decomposition of weighted gradients in a local neighborhood around the output location X_j:

$$G = \begin{bmatrix} \vdots & \vdots \\ f^{x_1}(X_i) & f^{x_2}(X_i) \\ \vdots & \vdots \end{bmatrix} = U_j S_j V_j \quad \forall X_i \in \mathrm{supp}\{w_i\}, \tag{2.18}$$

where $f^{x_1}(X_i)$ and $f^{x_2}(X_i)$ are the estimated gradients along x_1 and x_2 at the nearby point X_i, and w_i represents a local window function. The singular values s_1 and s_2 along the diagonal of S_j represent the energy along the dominant directions in the neighborhood defined by the local window function. Thus, the orientation angle is set such that the window is elongated orthogonal to the dominant gradient direction — that is, from the second column of V_j, $v = [v_1, v_2]^T$ as

$$\theta_j = \arctan\left(\frac{v_1}{v_2}\right). \tag{2.19}$$

The elongation parameter, σ_j, is then set according to

$$\sigma_j = \frac{s_1 + \lambda_1}{s_2 + \lambda_1}, \lambda_1 \geq 0, \tag{2.20}$$

where λ_1 is a regularization parameter. The scaling parameter, γ_j, is finally set according to

$$\gamma_i = \left(\frac{s_1 s_2 + \lambda_2}{M}\right)^{\alpha}, \tag{2.21}$$

where λ_2 is another regularization parameter, α is the structure sensitivity parameter, and M is the number of samples in the local neighborhood of the gradient analysis window w. Intuitively, the shape of the window support should be circular and relatively large in flat areas, elongated along edges, and small in textured areas to preserve detail. For all results presented in this chapter, we fix the regularization parameters to $\lambda_1 = 1$ and $\lambda_2 = 0.001$. The structure sensitivity parameter, α, is treated as a user-specified parameter, representing a trade-off between denoising and detail preservation. For a more detailed analysis of the choice of parameters, see Takeda et al. (2007) and Milanfar (2013). Note that since we are primarily interested in irregularly sampled data, we choose to adapt the window functions instead of the support of each observation (kernel) as in the work of Takeda et al. (2007).

2.4.5 COLOR CHANNEL CORRELATION

Modern demosaicing methods consider the correlation between color channels to improve CFA interpolation. Commonly used heuristics are that edges should match between color channels and that interpolation should be performed along and not across edges. For Bayer pattern CFA arrays, in which the green channel is sampled more densely than the red and blue channels, a common method is to first reconstruct the green channel, then use the gradients of the green channel when one is reconstructing the red and blue channels.

To take into account the correlation between color channels in our reconstruction, we adapt this idea to the irregularly sampled input data from multisensor camera systems. First we estimate the gradients for the green channel using the LPA with $M \geq 1$ and isotropic window supports. We then use the green channel gradients to locally adapt H_j as described in the previous section. Using the adapted H_j, we then compute the radiant power estimates for all color channels (red, green, and blue). This effectively forces the interpolation to be performed along and not across edges and preserves their location between the color channels. We refer to this reconstruction method as the color-adaptive local polynomial approximation (CALPA).

2.5 **EXAMPLE APPLICATIONS**

In this section we give an overview of two different applications, HDR reconstruction for multisensor systems and HDR reconstruction from image data captured spatially with varying per-pixel gain (ISO) with a Canon 5D consumer camera.

2.5.1 **MULTISENSOR HDR VIDEO CAPTURE**

The first application we consider is HDR video capture by multisensor imaging systems. To evaluate our algorithm, we compare its performance against other state-of-the-art methods. Since most other algorithms do not generalize to setups with arbitrary misalignments between the sensors, we also include comparisons with setups where the sensors are assumed to be perfectly aligned as well as setups where the sensors are only translationally offset from each other. We compare our algorithm against three other algorithms: first, the recent real-time multisensor reconstruction algorithm presented by Tocci et al. (2011); second, Debayer first — a method that performs demosaicing before HDR assembly, using heuristic linear radiometric weights; and third Debayer last — a method that performs demosaicing after HDR assembly, using radiometric weights inversely proportional to the variance estimate $\hat{\sigma}_f^2$ given by our sensor noise model (Section 2.3.1). For the above-mentioned methods we consider demosaicing using both bilinear interpolation and the gradient adaptive method presented by Malvar et al. (2004).

We base most of our comparisons on HDR image data generated from a camera simulation framework. In the simulator, the scene is represented by a virtually noise-free high-resolution HDR image, meticulously generated from a set of exposures captured 1 f-stop apart covering the dynamic range of the real scene. Each exposure is computed as an average of 100 frames, in this case captured with a Canon 5D SLR camera. The camera simulation framework allows us to generate images of arbitrary bit depth, sensor misalignments, exposure settings, and noise characteristics according to the model described in Section 2.3.1. For our experiments, we simulate different multisensor setups with different sensor misalignments, T_s, and different noise characteristics, given the simulation parameters g_s, n_s, t_s, and $\text{Var}[r_{s,i}]$. We focus our evaluation on simulated sensors with a difference of 3-4 f-stops between the sensor exposures as this is the commonest setup for multisensor HDR video systems in practice (Tocci et al., 2011).

In our first example, we demonstrate the LPA and CALPA using input data from three perfectly aligned sensors. Although this is not a fully realistic scenario in practice, it allows us to make comparisons with a wide range of previous methods which in many cases rely on this assumption. Fig. 2.7 displays two different examples comparing the LPA and the CALPA with the Debayer first algorithm and the Debayer last algorithm with two different debayering methods applied. The input data simulates three Canon 5D sensors captured with an exposure scaling of 4 f-stops apart. Compared with the Debayer first method, LPA reconstruction reduces color artifacts because of the joint CFA interpolation and HDR reconstruction. Another benefit of the LPA is the inherent denoising and robust smoothing and cross-sensor blending in regions where one or more of the sensors saturate. Compared with the Debayer last method, the LPA methods reduce color artifacts in high contrast regions (eg, around the light build filament). Compared with the LPA, the anisotropic window supports used in the CALPA reduce color artifacts in regions with high-frequency variations.

In Fig. 2.8, the LPA and the CALPA are compared with the real-time multisensor reconstruction method recently proposed by Tocci et al. (2011). The reconstructions were generated from three virtual

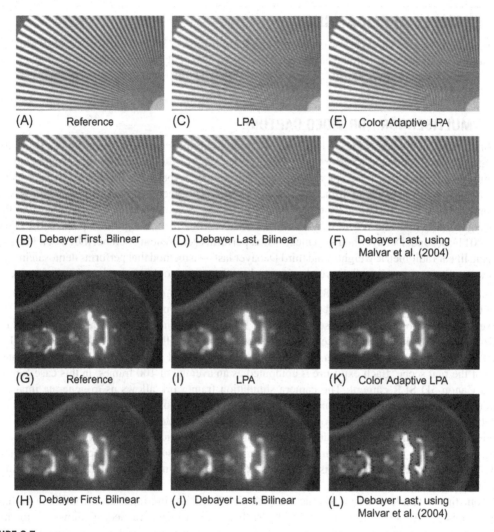

FIGURE 2.7

Reconstructed results from three perfectly aligned virtual Canon 5D sensors captured 3 f-stops apart. (A, G) Reference; (B, H) Debayer first, bilinear; (C, I) LPA; (D, J) Debayer last, bilinear; (E, K) CALPA; (F, L) Debayer Last, using the debayering method described by Malvar et al. (2004).

exposures of a simulated Kodak KAI-04050 sensor (used in our experimental HDR video system) captured 5 f-stops apart. The lowest-exposed and the highest-exposed sensors are perfectly aligned, while the middle sensor has a translational offset of [0.4, 0.45] in pixel units. This is similar to the alignment errors reported by Tocci et al. (2011) for their camera system. The method of Tocci et al. copes rather well with translational misalignments, because each reconstructed pixel uses information from only one sensor. However, the reconstruction suffers from noise. This is expected as the method of

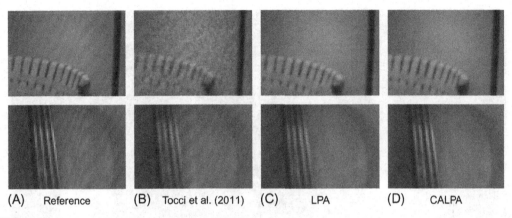

(A)   Reference   (B)   Tocci et al. (2011)   (C)   LPA   (D)   CALPA

FIGURE 2.8

Magnifications from reconstructions of a scene virtually exposed with three Kodak KAI-04050 sensors
5 f-stops apart. (A) Tone-mapped HDR reference images. (B) Reconstructions by the method proposed by
Tocci et al. (2011). (C) LPA $M = 1$, $h = 0.7$. (D) CALPA, $M = 1$, $h = 0.7$, $\alpha = 0.005$. In areas of sensor
transitions, the method of Tocci et al. (2011) exhibits severe noise as only one sensor is used at a time, as can
be seen in the magnification in the top row. In contrast, the LPA handles sensor transitions gracefully owing to
the incorporation of a sensor noise model and a statistical maximal likelihood fit to all sensor data. The CALPA
shows ever better results owing to the inherent denoising using adaptive filtering supports. Use of the
CALPA also helps in reducing unwanted color artifacts.

Tocci et al. considers only observations from the highest-exposed sensor at each point. In contrast, the
LPA and the CALPA are based on a maximum likelihood approach incorporating information from all
nonsaturated sensor observations. The CALPA effectively denoises the image by using larger filtering
supports in flat image areas. Compared with the other methods, the CALPA also reduces unwanted color
artifacts in areas of high spatial frequency. A robust multisensor reconstruction algorithm should handle
general subpixel misalignments between the sensors. Both the LPA and the CALPA can gracefully
handle any geometric misalignments between the sensors, including rotational errors.

Fig. 2.9 shows an example frame captured with the multisensor HDR video camera displayed in
Fig. 2.2. This camera system uses four high-quality Kodak KAI-04050 CCD sensors with a resolution
of 2336 × 1752 pixels and RGB Bayer pattern CFA sampling. The sensors receive different amounts
of light from an optical system with a common lens, a four-way beamsplitter, and four different ND
filters. All parts of the optical system except for a custom-built relay lens are off-the-shelf components.
The relay lens is used to extend the optical path to make room for the beamsplitters. The sensors
have 12-bit linear analog-to-digital conversion, and the exposure scalings cover a range of $1 : 2^{12}$,
yielding a dynamic range equivalent to $12 + 12 = 24$ bits of linear resolution, commonly referred to
as "24 f-stops" of dynamic range. The dynamic range can be extended further by variation of the per-
pixel gain between the sensors. The sensors are connected to a host computer through a CameraLink
interface. The system allows capture, processing, and off-line storage of up to 32 frames per second at
a resolution of four megapixels, amounting to around 1 GiB of raw data per second. Use of a graphics
processing unit, implementation of the multisensor LPA and CALPA enables real-time processing.

FIGURE 2.9

Left: A full resolution locally tone-mapped frame from a video sequence captured with the multisensor HDR video camera displayed in Fig. 2.2. The graphics processing unit implementation of the reconstruction algorithm reconstructs the image from input from four four-megapixel sensors at 26 frames per second. Right: Four gamma-mapped images 2 f-stops apart from the same HDR image.

2.5.2 SPATIAL MULTIPLEXING USING DUAL-ISO CAPTURE

Another application of the unified framework is HDR reconstruction based on single-shot RAW CFA images with varying gain (see, eg, Hajisharif et al., 2014). In this case, the input is a RAW image where the per-pixel gain, or ISO, setting varies between two different settings in a given configuration over the sensor. The high-gain pixels increase the signal-to-noise-ratio in darker areas of the scene, while the low gain setting enables capture of high-intensity regions without saturation. Fig. 2.10 illustrates a gain pattern configuration for Bayer pattern CFA sensors, where pairs of rows have different per-pixel gain. The input pixels are filtered to reconstruct the HDR output value at pixel location X_j. The different gains, g_1 and g_2, corresponding to, for example, no amplification and 16 times amplification of the analog readout, enable the capture of a wider range of intensities and extend the dynamic range in the final image. The HDR reconstruction is done in a similar way as for the multisensor case, with the difference that the reference, or target, grid is aligned with the input sensor. The HDR pixel value at a location X_j in the output image is estimated using the statistical approach described in Section 2.4 with either the LPA algorithm or CALPA algorithm. The noise model parameters are calibrated as described in Section 2.3.2.1.

The gain configuration illustrated in Fig. 2.10 (left) can be captured with off-the-shelf cameras. Fig. 2.11 shows an example HDR image captured with a Canon 5D mark III camera running the Magic

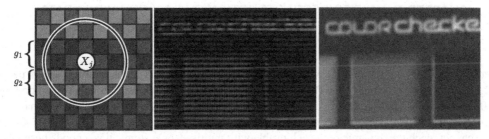

FIGURE 2.10

A gain pattern, with two different gain settings (ISO), for a sensor with a Bayer pattern CFA (left), RAW data where the different gain settings in pairs of rows are visible (middle), and the reconstructed HDR pixels for the zoomed-in region (right).

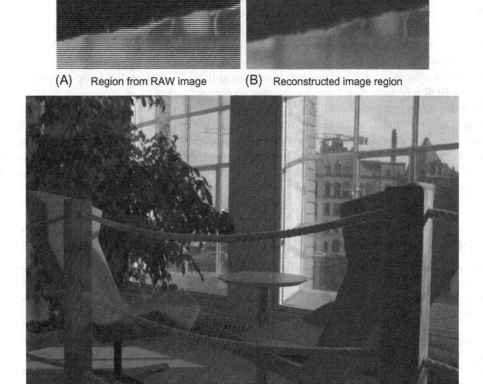

(A)   Region from RAW image   (B)   Reconstructed image region

(C)   Tone-mapped reconstructed HDR image

FIGURE 2.11

(A) A region from the RAW data from the sensor using dual-ISO capture on a Canon 5D camera.
(B) The corresponding region from the reconstructed image. (C) Tone-mapped image obtained with the noise-aware reconstruction method presented.

Lantern firmware with gain, or ISO, varying between pairs of rows to cover the CFA Bayer pattern. Fig. 2.11A displays a zoom-in on the RAW input data with simultaneous gain settings corresponding to ISO100 and ISO1600, and Fig. 2.11B displays the corresponding region in the reconstructed image. The bottom image is the final HDR image reconstructed by the LPA with $M = 1$. Although dual or multigain capture cannot reach the dynamic range achievable with multisensor systems, it is a good alternative in many cases. Spatial gain multiplexing is very robust both during capture and in the reconstruction, as it relies on a single sensor (geometric alignment is not necessary) and a single exposure time (each pixel exhibits the same motion blur). The technique is well suited for imaging in low-light environments as no light is lost because of optical filtering.

2.6 CONCLUSION

We have presented a unified framework for HDR reconstruction based on input from multisensor imaging systems and imaging systems where the exposure/gain is varying spatially over the sensor. A benefit of the method is that it performs all steps in the traditional HDR imaging pipeline in a single step. The reconstruction is based on spatially adaptive filtering and a noise-aware local likelihood estimation using isotropic and anisotropic filter supports taking into account the correlation between color channels. The method uses a sensor noise model to adapt the reconstruction to the characteristics of the camera system used. The framework is general in that the algorithms apply equally well to still-image HDR reconstruction from, for example, exposure bracketing.

There are still a number of challenges to be solved before HDR video can be found in consumer products. There are already a number of robust HDR capture methods — for example, use of multiple sensors or spatially varying gain, as described in this chapter. The solution could be a combination of the two. For multiple gain capture, further investigations are necessary regarding how many and which different gain setting to use, and how they should be best distributed over the sensor while ensuring easy implementation in hardware.

REFERENCES

Aggarwal, M., Ahuja, N., 2004. Split aperture imaging for high dynamic range. Int. J. Comput. Vis. 58 (1), 7–17.

Aguerrebere, C., Almansa, A., Gousseau, Y., Delon, J., Musé, P., 2014. Single shot high dynamic range imaging using piecewise linear estimators. In: IEEE International Conference on Computational Photography (ICCP), Santa Clara, CA.

Ajdin, B., Hullin, M.B., Fuchs, C., Seidel, H.P., Lensch, H.P.A., 2008. Demosaicing by smoothing along 1-D features. In: IEEE Conference on Computer Vision and Pattern Recognition (CVPR), Anchorage, AK, USA, pp. 1–8.

A1ex, 2013. Dynamic range improvement for canon dSLRs with 8-channel sensor readout by alternating ISO during sensor readout. Technical documentation, http://acoutts.com/a1ex/dual_iso.pdf.

Astola, J., Katkovnik, V., Egiazarian, K., 2006. Local Approximation Techniques in Signal and Image Processing. SPIE Publications, Bellingham, WA.

Debevec, P., Malik, J., 1997. Recovering high dynamic range radiance maps from photographs. In: SIGGRAPH '97 Proceedings of the 24th Annual Conference on Computer Graphics and Interactive Techniques. ACM Press/Addison-Wesley Publishing Co, New York, NY, pp. 369–378.

Foi, A., 2009. Clipped noisy images: heteroskedastic modeling and practical denoising. Signal Process. 89 (12), 2609–2629.

Froehlich, J., Grandinetti, S., Eberhardt, B., Walter, S., Schilling, A., Brendel, H., 2014. Creating cinematic wide gamut HDR-video for the evaluation of tone mapping operators and HDR-displays. In: IS&T/SPIE Electronic Imaging, pp. 90230X.

Granados, M., Ajdin, B., Wand, M., Theobalt, C., Seidel, H., Lensch, H., 2010. Optimal HDR reconstruction with linear digital cameras. In: IEEE Conference on Computer Vision and Pattern Recognition (CVPR), San Francisco, CA, USA, pp. 215–222.

Hajisharif, S., Kronander, J., Unger, J., 2014. HDR reconstruction for alternating gain (ISO) sensor readout. In: Eric Galin, M.W. (Ed.), Eurographics 2014 – Short Papers Proceedings. Eurographics Association, Strasbourg, France, pp. 25–28.

Hasinoff, S., Durand, F., Freeman, W., 2010. Noise-optimal capture for high dynamic range photography. In: IEEE Conference on Computer Vision and Pattern Recognition (CVPR), San Francisco, CA, USA, pp. 553–560.

Kronander, J., Gustavson, S., Bonnet, G., Unger, J., 2013. Unified HDR reconstruction from raw CFA data. In: IEEE International Conference on Computational Photography (ICCP), Cambridge, MA, USA, pp. 1–9.

Kronander, J., Gustavson, S., Bonnet, G., Ynnerman, A., Unger, J., 2014. A unified framework for multi-sensor HDR video reconstruction. Signal Process. Image Commun. 29 (2), 203–215.

Malvar, H.S., He, L.-W., Cutler, R., 2004. High-quality linear interpolation for demosaicing of Bayer-patterned color images. In: IEEE International Conference on Acoustics, Speech, and Signal Processing (ICASSP '04), vol. 3, Montreal, Quebec, Canada, pp. 485–488.

Manakov, A., Restrepo, J.F., Klehm, O., Hegedüs, R., Eisemann, E., Seidel, H.P., Ihrke, I., 2013. A reconfigurable camera add-on for high dynamic range, multi-spectral, polarization, and light-field imaging. ACM Trans. Graph. (Proc. SIGGRAPH 2013) 32 (4), 47:1–47:14.

McGuire, M., Hughes, J.F., 2007. Optical splitting trees for high-precision monocular imaging. IEEE Comput. Graph. Appl. 27, 32–42.

Milanfar, P. 2013. A tour of modern image filtering: new insights and methods, both practical and theoretical. IEEE Signal Process. Mag. 30 (1), 106–128.

Narasimhan, S.G., Nayar, S.K., 2005. Enhancing resolution along multiple imaging dimensions using assorted pixels. IEEE Trans. Pattern Anal. Mach. Intell. 27 (4), 518–530.

Nayar, S., Mitsunaga, T., 2000. High dynamic range imaging: spatially varying pixel exposures. In: IEEE Conference on Computer Vision and Pattern Recognition (CVPR), vol. 1, Hilton Head Island, SC, USA, pp. 472–479.

Schoberl, M., Belz, A., Seiler, J., Foessel, S., Kaup, A., 2012. High dynamic range video by spatially non-regular optical filtering. In: 19th IEEE International Conference on Image Processing (ICIP), Orlando, FL, USA, pp. 2757–2760.

Takeda, H., Farsiu, S., Milanfar, P., 2007. Kernel regression for image processing and reconstruction. IEEE Trans. Image Process. 16 (2), 349–366.

Tibshirani, R., Hastie, T., 1987. Local likelihood estimation. J. Am. Stat. Assoc. 82 (398), 559–567.

Tocci, M.D., Kiser, C., Tocci, N., Sen, P., 2011. A versatile HDR video production system. ACM Trans. Graph. (Proc. SIGGRAPH 2011) 30 (4), 41:1–41:10.

Unger, J., Gustavson, S., 2007. High-dynamic-range video for photometric measurement of illumination. In: Proceedings of SPIE – Electronic Imaging, Sensors, Cameras, and Systems for Scientific/Industrial Applications VIII. SPIE & IS&T, vol. 6501, Bellingham, Washington/Springfield, Virginia, USA.

Unger, J., Gustavson, S., Ollila, M., Johannesson, M., 2004. A real time light probe. In: Proceedings of the 25th Eurographics Annual Conference, Short Papers and Interactive Demos Eurographics Association, Grenoble, France, pp. 17–21.

Yasuma, F., Mitsunaga, T., Iso, D., Nayar, S.K., 2010. Generalized assorted pixel camera: postcapture control of resolution, dynamic range, and spectrum. IEEE Trans. Image Process. 19 (9), 2241–2253.

STACK-BASED ALGORITHMS FOR HDR CAPTURE AND RECONSTRUCTION

3

O. Gallo*, P. Sen[†]

*NVIDIA Research, Santa Clara, CA, United States**
University of California, Santa Barbara, CA, United States[†]

CHAPTER OUTLINE

3.1 INTRODUCTION

In this chapter, we examine approaches to capture high-dynamic-range (HDR) images and video using *conventional* digital cameras. This is in contrast to approaches with cameras that are specifically designed to capture a larger dynamic range in a single exposure (see Chapter 2). Since standard digital sensors can capture only a small fraction of the incident irradiance (see Chapter 1), approaches for capturing HDR images with a standard sensor must take a stack of N sequential images $Z_1, \ldots, Z_N$ with different exposure settings and combine their information together as a postprocess to reconstruct an HDR irradiance image, E. We refer to these as *stack-based* algorithms for HDR capture and reconstruction.

High Dynamic Range Video. http://dx.doi.org/10.1016/B978-0-08-100412-8.00003-6

FIGURE 3.1

The Great Wave, Sète, Gustave Le Gray, 1857. In one of the earliest examples of stack-based HDR imaging, French photographer Gustave Le Gray extended the dynamic range he could capture by taking two images with different exposure times and combining the two negatives into one.

From: www.metmuseum.org.

While the focus of this book is HDR video, a thorough discussion of HDR capture for still images is very important. First and foremost, a large portion of the methods proposed for HDR video use a similar strategy, in that they acquire a stream of differently exposed frames. Additionally, many of the topics we cover in this chapter are central also in the case of HDR video, even when the latter is captured with specialized sensors.

From a historical perspective, and although HDR imaging has only recently become widespread, the analog counterpart of today's stack-based approaches was introduced as early as the mid-1800s by French photographer Gustave Le Gray. To expand the limited dynamic range he could capture on film, he literally cut and pasted together multiple films, each measuring a different portion of the dynamic range. The resulting landscapes are simply breathtaking (see Fig. 3.1). The idea of taking multiple shots to extend the dynamic range a camera can capture reappeared in the context of digital photography over a 100 years later: two decades ago, Mann (1993)[1] and Madden (1993) proposed combining multiple low-dynamic-range (LDR) pictures into a single HDR image. Since then, stack-based HDR imaging has attracted growing interest from the research community. Today, most consumer cameras, and even some high-end DSLR cameras, offer HDR shooting modes generally based on this strategy.

The layout of this chapter roughly follows the steps involved in stack-based HDR imaging generation. First, we must determine how many pictures to take and what their exposure times should be to adequately capture the dynamic range of a given scene (Section 3.2). Once the images have been captured, we must merge their information to reconstruct the HDR result (Section 3.3). These approaches work well for static scenes captured with tripod-mounted cameras.

---

[1]The algorithmic details of the work of Mann were published in a later article (Mann and Picard, 1995).

Table 3.1 Notation Used in This Chapter	
N	number or exposures in the source image stack
$\{Z_i\}_{i=1:N}$	stack of N input LDR source images
$Z_i(p)$	value of pixel p in the ith exposure
$\{t_i\}_{i=1:N}$	exposure times for each of the N source images on the stack
$R_{i,j}$	exposure ratio between exposures j and i (if the exposure time is the only parameter changing, then $R_{i,j} = t_j/t_i$)
E	HDR irradiance image of the scene (W/m^2), which the algorithms in this chapter attempt to reconstruct from an input stack
$\widetilde{E}$	estimated scene irradiance
$\{X_i\}_{i=1:N}$	stack of N exposure images (J/m^2, computed as $X_i = E \cdot t_i$)
$f(\cdot)$	camera response function, which converts the pixel exposure X to the pixel value Z — that is, $Z_i(p) = f(X_i(p))$
$g(\cdot)$	inverse camera response function, which converts pixel values to pixel exposures — that is, $X_i(p) = g(Z_i(p))$ (note that $g(\cdot)$ is not an exact inverse of $f(\cdot)$)
$w_i(p)$	weight matrix indicating how well exposed each LDR pixel is (eg, for merging LDR images to form a final HDR result)
Z_{ref}	reference input LDR source, for algorithms that need a reference image from the stack to be specified

However, if the scene is dynamic or the camera is handheld, the slight differences between exposures in the stack will produce unacceptable ghostlike artifacts in the final result. Since this scenario is very common, a large body of research on stack-based HDR image reconstruction focuses on handling motion (Section 3.4), and two major kinds of methods have been developed: (1) methods that remove ghosting artifacts by rejecting information from images that contain motion (Section 3.4.2), and (2) methods that perform some kind of nonrigid registration to align the input images (Section 3.4.3).

Throughout the chapter, we will use the notation shown in Table 3.1, often rewriting equations from the different articles to match this notation for consistency and clarity.

3.2 METERING FOR HDR IMAGING

When a photographer presses the shutter button to take a picture, digital cameras analyze the scene content to determine the optimal capture parameters. Collectively, the algorithms designed to select these parameters are referred to as the three A's: autofocus, auto white balance, and autoexposure. The first two will not be discussed in this chapter as they do not have any specialized counterpart for the process of HDR capture. Autoexposure, also called metering, is the process of selecting the combination of exposure time, ISO setting, and aperture that optimally captures a specific scene on the basis of some criterion. The heuristics involved in metering algorithms range from considerations about motion blur to signal quality in terms of both the signal-to-noise ratio (SNR) and quantization. Additionally, they may include optimizations better suited to work with the algorithms used by the image signal processor in later stages.

In the absence of automatic methods, metering is the photographer's responsibility, requiring both technical and artistic skills. This process is particularly involved in the context of analog photography, because of the nonlinearity of the film's response. An example of a beautifully developed theory for metering is the Zone System by Ansel Adams and Fred Archer (Adams, 1948). In a nutshell, Adams and

Archer suggest dividing the range of gray levels that can be captured by the camera into ten segments, also called "zones." Metering then becomes the process of selecting an exposure time that assigns zones to the correct range of irradiance in the scene. For instance, the exposure time should be selected so that the fifth zone captures the values in the middle of the scene's irradiance distribution.

However, for the capture of the scene with a single exposure and a digital camera, an optimal autoexposure algorithm is expose to the right (ETTR)[2]; in essence, ETTR selects the longest possible exposure time that does not induce saturation (or blur for handheld cameras or dynamic scenes). The resulting image minimizes the impact of photon shot noise. Because of the discrete nature of light, the actual number of photons hitting a pixel in a given time can be modeled by a random Poisson process. Noting that for large numbers a Poisson process can be approximated by a Gaussian process, we have that the number of incoming photons is $n_p \sim \mathcal{N}(\mu, \sigma^2 = \mu)$, where μ is the average number of photons collected. Therefore, the SNR increases with a longer exposure time, as the latter increases the average number of photons collected: $\mathrm{SNR} = \mu/\sigma = \sqrt{\mu}$. A shorter exposure time also causes a larger quantization error: the analog signal from the sensor is linearly quantized by the analog-to-digital converter (ADC), which induces a mean square error of $\Delta^2/12$, where Δ, the size of the quantization bins, decreases linearly with the exposure time; see Fig. 3.2A. By encouraging a long exposure time, ETTR minimizes the quantization mean square error and increases the average number of photons collected, thus increasing the SNR.

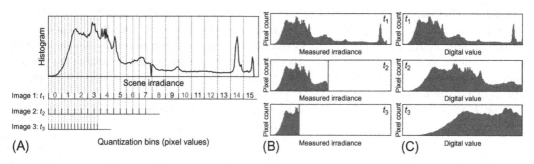

FIGURE 3.2

Changing the exposure time affects, among other things, the granularity of the quantization intervals and the shape of the histogram of the captured image. Specifically, the granularity is finer for longer exposure times: the irradiance distribution of a hypothetical scene and the quantization bins of three exposures separated by one stop ($t_3 = 2t_1 = 4t_1$) for a hypothetical four-bit sensor which measures pixel values from 0 to 15 are shown in (A). The last bin is indicated with a dotted line, signifying that it captures all the irradiance values from its left boundary to infinity. This causes saturation in the image, as is visible in (B), which shows the measured irradiance distribution for these exposure times. In the digital domain, the center of mass of these histograms shifts toward the right as the exposure time is lengthened (C); this is why the process of selecting the longest exposure that does not induce saturation is called expose to the right (ETTR) in this case, exposure time t_1. Note that all the graphs are normalized to show details, and the saturated pixel bins are clamped to fit in the graph.

---

[2]The name stems from the fact that longer exposure times push the center of mass of the brightness histogram toward the right; see also Fig. 3.2C.

Stack-based HDR imaging also requires three A's. Focus and white balance, as mentioned before, need no adaptation. In contrast, the metering strategy needs to account for the fact that different segments of the scene's irradiance must be sampled by different pictures in the stack. Note that the aperture setting affects the depth of field of the image, and thus should typically not change across the stack, leaving only the exposure time and ISO sensitivity as the main parameters that can be adjusted.[3]

Metering for HDR imaging is more involved than single-picture metering for several reasons. First, a metering algorithm needs to select the actual number of pictures required to completely sample the scene's irradiance, given the sensor's dynamic range. Note that this may be complicated by practical constraints: on the one hand, a large number of captures may be impractical for memory and computational requirements. A larger stack also requires a longer time for capture, making it likelier that the scene content will change or move. It may also degrade the user experience by forcing the photographer to wait while a long stream of pictures are captured. On the other hand, a sparser sampling of the range (ie, taking fewer pictures separated by a larger number of stops), may cause registration and merging issues. Second, and perhaps more obvious, it needs to select multiple exposure times. Several strategies have been proposed to perform metering for HDR imaging. We can roughly classify these methods in three main categories.

3.2.1 RANGE-AGNOSTIC METHODS

Use a standard autoexposure algorithm together with a simple progression of exposures, predefined or user-supplied (such as exposure values of 0, +1, −1[4]). This is the strategy most commonly found in commercial products since it is extremely efficient from a computational standpoint — no computation is really needed. However, these methods do not provide a guarantee that the scene's irradiance will be fully captured: overexposed or underexposed regions may still appear in the final result.

3.2.2 RANGE-AWARE METHODS

Select the exposure time by looking at some top and bottom percentile of pixel values in the image. By constraining the number of the pixels at both ends of the range, or their maximum and minimum brightness, methods of this class guarantee that the darkest and brightest regions of the scene be covered. The method proposed by Bilcu et al. (2008) is an example of this strategy: after metering for a single image, they select the exposure time of two more images to capture the highlights and the dark areas of the scene. To find the actual extent of the range, they iteratively change the exposure time while streaming the viewfinder frames. Because they always capture three images, the resulting stacks may be larger than necessary and suboptimal in terms of the noise characteristics. Gelfand et al. (2010) use a similar strategy but allow the number of exposures to vary on the basis of the actual range of a specific scene, although they limit the maximum number of stops between images.

---

[3]However, HDR reconstruction methods based on patch-based synthesis (see Section 3.4.3.2) can handle changes in aperture as well, as first shown by Sen et al. (2012).

[4]In general, the exposure value indicates a specific exposure level, corresponding to a set of different combinations of exposure time and aperture setting. It is also used in a relative sense to indicate power-of-2 increments of the exposure level, also called stops. Here we use the latter definition, where an exposure value of 0 corresponds to the exposure level obtained with standard autoexposure, and an exposure value of +1 indicates a picture that captures twice as many photons.

3.2.3 **NOISE-AWARE METHODS**

Model the noise characteristics of the camera system, sometimes even accounting for the scene's radiance distribution. For instance, the noise model proposed by Hasinoff et al. (2010) shows that use of a higher ISO setting is beneficial with regard to the SNR for a given time budget: the gain boosts the signal before quantization, thus reducing the effect of ADC noise. On the basis of this observation they propose an optimal, although scene-agnostic, selection of the exposure times for stack-based HDR. A closely related solution is the "HDR+" mode available on the Google NEXUS devices (Levoy, 2014). Rather than selecting different exposures, they take a burst of pictures with the same exposure time, selected to be as short as needed to avoid saturation. Because of the stochastic properties of photon shot noise and ISO noise, merging the different pictures yields a higher SNR in the dark regions of the scene. This strategy, however, does not seem to address the problem of quantization noise, which, as mentioned above, is a particularly pressing issue in the case of short exposures, where the size of the quantization bins is large.

Other noise-aware methods define the optimal sequence of exposures on the basis of the actual distribution of irradiance from the scene. Leveraging on this knowledge, these algorithms can produce a smaller stack with a higher SNR. Granados et al. (2010) performed an accurate analysis of the different sources of noise in the image formation process of linear cameras and greedily determined the optimal stack — in terms of exposure times and actual number of exposures — given a target SNR. To predict the SNR for a specific scene, they assume an a priori knowledge of the histogram of the scene irradiance. This is similar to the work of Chen and El Gamal (2002). The method of Gallo et al. (2012) extends these methods in two ways. First, it proposes a strategy to compute the actual HDR histogram of the irradiance of a specific scene. Second, it finds the *globally* optimal stack for a generic camera response function (CRF), making it possible to use merging strategies designed for nonlinear images. Fig. 3.3 shows a result from the method of Gallo et al. (2012).

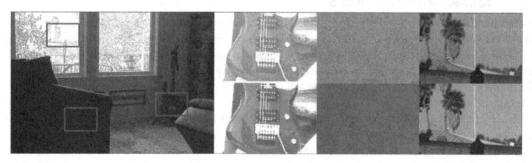

FIGURE 3.3

A naïve metering strategy, even one that prevents oversaturation and undersaturation in the final image, may lead to suboptimal results. In this example, Gallo et al. (2012) compare their method with one that captures the whole dynamic range uniformly (in this case with five pictures, each two stops apart). For the scene shown in the tone-mapped image on the left, their method selects only three images instead of the five images required for the uniform sampling. Nevertheless, as shown in the bottom row of the right, their method outperforms the uniform sampling method (top row on the right) in terms of noise.

Source: Images courtesy of Gallo et al. (2012).

When the metering process is completed, the selected images can be sequentially captured. In the following sections we discuss the processing involved in the combination of the resulting LDR images.

3.3 FROM LDR TO HDR

Once the necessary LDR images have been selected and captured, they need to be combined into a single irradiance map. In this section, we assume that the camera is steady, and that the captured scene is static during the acquisition of the stack; in other words, a given pixel p in the sensor measures the same irradiance E across the whole stack (these assumptions will be relaxed in Section 3.4). We will further assume that the only parameter that changes across the stack is the exposure time t_{exp} and, possibly, the ISO gain g. Without loss of generality, we subsume both with the variable $t = t_{exp} \cdot g$.

The measured energy density (J/m^2), often referred to as *exposure*, can then be modeled as

$$X_i(p) = E(p) \cdot t_i, \tag{3.1}$$

where i is the index of the specific LDR image in the stack. Eq. (3.1) is called the *reciprocity assumption* because it states that the exposure $X_i(p)$ can be kept constant when the irradiance changes by a factor k, provided that the exposure time t is also changed by a factor of $1/k$. This effect was first reported by Bunsen and Roscoe (1862).

There are two main approaches to the problem of combining information from the LDR images. The first works directly in the pixel's digital value, and never estimates the underlying irradiance map (these methods are often referred to as "exposure fusion" methods; see Section 3.3.2.3). The second approach works in the irradiance domain and computes an actual HDR map. The latter requires radiometric calibration, the process of determining the mapping between the digital value of a pixel and the corresponding irradiance (up to a scale factor), which we will discuss first. Later, we will describe different strategies to merge the LDR images into the final HDR irradiance map, and conclude this section by discussing exposure fusion techniques. Note that tone mapping, the process of compressing the dynamic range so that the image can be shown on a regular LDR display, will not be covered in this chapter, but is discussed in the second part of this book.

3.3.1 RADIOMETRIC CALIBRATION

Eq. (3.1) describes the relationship between the irradiance $E(p)$ (W/m^2) at pixel p and the corresponding energy density $X_i(p)$ (J/m^2). However, we cannot always access X. In analog cameras, the film's opacity is related to exposure via a highly nonlinear curve called the characteristic (or Hurter-Driffield) curve. In CCD and CMOS cameras, we can often access the RAW values, which are linearly related to the exposure X. However, manufacturers apply carefully-designed transfer functions that both compress the data and enhance the quality of the final image; see Fig. 3.4. These curves, combined with any other linear and nonlinear process applied by the rest of the image processing pipeline (eg, white balance), can be combined in a single function f, called the camera response function (CRF):

$$Z_i(p) = f(X_i(p)) = f(E(p) \cdot t_i), \tag{3.2}$$

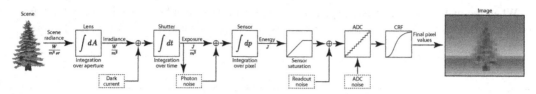

FIGURE 3.4

Imaging pipeline of a typical digital camera showing the different sources of noise. The radiant power of the scene is captured by the camera and integrated over the area of the lens aperture, over the time the shutter is open, and over the area of the pixel's footprint to be converted into energy. This signal could be cut off by the saturation of the sensor, which limits the dynamic range of the camera. The result is then quantized by an ADC, and the CRF is applied to get the final nonlinear digital pixel values.

Source: Diagram inspired by Figs. 1 in Debevec and Malik (1997) and Hasinoff et al. (2010). Tree model courtesy of Eezy (http://www.vecteezy.com).

where $Z_i(p)$ is the digital value associated with pixel p in the ith exposure. If we know the inverse of the CRF, we can estimate the irradiance at the pixel:

$$\widetilde{E}(p) = f^{-1}(Z_i(p))/t_i. \tag{3.3}$$

Strictly speaking, the CRF f is *not* invertible because of saturation, since all pixels whose irradiance is beyond a certain value are mapped to the highest digital value. Furthermore, the process of quantization maps a finite set of irradiance values to the same bin. Therefore, the function f is not one-to-one and cannot be inverted; after all, if it were invertible, the full irradiance could be recovered from a single image. Despite this observation, we follow conventional notation and say that radiometric calibration is the process of estimating the inverse of the CRF, $g = f^{-1}$. By "inverse function," we simply mean a lookup table that remaps the nonlinear values Z_i to values that are linearly related to the original irradiance $\widetilde{E}$, saturation and quantization aside.

Although different algorithms have been proposed for radiometric calibration, they generally assume that the CRF is fixed; it can then be sampled by the taking of multiple pictures of the same scene (same irradiance at each pixel) with different exposure times. The assumption that the CRF does not change across the pictures in a stack is paramount if its estimation is to be accurate. However, camera manufacturers exert great effort to optimize the visual quality of the final image, sometimes adapting the CRF to a specific scene to achieve this (Kim et al., 2012); this sets limits on the overall accuracy of the estimation process.

Camera manufacturers are often reluctant to share information about CRFs, which are their "secret sauce" necessary to deal with the low quality of the pictures that popular, cheap sensors produce. However, it is fairly safe to make a couple of assumptions when one is performing radiometric calibration. First and foremost, it is commonly assumed that f is monotonic. It is also natural to accept that f is spatially uniform.

The literature on radiometric calibration is vast. However, on the basis of the assumption they make about the shape of the CRF, most approaches can be classified into one of two classes: parametric and nonparametric methods. We describe a few representative methods from these two categories in Sections 3.3.1.1 and 3.3.1.2. A small number of methods explore the possibility of estimating the CRF from a single image; because this class is orthogonal to the previous class, we describe it separately in Section 3.3.1.3.

3.3.1.1 Parametric methods for radiometric calibration

While the CRF can differ from camera to camera, and even for the same camera from scene to scene, it is unlikely that its form is too exotic. On the basis of this observation, several methods assume a specific functional form for the CRF and attempt to estimate it using different strategies.

Farid (2001) assumes the CRF to be a simple gamma function, $Z = X^\gamma$, in which case the radiometric calibration process is reduced to estimating γ. He then observes that gamma compressing a signal introduces higher-order harmonics in the spectrum of the image. With that, he estimates the gamma as

$$\arg\min_{\gamma} \sum_{\omega_1,\omega_2 \in [0,2\pi)} |B(\omega_1,\omega_2)|, \tag{3.4}$$

where B is the bicoherence of the Fourier transform of Z, a measure of the correlation of harmonically-related frequencies (Farid, 2001).

In their work on HDR, Mann and Picard (1995) assume the CRF to be of a slightly more general form: $Z = \alpha + \beta X^\gamma$. To estimate α, essentially the black level of the camera, they use a picture captured with the lens cap on, usually referred to as dark frame. Then, assuming the images to be registered, they compute the cross-histogram of the intensity values of a pair of images. For eight-bit images, for example, this is a 256×256 two-dimensional histogram where bin (r,c) contains the number of pixels such that $Z_i(p) = c$ and $Z_j(p) = r$. The parameters (β,γ) can then be found by regression. Mann (2000) later extended this work by considering a number of different analytical forms for the CRF.

Mitsunaga and Nayar (1999) assume a polynomial form of the inverse of the CRF. Specifically,

$$X = f^{-1}(Z) = \sum_{k=0}^{K} c_k Z^k, \tag{3.5}$$

where K is the order of the polynomial. Given a rough estimate of the exposure time ratio $R_{i,i+1}$, the exact ratio and the inverse of the CRF can be found as

$$\arg\min_{\{c_n\},R_{i,i+1}} \sum_i \sum_p \left(\sum_k c_k(Z_i(p))^k - R_{i,i+1} \sum_k c_k(Z_{i+1}(p))^k \right)^2. \tag{3.6}$$

This is a straightforward least-squares optimization that can be solved iteratively for $R_{i,i+1}$ and the polynomial coefficients $\{c_n\}$, until convergence.

Grossberg and Nayar (2003b) relax the assumption on the CRF having a specific analytical form. They start by observing that, under the assumption that CRFs are monotonic, the space of the CRFs is convex, and therefore a linear combination of CRFs is still a CRF. After collecting a large number of real CRFs, $\{f_j\}_{j=1:J}$, they compute the first M eigenvectors of the covariance matrix, whose elements are defined as

$$C_{r,c} = \sum_j (f_j(X_r) - \bar{f}(X_r))(f_j(X_c) - \bar{f}(X_c)), \tag{3.7}$$

where $\bar{f} = 1/J \sum_{j=1}^{J} f_j$, X are the different exposure values, and (r, c) index the bin in the covariance matrix. They show that as few as $M = 3$ eigenvectors can capture 99.5% of the energy, while use of $M = 9$ eigenvectors produces curves that are visually indistinguishable from the ground truth. This approach is accurate and extremely efficient, which is why it is used by several methods, as we will see in later sections.

3.3.1.2 Nonparametric methods for radiometric calibration

Making explicit assumptions on the analytical form of the CRF is not necessary. After all, radiometric calibration can be reduced to computing the lookup table that maps digital values to irradiance (or exposure) estimates, while respecting some properties.

In one of the seminal articles that perhaps most popularized modern HDR stack-based imaging, Debevec and Malik (1997) use a least-squares formulation to recover the inverse of the CRF as well as the irradiance values, while imposing smoothness of the recovered response:

$$\arg \min_{E,g} \sum_{i,p} (E(p) - g(Z_i(p))/t_i) + \lambda \sum_{z=Z_{\min}}^{Z_{\max}} g''(z), \tag{3.8}$$

where $g = f^{-1}$. Essentially, the first term imposes the condition that Eq. (3.3) be satisfied, while the second encourages smoothness of the recovered CRF. Changing λ in Eq. (3.8) has a strong impact on the overall shape of the estimated CRF; for this reason their method can be seen as a means to convert the images in the stack to the same domain, rather than to perform accurate calibration.

Lee et al. (2013) make the acute observation that the estimates for the exposures X_i from the different images in the stack should be linearly dependent. More formally, the matrix formed with the N exposure images represented as column vectors, $[X_1, X_2, \ldots, X_N]$, should have rank 1. With the matrix $O = [Z_1, Z_2, \ldots, Z_N]$, where the columns are the observed images, the inverse of the CRF g can then be found as

$$g = \arg \min_{g} \mathrm{rank}(g \otimes O), \tag{3.9}$$

where the operator $\otimes$ represents an element-wise application of a function. For numerical considerations, rather than minimizing the rank of the matrix, Lee et al. suggest minimizing the ratio of the first two singular values of $g \otimes O$ (this is also called the condition number). They propose solving

$$g = \arg \min_{g} \phi(g \otimes O) + \lambda \sum_{Z_i} H\left(-\left.\frac{\partial g}{\partial Z}\right|_{Z_i}\right), \tag{3.10}$$

where $\phi(\cdot)$ is the first condition number of a matrix, and $H(\cdot)$ is the Heaviside step function, which is 1 if its argument is nonnegative, and 0 otherwise. The second term encourages monotonicity: it adds a penalty that is proportional to the number of points where g is decreasing.

Another interesting approach is the work of Kim et al. (2012). On the basis of their analysis of a large database of JPEG+RAW[5] images taken with different cameras and settings, they propose that a single CRF, as traditionally defined and estimated, is not sufficient to explain all of the in-camera

---

[5]RAW images are to a first approximation linear with the exposure X, and can therefore be used as ground truth to estimate the irradiance impinging on the sensor.

processing steps. They observe that cameras perform a gamut mapping that is a function of the scene, or, more specifically, of the "picture style." Therefore, they propose the following image formation model:

$$Z = f(h(T_s T_w X)), \tag{3.11}$$

where $h(\cdot)$ is the gamut mapping, T_s is the matrix for convection to sRGB, and T_w is the white balance matrix. The key observation is that the gamut mapping, carefully adapted to the scene's type to improve the visual quality of the JPEG image, can be detrimental to the estimation of the inverse of the CRF; however, they hypothesize that the pixels that are affected the most by gamut mapping are the ones that are highly saturated (ie, in the HSV color space they have a high saturation value), and show that removing them from the computation of the CRF allows a more accurate result to be obtained.

Virtually all the methods described in this section require that the input stack be perfectly registered; in other words, they assume that the underlying irradiance at pixel p is the same across the stack. Grossberg and Nayar (2003a) propose overcoming this constraint by estimating intensity mapping functions (IMFs). The idea is similar to that of comparagrams (Mann and Picard, 1995): IMFs capture how the brightness values change between two images in the stack. However, rather than building the cross-histogram of two images, which implicitly assumes registration, they look at the cumulative histogram of brightness: $C(\widetilde{Z}) = \sum_{Z=0}^{\widetilde{Z}} H(Z)$, where H is the histogram of the image. The advantage of the cumulative histogram of an image is that it is robust to small motions in the scene. Given the cumulative histogram of two images C_1 and C_2, it is straightforward to compute the IMF $\tau_{1,2}$:

$$\tau_{1,2}(Z_1) = C_2^{-1}(C_1(Z_1)). \tag{3.12}$$

More recently, Badki et al. (2015) proposed an algorithm specifically designed to tackle the problem of radiometric calibration for scenes with significant motion. Their approach builds upon both the work of Grossberg and Nayar (2003a) and the method for radiometric calibration by rank-minimization of Lee et al. (2013). First, inspired by the method of Hu et al. (2012), they extend the method of Grossberg and Nayar to large motions by proposing a new random sample consensus (RANSAC)-based method for computing the IMFs that is robust to such motions. Second, they replace the least-squares optimization for solving for the CRF in Grossberg and Nayar with the rank-minimization scheme of Lee et al. However, the original method of Lee et al. uses artifact-prone, pixel-wise correspondences in the optimization, so Badki et al. reformulated the optimization to replace these correspondences with IMFs. The result is an algorithm that can solve for the CRF even in cases of significant camera and scene motion.

3.3.1.3 Single-image methods for radiometric calibration

The methods we have described thus far assume that multiple images Z_i of the same scene are available, essentially allowing one to measure different segments of the CRF. However, radiometric calibration can also be performed on a single image, although with potentially lower accuracy. Single-image methods can be beneficial in the context of stack-based HDR imaging when the assumption that the CRF is the same across the stack is not valid.

Matsushita and Lin (2007) leverage the fact that the different sources of noise in a camera system can be modeled with symmetric distributions: the cumulative noise distribution should then be symmetric as well, and any deviation from the overall symmetry results only from the nonlinearity of the CRF.

Therefore, using existing methods to estimate the noise distribution from a single image, they frame the problem of radiometric calibration as

$$g = \arg\min_{g} \xi_\eta, \qquad (3.13)$$

where ξ_η is a measure of the skewness induced by f to the noise distribution η. Eq. (3.13) essentially states that g should restore the symmetry of the noise distribution that is expected before the CRF is applied to the image.

Lin et al. (2004) propose another single-image method. They observe that, owing to the finite size of the pixels, the irradiance of pixels at the boundary between two uniform regions is a linear combination of the values on either side of the edge. Moreover, moving along a direction orthogonal to the edge in image space should correspond to moving along a line in RGB space only if the CRF is linear. However, if a nonlinear CRF is applied to the pixel values, these linear segments become curved. Therefore, they use the method of Grossberg and Nayar (2003b) to parameterize the CRF, and formulated an optimization problem where the solution for g maximizes the linearity in the RGB space of several segments that cross image edges.

Lin et al. (2004) extended this method to a single, grayscale image based on the same idea: the irradiance values of edge pixels should be a linear combination of the irradiance values of the regions on either side of the edge. However, since color is not available, they look at the histograms of the intensities in patches lying across edges. These histograms should be roughly uniform, because the point spread function, together with the finite pixel size, turns a sharp edge into a smooth gradient. Once again, they formulate an optimization problem where the inverse CRF is the function that maximizes the uniformity of histograms of different "edge" patches.

The methods described in this section are not intended to serve as a complete survey of the range of radiometric calibration methods; rather they are meant to offer insight into the strategies most commonly used. Many other relevant methods have been proposed, such as methods that model and estimate the noise characteristics of the sensor (Tsin et al., 2001; Granados et al., 2010), approaches based on a probabilistic framework (Xiong et al., 2012), and algorithms working with video sequences, where the CRF may also vary from frame to frame (Grundmann et al., 2013).

3.3.2 MERGING MULTIPLE LDR IMAGES INTO THE FINAL HDR RESULT

The process of radiometric calibration we described in Section 3.3.1 essentially maps images captured with different exposure times, and processed with nonlinear operators, to the same linear domain. In this domain, an estimate of the irradiance $\widetilde{E}(p)$ can be computed as a linear combination of the values of the corresponding pixels across the stack:

$$\widetilde{E}(p) = \frac{\sum_i w_i(\cdot) \cdot X_i(p)/t_i}{\sum_i w_i(\cdot)}, \qquad (3.14)$$

where we do not make the dependency of the weights $w_i(\cdot)$ explicit because, for different methods, they can be a function of the pixel value $Z_i(p)$ or the exposure $X_i(p)$. Eq. (3.14) is at the heart of most methods that merge multiple LDR images into a single HDR image, with the difference between the various methods lying in the actual definition of the weights w. These weights can have a big impact on the quality of the final irradiance estimate because the different images in the stack will, in general, be affected by different amounts of quantization noise, photon shot noise, thermal noise, etc.

Debevec and Malik (1997) observed that the nonlinearity induced by clipping (saturation or underexposure) limits the accuracy with which the true exposure X of these pixels can be recovered. Therefore, they empirically defined a simple triangle function for w that attenuates the contribution of pixels whose exposure is close to either end of the range:

$$w_{DM}(Z) = \min(Z - Z_{min}, Z_{max} - Z), \tag{3.15}$$

where $[Z_{min}, Z_{max}]$ is the range of the pixel values. Mann and Picard (1995) adopted a similar solution, but quantified more accurately the quality of the irradiance estimate offered by each image in the stack. Specifically, they propose considering the granularity of the quantization induced by the CRF. Where the CRF is steeper, the mapping from X to digital value Z produces a lower quantization error; conversely, where the CRF is flatter, larger ranges of the exposure axis are mapped to the same digital value. Therefore, they define the weights as

$$w_{MP}(X) = f'(X). \tag{3.16}$$

Note that Eq. (3.16) only accounts for quantization noise and ignores the other sources of noise. Mitsunaga and Nayar (1999) extended the work by Mann and Picard by explicitly considering the SNR in the weight computation:

$$w_{MN}(X) = SNR_X \cdot w_{MP}(X) = \frac{X}{\sigma_X} \cdot f'(X) = \frac{g(Z)}{\sigma_X} \cdot \frac{1}{g'(Z)} \approx \frac{g(Z)}{g'(Z)}, \tag{3.17}$$

where, again, $g = f^{-1}$ and, in the last step of the equation, the noise σ_X is assumed to be independent of the level itself, and is therefore dropped. As pointed out by Granados et al. (2010), for linear cameras we can write $w_{MN}(X) = t$, since the rest of the terms are the same in every LDR.

Robertson et al. (2003) use a weighted least squares approach, where the contribution of each pixel to the error is weighted with Mann and Picard's weight, w_{MP}, from Eq. (3.16):

$$Error = \sum_{i,p} w_{MP}(X_i(p)) \left(X_i(p) - t_i \widetilde{E}(p)\right)^2, \tag{3.18}$$

which can be minimized, leading to the weights:

$$w_R = w_{MP} \cdot t^2. \tag{3.19}$$

3.3.2.1 Maximum likelihood estimation

A more theoretically founded approach is to compute the maximum likelihood (ML) estimate of the irradiance $\widetilde{E}(p)$ (Tsin et al., 2001; Granados et al., 2010). Given two irradiance estimates $E_i(p) = X_i(p)/t_i$ from two different images in the stack, we seek to compute:

$$\widetilde{E} = \arg\max_E p(\widetilde{E} \mid E_1, E_2), \tag{3.20}$$

where we omitted the dependence on the pixel p for clarity. We can assume that the observations are drawn from two independent Gaussian distributions $\mathcal{N}(E_i, \sigma_i)$. We can then write

$$p(\widetilde{E} \mid E_1, E_2) = \frac{p(E_1, E_2 \mid \widetilde{E})p(\widetilde{E})}{p(E_1, E_2)}$$
$$\propto p(E_1 \mid \widetilde{E}) \cdot p(E_2 \mid \widetilde{E}), \tag{3.21}$$

where we made the common assumption of a uniform prior distribution. Plugging Eq. (3.21) into Eq. (3.20) and taking the logarithm, we can write

$$\widetilde{E} = \arg\max_{\widetilde{E}} p(E_1 \mid \widetilde{E}) \cdot p(E_2 \mid \widetilde{E})$$
$$= \arg\min_{\widetilde{E}} \frac{(E_1 - E)^2}{\sigma_1^2} + \frac{(E_2 - E)^2}{\sigma_2^2}. \tag{3.22}$$

We can find the ML estimate for $\widetilde{E}$ by setting the derivative with respect to $\widetilde{E}$ in Eq. (3.22) to zero:

$$\widetilde{E} = \frac{\sigma_2^2}{\sigma_1^2 + \sigma_2^2} E_1 + \frac{\sigma_1^2}{\sigma_1^2 + \sigma_2^2} E_2$$
$$= \frac{1/\sigma_1^2 E_1 + 1/\sigma_2^2 E_2}{1/\sigma_1^2 + 1/\sigma_2^2}, \tag{3.23}$$

from which we can see that $w_{\text{ML}} = 1/\sigma_i^2$. Several methods build on this result by observing that weights should at least account for the uncertainty of the pixel's value. The first attempt in this direction was the work of Tsin et al. (2001). After modeling white balance as an affine transformation of the exposure X, and calibrating the sensor for photon shot noise and thermal noise, they define the weights as

$$w_{\text{T}} = \frac{1}{\sigma(Z)}, \tag{3.24}$$

where $\sigma(Z)$ is the standard deviation of the signal, measured from the residuals of the irradiance estimation. Kirk and Andersen (2006) used the ML weights as well:

$$w_{\text{KA}} = \frac{1}{\sigma(X_i/t_i)} = \frac{t_i^2}{\sigma(X_i)} \approx \frac{t_i^2}{\sigma(Z_i)g'(Z_i)^2}. \tag{3.25}$$

Arguably, the most complete noise model was proposed by Granados et al. (2010). They too use the ML weights, but improved on previous work by considering both spatial and temporal noise, the latter being also modeled more accurately than in other approaches. Moreover, they precalibrate the camera noise parameters to avoid polluting the irradiance estimate with the uncertainty of the noise estimation.

3.3.2.2 Winner-take-all merging schemes

A few researchers have proposed a different approach to the generation of an HDR map from a stack of LDR images. Such work is based on the observation that the picture with the longest exposure in the stack is also the one with the smallest quantization noise, and the one impacted by the least photon shot noise (see also Section 3.2). Following this logic, in his work preceding the work of Mann and

Picard (1995), Madden (1993) suggested combining the different images in an HDR stack by using the longest, nonsaturated exposure available for each pixel p. A similar approach was proposed by Tocci et al. (2011); however, they also suggest blending the irradiance estimates at the very top and bottom of the useful range of each LDR image to prevent banding artifacts in the transition areas. Additionally, Tocci et al. worked in the Bayer domain to prevent artifacts due to demosaicing when a subset of the color channels saturate, and assessed the reliability of a pixel's estimate also based on its neighborhood.

3.3.2.3 Exposure fusion methods

The approach to HDR we have described so far consists of radiometric calibration, followed by a merging process that produces the final HDR result. To be displayed on regular monitors, the HDR map needs to be tone mapped. An orthogonal approach is that of *fusing* the images directly in the nonlinear brightness domain. The most popular method in this category is *exposure fusion* by Mertens et al. (2007). Their simple and effective method sidesteps the estimation of the exposure values X_i, and blends the digital values Z_i directly. To reflect the quality of a pixel value, they define

$$w_{EF} = w_s \cdot w_c \cdot w_e, \tag{3.26}$$

where w_s, the color saturation weight, encourages more vivid colors, w_c, the contrast weight, penalizes low contrast, and w_e, the well-exposedness weight, prefers pixels close to the middle of the range. A naïve application of the method directly to the image may cause visible seams due to abrupt changes in the values of the weights of neighboring pixels. To prevent such artifacts, Mertens et al. decompose the image into a Laplacian pyramid, and combine it with a Gaussian pyramid decomposition of the weight maps to create the final image. Once again, unlike Eq. (3.14), w_{EF} is used in the weighted average of the digital values Z_i. Merging images directly in the nonlinear brightness domain has advantages and disadvantages. In general, it creates natural-looking results, whereas the tone mapping procedure required for the standard HDR pipeline often produces unnaturally contrasted pictures. Moreover, artifacts due to misregistration are often attenuated by the weighting process. At the same time, it never produces an actual HDR irradiance map, which can be beneficial for computer vision tasks. Finally, when the difference in brightness between the images in the stack is too large, it can introduce artifacts caused by the Gaussian pyramid decomposition of the weights. Several methods build on this idea to increase computational efficiency (Gelfand et al., 2010), to embed deghosting (Zhang and Cham, 2010; An et al., 2011; Gallo et al., 2015), or simply to propose different weights (Shen et al., 2011).

3.4 HANDLING ARTIFACTS FROM MOTION DURING HDR RECONSTRUCTION

The algorithms described in the previous section assume the scene to be static and the camera to be steady. However, when the stack of LDR images is captured in the presence of camera or scene motion, the misalignment between different exposures produces ghostlike artifacts in the final HDR result (see Fig. 3.5B). Since this is a common scenario, addressing motion artifacts is an important problem for practical HDR capture. Indeed, there is a large body of research on the subject, some of which we will survey here. These methods are often known as HDR "deghosting" algorithms, because they deghost

(A)   Low-dynamic-range inputs   (B) HDR Result

FIGURE 3.5

Example of ghosting with stack-based HDR imaging for dynamic scenes. (A) Stack of input images $Z_1, \ldots, Z_5$. Some of the input images were captured while people were in the scene. (B) HDR result from traditional merging (Section 3.3), with objectionable ghosting artifacts.

Source: Images courtesy of Gallo et al. (2009).

(or remove ghosting artifacts from) the final HDR result. Readers seeking detailed explanations of the individual algorithms or thorough comparisons are referred to the original articles cited, as well as survey articles in the field (Srikantha and Sidibe, 2012; Hadziabdic et al., 2013; Tursun et al., 2015).

Before we begin, we note that stack-based approaches to HDR reconstruction cannot always recover the actual HDR image when the scene is dynamic, at least not like an actual HDR camera would. For example, consider the situation shown in Fig. 3.6, where the volleyball in front of the bright window occupies different positions across the two-image stack. In the long exposure, which has been selected as the reference, the window is almost entirely saturated and offers no useful detail. Ideally, we would recover this information from the short exposure, which properly captures the scene outside the window. Unfortunately, in the second frame, the ball has moved and blocks part of the view of the window, making it impossible to capture the scene behind it. Because this information is not available in any picture of the stack, we cannot reconstruct an HDR image that would exactly reproduce the structure of the scene as it was when the reference image was captured, as shown in Fig. 3.6C. However, some of the deghosting algorithms we will discuss are able to reconstruct *plausible* HDR results, even in extreme cases. Furthermore, they offer the only practical way to capture HDR images of dynamic scenes with conventional digital cameras.

Previous deghosting work can be divided into two major categories: (1) *rejection-based* algorithms and (2) *alignment* algorithms. Rejection-based algorithms assume the scene to be mostly static and use a rejection technique to eliminate motion artifacts, while alignment algorithms perform some kind of nonrigid registration between the images so that they can be merged to produce the final HDR result. Each kind of algorithm has advantages and disadvantages, which we will discuss at a high level below. Before we begin discussing the two major kinds of deghosting algorithms, however, we note that either approach can first address artifacts from small camera motions through simple, rigid-alignment approaches as described in the next section.

(A) Long Exposure (reference)   (B)   Short Exposure   (C)   Actual HDR scene

FIGURE 3.6

The problem with stack-based HDR imaging when the scene is dynamic. The region marked in the reference image (A) is occluded in the exposure that captures the highlights (B), making it impossible to reconstruct the actual content of the scene, which is shown in (C).

3.4.1 SIMPLE RIGID-ALIGNMENT METHODS

A simple rigid-alignment preprocess (eg, one using a rotation, translation, or homography matrix to align the images) can often eliminate many of the artifacts from small camera motions, making it easier to deghost images that contain mostly static objects. Of course, such rigid registrations do not address the problem of parallax (caused by camera translation), or artifacts caused by highly dynamic scenes. However, they usually work reasonably well when the camera motion is relatively small and the scene does not undergo significant changes.

To our knowledge, the first method which performed a simple rigid-alignment preprocess is the work of Bogoni (2000), who applied a global affine alignment before his optical flow alignment (we will discuss this method in more detail in Section 3.4.3.1). Another early method is the work of Ward (2003), which targets artifacts from camera translations. To compare the differently exposed images, Ward proposes first converting them into median threshold bitmaps, which are binary images with 1's for pixels greater than the images' median. This strategy stems from the observation that median threshold bitmaps from differently exposed images resemble each other more closely than when other potential transformations, such as edge operators, are applied to the images.

One can use median threshold bitmaps to measure the registration quality by simply XORing the pixels of the median threshold bitmaps to see where they are different. The optimal translation is the one that maximizes the number of 1's in the median threshold bitmap. To minimize the impact of noise induced by the pixels close to the median threshold, Ward excludes the pixels whose distance from the threshold is within the noise tolerance. This process can be accelerated with a pyramidal approach, where the translational alignment is computed on coarse versions of the images and then refined at higher resolutions. This multiscale approach also reduces the chance of convergence to a local minimum.

Tomaszewska and Mantiuk (2007) proposed a different method of rigid alignment by using SIFT to extract key points in each image and finding correspondences between them. They then eliminate spurious matches using RANSAC to estimate the homography that can be used to prewarp the images. These warped images can then be merged by any of the methods described in Section 3.3. In the end, different flavors of methods like these are common preprocessing steps for more advanced algorithms, as we will see in the next section.

3.4.2 REJECTION ALGORITHMS FOR HDR DEGHOSTING

Rejection-based algorithms assume minimal scene motion and a static camera so that only few pixels actually exhibit motion. If the camera shakes slightly, a simple rigid registration process, such as one of those described in the previous section, can be applied as a preprocess to align the images and satisfy this assumption. Since most of the pixels will exhibit no motion under these assumptions, most of the final HDR images can be computed with the standard HDR merging process for static scenes described in Section 3.3. To prevent artifacts at the pixels affected by motion, only the images that are deemed to be static at those locations are combined.

The challenge for these rejection algorithms, therefore, is to detect pixels affected by motion and select from the stack the pixels that can be used in the corresponding locations. These algorithms are usually easy to implement and fairly fast, as they only have to detect motion pixels that deviate from the predicted value. Furthermore, because of their design, they are usually successful at completely removing ghosting artifacts, but sometimes have to compromise on the extent of the dynamic range reconstructed in certain regions.

However, rejection algorithms do have serious shortcomings. Perhaps most importantly, these methods cannot handle moving HDR content since they typically discard from the stack any pixels that contain motion. For example, consider a scene with a moving object whose radiance has a dynamic range too high to be captured by a single image (eg, a moving person who is partly in the shadows and partly under direct sunlight). Rejection-based techniques cannot reconstruct this HDR image correctly, as these methods only merge *corresponding* pixels across the stack of images, rather than compensating for motion (ie, they do not move content around). In these cases, different portions of the HDR irradiance range may be measured by nonoverlapping regions of the images in the stack, and therefore the values from a single pixel across the stack cannot be combined to get a proper HDR result. Therefore, in general, rejection algorithms have not been as effective in reconstructing HDR results from complex dynamic scenes as the registration-based algorithms we will examine in Section 3.4.3.

Nevertheless, it is useful to study rejection-based techniques because the results they produce are generally not affected by motion artifacts. We can classify rejection-based methods into two categories, which we will describe in the subsequent sections: (1) those that do not select a reference image and try to use information from all images equally (often producing an image from only the static parts of the scene), and (2) those that select an image in the stack as the reference (with the goal of producing an HDR result that resembles this image).

3.4.2.1 Rejection methods without a reference image

Rejection methods that do not define a single reference image are based on the observation that small moving objects tend to affect different regions of the images across the stack. Therefore, if the stack of LDR images is large enough (usually five or more images), a pixel p is likely to capture the irradiance from the static parts of the scene in most of the pictures. Methods from this category then propose a model for how pixel p should behave across the stack if it represented a static object, and discard the values $Z_i(p)$ across the stack that do not follow this model, as they are likely to be affected by motion. However, these methods run into the problem that neighboring pixels may come from a different subset of exposures where objects might be in different positions, which would introduce visible discontinuities. To minimize these effects, these algorithms generally identify clusters or groups of pixels that can be drawn coherently from one (or more) of the input LDR images.

One of the first methods to do this was described in Section 4.7 of the book by Reinhard et al. (2005). In this method, the CRF is assumed to be known so that the LDR images Z_i can be converted into their corresponding irradiance images E_i. The different images E_i should theoretically be the same, except for noise, saturation, and motion, which may alter some of the pixels' values from image to image. Therefore, Reinhard et al. propose computing the *weighted normalized variance* of the values at each pixel p to determine which pixels are affected by motion:

$$\sigma^2(p) = \frac{\sum_{i=0}^{N} w_i(p)E_i(p)^2 / \sum_{i=0}^{N} w_i(p)}{\left(\sum_{i=0}^{N} w_i(p)E_i(p)\right)^2 / \left(\sum_{i=0}^{N} w_i(p)\right)^2} - 1. \tag{3.27}$$

This equation, explained only verbally by Reinhard et al. (2005) and later presented mathematically by Jacobs et al. (2008), uses weights w_i to exclude overexposed or underexposed pixels from the computation as their divergence from the true irradiance may bias the estimate of the variance. Note that unlike traditional variance, the variance in Eq. (3.27) is normalized to the actual size of the signal.

The key observation is that when one is looking across the image stack, pixels that are not affected by motion should have a smaller variance than those that measure irradiance from different objects. Of course, one could set a simple threshold for this variance to distinguish between these two cases. However, this naïve approach has the problem that the image would suffer from discontinuity artifacts when neighboring pixels are selected from different images with different objects.

To avoid this problem, rejection methods that do not define a reference image must group pixels together into larger clusters, where all the pixels in a cluster are drawn coherently from the same image in the stack. In the particular case of Reinhard et al. (2005), morphological operators such as erosion and dilation are used to grow the binary image after thresholding of the variance to create larger, contiguous regions that are identified to have motion. To decide which exposure to use in each region, they generate a histogram of irradiance values in each region and find the maximum value that is not in the top 2%, which they consider to be outliers. They then find the longest exposure that still includes this maximum value within its valid range, and interpolate between this exposure and the original HDR result using the per-pixel variance as a mixing coefficient. In this way, pixels with lower variance across the stack will use the original HDR result, while pixels with larger variance will use the single exposure. This algorithm is able to produce deghosted images and, at the same time, ensure that each region is drawn coherently from one exposure.

In another method, Eden et al. (2006) first use a SIFT-based feature registration technique to align the input images in the presence of varying exposure levels. Once the stack has been aligned, they map the images to the irradiance domain, where they draw each pixel of the final composite from one of the input images. This is done in two steps. In the first step, they use a subset of the aligned input images to create a reference panorama that covers the full angular extent of all the inputs using graph cuts (Boykov et al., 2001). However, because of overexposure or underexposure this reference image could have areas of missing information, so they introduce detail from images that are better exposed while solving for a smooth transition between regions in a second pass. This problem is minimized via max-flow graph cut to produce the final result, which can be smoothed out to remove any remaining seams.

The approach of Khan et al. (2006) attempts to compute a ghost-free image through several iterations of kernel density estimation that modify the blending weights w_i of Eq. (3.14), by assuming that background (static) pixels are the most common. Essentially, they compute the probability that a given pixel is part of the background, and use this weight when blending so pixels from dynamic objects (and not the background) get a smaller weight. To do this, they represent each pixel in the stack of images with a five-dimensional vector $\mathbf{x}_i(p)$, where i is the index of the image in the stack and p is the pixel location. This vector contains the three LDR color channels of the pixel value (in Lab space) as well as the coordinates of the pixel on the image.

For a given pixel p, they select all pixels $\mathbf{y}_j(q)$ in its 3×3 neighborhood over all the images in the stack, denoted by $\mathcal{N}(p)$. Note that the pixels at position p across the stack are not included in this neighborhood. They begin by assuming that all $\mathbf{y}_j(q)$ are equally likely to be part of the background. The probability that a pixel p belongs to the background B (given by $P(\mathbf{x}_i(p)\,|\,B)$) can then be calculated with a kernel density estimator:

$$P(\mathbf{x}_i(p)\,|\,B) = \frac{\sum\limits_{j,q \in \mathcal{N}(p)} w_{j,q} K_H(\mathbf{x}_i(p) - \mathbf{y}_j(q))}{\sum\limits_{j,q \in \mathcal{N}(p)} w_{j,q}}, \tag{3.28}$$

where the kernel K_H is a five-dimensional multivariate Gaussian density function, and the weight $w_{j,q}$ indicates the probability of the pixel belonging to the background. For the first iteration, these weights are initialized to a "hat" function similar in spirit to that of Debevec and Malik (1997). For subsequent iterations, the value of the weights can be set to the probability that the pixel belongs to the background, as computed by Eq. (3.28) in the previous iterations. However, each time the newly computed weights are multiplied by the initial weights from the hat function to continually diminish the probability that pixels that are overexposed or underexposed are used in the final estimates. On convergence, the weights are plugged into Eq. (3.14) to merge the LDR images into an HDR result.

Jacobs et al. (2008) extended the deghosting algorithm of Reinhard et al. (2005) in several ways. First, they prealign the images as in the earlier work of Ward (2003), but in this case they iteratively solve for the translation *and* rotation that maximizes the XOR score between the two median threshold bitmaps. In the second stage, they replace the variance metric of Eq. (3.27) with a local entropy measure that indicates movement in the scene. Specifically, they measure the local entropy at each pixel in the LDR image Z_i by looking at the pixel values z within a two-dimensional window around pixel p:

$$H_i(p) = -\sum_z P(Z = z) \log(P(Z = z)), \tag{3.29}$$

where the probability function $P(Z = z)$ is computed from the normalized histogram of the intensity values of pixels within the window. Using these entropies, they compute an uncertainty image U, which is the local weighted entropy difference between the images:

$$U(p) = \sum_{i=1}^{N-1} \sum_{j=0}^{i-1} \frac{v_{ij}}{\sum_{i=1}^{N-1} \sum_{j=0}^{i-1} v_{ij}} |H_i(p) - H_j(p)|, \tag{3.30}$$

where $v_{ij} = \min(w_i(p), w_j(p))$, and weights $w_i(p)$ and $w_j(p)$ are computed with Debevec and Malik's triangle function in Eq. (3.15), with $Z_{\min} = 0.05$ and $Z_{\max} = 0.95$. The intuition is that static regions would have similar local entropy measures across the LDR images, even if they are near edges, which might increase the variance because of slight camera motions. This method also does not need a priori knowledge of the CRF, as the entropy measurement can be done in the LDR domain. As with previous methods, this uncertainty image is thresholded and the resulting binary image is eroded and dilated to produce contiguous regions that are affected by motion. At this point, each region is filled with values from one of the irradiance images E_i that is not overexposed or underexposed in that region and is blended with the original HDR value to avoid artifacts at the borders.

Sidibe et al. (2009) observed that the value of pixel p across the stack should increase with the exposure time, since the camera response curve is monotonically increasing: $Z_i(p) \leq Z_j(p)$ if $t_i < t_j$. Therefore, they propose identifying regions where this order relation is broken at least once as ghosted regions. Of course, there might be motions that preserve this order which would not be detected. In the ghosted regions, they use the input images that they deem to have captured the background, which is assumed to appear in most of the images. To do this, they effectively compute the histogram of irradiance values at each ghosted pixel and compute the mode of this distribution, which is the value that appears the most often. The mode is assumed to be the background and the values are merged (ignoring pixel values close to saturation or zero) to form the final HDR image. To have enough samples at each pixel to compute the mode, they require at least five images in the stack.

In another approach, Pece and Kautz (2010) first compute median threshold bitmaps for each image in the stack as proposed by Ward (2003) and accumulate these binary maps for each pixel over all the exposures. Values that are neither 0 nor N are considered motion, and the morphological operators of dilation and erosion are applied to this result to generate the final motion map. Pece and Kautz present results they obtained using exposure fusion (Mertens et al., 2007), where they select the best available exposure for each of the clusters in the motion map to produce their results.

Zhang and Cham (2012) presented a technique similar to exposure fusion (Mertens et al., 2007) (see Section 3.3.2.3) because they fuse the images without generating an HDR image first, but use a novel consistency metric that uses the image gradient to detect movement. To begin, they compute the magnitude $M_i(p)$ and direction $\theta_i(p)$ of the gradient around every pixel of each image in the stack. Next, they observe that the magnitude of the gradient can be used to determine saturated or underexposed pixels, as these regions typically have lower gradient magnitude. Therefore, they propose a *visibility* measure that indicates how well exposed and visible a particular pixel is:

$$V_i(p) = \frac{M_i(p)}{\sum_{i=1}^{N} M_i(p) + \varepsilon}, \tag{3.31}$$

where the ε is a small value (eg, 10^{-25}) to avoid division by zero. Finally, they observe that the gradient direction can serve as a consistency measure to detect motion across the exposure stack because of its invariant property over different exposures. Therefore, they compute the gradient direction difference of the ith image with respect to the jth image as follows:

$$d_{ij}(p) = \frac{\sum_{k \in \mathcal{N}} |\theta_i(p+k) - \theta_j(p+k)|}{M^2}, \tag{3.32}$$

where $\mathcal{N}(p)$ is the set of offsets of the pixels in an $M \times M$ square neighborhood around pixel p. With this, a consistency score S_i can be computed for every image. One does this by accumulating a Gaussian weight for each pixel based on the difference of its gradient direction across the stack:

$$S_i(p) = \sum_{j=1}^{N} \exp\left(-\frac{d_{ij}(p)^2}{2 \cdot 0.2}\right). \tag{3.33}$$

Given these scores, a consistency score for each pixel p in the stack image i can then be calculated as

$$C_i(p) = \frac{S_i(p) \cdot \alpha_i(p)}{\sum_{j=1}^{N} S_j(p) \cdot \alpha_j(p) + \varepsilon}, \qquad (3.34)$$

where $\alpha_i(p)$ is simply 1 if the pixel is well exposed and 0 if it is not. Here, we use the term "well exposed" to define a pixel whose value is in the middle of its range — say, between 0.1 and 0.9 in a normalized pixel value range. These consistency scores can then be used to compute the final weights for the fusion process (Eq. 3.26):

$$w_{\text{EF}}(p) = \frac{V_i(p) \cdot C_i(p)}{\sum_{j=1}^{N} V_j(p) \cdot C_j(p) + \varepsilon}. \qquad (3.35)$$

The final image can then be fused without the need for tone mapping, but does not produce a true HDR result.

Granados et al. (2013) propose using a noise-aware model to determine whether the image stack values for a particular pixel are *consistent*, which means that they measure the same static irradiance. They observe that for a pixel in a static region, the exposure values across the stack should all be within an error margin based on the noise of the imaging system. Therefore, rather than using an arbitrary threshold to detect motion, they characterize the noise in the imaging system (both shot noise and readout noise) as a Gaussian distribution, which enables them to determine the probability that the difference between two pixel values is caused by scene motion or noise. This idea can be extended to the N images in the stack to produce *consistent* subsets, which will not introduce ghosting artifacts when combined.

Once these consistent subsets have been identified, the next challenge is to ensure that neighboring pixels are drawn coherently from the subsets to avoid artifacts. To do this, they pose the irradiance reconstruction problem as a labeling problem, solved by minimizing an energy function with two terms. The first, a consistency term, encourages the pixels to be selected from consistent subsets of the image to reduce ghosting. The second is a prior term that penalizes incoherency across neighboring pixels by enforcing that neighboring pixels should be drawn from the same consistent subset. They solve this labeling problem using the expansion-move graph-cuts algorithm and then merge the consistent sets at each pixel to produce the final HDR result. However, despite this graph-cut optimization, their method still cannot always guarantee a semantically consistent result, and thus it requires a manual intervention to resolve remaining issues.

Finally, Oh et al. (2015) recently proposed a clever rank minimization strategy to solve for the final HDR image. They begin by assuming that there are two kinds of motion between the images in the stack. The first is global motion due to camera movement, which they assume can be modeled with a homography. The second is local motion, which they want to eliminate, and is caused by the nonrigid movement of objects in the scene. Their key observation is that if global motion is accounted for, the stack of exposure images $X_1, \ldots, X_N$ should be linearly dependent. In other words, barring local motion, saturation, or noise, the globally aligned exposure images would simply be scaled versions of E (ie, $X_i = E \cdot t_i$). Therefore, they attempt to eliminate motion artifacts by enforcing that the matrix whose columns are the input LDR images should be of rank 1 (ie, all columns should be linearly dependent).

Oh et al. first account for the global motion by modeling the hypothetical process of capturing "globally aligned" LDR source images, as if the camera were not moving. This can be written as $\widetilde{Z}_i = f(\widetilde{X}_i + \eta_i)$, where $\widetilde{Z}_i$ are the LDR images that would have been taken with a static camera, $\widetilde{X}_i$ is the ideal exposure image that contains only static scene information, and η_i is a "noise" term representing the local motion in the scene. Since we can apply a homography operator $\circ h_i$ to perform global alignment on each of the inputs Z_i, we can write $\widetilde{Z}_i = Z_i \circ h_i$. Once the camera has been calibrated so that its response curve is linear (see Section 3.3.1), the capture process can be modeled as

$$\widetilde{Z}_i = Z_i \circ h_i = a\widetilde{X}_i + a\eta_i. \tag{3.36}$$

We can then vectorize the terms in this equation and combine them into matrices using all of the N captured images: $\mathbf{Z} \circ \mathbf{h} = \mathbf{X} + \boldsymbol{\eta}$. Since all the columns of $\mathbf{X}$ are simply scaled versions of the static scene irradiance E, it is a rank 1 matrix. At the same time, $\boldsymbol{\eta}$ is sparse if we assume that most of the scene is static and only a few areas are affected by motion. Therefore, the problem of removing motion artifacts from the HDR image is equivalent to the problem of solving for a rank 1 matrix $\mathbf{X}$ and sparse matrix $\boldsymbol{\eta}$ through the following optimization:

$$\mathbf{X}^*, \boldsymbol{\eta}^*, \mathbf{h}^* = \arg\min_{\mathbf{X}, \boldsymbol{\eta}, \mathbf{h}} \; p_2(\mathbf{X}) + \lambda \|\boldsymbol{\eta}\|_1$$

$$\text{subject to } \mathbf{Z} \circ \mathbf{h} = \mathbf{X} + \boldsymbol{\eta}. \tag{3.37}$$

Here, $p_2(\mathbf{X}) = \sum_{i=2}^{N} \sigma_i(\mathbf{X})$ is the sum of the singular values from the second to the last[6] which measures the rank of the matrix. The L1 norm $\| \cdot \|_1$ is a measure of sparsity, and the weighting coefficient λ balances the contribution of the two terms. This constrained optimization problem can be solved with augmented Lagrange multipliers (Peng et al., 2012), where the problem is divided into three different subproblems for $\mathbf{X}$, $\boldsymbol{\eta}$, and $\mathbf{h}$ and minimized iteratively.

As discussed earlier, rejection-based algorithms have their drawbacks, but this subset of algorithms that do not specify a reference image have additional problems. For example, they can often produce images that contain duplicate objects or other artifacts, because the semantic meaning of objects is lost when the consistent sets computed in neighboring pixels are not coherent. These artifacts typically require a manual correction. Furthermore, since they typically strive to use only "background" pixels from each image, rejection methods of this type will suppress dynamic objects from the HDR result.

Finally, because these algorithms produce images that do not adhere to a ground truth reference (ie, an HDR picture taken at a specific moment in time), they cannot be easily extended to the capture of HDR video. The reason for this is twofold. First, they do not guarantee temporal continuity since each frame is individually computed, and may use a pixel cluster that is not temporally coherent with the neighboring frames. Second, even if temporal coherency could be enforced, the fact that dynamic objects are usually suppressed defeats the purpose of taking a video in the first place.

3.4.2.2 Reference-based rejection methods

The algorithms in this category select a single image from the stack as the *reference* and use it as the foundation of the final image. In other words, the HDR result will be geometrically consistent with this

---

[6]Assuming that the number of images N is smaller than the number of pixels in each image.

reference, at least in the parts where it is well exposed. The other images in the stack will be tested against the reference, and pixels deemed to have been affected by motion will be rejected. For regions where all the images in the stack are rejected, the HDR result would be reconstructed with use of only the reference.

One of the first examples of these algorithms is the work of Grosch (2006), which takes two differently exposed images and first aligns them using a variant of the method proposed by Ward (2003), extended to consider both translation and rotation. He then computes the CRF on the largely aligned images using the method of Grossberg and Nayar (2003a), and uses the first image (the reference) to predict the estimated values in the second:

$$\widetilde{Z}_2(p) = f\left(\frac{t_2}{t_1} g(Z_1(p))\right). \tag{3.38}$$

If the predicted color $\widetilde{Z}_2(p)$ is beyond a threshold from the actual color in the second image (ie, $|\widetilde{Z}_2(p) - Z_2(p)| > \varepsilon$), the algorithm assumes that $Z_2(p)$ would introduce motion artifacts, and falls back to using only the first image at these locations. This produces an artifact-free result because it largely follows the reference, and has the advantage that it does not need a priori knowledge of the CRF. However, if the scene contains large moving objects, then the radiometric calibration step could fail as well, unless a more robust calibration procedure, such as the algorithm of Badki et al. (2015), is used (see Section 3.3.1.2).

Gallo et al. (2009) propose a similar approach. They first define the reference as the image in the stack with the fewest overexposed and underexposed pixels; they then compare the values of the pixels of the different images in the stack against it. They perform the comparison in the log-irradiance domain, where the following relationship holds:

$$\ln(X_{\text{ref}}) = \ln(X_i) + \ln(t_i/t_{\text{ref}}), \tag{3.39}$$

where the dependence on pixel p is omitted for clarity. Pixels whose exposure $X_i(p)$ is farther than a threshold from the value predicted by Eq. (3.39) belong to moving objects. However, for increased robustness, rather than working directly with pixels, Gallo et al. propose working with patches; a patch from the ith image in the stack is merged with the corresponding patch in the reference if the number of its pixels obeying Eq. (3.39) is above the threshold. The patches are defined on a regular grid; because two neighboring patches in the reference image can be merged with a different subset of patches from the stack, visible seams may exist at the patches' boundaries. To address this issue, the patches are blended with a Poisson solver (Pérez et al., 2003).

Raman and Chaudhuri (2011) extended the work of Gallo et al. (2009) by replacing squared patches with superpixels, which are inherently more edge aware. However, rather than computing the HDR irradiance map, they fuse the images directly using their nonlinear digital pixel values. To begin, they compute the weighted variance proposed by Reinhard et al. (2005), see Eq. (3.27), to identify the pixels that may have measured irradiance from moving objects. Then, using only the pixels that are deemed to have captured static objects, they fit fourth-order polynomials to create a set of $N-1$ IMFs that map the pixel values of each exposure in the stack to the reference.

In the next step, all the images but the reference are segmented into superpixels with homogeneous color and texture. The idea is to blend the superpixels that are static with respect to the reference with the well-exposed reference information. To identify a superpixel as static, the authors use the IMF and compare its pixels with those of the superpixel in the reference; to make the process more robust to noise they also threshold the distance of each pixel from the predicted value. If 90% of the pixels are within this threshold, then the superpixel is considered to be static with respect to the reference. These static superpixels are then decomposed into 6×6 patches with an overlap of one pixel on each edge. The patches with more than 90% of pixels within the static superpixel are considered static as well, and their gradients are merged by use of a Gaussian weighting function based on exposure. Finally, a Poisson solver is used to reconstruct the final color information from the gradients (Pérez et al., 2003).

Wu et al. (2010) propose a set of criteria for detecting moving pixels. First, they use a criterion that ensures that the pixel values are monotonically increasing as the exposure time is lengthened, similar to the earlier work of Sidibe et al. (2009). Next, they use a criterion similar to that of Grosch (2006) that compares a pixel's value with that predicted from another exposure after compensation for the CRF and the exposure time ratio. If a pixel violates any of these criteria, then it is considered to be affected by motion. The final motion map is generated by use of the morphological operators, such as opening and closing. Once the pixels affected by motion have been identified, Wu et al. proceed to compute the final HDR image. Specifically, they select image k as a reference and use it to fill in the pixels affected by motion in the neighboring images $k - 1$ and $k + 1$ with the value predicted by the camera response curve as in Eq. (3.38). These new images are then used to predict the next images, and so on until the entire stack has been processed. Finally, they correct boundary artifacts near the edges of the regions in the motion map by convolving the images with a low-pass kernel, and using the result to replace the values calculated originally in these regions. The HDR image is then computed with the standard merging equation (Eq. 3.14).

Heo et al. (2010) first globally aligned the images in the stack with the reference image using a homography estimated with SIFT features using RANSAC. Next, $N - 1$ joint histograms are computed between the values in the reference and those in the other images. These histograms are then converted into smooth joint probability distribution functions through a Parzen windowing process using a 5×5 Gaussian filter followed by a normalization to enforce that the subtended area sums to 1. Pixels in the other images in the stack with a joint probability less than a fixed threshold are labeled as ghost pixels. This simple thresholding of the joint probability to determine ghost regions can be very noisy, however, so Heo et al. further refine the ghost regions using an energy minimization that enforces smoothness between neighboring pixels, and which is solved with use of graph cuts.

The refined ghost regions can be used to compute new joint histograms that are not affected by motion artifacts; therefore, the algorithm iteratively alternates between computing the joint probability distribution functions and detecting the ghost regions. The pixels not affected by motion are then used to compute the CRF with the method of Debevec and Malik (1997). To further reduce artifacts, this CRF is used to refine the radiance values of all the pixels in the other images in order to make their values more consistent with the reference image. Finally, the different exposures are blended to generate the final HDR result with a weighted filtering step. These weights are computed by application of a bilateral filter (Tomasi and Manduchi, 1998) to all the samples in a patch around a pixel, with use of a global intensity transfer function to compare the differently exposed pixel values.

Rejection-based methods that use a single reference image generally reduce or completely remove ghosting artifacts from the final HDR image. They do, however, have some of the shortcomings of all the rejection-based methods we discussed earlier, such as not being able to handle dynamic HDR content. Furthermore, if the regions where the reference is overexposed or underexposed are large, these algorithms could have problems recovering the full dynamic range because of their heavy reliance on the reference; see Fig. 3.7.

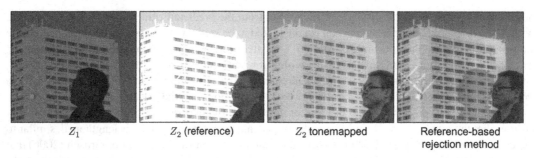

| Z_1 | Z_2 (reference) | Z_2 tonemapped | Reference-based rejection method |

FIGURE 3.7

Rejection-based HDR reconstruction methods cannot "move" information around the image. Here, the façade of the building is completely saturated in the reference image, as seen in the tone-mapped version of Z_2. A reference-based rejection method, such as that shown in the last image, produces a gray halo in the final result because it falls back to the reference when motion is detected. Since the reference is saturated in this region, the measured irradiance is much lower than the actual irradiance measured by the low exposure, resulting in the artifact visible in the rightmost image.

Source: Images courtesy of Sen et al. (2012).

3.4.3 NONRIGID REGISTRATION ALGORITHMS FOR HDR DEGHOSTING

Rather than simply rejecting content that could generate ghosting, one can compensate for motion by means of nonrigid registration. To do this, two kinds of algorithms have been proposed: (1) algorithms based on a flavor of optical flow to align the images, and (2) algorithms based on patch-based synthesis. Note that while nonrigid registration algorithms have the potential to preserve a larger dynamic range from the stack, they tend to introduce objectionable artifacts when the estimation of the displacement between the images fails. This is particularly true for flow-based algorithms, which we will discuss first.

3.4.3.1 Optical flow and correspondence registration methods

Bogoni (2000) presented perhaps the earliest known method to register a stack of images for HDR reconstruction. First, he applies an affine motion estimation to globally align the images. This process, based on earlier work on registration for image mosaics (Hansen et al., 1994), operates in a multiresolution fashion from coarse to fine, using a Laplacian pyramid scheme. At each iteration, the optical flow field is computed from one image to another by means of local cross-correlation analysis, and then an affine motion model is fit to the flow field by weighted least-squares regression. The affine transform is then used to warp each image to align it roughly with the reference. At this point, a second step performs unconstrained motion estimation with optical flow between each source image and a predefined reference. This resulting field is used to warp the individual sources to compute the final registration with the reference.

Jinno and Okuda (2008) propose addressing the problem of ghosting with Markov random fields. After selecting the reference image, they estimate three arrays (the same size as the images) for each of the other images in the stack. The first is a displacement field **d**, and the second is a binary occlusion field **o** that indicates the parts of the reference that are occluded in the second image. This is computed by thresholding the maximum search distance for the displacement field: if a pixel cannot be found in a neighborhood $\mathcal{N}(p)$ around a given pixel that has a luminance within a specific threshold, then pixel p is considered occluded. The third is a saturation field, a binary mask that keeps track of the regions where the second image is overexposed or underexposed. Since these arrays are spatially coherent, they can be modeled as Markov random fields and computed with use of Bayes rule as an estimation problem that finds the most probable fields $\mathbf{d}$, $\mathbf{o}$, and $\mathbf{s}$ given observed images Z_{ref} and Z_i:

$$
\begin{aligned}
\max_{\mathbf{d,o,s}} P(\mathbf{d,o,s} \mid Z_{\text{ref}}, Z_i) &= \max_{\mathbf{d,o,s}} \frac{P(Z_{\text{ref}} \mid \mathbf{d,o,s}, Z_i) P(\mathbf{d,o,s} \mid Z_i)}{P(Z_{\text{ref}})} \\
&= \max_{\mathbf{d,o,s}} \frac{P(Z_{\text{ref}} \mid \mathbf{d,o,s}, Z_i) P(\mathbf{d} \mid \mathbf{o,s}, Z_i) P(\mathbf{o} \mid \mathbf{s}, Z_i) P(\mathbf{s} \mid Z_i)}{P(Z_{\text{ref}})}.
\end{aligned} \tag{3.40}
$$

This problem is analogous to that of finding $\max_{\mathbf{d,o,s}} P(Z_{\text{ref}} \mid \mathbf{d,o,s}, Z_i) P(\mathbf{d} \mid \mathbf{o,s}, Z_i) P(\mathbf{o} \mid \mathbf{s}, Z_i) P(\mathbf{s} \mid Z_i)$, which they approximate by first finding $\mathbf{s}$ through thresholding, and then iteratively solving for $\mathbf{d}$ and $\mathbf{o}$. Once they have these fields, they can use them during the merging stage to produce the final HDR result.

Zimmer et al. (2011) align images in the stack with a specified reference using an energy-based optical flow optimization that is more tolerant to changes in exposure. To achieve this invariance, they define an energy function that leverages the gradient constancy, similar to that of Brox et al. (2009) and that of Brox and Malik (2011). Specifically, for each image i in the stack they compute a dense displacement field $\mathbf{u}_i(p) = [u_i(p), v_i(p)]^\mathsf{T}$ that specifies an offset at every pixel by minimizing an energy function of the form

$$
E(\mathbf{u}_i(p)) = \sum_{p \in \Omega} D(\mathbf{u}_i(p)) + \lambda S(\nabla \mathbf{u}_i(p)), \tag{3.41}
$$

where $D(\mathbf{u}_i(p))$ is the data term that tries to align the image with the reference, and S is the smoothness term (regularizer) that encourages smooth flow in places where the reference image is unreliable (ie, overexposed or underexposed). Because the brightness constancy across the stack is violated in this application, they propose that the data term $D(\mathbf{u}_i(p))$ should try to match the gradient of the offset region in image Z_i with that of the reference:

$$
D(\mathbf{u}_i(p)) = \Psi \left(\frac{1}{n_x} \left| \frac{\partial}{\partial x} Z_i(p + \mathbf{u}_i(p)) - \frac{\partial}{\partial x} Z_{\text{ref}}(p) \right|^2 + \frac{1}{n_y} \left| \frac{\partial}{\partial y} Z_i(p + \mathbf{u}_i(p)) - \frac{\partial}{\partial y} Z_{\text{ref}}(p) \right|^2 \right), \tag{3.42}
$$

where Ψ is regularized L_1 norm $\Psi(s^2) = \sqrt{s^2 + 0.001^2}$ and n_x and n_y are normalization factors. For the smoothness term $S(\nabla \delta p_i(p))$, they use a regularizer based on total variation:

$$
S(\nabla \mathbf{u}_i(p)) = \Psi(|\nabla u_i(p)|^2 + |\nabla v_i(p)|^2). \tag{3.43}
$$

The energy equation (Eq. 3.41) is then optimized by a semi-implicit gradient descent scheme, and the final flows are used to warp each of the input images in the stack, which are then merged with use of the method of Robertson et al. (1999).

Later, Hu et al. (2012) proposed using the patch-based, nonrigid dense correspondence method of HaCohen et al. (2011) to compute dense correspondences between the reference image and the other images in the stack, called the *source* images. They then use this correspondence field to warp pixels in the source images to match the appearance of the reference. However, because of occlusions and disocclusions, as well as brightness changes, the correspondences are generally incomplete; this can result in "holes" (ie, regions where the pixels' values are undefined).

To address this problem, Hu et al. first propose a robust strategy to estimate the IMFs (Grossberg and Nayar, 2003a) using the known pixel correspondences. For each hole in the warped source, they then attempt to paste the pixels from the original source image; however, to compensate for motion, they first apply a local projective transformation to align them with those in the reference. To ensure that the pasted pixels cause no artifacts, Hu et al. take a bounding box larger than the hole to be transformed and pasted. If the pixels within the box, but outside the hole, do not match, the region is considered to be affected by motion. In these cases, the pixels are pasted directly from the reference after their brightness values have been appropriately corrected with the estimated IMF.

Another method based on a flavor of optical flow is the approach of Gallo et al. (2015). The method is based on the observation that modern cameras offer fast bursts modes, which make arbitrarily large displacements unlikely. Therefore, instead of computing the optical flow at each pixel, they suggest computing it only at sparse locations, and then to propagate it to the rest of the pixels. Specifically, there are four stages to the algorithm.

Gallo et al. first describe a novel method to find and match corners across two images in the stack, one chosen to be the reference and the other being the source image. Their corners are based on the changes of average brightness around different pixels, which can be computed efficiently with integral images. The second stage identifies and removes matches that are either incorrect or belong to structures that move in a highly nonrigid fashion. To achieve this, Gallo et al. observe that good matches should be locally consistent with a homography, and propose a modification of the RANSAC algorithm to isolate those that are not. The set of matches that are both correct and rigid offers an estimate of the flow at sparse locations, which can be propagated to the rest of the pixels with use of the reference image as a guide to an edge-aware diffusion algorithm. With the dense flow, the source image can be warped to be geometrically consistent with the reference. However, to account for possible errors in the flow propagation, Gallo et al. modify the algorithm of Mertens et al. (2007), see Section 3.3.2.3, by adding a fourth weight. Specifically, they use the structural similarity index proposed by Wang et al. (2004) to account for the quality of the registration at different locations, and reduce the contribution of regions that are not correctly registered. For a pair of five-megapixel images, Gallo et al. report execution times of less than 150 ms on a desktop computer and less than 700 ms on a commercial tablet computer. For reference, the methods described in Section 3.4.3.2 are several orders of magnitude slower. As mentioned before, this large speedup is possible thanks to the observation that arbitrarily large displacements are unlikely when the stack is captured in a fast burst.

Compared with rejection-based approaches for HDR reconstruction, these alignment methods, which rely on correspondences between the different images in the stack, have the advantage that they can move content around. This allows them to handle dynamic objects with HDR illumination. However, finding reliable correspondences, especially in cases of complex motion and deformation, is quite difficult and can introduce new artifacts. These problems can largely be resolved by use of the patch-based synthesis methods discussed next.

3.4.3.2 Patch-based synthesis methods

The most successful kind of HDR deghosting algorithms are perhaps those that align the stack of images together by using *patch-based synthesis* to generate plausible images that are registered to the reference (Sen et al., 2012; Hu et al., 2013; Kalantari et al., 2013). Indeed, a recent state-of-the-art report on HDR deghosting techniques has shown that these algorithms produce the best results for general scenes (Tursun et al., 2015). These methods can also be considered a new kind of algorithm (different from the rejection and registration algorithms we have discussed) because they can solve for the aligned images and the HDR reconstruction simultaneously. Although patch-based synthesis had previously been shown to be very powerful for various computational imaging tasks (such as image hole filling (Wexler et al., 2007), image summarization and editing (Simakov et al., 2008; Barnes et al., 2010), morphing (Shechtman et al., 2010), and finding dense correspondences between images (HaCohen et al., 2011)), these new methods apply it to HDR reconstruction by posing the problem as an energy optimization.

To use patch-based synthesis for HDR reconstruction, the two independent methods of Sen et al. (2012) and Hu et al. (2013) make similar observations: after registration, each image Z_i from the stack should look as if it were taken at the same time as the reference Z_{ref}, but should be photometrically consistent with the original Z_i, thereby capturing all of the additional dynamic range information contained in the original image.

Sen et al. (2012) propose doing this using a new optimization equation that codifies the objective of reference-based HDR reconstruction algorithms: (1) to produce an HDR image that resembles the reference in portions where the reference is well exposed, and (2) to leverage well-exposed information from other images in the stack in places where the reference is not well exposed. This results in what they call the *HDR image synthesis equation*, which contains two terms:

$$\text{Energy}\,(E) = \sum_{p \in \text{pixels}} \left[\alpha_{ref}(p) \cdot \left(g(Z_{ref}(p))/t_{ref} - E(p) \right)^2 + (1 - \alpha_{ref}(p)) \cdot E_{MBDS}(H \,|\, L_1, \ldots, L_N) \right]. \quad (3.44)$$

The first term states that the ideal HDR image E should be close in an L_2 sense to the LDR reference Z_{ref} mapped to the linear irradiance domain. This should be done only for the pixels where the reference is properly exposed, as given by the $\alpha_{ref}(p)$ term, which is a trapezoid function in the pixel intensity domain that favors intensities near the middle of the pixel value range.

In the parts where the reference image Z_{ref} is poorly exposed as indicated by the $1 - \alpha_{ref}(p)$ term, the algorithm draws information from the other images in the stack using a novel multisource bidirectional equation E_{MBDS} that extends the bidirectional similarity metric of Simakov et al. (2008):

$$\text{BDS}\,(T \,|\, S) = \frac{1}{|S|} \sum_{P \in S} \min_{Q \in T} d(P, Q) + \frac{1}{|T|} \sum_{Q \in T} \min_{P \in S} d(Q, P). \quad (3.45)$$

The original function of Simakov et al. takes a pair of images (source S and target T) and ensures that all of the patches (small blocks of pixels) in S can be found in T (first term, called "completeness") and vice versa (second term, called "coherence"). Note that the coherence term ensures that the final target does not contain objectionable artifacts, as these artifacts are not found in the original source.

However, Eq. (3.45) does not work for HDR reconstruction directly; sometimes content that should be visible in the ith exposure when "aligned" with the reference exposure might be occluded in Z_i and needs to be drawn from a different image. So rather than using a pairwise bidirectional similarity metric, Sen et al. introduce a multisource bidirectional similarity metric E_{MBDS} that draws information from all the images in the stack simultaneously.

To optimize Eq. (3.44), Sen et al. introduce auxiliary variables $\tilde{Z}_i$ that represent the different LDR images in the stack after they have been aligned with the reference. This equation can then be solved with an iterative, two-stage algorithm that solves for the $\tilde{Z}_1, \ldots, \tilde{Z}_N$ and E simultaneously:

Stage 1: The algorithm first solves for the aligned LDR images $\tilde{Z}_1, \ldots, \tilde{Z}_N$ with a bidirectional search-and-vote process (Simakov et al., 2008) accelerated by PatchMatch (Barnes et al., 2009). This process draws information into each of the aligned LDR images from the entire stack, which has been adjusted to match the corresponding exposure level. To produce images aligned with the reference, the irradiance image E from the previous iteration (which has been injected with the reference in step 2) is used as the initial target for the search-and-vote process.

Stage 2: Next, the algorithm optimizes for E by merging the aligned images $\tilde{Z}_1, \ldots, \tilde{Z}_N$ using a standard HDR merging process (Section 3.3) and then injects the portions of the reference image where it is well exposed into the result.

Once the new E has been computed, it is used to extract the new image targets for the next iteration, and the algorithm goes back to stage 1. These two stages are performed at every iteration of the algorithm until it converges. Furthermore, as is common for patch-based methods such as this (eg, Simakov et al., 2008), this core algorithm is performed at multiple scales, starting at the coarsest resolution and working to the finest. Once the algorithm has converged, it returns both the desired HDR image E and the "aligned" images at each exposure $\tilde{Z}_1, \ldots, \tilde{Z}_N$. A result produced with this algorithm is shown in Fig. 3.8. This algorithm was later extended by Kalantari et al. (2013) to reconstruct HDR video, as described in Chapter 4.

Hu et al. (2013) propose a different patch-based synthesis algorithm, which, unlike the algorithm of Sen et al., does not require the camera calibration curve to be known a priori. Specifically, they calculate the aligned images $\tilde{Z}_i$ as

$$\tilde{Z}_i = \arg \min_{\tilde{Z}_i, \tau, \mathbf{u}} \left(C_r(\tilde{Z}_i, Z_{\text{ref}}, \tau) + C_t(\tilde{Z}_i, Z_i, \mathbf{u}) \right), \tag{3.46}$$

(A)   Low-dynamic-range inputs   (B)   Aligned intermediate images   (C) HDR result

FIGURE 3.8

Sample result of the HDR reconstruction algorithm of Sen et al. (2012). (A) Input LDR images (first, third, and fifth images shown from of a five-image input stack: Z_1, Z_3, Z_5). (B) Corresponding aligned images ($\tilde{Z}_1, \tilde{Z}_3, \tilde{Z}_5$), computed by the algorithm. (C) Tone-mapped HDR result after the reconstruction.

Source: Images courtesy of Sen et al. (2012).

where $\mathbf{u}$ is the displacement field that "warps" image Z_i to match the geometric appearance of the reference, and τ is the IMF between the source image Z_i and the reference Z_{ref}. In Eq. (3.46), the first term, C_r for "radiance consistency," encourages the aligned image $\widetilde{Z}_i$ to be geometrically consistent with the reference, Z_{ref}:

$$C_r(\widetilde{Z}_i, Z_{\text{ref}}, \tau) = \sum_p \left(\|\widetilde{Z}_i(p) - \tau(Z_{\text{ref}}(p))\|^2 + \alpha \, \|\nabla \widetilde{Z}_i(p) - \nabla \tau(Z_{\text{ref}}(p))\|^2 \right). \tag{3.47}$$

Note that both the images and their gradients are accounted for in Eq. (3.47). The second term in Eq. (3.46), C_t, is what Hu et al. call the "texture consistency" term:

$$C_t(\widetilde{Z}_i, Z_i, \mathbf{u}) = \frac{1}{k} \sum_p \left(\|P_{\widetilde{Z}}(p) - P_{Z_i}(p + \mathbf{u}(p))\|^2 + \alpha \, \|\nabla(P_{\widetilde{Z}}(p) - \nabla P_{Z_i}(p + \mathbf{u}(p))\|^2 \right), \tag{3.48}$$

where k is a normalization factor. The texture consistency term enforces similarity between the patch around pixel p in the warped source, $P_{\widetilde{Z}}(p)$, and the corresponding patch in the source image, $P_{Z_i}(p + \mathbf{u}(p))$. This helps enforce that the synthesized content is plausible and free of artifacts.

Hu et al. tackle this optimization iteratively using a coarse-to-fine approach, which helps in two ways. First, it prevents the optimization from falling in a local minimum. Second, it allows the algorithm to deal with large oversaturated or undersaturated regions: a patch that is entirely saturated at the finest level could include information from neighboring nonsaturated pixels at coarser levels, thus allowing information to propagate inward. To do this, they propose an iterative, three-stage algorithm:

Stage 1: First, they estimate τ using the intensity histograms of the images (Grossberg and Nayar, 2003a) at the coarsest level of the pyramid and initialize $\widetilde{Z}_i = \tau(Z_i)$ for the same level. The displacement $\mathbf{u}$, which appears only in C_t, can then be estimated with PatchMatch (Barnes et al., 2009).

Stage 2: In a second step, they propose refining the estimate of $\widetilde{Z}_i$ by minimizing C_r. However, for the areas where the reference image is overexposed or underexposed, they average $\tau(Z_{\text{ref}}(p))$ with the corresponding location in the source image $Z_i(p + \mathbf{u}(p))$ with a weight that accounts for how likely the latter is to become overexposed or underexposed in the reference image.

Stage 3: In the third and last step, with the new $\widetilde{Z}_i$, they refine the IMF, τ. Moving to the next finer level, they leave τ unchanged and linearly interpolate u. The latent image $\widetilde{Z}_i$, instead, is initialized with a weighted average of $\tau(Z_{\text{ref}})$ and $Z_i(p + u(p))$.

Results from this approach can be seen in Fig. 3.9.

As discussed earlier, these patch-based synthesis methods have the advantage that they work very well for scenes with complex, arbitrarily large motion where other algorithms would normally fail. However, they are expensive and require considerable time and hardware resources for evaluation: the reference implementations provided by Hu et al. take more than 1 min for VGA images. Furthermore, although they can produce plausible results, they are only *hallucinating* the final result as compared with the true HDR result that would have been captured by a hypothetical HDR camera.

| (A) | Low-dynamic-range inputs | (B) | Aligned intermediate results | (C) | HDR Result |

FIGURE 3.9

Sample result of the HDR reconstruction algorithm of Hu et al. (2012). (A) Input LDR images Z_1, Z_2, Z_3. (B) Corresponding aligned images $(\tilde{Z}_1, \tilde{Z}_2, \tilde{Z}_3)$ computed by the algorithm. (C) Tone-mapped HDR result after the reconstruction.

Source: Images courtesy of Hu et al. (2013).

3.5 CONCLUSION

In this chapter, we have examined approaches to capture HDR images and video by taking a stack of multiple images at different exposure settings. We began by studying algorithms for metering, which set the exposure levels for the different images in the stack. Next, we studied the process of merging the LDR images into a final HDR result, which included a radiometric calibration process (to compute the irradiance images from the original pixel values) and merging schemes (which compute the weights of the different irradiance images to compute the final HDR). Finally, we examined algorithms developed to handle artifacts from motion when one is capturing stack-based HDR images, which included rejection algorithms and registration algorithms.

ACKNOWLEDGMENTS

P. Sen was partially funded by National Science Foundation grants IIS-1342931 and IIS-1321168. The authors also thank the many researchers within the computational imaging community who published the work described in this chapter. Without their innovations, none of this would have been possible.

REFERENCES

Adams, A., 1948. The Negative: Exposure and Development, vol. 2. Morgan and Lester, New York.

An, J., Lee, S., Kuk, J., Cho, N., 2011. A multi-exposure image fusion algorithm without ghost effect. In: The IEEE International Conference on Acoustics, Speech, and Signal Processing (ICASSP).

Badki, A., Kalantari, N.K., Sen, P., 2015. Robust radiometric calibration for dynamic scenes in the wild. In: The IEEE International Conference on Computational Photography (ICCP).

Barnes, C., Shechtman, E., Finkelstein, A., Goldman, D.B., 2009. PatchMatch: a randomized correspondence algorithm for structural image editing. ACM Trans. Graph. (Proc. SIGGRAPH) 28 (3), 24:1–24:11.

Barnes, C., Shechtman, E., Goldman, D.B., Finkelstein, A., 2010. The generalized PatchMatch correspondence algorithm. In: The European Conference of Computer Vision (ECCV).

Bilcu, R.C., Burian, A., Knuutila, A., Vehviläinen, M., 2008. High dynamic range imaging on mobile devices. In: The IEEE International Conference of Electronics, Circuits, and Systems (ICECS).

Bogoni, L., 2000. Extending dynamic range of monochrome and color images through fusion. In: The IEEE International Conference of Pattern Recognition (ICPR).

Boykov, Y., Veksler, O., Zabih, R., 2001. Fast approximate energy minimization via graph cuts. IEEE Trans. Pattern Anal. Mach. Intell. 23 (11), 1222–1239.

Brox, T., Malik, J., 2011. Large displacement optical flow: descriptor matching in variational motion estimation. IEEE Trans. Pattern Anal. Mach. Intell. 33 (3), 500–513.

Brox, T., Bregler, C., Malik, J., 2009. Large displacement optical flow. In: The IEEE Conference on Computer Vision and Pattern Recognition (CVPR).

Bunsen, R.W., Roscoe, H.E., 1862. Photochemical researches — Part V. On the measurement of the chemical action of direct and diffuse sunlight. Proc. R. Soc. Lond. 12, 306–312.

Chen, T., El Gamal, A., 2002. Optimal scheduling of capture times in a multiple-capture imaging system. In: The International Society for Optics and Electronics (SPIE).

Debevec, P.E., Malik, J., 1997. Recovering high dynamic range radiance maps from photographs. In: ACM SIGGRAPH.

Eden, A., Uyttendaele, M., Szeliski, R., 2006. Seamless image stitching of scenes with large motions and exposure differences. In: The IEEE Conference on Computer Vision and Pattern Recognition (CVPR).

Farid, H., 2001. Blind inverse gamma correction. IEEE Trans. Image Process. 10 (10), 1428–1433.

Gallo, O., Gelfand, N., Chen, W.C., Tico, M., Pulli, K., 2009. Artifact-free high dynamic range imaging. In: The IEEE International Conference on Computational Photography (ICCP).

Gallo, O., Tico, M., Manduchi, R., Gelfand, N., Pulli, K., 2012. Metering for exposure stacks. In: Eurographics, vol. 31, pp. 479–488.

Gallo, O., Troccoli, A., Hu, J., Pulli, K., Kautz, J., 2015. Locally non-rigid registration for mobile HDR photography. The IEEE Conference on Computer Vision and Pattern Recognition Workshops (CVPRW).

Gelfand, N., Adams, A., Park, S.H., Pulli, K., 2010. Multi-exposure imaging on mobile devices. In: The ACM Conference on Multimedia (MM).

Granados, M., Ajdin, B., Wand, M., Theobalt, C., Seidel, H.P., Lensch, H.P.A., 2010. Optimal HDR reconstruction with linear digital cameras. In: The IEEE Conference on Computer Vision and Pattern Recognition (CVPR).

Granados, M., Kim, K.I., Tompkin, J., Theobalt, C., 2013. Automatic noise modeling for ghost-free HDR reconstruction. ACM Trans. Graph. (Proc. SIGGRAPH Asia) 32 (6), 201:1–201:10.

Grosch, T., 2006. Fast and robust high dynamic range image generation with camera and object movement. In: The International Symposium on Vision, Modeling and Visualization.

Grossberg, M.D., Nayar, S.K., 2003a. Determining the camera response from images: what is knowable? IEEE Trans. Pattern Anal. Mach. Intell. 25 (11), 1455–1467.

Grossberg, M.D., Nayar, S.K., 2003b. What is the space of camera response functions? In: The IEEE Conference on Computer Vision and Pattern Recognition (CVPR).

Grundmann, M., McClanahan, C., Kang, S.B., Essa, I., 2013. Post-processing approach for radiometric self-calibration of video. In: The IEEE International Conference on Computational Photography (ICCP).

HaCohen, Y., Shechtman, E., Goldman, D.B., Lischinski, D., 2011. Non-rigid dense correspondence with applications for image enhancement. ACM Trans. Graph. (Proc. SIGGRAPH) 30 (4), 70:1–70:10.

Hadziabdic, K.K., Telalovic, J.H., Mantiuk, R., 2013. Comparison of deghosting algorithms for multi-exposure high dynamic range imaging. In: The ACM Spring Conference on Computer Graphics.

Hansen, M., Anandan, P., Dana, K., Van der Wal, G., Burt, P., 1994. Real-time scene stabilization and mosaic construction. In: The IEEE Workshop on Applications of Computer Vision.

Hasinoff, S.W., Durand, F., Freeman, W.T., 2010. Noise-optimal capture for high dynamic range photography. In: The IEEE Conference on Computer Vision and Pattern Recognition (CVPR).

Heo, Y., Lee, K.M., Lee, S.U., Moon, Y., Cha, J., 2010. Ghost-free high dynamic range imaging. In: The Asian Conference of Computer Vision (ACCV).

Hu, J., Gallo, O., Pulli, K., 2012. Exposure stacks of live scenes with hand-held cameras. In: The European Conference of Computer Vision (ECCV).

Hu, J., Gallo, O., Pulli, K., Sun, X., 2013. HDR deghosting: How to deal with saturation? In: The IEEE Conference on Computer Vision and Pattern Recognition (CVPR).

Jacobs, K., Loscos, C., Ward, G., 2008. Automatic high-dynamic-range image generation for dynamic scenes. IEEE Comput. Graph. Appl. 28 (2), 84–93.

Jinno, T., Okuda, M., 2008. Motion blur free HDR image acquisition using multiple exposures. In: The IEEE International Conference on Image Processing (ICIP).

Kalantari, N.K., Shechtman, E., Barnes, C., Darabi, S., Goldman, D.B., Sen, P., 2013. Patch-based high dynamic range video. ACM Trans. Graph. (Proc. SIGGRAPH Asia) 32 (6), 202:1–202:8.

Khan, E.A., Akyüz, A.O., Reinhard, E., 2006. Ghost removal in high-dynamic-range images. In: The IEEE International Conference on Image Processing (ICIP).

Kim, S.J., Lin, H.T., Lu, Z., Süsstrunk, S., Lin, S., Brown, M.S., 2012. A new in-camera imaging model for color computer vision and its application. IEEE Trans. Pattern Anal. Mach. Intell. 34 (12), 2289–2302.

Kirk, K., Andersen, H.J., 2006. Noise characterization of weighting schemes for combination of multiple exposures. In: The British Machine Vision Conference (BMVC).

Lee, J.Y., Matsushita, Y., Shi, B., Kweon, I.S., Ikeuchi, K., 2013. Radiometric calibration by rank minimization. IEEE Trans. Pattern Anal. Mach. Intell. 35 (1), 144–156.

Levoy, M., 2014. HDR+: Low light and high dynamic range photography in the Google Camera App. http://googleresearch.blogspot.com/2014/10/hdr-low-light-and-high-dynamic-range.html (Accessed on Jan. 1, 2015).

Lin, S., Gu, J., Yamazaki, S., Shum, H.Y., 2004. Radiometric calibration from a single image. In: The IEEE Conference on Computer Vision and Pattern Recognition (CVPR).

Madden, B.C., 1993. Extended intensity range imaging, Technical report, University of Pennsylvania.

Mann, S., 1993. Compositing multiple pictures of the same scene: generalized large-displacement 8-parameter motion. In: The Society for Imaging Science and Technology (IS&T).

Mann, S., 2000. Comparametric equations with practical applications in quantigraphic image processing. IEEE Trans. Image Process. 9 (8), 1389–1406.

Mann, S., Picard, R., 1995. Being 'undigital' with digital cameras: extending dynamic range by combining differently exposed pictures. In: The Society for Imaging Science and Technology (IS&T).

Matsushita, Y., Lin, S., 2007. Radiometric calibration from noise distributions. In: The IEEE Conference on Computer Vision and Pattern Recognition (CVPR).

Mertens, T., Kautz, J., Van Reeth, F., 2007. Exposure fusion. In: The Pacific Conference on Computer Graphics and Applications.

Mitsunaga, T., Nayar, S.K., 1999. Radiometric self calibration. In: The IEEE Conference on Computer Vision and Pattern Recognition (CVPR).

Oh, T.H., Lee, J.Y., Kweon, I.S., 2015. Robust high dynamic range imaging by rank minimization. IEEE Trans. Pattern Anal. Mach. Intell. 37 (6), 1219–1232.

Pece, F., Kautz, J., 2010. Bitmap movement detection: HDR for dynamic scenes. In: The Conference on Visual Media Production (CVMP).

Peng, Y., Ganesh, A., Wright, J., Xu, W., Ma, Y., 2012. RASL: Robust alignment by sparse and low-rank decomposition for linearly correlated images. IEEE Trans. Pattern Anal. Mach. Intell. 34 (11), 2233–2246.

Pérez, P., Gangnet, M., Blake, A., 2003. Poisson image editing. ACM Trans. Graph. (Proc. SIGGRAPH) 22 (3), 313–318.

Raman, S., Chaudhuri, S., 2011. Reconstruction of high contrast images for dynamic scenes. Vis. Comput. 27 (12), 1099–1114.

Reinhard, E., Ward, G., Pattanaik, S., Debevec, P., 2005. High Dynamic Range Imaging: Acquisition, Display, and Image-Based Lighting (The Morgan Kaufmann Series in Computer Graphics). Morgan Kaufmann Publishers Inc., San Francisco, CA.

Robertson, M.A., Borman, S., Stevenson, R.L., 1999. Dynamic range improvement through multiple exposures. In: The IEEE International Conference on Image Processing (ICIP).

Robertson, M.A., Borman, S., Stevenson, R.L., 2003. Estimation-theoretic approach to dynamic range enhancement using multiple exposures. J. Electron. Imaging 12 (2), 219–228.

Sen, P., Kalantari, N.K., Yaesoubi, M., Darabi, S., Goldman, D.B., Shechtman, E., 2012. Robust patch-based HDR reconstruction of dynamic scenes. ACM Trans. Graph. (Proc. SIGGRAPH Asia) 31 (6), 203:1–203:11.

Shechtman, E., Rav-Acha, A., Irani, M., Seitz, S., 2010. Regenerative morphing. In: The IEEE Conference on Computer Vision and Pattern Recognition (CVPR).

Shen, R., Cheng, I., Shi, J., Basu, A., 2011. Generalized random walks for fusion of multi-exposure images. IEEE Trans. Image Process. 20 (12), 3634–3646.

Sidibe, D., Puech, W., Strauss, O., 2009. Ghost detection and removal in high dynamic range images. In: The European Signal Processing Conference (EUSIPCO).

Simakov, D., Caspi, Y., Shechtman, E., Irani, M., 2008. Summarizing visual data using bidirectional similarity. In: The IEEE Conference on Computer Vision and Pattern Recognition (CVPR).

Srikantha, A., Sidibe, D., 2012. Ghost detection and removal for high dynamic range images: recent advances. Signal Process. Image Commun. 27 (6), 650–662.

Tocci, M.D., Kiser, C., Tocci, N., Sen, P., 2011. A versatile HDR video production system. ACM Trans. Graph. 30 (4), 41:1–41:10.

Tomasi, C., Manduchi, R., 1998. Bilateral filtering for gray and color images. In: The IEEE International Conference on Computer Vision (ICCV).

Tomaszewska, A., Mantiuk, R., 2007. Image registration for multi-exposure high dynamic range image acquisition. In: The International Conference in Central Europe on Computer Graphics, Visualization and Computer Vision (WSCG).

Tsin, Y., Ramesh, V., Kanade, T., 2001. Statistical calibration of CCD imaging process. In: The IEEE International Conference on Computer Vision (ICCV).

Tursun, O.T., Akyz, A.O., Erdem, A., Erdem, E., 2015. The state of the art in HDR deghosting: a survey and evaluation. In: Eurographics STAR Reports.

Wang, Z., Bovik, A.C., Sheikh, H.R., Simoncelli, E.P., 2004. Image quality assessment: from error visibility to structural similarity. IEEE Trans. Image Process. 13 (4), 600–612.

Ward, G., 2003. Fast, robust image registration for compositing high dynamic range photographs from hand-held exposures. J. Graph. Tools 8 (2), 17–30.

Wexler, Y., Shechtman, E., Irani, M., 2007. Space-time completion of video. IEEE Trans. Pattern Anal. Mach. Intell. 29 (3), 463–476.

Wu, S., Xie, S., Rahardja, S., Li, Z., 2010. A robust and fast anti-ghosting algorithm for high dynamic range imaging. In: The IEEE International Conference on Image Processing (ICIP).

Xiong, Y., Saenko, K., Darrell, T., Zickler, T., 2012. From pixels to physics: probabilistic color de-rendering. In: The IEEE Conference on Computer Vision and Pattern Recognition (CVPR).

Zhang, W., Cham, W.K., 2010. Gradient-directed composition of multi-exposure images. In: The IEEE Conference on Computer Vision and Pattern Recognition (CVPR).

Zhang, W., Cham, W.K., 2012. Reference-guided exposure fusion in dynamic scenes. J. Vis. Commun. Image Represent. 23 (3), 467–475.

Zimmer, H., Bruhn, A., Weickert, J., 2011. Freehand HDR imaging of moving scenes with simultaneous resolution enhancement. Eurographics.

MULTIVIEW HDR VIDEO SEQUENCE GENERATION

4

R.R. Orozco*, C. Loscos†, I. Martin*, A. Artusi*

University of Girona, Girona, Spain*
University of Reims Champagne-Ardenne, Reims, France†

CHAPTER OUTLINE

4.1 INTRODUCTION

There is a huge gap between the range of lighting perceived by the human visual system and what conventional digital cameras and displays are able to capture and represent respectively. High dynamic range (HDR) imaging aims at reducing this gap by capturing a more realistic range of light values, such that dark and bright areas can be recorded in the same image or video. Visually, this avoids underexposure and overexposure in such areas.

The combination of the latest advances in different digital imaging areas, such as 4K color video resolution, stereo- and multiscopic video, and HDR imaging, promises an unprecedented video experience for users. However, big challenges of different nature remain to be overcome before such technologies converge. In particular, the whole pipeline of acquisition, compression, transmission, and

High Dynamic Range Video. http://dx.doi.org/10.1016/B978-0-08-100412-8.00004-8

display of HDR imagery presents unsolved problems to which solutions need to be found before we can enjoy HDR films on a TV at home. Just one example is enough to illustrate such challenges: 1 min of HDR video at 30 frames per second full high-definition resolution requires 42GB of storage compared with four times less for traditional digital low dynamic range (LDR) video (Chalmers and Debattista, 2011). Among such challenges, the extension from HDR images to HDR video and stereo HDR video has a very important role.

Techniques for HDR video acquisition are evolving. There are two main approaches to capture HDR images: HDR sensors or multiple exposure with LDR sensors. Some HDR camera prototypes have been lately presented to the research community (Nayar and Branzoi, 2003; Chalmers et al., 2009; Tocci et al., 2011), but are not available yet for commercial use. Their commercial counterparts, such as the Viper camera (ThomsomGrassValley, 2005), the Phantom HD camera (Research, 2005), and the Red Epic camera (RedCompany, 2006), are only few in number, the range addressed remains limited (below 16 f-stops), and they are still unaffordable for most users. Diverse alternatives have been presented to acquire HDR values with conventional cameras. Chapter 2 discussed some techniques based on RAW data acquired directly from the camera sensor and hardware prototypes designed to increase the range of light captured in each frame.

Chapter 3 introduced some computational procedures to recover HDR images from different LDR exposures. There is a common problem to merge the different exposures in HDR video and stereoscopic HDR video when LDR images do not superpose exactly. Pixels are misaligned: pixels corresponding to the same object in the scene do not correspond to the same position in all the images. A matching process is thus required to find correspondences between pixels in the different images that represent the same object in the scene.

In this chapter we present the various solutions that are proposed to go from images to sequences with changes in content and/or viewpoint. One key issue is pixel registration in differently exposed sequences, with the possibility of creating HDR videos or multiview (including stereo) image sequences.

The chapter is divided into three main sections. In Section 4.2, we provide a classification of the different natures of possible misalignment according to the position of the camera and the scene content. Section 4.3 is dedicated to multiexposure HDR video generation from a free-path camera. Section 4.4 is dedicated to multiscopic HDR video.

4.2 HDR AND STEREO HDR VIDEO ACQUISITION

Chapter 3 presented the idea that merging multiple exposures allows one to recover HDR images. Multiple exposures can be acquired in different ways. One is exposure time variation for multiple exposures of the same scene sequence contributing to each video frame. Use of deghosting algorithms to produce an HDR per frame and obtain HDR video is not straightforward. Several constraints need to be considered to decide if an algorithm for HDR image generation can be extended to multiple exposure video sequences. Even the kind of camera we use is important, because some camera application programming interfaces do not allow exposure times to be adjusted. This section is dedicated to analyzing the two main aspects that need to be considered: temporal and spatial issues in HDR acquisition.

4.2.1 **TEMPORAL CONSIDERATIONS**

The typical frame rate to play video is around 24 frames per second. This means that we need to create at least 24 HDR frames per second. Unlike still HDR images, we cannot just capture long exposures to try to increase the dynamic range. The sum of the exposure times for each frame must be small enough to guarantee video frame rates. The difference in exposure times between consecutive frames is limited for temporal coherence not to be compromised. Such limitations must be considered before we can extend methods designed for image deghosting. Some of them are suitable for generating only one HDR image for each multiexposed sequence. In the video context, the use of long exposure times to increase the dynamic range is not always possible without compromising video frame rates.

Besides, stereo HDR video requires at least two views per frame, while some multiscopic displays may accept more than nine different views. Computational approaches in this case are preferred to the use of multiple expensive HDR cameras. To achieve HDR in stereoscopy, we need to generate one HDR image per view for each stereo HDR frame. Most proposed approaches recover HDR values from differently exposed stereo LDR images. Synchronization in capturing the different views is one supplementary key consideration.

Thus, four solutions can be considered:

1. All views have the same exposure time at each take. The exposure varies from one take to another. While synchronization is simplified here, the difficulty is to reconstruct temporarily coherent HDR information in overexposed and underexposed areas.
2. Each view has a different exposure time. While finding HDR information in all areas of the images is ensured, synchronization is problematic: either each objective waits for the others before taking the next frame or frames need to be synchronized afterward.
3. Bonnard et al. (2012) proposed an alternative to the first two solutions. It involves placing neutral density filters on the camera objectives in order to simulate different exposure times. An advantage of this is that all objectives use the same exposure time even if the resulting images simulate different exposures. Synchronization is thus reduced to synchronizing the different objectives. A major drawback arises from the fact that each view takes the same exposure at each frame. Underexposed or overexposed areas will remain as such through the entire video.
4. Other alternatives have been proposed to acquire for each objective, several exposures at once. This could be done, for example, through a beam splitter (Tocci et al., 2011) or a spatially varying mask (Nayar and Branzoi, 2003). Optical elements can split light beams onto different sensors with different exposure settings (Tocci et al., 2011).

4.2.2 **SPATIAL CONSIDERATIONS**

While LDR images are being taken for reconstruction of HDR information, spatial variation can occur. In the following we classify these variations according to their cause and the misalignment they produce (Section 4.2.2.1). The types of misalignment are discussed in Section 4.2.2.2. Finally, we review the different approaches specifically designed to manage misalignment of multiexposed LDR images for HDR generation (Section 4.2.2.3).

Table 4.1 Different Configurations of the Camera and Scene

	Camera	Scene	Misalignment	HDR Video
1	Static	Static	–	Time lapse
2	Static	Dynamic	Local	Camera constrained
3	Free path	Static	Global	Scene constrained
4	Free path	Dynamic	Local and global	General case
5	Stereo/multiscopic	Static/dynamic	Constrained global	Stereo/multiscopic

4.2.2.1 Camera versus scene movement

Misalignment can be classified into different categories according to the kind of movement of the camera and the objects in the scene. Table 4.1 shows the types of possible misalignment present in multiple exposures for HDR generation.

A *static camera* refers to a camera fixed to a tripod or any other support that keeps the objective still during the acquisition. The *free path* classification considers camera movement because the camera is either handheld or supported by an articulated device, therefore following an arbitrary trajectory. Stereo or multiscopic acquisition includes stereo pairs of cameras or camera rigs composed of three or more positions of one or more cameras horizontally aligned, to capture respectively two views or more of the same scene.

The scene is classified as *dynamic* if any object moves during the acquisition, no matter what the amount of movement or the size of the object is. Otherwise the scene is considered as *static*.

In every case it is possible to produce an HDR video. For static images (first row in Table 4.1), it is possible to repeat the acquisition for a given time step and combine the HDR video into a time-lapse video. Time lapse (also known as slow motion) is a technique for capturing frames significantly slower than video frame rates. When played at video frame rates, time appears to be faster. This is often used to capture a natural phenomenon such sunrise or sunset, or in the animation industry. Camera- and scene-constrained sequences are, in general, easier to align than the general case since only one kind of misalignment takes place. The three cases (second, third, and fourth rows in Table 4.1) are used in film production. The last case (fifth row in Table 4.1), concerning stereo and multiscopic cameras, has become very popular since stereo and autostereoscopic displays have appeared on the market.

4.2.2.2 Local and/or global misalignment

Misalignments as defined in the previous section can be categorized in three different types:

1. *Global* misalignment is the consequence of camera motion (change in position) and affects all pixels in the image (third and fifth rows in Table 4.1). It is common in exposure sequences acquired with handheld cameras, although it is possible to find misalignment even for still sequences acquired with tripods (camera shaking with the mechanism activation). Between consecutive pairs of images, it is generally a small movement corresponding to translations or rotations. It may cause ghosting in the resulting HDR image, but some efficient techniques have been proposed to correct this misalignment. However, even for small movements, problems of object occlusion and parallax could be difficult to solve.

2. *Local* misalignment is produced by dynamic objects in the scene and affects only certain areas inside the image (second row in Table 4.1). Capturing a set of LDR images takes at least the sum of the shutter speeds of each picture. This time is enough to introduce differences in the positions of a dynamic object in the scene. In this case some areas occluded in some images may be visible in others. Depending on the speed of the dynamic object and the kind of the movement, it may produce important differences between the inputs.

3. *Local and global* misalignment combines the two previous types and concerns the fourth row in Table 4.1. When a camera follows a free path to record a dynamic scene, each frame contains both local and global misalignment. Pixels in the image are affected by transformations of different nature.

Fig. 4.1 shows an example of movement in a common multiexposure sequence. Notice that in stereo pairs (Fig. 4.1A and B), both images were acquired at the same time by two different sensors. Even if

(A)   Low exposure   (B)   High exposure

(C)   Low exposure   (D)   High exposure

FIGURE 4.1

Different exposure frames of a stereo pair (A, B) from Middlebury (2006) and a dynamic video sequence captured with a handheld camera (C, D).

the scene is dynamic, both images correspond to the same time and no local misalignment is possible. The only misalignment possible is global, due to changes in the perspective from the two points of view.

4.2.2.3 Generated HDR content

If both the scene and the camera remain static during the acquisition, the multiple exposures are aligned. The only possible result is an HDR image. However, even in this case, time-lapsed HDR video can be produced combining HDR frames acquired from static multiexposed LDR images. In such a case, any of the existing techniques for static images (Mann and Picard, 1995; Debevec and Malik, 1997; Mitsunaga and Nayar, 1999) can be used to recover radiance values and merge them into HDR images.

In cases where the camera remains static recording a dynamic scene, it is possible to detect the areas affected by dynamic objects in the scene and treat them locally. Several techniques have been presented for motion detection in the context of HDR image generation (Khan et al., 2006; Jacobs et al., 2008; Gallo et al., 2009; Pece and Kautz, 2010; Heo et al., 2011; Orozco et al., 2012, 2013, 2014), many of them already discussed in Chapter 3.

In the opposite case (static scene and dynamic camera), the movement between consecutive frames is very small. Some computationally efficient methods were proposed to solve such misalignment (Ward, 2003; Grosch, 2006; Skala, 2007). The most difficult case is when both the camera and the scene move. In such a case, dense correspondences between frames are required.

In a handheld video sequence (Fig. 4.1C and D), the images correspond to different time instants and every pixel in the image is affected by global misalignment due to changes in the position of the camera, and some pixels are also affected by local misalignment due to dynamic objects like the boat in this figure.

The rest of this chapter is dedicated to analyzing the most important approaches to solve the misalignment described in Table 4.1. Section 4.3 focuses on cases corresponding to the second to fourth, whereas Section 4.4 analyzes existing solutions for the case described in the fifth row.

4.3 FREE-PATH SINGLE CAMERA

Most approaches for deghosting in HDR imaging are based on selecting the best exposure as a reference image and solving the misalignment only for the reference image. The extension of such algorithms for HDR video by repeating the process, taking as a reference each frame in a multiexposed video sequence (see Fig. 4.2) is not always possible. Most of these methods fail if the reference contains large overexposed or underexposed regions and they do not pay attention to temporal coherence. This is an important drawback for extending such methods to HDR video generation. It is fundamental to produce HDR frames free of ghosting in the resulting sequence and they must look alike. Otherwise, flickering will be noticeable in the resulting HDR video.

4.3.1 MULTIEXPOSURE ACQUISITION SETUP

In HDR images, the number of exposures and the shutter speed used for each LDR image vary depending on the scene conditions. It is common to take several exposures and to use long shutter speeds to get details in shadow areas of dark scenes. In contrast, for brighter scenes, less light exposure

FIGURE 4.2

Multiexposure video sequence alternating three different exposures.

and thus faster shutter speeds are preferable. However, as discussed in Section 4.2, it is difficult to implement this approach in the HDR video context because of timing constraints.

Similarly to auto exposure control of digital cameras, digital video cameras have a function called auto gain control (AGC) in charge of measuring the brightness distribution of the scene and calculating the best exposure time for the scene conditions. This function provides an optimal exposure value for the scene.

The ratios of light exposure are measured in f-stops. Each f-stop means a factor of 2 (double or half, added or subtracted exposure value). The most extended idea is to capture additional exposures at fixed multiples of this medium exposure value (eg, ±2 f-stops) to obtain high and low exposures. Many authors use only two exposures (low and high) to generate three frames of HDR video (Kang et al., 2003; Sand and Teller, 2004; Mangiat and Gibson, 2010), while the most recent approach (Kalantari et al., 2013) uses three (low, medium, and high) exposures to generate the same number of HDR frames (see Fig. 4.2).

4.3.2 PER-FRAME HDR VIDEO GENERATION

To our knowledge, Kang et al. (2003) proposed the first method to extend multiple exposure image methods to video sequences. Their system uses a programmed video camera that temporally alternates long and short exposures. On the basis of the AGC function, this method calculates adaptive exposure values at each frame depending on the scene lighting conditions. The ratio between exposures of consecutive frames can range from 1 f-stop to a maximum of 16 f-stops.

Every HDR frame for a given time t_i is generated with use of information from adjacent frames t_{i-1} and t_{i+1}. They reexpose the short-exposure frame with the long-exposure value. Once images have been transformed to the same exposure, motion estimation is performed for the two adjacent images. It consists of two steps:

1. Global registration by estimation of an affine transform between them.
2. Gradient-based optical flow to determine dense motion field for local correction.

In the regions where the current frame is well exposed, images are merged by a weighted function to prevent ghosting. For the overexposed or underexposed regions, the previous/next frames are bidirectionally interpolated with use of optical flow and a hierarchical homography algorithm.

Despite the novelty of this work and the promising results for some scenes, gradient-based optical flow is not accurate enough to find forward/backward flow fields. Boosting the short exposure to the long one will increase the noise, details such as edges may be lost, and slight variations of brightness may persist because of inaccuracies in the camera response function (CRF). This may produce ghosting and errors in registration for fast nonrigid moving objects.

Sand and Teller (2004) proposed an algorithm to register two different video sequences of the same scene. HDR video is one of the most direct applications of their method. Differently exposed videos can be matched with their method. The matching search starts by selecting feature points and comparing the selected features of the two images. A matching cost is evaluated in some feature points by use of two terms:

1. Pixel consistency, instead of comparing equal pixels or patches in two images, they compare a single pixel in the reference image with a 3×3 patch in the source image. Correspondence is evaluated within a window around each pixel and pixel matching probabilities are assigned.
2. Motion regression and consistency, to determine how well a particular correspondence is consistent with its neighbors.

The matchings obtained are used to find regression predictions that are improved in a regression process. After high likelihood correspondences have been found, a locally weighted linear regression method is used to interpolate and extrapolate correspondences for the rest of pixels, with a dense correspondence field being obtained. This scheme is extended to all frame pairs of the video sequence. This method offers very good results for highly textured scenes but poor results otherwise. Processing each pair of frames might take up to 1.31 s, and full video matching might take several minutes for each second of video input.

Mangiat and Gibson (2010) proposes improving the method of Kang et al. (2003) by using a block-based motion estimation algorithm. They also worked with a video sequence that alternates two exposure values. They used a CRF recovered with a sequence of 12 static exposures using the method presented by Debevec and Malik (1997). The short exposure is boosted by the CRF to match the long exposure.

They used software (Sühring, 2008) to calculate block-based forward and backward motion estimation vectors for each frame with respect to the adjacent ones. However, such estimation is likely to fail in saturated areas. A second step of bidirectional motion estimation is performed to fill in the saturated areas with information from the previous and next frames. The cost function is the sum of absolute differences (SAD), adding a cost term that relates the motion vector estimated for adjacent frames. Block-based motion estimation is prone to artifacts such as discontinuities at block boundaries. Differences between the images in the radiance domain are detected and assumed to be artifacts. Such pixels are considered to be like holes and are replaced by pixels in the contour of such areas. Nevertheless, poorly registered pixels may pass to the HDR merging step. They propose the use of a cross-bilateral filter to treat the tone-mapped HDR image using edge information at each frame. Despite the different attempts to avoid artifacts, the problem of fast motion (eg, eyes blinking) remains unsolved. The filtering step executed in the tone-mapped images cannot be used for HDR displays.

Sen et al. (2012) recently presented a method based on a patch-based energy minimization that integrates alignment and reconstruction in a joint optimization for HDR image synthesis. Their method relies on a patch-based nearest-neighbor search proposed by Barnes et al. (2009) and a multisource bidirectional similarity measure inspired by Simakov et al. (2008). This method allows the production of an HDR result that is aligned to one of the exposures and contains information from all the remaining exposures. The results are very accurate but dependent on the quality of the reference image. Artifacts may appear if the reference image has large underexposed or saturated areas.

Kalantari et al. (2013) proposed a new approach for HDR reconstruction from multiexposure video sequences built on patch-based synthesis for HDR images from (Sen et al., 2012) combined with optical flow. Instead of alternating two exposures, they propose using three different exposure values. The HDR video reconstruction is guided by an energy function that includes terms for mapping the LDR images in the radiance domain, ensures the similitude of the resulting HDR values with the LDR reference, and reinforces temporal coherence. They used the sum of squared distances (SSD) to compare two patches.

Their algorithm uses optical flow for a first motion estimation step. This estimation helps to compute a window size to constrain the patch-match search and vote step. The results are very accurate for detailed regions in the scene. However, slight flickering appears in large areas of the same colors.

4.4 MULTISCOPIC HDR VIDEO

In this section we address the generation of HDR images for more than two views. We first review the basics of stereoscopic imaging (Section 4.4.1) and epipolar geometry (Section 4.4.2), before we discuss recent contributions to the generation of multiscopic HDR images (Section 4.4.3).

4.4.1 STEREOSCOPIC IMAGING

Apart from a huge number of colors and huge amount of fine detail, our visual system is able to perceive depth and 3D shape of objects. Traditional images offer a representation of reality projected in two dimensions. We can guess the distribution of objects in depth because of monoscopic cues such as perspective, but we cannot actually perceive depth. Our brain needs to receive two slightly different images to actually perceive depth.

Stereoscopy enables depth perception. Stereo images refer to a pair of images acquired with two cameras horizontally aligned and separated at a scalable distance similar to the distance between our eyes. Stereoscopic display systems project them in a way such that each eye perceives only one of the images. In recent years, stereoscopic video technologies such as stereoscopic video cameras and stereoscopic displays have become available to consumers (Mendiburu et al., 2012; Dufaux et al., 2013; Urey et al., 2011). Stereo video requires the recording of at least two views of a scene, one for each eye. Some autostereoscopic displays render more than nine different views for an optimal experience (Lucas et al., 2013).

Some prototypes have been proposed to acquire stereo HDR content from two or more differently exposed views. Most approaches (Troccoli et al., 2006; Lin and Chang, 2009; Sun et al., 2010; Rufenacht, 2011; Bätz et al., 2014; Akhavan et al., 2014) are based on a rig of two cameras placed like a conventional stereo configuration that captures different exposed images. Section 4.4.3 focuses on analyzing the different existing approaches for multiscopic HDR acquisition.

4.4.2 EPIPOLAR GEOMETRY

Stereo images permit a viewer to see depth. The geometry that relates 3D objects to their 2D projection in stereo vision is known as epipolar geometry. Depth can be retrieved mathematically from a pair of images with use of epipolar geometry. Fig. 4.3 describes the main components of epipolar geometry. A point x in the 3D world coordinates is projected onto the left and right images I_L and I_R respectively. c_L and c_R are the two centers of projection of the cameras; the plane formed by them and the point x is known as the epipolar plane. x_L and x_R are the projections of x in I_L and I_R, respectively.

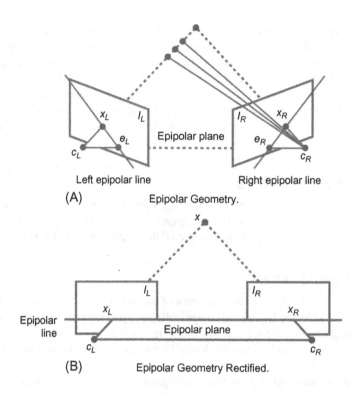

FIGURE 4.3

Main elements of the epipolar geometry: (A) epipolar geometry; (B) epipolar geometry rectified.

For any point x_L in the left image, the distance to x is unknown. According to the epipolar geometry, the corresponding point x_R is located somewhere on the right epipolar line. Epipolar geometry does not mean direct correspondence between pixels. However, it reduces the search for a matching pixel to a single epipolar line.

If the image planes are aligned and their optical axes are made parallel, the two epipolar lines (left and right) converge. This process is known as "rectification." After rectification, the search space for a pixel match is reduced to the same image row.

To the best of our knowledge, all methods concerned with stereoscopic HDR content take advantage of the epipolar constraint during the matching process. Rectified image sets are available on the Internet for testing purposes (Middlebury, 2006).

4.4.3 MULTIPLE-EXPOSURE STEREO MATCHING

4.4.3.1 Problem formulation

The use of stereo matching or disparity estimation for pixel matching on differently exposed stereo/multiview images is not straightforward. Stereo matching or disparity estimation is the process of finding the pixels in the multiscopic views that correspond to the same 3D point in the scene. The rectified epipolar geometry simplifies this process of finding correspondences on the same epipolar line.

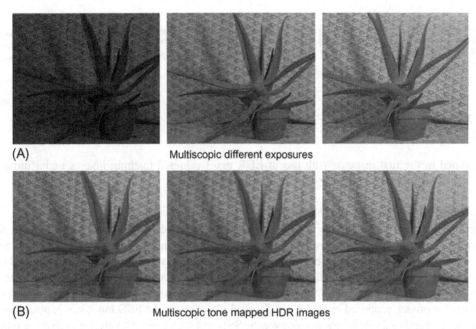

(A)   Multiscopic different exposures

(B)   Multiscopic tone mapped HDR images

FIGURE 4.4

"Aloe" set of LDR multiview images from Middlebury (2006): (A) multiscopic different exposures;
(B) multiscopic tone-mapped HDR images.

It is not necessary to calculate the 3D point coordinates to find the corresponding pixel on the same row of the other image. The disparity is the distance d between a pixel and its horizontal match in the other image. Akhavan et al. (2013, 2014) compared the different ways of obtaining disparity maps from HDR, LDR, and tone-mapped stereo images. A useful comparison among them is offered, illustrating that HDR input can have a significant impact on the quality of the result.

Fig. 4.4 shows an example of a differently exposed multiview set corresponding to one frame in a multiscopic system of three views. The main goal of stereo matching is to find the correspondences between pixels to generate one HDR image per view for each frame.

Correspondence methods rely on matching cost functions for computing the similarity of images. It is important to consider that even when one uses radiance space images, there might be brightness differences. Such differences may be introduced by the camera because of image noise or slightly different settings or vignetting. For good analysis and comparison between existing matching costs and their properties, see (Scharstein and Szeliski, 2002; Hirschmuller and Scharstein, 2009; Bonnard et al., 2014).

Many approaches have been presented to recover HDR from multiview and multiexposed sets of images. Some of them (Troccoli et al., 2006; Lin and Chang, 2009; Sun et al., 2010) share the same pipeline as in Fig. 4.5. All the work mentioned takes as input a set of images with different exposures acquired with a camera with unknown response function. In such cases, the disparity maps need to

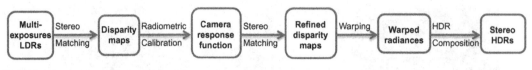

FIGURE 4.5

General multiexposed stereo pipeline for stereo HDR. Proposed by Troccoli et al. (2006), used by Sun et al. (2010) and Lin and Chang (2009), and modified later by Bätz et al. (2014).

be calculated in the first instance with use of LDR pixel values. Matching images under important differences of brightness is still a big challenge in computer vision.

4.4.3.2 Per frame CRF recovery methods

To our knowledge, Troccoli et al. (2006) proposed the first technique for HDR recovery from multiscopic images of different exposures. They observed that the normalized cross-correlation (NCC) is approximately invariant to exposure changes when the camera has a gamma response function. Under such an assumption, they used the algorithm described by Kang and Szeliski (2004) to compute the depth maps that maximize the correspondence between one pixel and its projection in the other image. The original approach proposed by Kang and Szeliski (2004) used the SSD but it was replaced by the NCC. Eqs. (4.1) and (4.2) show how to calculate the SSD and NCC, respectively, for two image patches of N pixels centered in p and q of images I_L and I_R.

Images are warped to the same viewpoint with the depth map. Once pixels have been aligned, the CRF is calculated by the method proposed by Grossberg and Nayar (2003) over a selected set of matches. The problem that arises from the use of the NCC is that this introduces ambiguity within the equivalence gamma response. With the CRF and the exposures, all images are transformed to radiance space and the matching process is repeated, this time with the SSD. The new depth map improves the previous one and helps to correct artifacts. The warping is updated and HDR values are calculated by a weighted average function.

$$\text{SSD} = \sum_{q,p \in N} (I_L(q) - I_R(p))^2, \tag{4.1}$$

$$\text{NCC} = \frac{\sum\limits_{q,p \in N} I_L(q) \cdot I_R(p)}{\sqrt{\sum\limits_{q,p \in N} I_L(q)^2 \cdot \sum\limits_{q,p \in N} I_R(p)^2}}. \tag{4.2}$$

The same problem was addressed by Lin and Chang (2009). Instead of the NCC, they use scale-invariant feature transform (SIFT) descriptors to find matches between LDR stereo images. SIFT is not robust for different-exposure images. Only the matches that are coherent with the epipolar and exposure constraints are selected for the next step. The selected pixels are used to calculate the CRF.

The stereo matching algorithm they propose is based on previous work (Sun et al., 2003). Belief propagation is used to calculate the disparity maps. The stereo HDR images are calculated by means of a weighted average function. Even with the best results, SIFT is not robust enough if there are significant exposure variations.

A ghost removal technique is used afterward to treat the artifacts due to noise or stereo mismatches. The HDR image is exposed to the best exposure. The difference between them is calculated and pixels over a threshold are rejected, considering them like mismatches. This is risky because HDR values in areas underexposed and overexposed in the best exposure may be rejected. In this case the problem of ghosting would be solved but LDR values may be introduced in the resulting HDR image.

Sun et al. (2010) (inspired by Troccoli et al., 2006) also follow the pipeline described in Fig. 4.5. They assume that the disparity map between two rectified stereo images can be modeled as a Markov random field. The matching problem is presented like a Bayesian labeling problem. The optimal label (disparity) values are obtained by minimization of an energy function. The energy function they use is composed of a pixel dissimilarity term (NCC in their solution) and a disparity smoothness term. It is minimized by the graph cut algorithm to produce initial disparities. The best disparities are selected to calculate the CRF with the algorithm proposed by Mitsunaga and Nayar (1999). Images are converted to radiance space and another energy minimization is performed to remove artifacts. This time the pixel dissimilarity cost is computed with the Hamming distance between candidates.

The methods presented so far have a high computational cost. Calculating the CRF from nonaligned images may introduce errors because the matching between them may not be robust. Two exposures are not enough to obtain a robust CRF with existing techniques. Some of them execute two passes of the stereo matching algorithm, the first one to detect matches for the CRF recovery and the second one to refine the matching results. One might avoid this by calculating the CRF in a previous step using multiple exposures of static scenes. Any of the available techniques (Mann and Picard, 1995; Debevec and Malik, 1997; Mitsunaga and Nayar, 1999; Grossberg and Nayar, 2003) can be used to get the CRF corresponding to each camera. The curves help to transform pixel values into radiance for each image, and the matching process is executed in radiance space images. This avoids one stereo matching step and prevents errors introduced by disparity estimation and image warping.

4.4.3.3 Offline CRF recovery methods

Bonnard et al. (2012) proposed a method to create content that combines depth and HDR video for autostereoscopic displays. Instead of varying the exposure times, they use neutral density filters to capture different exposures. A camera with eight synchronized objectives and three pairs of 0.3, 0.6, and 0.9 filters plus two nonfiltered views provides eight views with four different exposures of the scene stored in 10-bit RAW files. They used a geometry-based approach to recover depth information from epipolar geometry. Depth maps drive the pixel matching procedure.

Bätz et al. (2014) presented a workflow for disparity estimation divided into the following steps:

- *Cost initialization* consists in evaluating the cost function for all values within a disparity search range. They use zero-normalized cross correlation (ZNCC), defined in Eq. (4.3):

$$\text{ZNCC} = \frac{\sum\limits_{q,p \in N} (I_L(q) - \bar{I}_L(q)) \cdot (I_R(p) - \bar{I}_R(p))}{\sqrt{\sum\limits_{q,p \in N} (I_L(q) - \bar{I}_L(q))^2 \cdot \sum\limits_{q,p \in N} (I_R(p) - \bar{I}_R(p))^2}}. \qquad (4.3)$$

The matching is performed on the luminance channel of a radiance space image with patches of 9×9 pixels. The result of searching for disparities is the disparity space image (DSI), a matrix of $m \times n \times d + 1$ for an image of $m \times n$ pixels, with $d + 1$ being the disparity search range.

- *Cost aggregation* is done to smooth the DSI and find the actual disparity of each pixel in the image. They use an improved version of the cross-based aggregation method described by Mei et al. (2011). This step is performed over the actual RGB images, not in the luminance channel as in the previous step.

- *Image warping* is responsible for actually shifting all pixels according to their disparities. How to deal with occluded areas between the images is the main challenge. Bätz et al. (2014) propose doing the warping in the original LDR images, which adds a new challenge: dealing with underexposed and overexposed areas. A backward image warping is chosen to implicitly ignore the saturation problems. The algorithm produces a new warped image with the appearance of the reference one by using the target image and the corresponding disparity map. Bilinear interpolation is used to retrieve values at subpixel precision.

Selmanovic et al. (2014) propose generating stereo HDR video from a pair of HDR and LDR videos, using an HDR camera (Chalmers et al., 2009) and a traditional digital camera (Canon 1Ds Mark II) in a stereo configuration. Their work is an extension to video of previous work (Selmanović et al., 2013) focused only on stereo HDR images. In this case, one HDR view needs to be reconstructed from two completely different sources.

Their method proposes three different approaches to generate the HDR video:

1. *Stereo correspondence* is computed to recover the disparity map between the HDR and LDR images. The disparity map allows the HDR values to be transferred to the LDR image. The SAD (Eq. 4.4) is used as a matching cost function. Both images are transformed to Lab color space, which is perceptually more accurate than RGB.

$$\text{SAD} = \sum_{q,p \in N} |I_L(q) - I_R(p)|. \tag{4.4}$$

   The selection of the best disparity value for each pixel is based on the *winner takes all* technique. The lower SAD is selected in each case. An image warping step based on the work of Fehn (2004) is used to generate a new HDR image corresponding to the LDR view. The SAD stereo matcher can be implemented to run in real time but the resulting disparity maps could be noisy and not accurate. The overexposed and underexposed pixels may end up in the wrong position. In large areas of the same color and hence same SAD cost, the disparity will be constant. Occlusions and reflective or specular objects may cause some artifacts.

2. The *expansion operator* could be used to produce an HDR image from the LDR view. Detailed state-of-the-art reports on LDR expansion have been presented by Banterle et al. (2009) and Hirakawa and Simon (2011). However, in this case, we need the expanded HDR to remain coherent with the original LDR. Inverse tone mappers are not suitable because the resulting HDR image may be very different from the acquired one, producing results that are not possible to fuse through a common binocular vision.

   Selmanovic et al. (2014) propose an expansion operator based on a mapping between the HDR and the LDR image using the first one as a reference. A reconstruction function maps LDR to HDR values (Eq. 4.5) based on an HDR histogram with 256 bins putting the same number of HDR values in each bin as there are in the LDR histogram.

$$RF = \frac{1}{\text{Card}(\Omega_c)} \sum_{i=M(c)}^{M(c)+\text{Card}(\Omega_c)} c_{\text{hdr}}(i). \tag{4.5}$$

In Eq. (4.5), $\Omega_c = \{j = i..N : c_{\text{ldr}}(j) = c\}$, $c = 0..255$ is the index if a bin Ω_c, $\text{Card}(\cdot)$ returns the number of elements in the bin, N is the number of pixels in the image, $c_{\text{ldr}}(j)$ are the intensity values for pixel j, $M(c) = \sum_0^c \text{Card}(\Omega_c)$ is the number of pixels in the previous bin, and c_{hdr} are the intensities of all HDR pixels sorted in ascending order. RF is used to calculate the look-up table and afterward expansion can be performed directly, assigning the corresponding HDR value to each LDR pixel.

The expansion runs in real time, is not view dependent, and avoids stereo matching. The main limitation is again on saturated regions.

3. The *hybrid method* combines the two previous methods. Two HDR images are generated by the previous approaches (stereo matching and expansion operator). Pixels in well-exposed regions are expanded by the first method (expansion operator), whereas matches for pixels in underexposed or overexposed regions are found by SAD stereo matching with the addition of a correction step. A mask of undersaturated and oversaturated regions is created by use of a threshold for pixels over 250 or below 5. The areas out of the mask are filled in with the expansion operator, and the underexpose or overexposed regions are filled in with an adapted version of the SAD stereo matching to recover more accurate values in overexposed or underexposed regions.

Instead of having the same disparity over the whole underexposed or overexposed region, this variant interpolates disparities from well-exposed edges. Edges are detected by a fast morphological edge detection technique described by Lee et al. (1987). Nevertheless, some small artifacts may still be produced by the SAD stereo matching in such areas.

Orozco et al. (2015) presented a method to generate multiscopic HDR images from LDR multiexposure images. They adapted a patch match approach to find matches between stereo images using epipolar geometry constrains. This method reduces the search space in the matching process and improves the incoherence problem of the patch match. Each image in the set of multiexposed images is used as a reference, looking for matches in all the remaining images. These accurate matches allow one to synthesize images corresponding to each view which are merged into one HDR per view that can be used in autostereoscopic displays.

4.5 CONCLUSIONS

We have presented a selection of the most significant methods to recover HDR values in misaligned multiple-exposure sequences. These approaches are based on the type of misalignment affecting the sequence of LDR images.

Despite the amount of research focused on this topic, there is not yet a fully robust solution for the two main cases we have analyzed in this chapter: free-path camera and multiscopic different views.

The problem of large saturated areas remains unsolved and temporal coherence problems persist in both HDR and stereo HDR video. However, important advances have been made lately in multiscopic and HDR video and such progress makes it likely we will soon have a robust method that includes large saturated areas.

REFERENCES

Akhavan, T., Kapeller, C., Cho, J.H., Gelautz, M., 2014. Stereo HDR disparity map computation using structured light. In: HDRi2014 Second International Conference and SME Workshop on HDR Imaging.

Akhavan, T., Yoo, H., Gelautz, M., 2013. A framework for HDR stereo matching using multi-exposed images. In: Proceedings of HDRi2013 First International Conference and SME Workshop on HDR Imaging, Paper no. 8. The Eurographics Association and Blackwell Publishing Ltd., Oxford/Malden.

Banterle, F., Debattista, K., Artusi, A., Pattanaik, S., Myszkowski, K., Ledda, P., Chalmers, A., 2009. High dynamic range imaging and low dynamic range expansion for generating HDR content. Comput. Graph. Forum 28 (8), 2343–2367.

Barnes, C., Shechtman, E., Finkelstein, A., Goldman, D.B., 2009. Patchmatch: a randomized correspondence algorithm for structural image editing. ACM Trans. Graph. (Proc. SIGGRAPH) 28 (3).

Bätz, M., Richter, T., Garbas, J.U., Papst, A., Seiler, J., Kaup, A., 2014. High dynamic range video reconstruction from a stereo camera setup. Signal Process. Image Commun. 29 (2), 191–202. Special Issue on Advances in High Dynamic Range Video Research.

Bonnard, J., Loscos, C., Valette, G., Nourrit, J.M., Lucas, L., 2012. High-dynamic range video acquisition with a multiview camera. In: Proc. SPIE 8436, Optics, Photonics, and Digital Technologies for Multimedia Applications II, 84360A.

Bonnard, J., Valette, G., Nourrit, J.M., Loscos, C., 2014. Analysis of the consequences of data quality and calibration on 3D HDR image generation. In: European Signal Processing Conference (EUSIPCO), Lisbonne, Portugal.

Chalmers, A., Debattista, K., 2011. HDR video: capturing and displaying dynamic real-world lighting. In: 19th Color Imaging Conference: Color Science and Engineering Systems, Technologies, and Applications, CIC19, pp. 177–180.

Chalmers, A., Bonnet, G., Banterle, F., Dubla, P., Debattista, K., Artusi, A., Moir, C., 2009. High-dynamic-range video solution. In: ACM SIGGRAPH ASIA 2009 Art Gallery & Emerging Technologies: Adaptation, SIGGRAPH ASIA '09. ACM, New York, NY, pp. 71–71.

Debevec, P., Malik, J., 1997. Recovering high dynamic range radiance maps from photographs. In: Proceedings of ACM SIGGRAPH (Computer Graphics), vol. 31, pp. 369–378.

Dufaux, F., Pesquet-Popescu, B., Cagnazzo, M., 2013. Emerging Technologies for 3D Video: Creation, Coding, Transmission and Rendering. John Wiley & Sons, New York, NY.

Fehn, C., 2004. Depth-image-based rendering (DIBR), compression, and transmission for a new approach on 3D-TV. In: Proc. SPIE, vol. 5291, pp. 93–104.

Gallo, O., Gelfand, N., Chen, W.C., Tico, M., Pulli, K., 2009. Artifact-free high dynamic range imaging. In: IEEE International Conference on Computational Photography (ICCP).

Grosch, T., 2006. Fast and robust high dynamic range image generation with camera and object movement. In: Vision, Modeling and Visualization, RWTH Aachen, pp. 277–284.

Grossberg, M.D., Nayar, S.K., 2003. Determining the camera response from images: what is knowable? IEEE Trans. Pattern Anal. Mach. Intell. 25 (11), 1455–1467.

Heo, Y.S., Lee, K.M., Lee, S.U., Moon, Y., Cha, J., 2011. Ghost-free high dynamic range imaging. In: Proceedings of the 10th Asian Conference on Computer vision, Volume Part IV, ACCV'10. Springer-Verlag, Berlin/Heidelberg, pp. 486–500.

Hirakawa, K., Simon, P., 2011. Single-shot high dynamic range imaging with conventional camera hardware. In: IEEE International Conference on Computer Vision (ICCV), 2011, pp. 1339–1346.

Hirschmuller, H., Scharstein, D., 2009. Evaluation of stereo matching costs on images with radiometric differences. IEEE Trans. Pattern Anal. Mach. Intell. 31 (9), 1582–1599.

Jacobs, K., Loscos, C., Ward, G., 2008. Automatic high-dynamic range generation for dynamic scenes. IEEE Comput. Graph. Appl. 28, 24–33.

Kalantari, N.K., Shechtman, E., Barnes, C., Darabi, S., Goldman, D.B., Sen, P., 2013. Patch-based high dynamic range video. ACM Trans. Graph. 32 (6), 202:1–202:8.

Kang, S.B., Szeliski, R., 2004. Extracting view-dependent depth maps from a collection of images. Int. J. Comput. Vis. 58 (2), 139–163.

Kang, S.B., Uyttendaele, M., Winder, S., Szeliski, R., 2003. High dynamic range video. ACM Trans. Graph. 22 (3), 319–325.

Khan, E.A., Akyüz, A.O., Reinhard, E., 2006. Ghost removal in high dynamic range images. IEEE International Conference on Image Processing, pp. 2005–2008.

Lee, J., Haralick, R., Shapiro, L., 1987. Morphologic edge detection. IEEE J. Robot. Autom. 3 (2), 142–156.

Lin, H.Y., Chang, W.Z., 2009. High dynamic range imaging for stereoscopic scene representation. In: 16th IEEE International Conference on Image Processing (ICIP), 2009, pp. 4305–4308.

Lucas, L., Loscos, C., Remion, Y., 2013. 3D Video From Capture to Diffusion. Wiley-ISTE, New York, NY.

Mangiat, S., Gibson, J., 2010. High dynamic range video with ghost removal. Proc. SPIE 7798, 779812–779812-8.

Mann, S., Picard, R.W., 1995. On Being Undigital With Digital Cameras: Extending Dynamic Range by Combining Differently Exposed Pictures. Perceptual Computing Section, Media Laboratory, Massachusetts Institute of Technology, Cambridge, MA.

Mei, X., Sun, X., Zhou, M., Jiao, S., Wang, H., Zhang, X., 2011. On building an accurate stereo matching system on graphics hardware. In: IEEE International Conference on Computer Vision Workshops (ICCV Workshops), 2011, pp. 467–474.

Mendiburu, B., Pupulin, Y., Schklair, S. (Eds.), 2012. 3D {TV} and 3D Cinema. Focal Press, Boston.

Middlebury, 2006. Middlebury stereo datasets. http://vision.middlebury.edu/stereo/data/.

Mitsunaga, T., Nayar, S., 1999. Radiometric self calibration. In: IEEE International Conference on Computer Vision and Pattern Recognition, vol. 1, pp. 374–380.

Nayar, S.K., Branzoi, V., 2003. Adaptive dynamic range imaging: optical control of pixel exposures over space and time. In: Proceedings of the Ninth IEEE International Conference on Computer Vision, Volume 2, ICCV '03. IEEE Computer Society, Washington, DC, p. 1168.

Orozco, R.R., Martin, I., Loscos, C., Artusi, A., 2015. Multiscopic HDR image sequence generation. In: WSCG, 23rd International Conference in Central Europe on Computer Graphics, Visualization and Computer Vision 2015, Plzen, Czech Republic.

Orozco, R.R., Martin, I., Loscos, C., Artusi, A., 2014. Génération de séquences d'images multivues hdr: vers la vidéo hdr. In: 27es journées de l'Association française d'informatique graphique et du chapitre français d'Eurographics, Reims, France, November 2014.

Orozco, R.R., Martin, I., Loscos, C., Artusi, A., 2013. Patch-based registration for auto-stereoscopic HDR content creation. In: HDRi2013—First International Conference and SME Workshop on HDR Imaging, Oporto Portugal, April 2013.

Orozco, R.R., Martin, I., Loscos, C., Vasquez, P.P., 2012. Full high-dynamic range images for dynamic scenes. In: Proc. SPIE 8436, Optics, Photonics, and Digital Technologies for Multimedia Applications II, 843609.

Pece, F., Kautz, J., 2010. Bitmap movement detection: HDR for dynamic scenes. In: 2010 Conference on Visual Media Production (CVMP), pp. 1–8.

RedCompany, 2006. Read one. http://www.red.com.

Research, V., 2005. Phantom HD. http://www.visionresearch.com.

Rufenacht, D., 2011. Stereoscopic High Dynamic Range Video. Ph.D. thesis, Ecole Polytechnique Fédérale de Lausanne (EPFL), Switzerland.

Sand, P., Teller, S., 2004. Video matching. ACM Trans. Graph. 23 (3), 592–599.

Scharstein, D., Szeliski, R., 2002. A taxonomy and evaluation of dense two-frame stereo correspondence algorithms. Int. J. Comput. Vis. 47 (1), 7–42.

Selmanović, E., Debattista, K., Bashford-Rogers, T., Chalmers, A., 2013. Generating stereoscopic HDR images using HDR-LDR image pairs. ACM Trans. Appl. Percept. 10 (1), 3:1–3:18.

Selmanovic, E., Debattista, K., Bashford-Rogers, T., Chalmers, A., 2014. Enabling stereoscopic high dynamic range video. Signal Process. Image Commun. 29 (2), 216–228. Special Issue on Advances in High Dynamic Range Video Research.

Sen, P., Kalantari, N.K., Yaesoubi, M., Darabi, S., Goldman, D.B., Shechtman, E., 2012. Robust patch-based HDR reconstruction of dynamic scenes. ACM Transactions on Graphics (Proceedings of SIGGRAPH Asia 2012) 31 (6), 203:1–203:11.

Simakov, D., Caspi, Y., Shechtman, E., Irani, M., 2008. Summarizing visual data using bidirectional similarity. In: IEEE Conference on Computer Vision and Pattern Recognition 2008 (CVPR'08).

Skala, P.V., (Ed.) 2007. Image Registration for Multi-Exposure High Dynamic Range Image Acquisition. University of West Bohemia.

Sühring, K., 2008. H.264/avc reference software. http://iphome.hhi.de/suehring/tml/.

Sun, J., Zheng, N.N., Shum, H.Y., 2003. Stereo matching using belief propagation. IEEE Trans. Pattern Anal. Mach. Intell. 25 (7), 787–800.

Sun, N., Mansour, H., Ward, R., 2010. HDR image construction from multi-exposed stereo LDR images. In: Proceedings of the IEEE International Conference on Image Processing (ICIP), Hong Kong.

ThomsomGrassValley, 2005. Viper filmstream. http://www.ThomsomGrassValley.com.

Tocci, M.D., Kiser, C., Tocci, N., Sen, P., 2011. A versatile HDR video production system. ACM Trans. Graph. 30 (4), 41:1–41:10.

Troccoli, A., Kang, S.B., Seitz, S., 2006. Multi-view multi-exposure stereo. In: Third International Symposium on 3D Data Processing, Visualization, and Transmission, pp. 861–868.

Urey, H., Chellappan, K.V., Erden, E., Surman, P., 2011. State of the art in stereoscopic and autostereoscopic displays. Proc. IEEE 99, 540–555.

Ward, G., 2003. Fast, robust image registration for compositing high dynamic range photographs from handheld exposures. J. Graph. Tools 8, 17–30.

HDR, CINEMATOGRAPHY, AND STEREOSCOPY

Y. Pupulin, O. Letz, C. Lèbre
Binocle, Paris, France

CHAPTER OUTLINE

5.1 INTRODUCTION

New technologies in the field of digital imaging, such as stereoscopy and high dynamic range (HDR) imaging, provide novel means for artistic expression, which will be investigated in this chapter. Our aim is to look for new possibilities that these technologies offer in cinematography. Particularly in the context of stereoscopy, the goal is to find better photographic techniques delivering the most expressive content. Such techniques are often very different from those used in 2D cinematography — for example, if high contrast or fast motion is acceptable in 2D cinematography, it is not desirable in stereoscopy.

A wide contrast range was sought in 2D movies for low-contrast shooting used for optical effects. During the laboratory processes, contrast enhancement was frequently observed, which was making the plates difficult to integrate. Plates were mainly used for special effects on set, such as rear projection, where contrast was further emphasized by the image-projection system. Low-speed daylight films were treated with "low contrast" in the laboratory, thus enabling an exposure value of 13 to be obtained,[1] allowing better integration between lit actors and projected images.

Apart from shots with old-style special effects, a low-contrast image, even though it has been cleverly used in certain movie sequences, is not very flattering to the brain, which usually prefers highly contrasted pictures, analyzing them as sharp images. How the brain functions is thus a primary focus of artistic practices in photography. One can wonder where the artistic interest is to systematically create HDR images and to render all the details of image regions legible, irrespective of their illumination level.

In the field of television and cinematography, which are gradually converting to fully digital, four areas of research are attracting great interest among professionals: HDR, wide color gamut (WCG), ultra high definition (UltraHD), and high frame rate (HFR). Former standards are being reconsidered — for example, HFR with respect to acquisition and projection frequencies that remained unchanged since the invention of talking pictures. The same observation also applies to television. Those frequencies represented reasonable comfort for viewers and even though it was known, as demonstrated by Showscan,[2] that it would be better to increase acquisition, broadcasting, and projection frequencies, such a change was not economically feasible before the widespread use of digital technologies.

Likewise, contrast, color, or resolution could change only in specific and limited ranges until the advent of digital technologies that resulted in a major step forward in the possibilities enabled by these four domains of research: HDR, WCG, UHD, and HFR.

With 2D cinematography, some operators or film makers still prefer film to digital format whenever a chain involving visual effects is not involved. This is because of certain digital specificities and essentially to the aspect of dematerialization that is perceived as a loss of grain and know-how. As an example and though paradoxical, it is notable that the digital technique reveals and freezes motion blurs which up to now were abstract and moving as they were linked to film grain mobility.

If the digital world remains controversial for cinematography, the same is not true for stereoscopy. The development of digital imaging and motion control cameras is the basis for sustained control in staging and visual comfort. Therefore, stereoscopy has been the frontrunner bringing about the digital revolution as well as its broad acceptance in cinematography.

In this context and in times of transitional but real regression of stereoscopy, many industrial companies are prone to present the different opportunities opened up by digital technologies in terms of dynamic range, color space, frequency, or resolution as remarkable revolutions. For example, 4K resolution becomes an innovation that might supersede both high-definition (HD) television, which is seen as too "limited," and stereoscopy, which is considered to be too "complicated" in its implementation and control.

---

[1]A difference of one exposure value corresponds to one f-stop.

[2]Showscan is a cinematic process developed by Douglas Trumbull: shooting with 65 mm film at 60 frames per second and projection with 70 mm film at 60 frames per second.

The same perspective also holds for HDR, HFR, or WCG. But if we can imagine 4K, HDR, HFR, and WCG channels in the future, it is obvious that combining them will be impossible to manage in terms of bandwidth in the near future.

So what is the meaning of these sometimes overlapping research directions in which we are engaged as operators or stereographers? The digital revolution offers the opportunity for new standards that might go beyond those of cinematography or television in order to resolve stereoscopy-related challenges. In particular, the overcoming of conflicts between 2D and 3D cinematography can be foreseen. For instance, HDR would allow the mitigation of discomfort relating to the strong stereoscopic effect and high contrast. Every stereographer knows this problem with resulting ghost images. Moreover, sudden movements by the subject or the camera will only worsen this effect, even though the issue can possibly be resolved with HFR. This shows the extent these research directions are interlinked.

The HDR benefit may be to open up new horizons for staging by extension of the range of 2D and, especially, 3D pictures which can be produced, with a contrast management detached from the real subject. In this field, as in other fields of investigation, we realize that artistic creation may be enhanced by digital-related research, whether it is for 2D or 3D movie making and in spite of loss due to film grain.

5.2 EXPERIMENTS WITH THE HDR TECHNIQUE

With our experience in special effects and stereoscopy, we participated in the NEVEx R&D project addressing the fields of HDR for 2D and 3D pictures.[3]

The following description is a summary of the last shooting campaign conducted in the framework of this project.

5.2.1 EXPOSURES

Still cameras have featured HDR functions for a long time. In this case, a slight delay in exposure times is not a problem. However, application of the same principle to motion picture cameras — that is, two pictures are taken successively at different exposure levels — creates a ghosting effect of the subject within a single image, as movements are no longer synchronized. Although temporal double exposure can be used in art or experimental films, or even in commercials or video clips, its application to conventional media is still a concern.

As experts in stereoscopy, our plan was to perform shooting with the help of a rig that enables double exposure of the views, as required in the creation of fully synchronized HDR images. However, the downside of the use of a rig lies in the loss of one f-stop caused by the semireflective mirror.

Exposure differences thus depend on the subject but also on the choices for enlarging or reducing overlap areas corresponding to identical exposures. During the various shoots, use of the same subject allowed for various HDR exposures when its contrast was sufficiently wide.

---

[3]The research consortium comprised AcceptTV, Binocle, DxO Labs, DxO SIG, Institut de Recherche en Communications et Cybernétique de Nantes (IRCCyN)/University of Nantes, Institut de Recherche en Informatique et Systèmes Aléatoires (IRISA)/University of Rennes 1, Polymorph Software, Technicolor, Télécom ParisTech, TF1, Thomson Video Networks, and Transvideo.

As the project was dedicated to television, it was initially decided to use HD cameras (1920×1080) fitted with 2/3-inch sensors. The recordings from the first cycles could not be exploited as the noise level in low-light conditions was not compatible with the requirements. Therefore, it was concluded that the success of HDR was foremost determined by the sensitivity of the sensors in low-light conditions. This conclusion led us to use cameras with larger, 35 mm sensors, but still with HD resolution suitable for television. This time, the resolution was the limiting factor, due to the excessive geometric variation of the objects exposed in low-light and high-light conditions. As a consequence, we decided to abandon the HD cameras in favor of motion picture cameras with 4K resolution and high sensitivity in low-light conditions.

The resulting shooting experiments are described hereafter.

5.2.2 CAMERAS

We chose a Sony F65 camera for its high resolution and sensitivity in low-light conditions, as illustrated in Fig. 5.1.

This camera comprises the following elements and parameters:

- Super 35 CMOS sensor supporting 4K resolution (4096×2160 pixels).
- Recording on an SRMemory card, which uses 16 bits per sample in 4:4:4[4] and provides all color information.

FIGURE 5.1

Sony F65 camera.

---

[4]Chroma subsampling is commonly expressed as a three part ratio $x : y : z$ (eg, 4:2:2), that describe the number of luminance and chrominance samples in a conceptual region that is 'x' pixels wide, and 2 pixels high. The parts are (in their respective order):

x: horizontal sampling reference (width of the conceptual region). Usually, 4.

y: number of chrominance samples (Cr, Cb) in the first row of 'x' pixels.

z: number of changes of chrominance samples (Cr, Cb) between first and second row of 'x' pixels.

4:4:4 means there is no chroma subsampling.

- Mechanical shutter, acting as a global shutter to limit ghosting. Many cameras use a rolling shutter, which is not suitable for merging two images captured in transparency and reflection, respectively, from a half-silvered mirror.
- Perfect synchronization of images via Genlock tri-level (generator locking using tri-level synchronising pulses).
- Built-in neutral density filters, largely avoiding the use of external filters, which are more difficult to use.

5.2.3 LENSES

Two types of lenses with a PL mount were used in the project:

- Optimo DP 16–42 mm (aperture T2.8) lenses, as illustrated in Fig. 5.2, developed by Angénieux and motorized by Binocle. The key benefit is that they have been specifically designed for stereoscopic shoots, thanks to improved compatibility of focal distances, focusing paths, f-stops, and optical centers with adjustable tracking.
- Zeiss Ultra Prime fixed lenses, as illustrated in Fig. 5.3, which are not primarily designed for stereoscopic shoots, but offer a larger lens aperture of T1.9. Matching of lenses, particularly in terms of scale, was achieved with the help of technicians from the Optical Division of Panavision Alga Paris.

5.2.4 FILTERS

To keep the same features between the two captured images — depth of field, noise level, and shutter speed — for their subsequent fusion, an exposure difference between the two cameras was obtained by the addition of neutral density filters to the camera capturing high-light levels:

- Sony F65 1/8, 1/16, 1/32, and 1/64 neutral density filters were primarily used. They are expressed in camera aperture fractions and correspond to ND0.9, ND1.2, ND1.5, and ND1.8 in that order, or in other words to differences in aperture of three, four, five, and six f-stops.

FIGURE 5.2

Optimo lenses from Angénieux. Optimo red lenses are dedicated to stereoscopy.

FIGURE 5.3

Zeiss Ultra Prime range.

- Where applicable, external Schneider ND0.3 and ND0.6 filters enabled further optimization of the exposure difference by one or two additional f-stops. But, as for stereoscopy, the added filters create greater geometric and photometric disparities in the picture, and their positioning is tricky. Schneider filters were used only when the Sony F65 camera's built-in filters were insufficient.

5.2.5 **RIG**

The Brigger III rig from Binocle, as illustrated in Fig. 5.4, is designed to receive cameras with weights and dimensions corresponding to Super 35 cameras and lenses. The rig is fitted with adjustable settings

FIGURE 5.4

Brigger III rig from Binocle.

required to make the optical axes of both cameras coincide: altitude, tilt, and axis, but also convergence and interaxial distance for stereoscopic setting. To obtain the best coincidence of the optical axes, the interaxial distances are zero and convergences form a flat angle, as usual when a stereoscopic rig is initialized. The rig includes a half-silvered mirror and a quarter-wave retarder filter.

5.2.6 **EXPOSURE TIME FOR IMAGES**

Two tools are used to control shooting:

- Transvideo CineMonitorHD 3DView Evolution, 12 inch, as illustrated in Fig. 5.5. Dedicated to stereoscopic imaging, it allows the analysis of two video streams both for synchronization and for viewing the two superimposed images. Transvideo upgraded it to ensure display of exposure values common to both images and specific areas relative to high-light and low-light areas.
- Tagger Movie from Binocle, as illustrated in Fig. 5.6, which enables the analysis of two images in terms of alignment, with one-pixel precision. It is also fitted with stereoscopic correction software that simplifies postproduction by ensuring, from the start of shooting, that both views can be fully blended. These tools were upgraded for the project insofar as the disparity indicators did not handle identical images with a coincident optical axis.

FIGURE 5.5

Transvideo CineMonitorHD 3DView Evolution, 12 in.

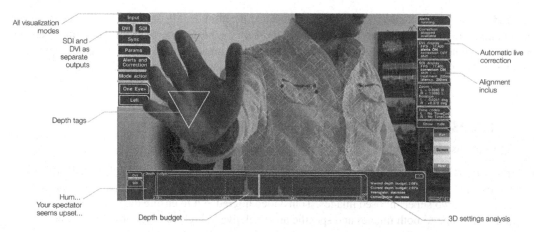

FIGURE 5.6

TaggerMovie from Binocle.

FIGURE 5.7

Density pattern called "noise pattern."

5.2.7 PREPARATION FOR SHOOTING

Besides standard equipment testing as performed before each 2D or stereoscopic shoot, response curves were characterized for every camera to ensure subsequent fusion of images.

We characterized each Sony F65 camera sensor by filming density patterns, as illustrated in Fig. 5.7, at different exposures. Fig. 5.8 shows an example of the contrast curve obtained by this procedure. The spectral response of the Sony F65 camera sensor is depicted in Fig. 5.9.

Half-silvered mirror reflection and transmission characteristics are given in Table 5.1.

5.2.8 SELECTION OF THE LOCATION

Lectoure, a medieval city in the south of France, was selected for these series of HDR shoots. During August, favorable sunshine is well suited for HDR imaging. We shot different scenes, as illustrated in Fig. 5.10:

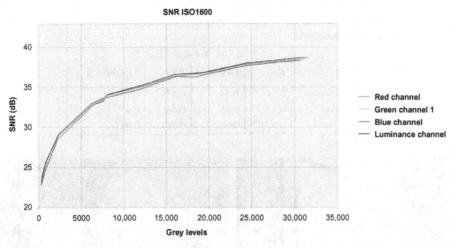

FIGURE 5.8

As an example, contrast curve plotted from noise pattern capture. *SNR*, signal-to-noise ratio.

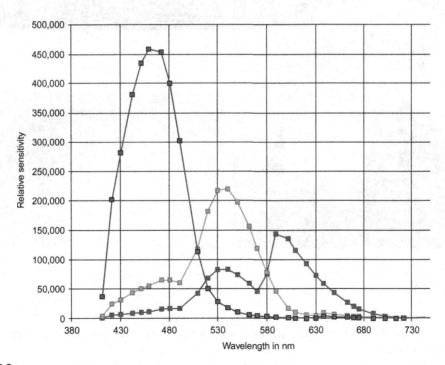

FIGURE 5.9

Spectral response of the Sony F65 camera sensor.

Table 5.1 Variations of Half-Silvered Mirror Reflection and Transmission

	Transmission			
	Red	Green	Blue	Luminance
Mirror transmission	0.54	0.53	0.55	0.53
Mirror reflection	0.44	0.45	0.44	0.45
Loss	0.979	0.982	0.987	0.982

FIGURE 5.10

Lectoure: (A) narrow lane, (B) church, (C) high-level view, and (D) air balloon.

- Narrow lanes with high contrast between shady and sunny areas.
- A church where the sun heavily comes through stained-glass choir windows, whereas aisles are left in deep darkness.
- A high-ground site offering a 360° view of the surrounding valleys for sunrise and sunset.
- A firework festival preceded by the ascent of air balloons as the sun sets.

5.2.9 **SHOOTING**

In parallel to HDR shoots with the Brigger III rig, a third Sony F65 camera captured a low dynamic range (LDR) reference with an exposure chosen in order to use most of its dynamic range. This setting could be different from the one used for the high-light-condition and low-light condition cameras of the HDR setup. Both devices are illustrated in Fig. 5.11. It is important to underline that exposure time results from an artistic intention that corresponds to the scene content and that in cinematography technology no exposure time is meaningfully better than any other. The directors and operators are the only ones to decide on the exposure time for a subject reproduced as a highlighted object.

For the selection of exposure times in high-light and low-light conditions, the director of photography uses standard tools (cells and zebra camera) as well as Transvideo functions that allow a double viewing of oscilloscopes and histograms.

Besides artistic choices, it is important to clarify two points in relation to exposure times.

In the LDR process, the operator determines the exposure time according to high-light levels, as the "burned" part of the image is definitively lost.

To get a successfully fused HDR image from two captured images, the low-light image must be as noise free as possible. The director of photography primarily determines the exposure time for low-light levels and then adjust the exposure time for high-light levels according to the possibilities linked to the difference in brightness of the subject.

For different subjects with a similar contrast level, the exposure time differences between two cameras are not necessarily the same. According to the quality and range required for the common parts of the exposure, the dynamic range of the system is more or less extended. As an example, one branch with backlit leaves will require a more extended common exposure area than the mass of dark trees against the same sky. In this example, details of the contours of the foreground leaves must be preserved from size variation in relation to exposure differences and this thus determines the global dynamic range of the system. Fig. 5.12 illustrates the selection of the exposure. Fig. 5.13 shows an example of the shooting environment.

FIGURE 5.11

Both devices facing the sunset.

FIGURE 5.12

Setting of HDR double exposure (foreground) and LDR camera (background).

FIGURE 5.13

Shooting environment.

5.3 **POSTPRODUCTION**

5.3.1 **ISSUES RELATED TO DOUBLE EXPOSURE**

HDR double exposure implies that the two images coincide by use of a half-silvered mirror and a rig. Each camera is adjusted to obtain high-light or low-light information. The camera will be preferably used in transmission for low-light images, which are more sensitive to deterioration.

To produce a good-quality HDR image, it is necessary to process the two views to remove geometric disparities, which always appear between two pictures in spite of the operators taking care with the settings. As in stereoscopy, disparities are due to imperfections of electronic, mechanical, and optical systems. Accordingly, it is necessary to use correction software. For this purpose, we apply our software Disparity Killer,[5] developed with the support of the Institut National de Recherche en Informatique et en Automatique (INRIA).

Fusion of images requires that streams are linearized. To capture the maximum dynamic range, cameras always use a gamma function, or the equivalent, which results in brightness dynamic compression. Furthermore, recording and reading of recorded data streams may add several nonlinear functions. For these reasons, acquired images from both streams are nonlinear, and the reverse function corresponding to the cumulative curves of the cameras and recorders will have to be applied to streams resulting from the whole processing chain.

To produce the HDR image, global processing of both streams is done in three steps, as illustrated in the workflow in Fig. 5.14:

1. Linearization of both images
2. Size and color correction for matching both images
3. Fusion itself

5.3.2 **LINEARIZATION**

For historical and technical reasons, the pixel intensity of a digital camera is not encoded in proportion to the amount of incoming light. Conversely, most postproduction algorithms, and even more so for HDR image fusion algorithms, assume that the captured luminous intensities are encoded in a linear way. Thus, to maximize HDR rendering quality, it is imperative to transform the video streams which are produced. This phase is called linearization. In a linearized video, white level is encoded to 100%, black to 0%, and gray to 50% — that is, 50% white plus 50% black. Perceptual gray is estimated to be around 20%. It is defined according to a sample population as being equivalent to 50% white and 50% black. Linear and gamma-encoded gradients are illustrated in Fig. 5.15.

The linearization must be achieved after the creation of the reverse curve corresponding to all the operations resulting from the whole chain. Examples of linear and gamma response curves are shown in Fig. 5.16. The key parameter to validate linearization is the ratio of intensities between both images. If the images are linear, the ratio is constant.

---

[5]Disparity Killer is the name of the first stereoscopic correction software. It was developed by INRIA and Binocle to ensure cerebral comfort for shooting crews and viewers. It allows adjustment of images at a subpixel level. It works in real time in Tagger Live for television shoots and in Tagger Movie for cinema shoots.

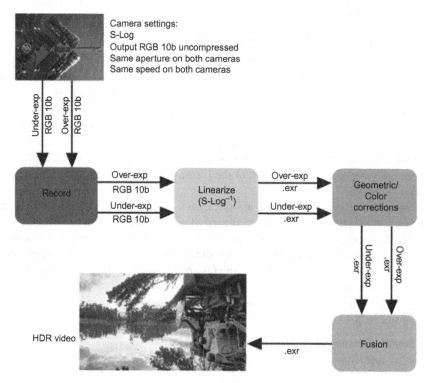

FIGURE 5.14

Postproduction workflow.

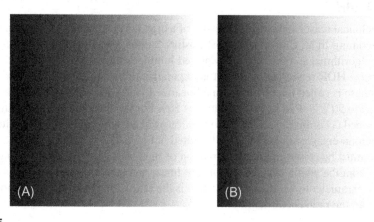

FIGURE 5.15

(A) Linear gradient and (B) gamma-encoded gradient.

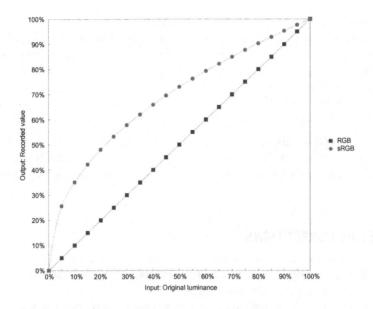

FIGURE 5.16

Linear/gamma response curves.

Not only the cameras used do not provide such images but, furthermore, each step of the postproduction process (recorder, codec) tends to modify the curve. The response curve from cameras and from other elements of the chain must be characterized to establish the function between perceived light intensity and its related digital value.

The main concern is to get inverse algorithms from those used by the manufacturers for cameras and other elements of the chain. It is therefore necessary to analyze images to generate the response curve.

This has been achieved through specific capture in the overall conditions of taking, recording, and transferring intended in the various shoots. The acquisition process in the laboratory must be identically reproduced at the shooting scene. On the basis of camera characterization, lookup tables are calculated for linearization of video contents. The reverse function can be then applied to modify the curves of the images produced.

5.3.3 CORRECTIONS

We have seen that imaging conditions are met by use of a stereoscopic rig fitted with a mirror, two cameras equipped with paired lenses, and one filter allowing differentiation high-light and low-light exposures.

As in stereoscopy, we have to face geometric and colorimetric distortions linked to the different system components:

- Sensors and electronics cannot be strictly identical.
- Cameras-lens pairs cannot be strictly similar and, because of mechanical assembly, the optical axes cannot be coincident.

- Optical axes are shifted in a different way for each camera.
- Semireflective mirrors do not have equivalent reflection and transmission features for all the wavelength range and sustain distortions resulting from gravity and mechanical stresses.
- Neutral density filters which are only basically neutral and which involve colorimetric and/or optical aberrations.
- Typically any mechanical, optical, or electronic appliance is imperfect and induces cumulative aberrations that can be corrected only by digital technology in stereoscopy.

The same applies to our HDR experience, with the result being that we have to correct both images to remove vertical and scale disparities and also horizontal disparities, which, usually, are not a problem in stereoscopic use because the visual illusion is the result of horizontal disparity varying in visual depth perception.

5.3.4 GEOMETRIC CORRECTIONS

Geometric correction of two captured streams is even more critical than in stereoscopy for their geometric alignment according to pixel precision, even though the result affects not the spectator's brain but only the structure and resolution of images.

Disparity Killer has been accordingly modified to take into account two superimposed images. This is never applied in stereoscopy because in this case the difference between viewpoints will influence the visual analysis of the image, the interpretation of corrections to be made, and the corrections themselves.

However, there is an outstanding problem the solution of which remains unknown: differences in the exposure time of sensors will involve different sizes of the same subject either underexposed or overexposed, as for a film.

5.3.5 COLORIMETRIC CORRECTIONS

Colorimetric correction between two images serves to overcome variations resulting from the sensor, lenses, mirror, and filters. The perfect semitranslucent mirror supposed to transmit and reflect 50% of light without affecting colors does not exist. Similarly, neutral density filters always show a slight variation. Therefore, it is necessary to calibrate both images so that same subject ranges may have equivalent chrominance.

This calibration does not prevent calibration of the different shots of the fused HDR sequence as would be done in any audiovisual presentation.

5.3.6 FUSION

The fusion process combines linearized and corrected contents. The purpose of fusion is to create a single image containing all the details of each input image by use of the underexposed images for light parts, the overexposed images for dark parts, and blending of both images for intermediate parts of the scene. The resulting image is represented in floating points and can be displayed only on an HDR monitor. A tone mapping process is required to display the image on conventional monitors or to distribute it on various media.

5.3.7 EXAMPLES

Examples of image pairs with different exposures are shown in Figs. 5.17 and 5.18, along with the corresponding linear and color-graded HDR images.

5.4 HDR: ENHANCED ARTISTIC PALETTE AVAILABLE FOR DIRECTORS OF PHOTOGRAPHY AND DIRECTORS

Use cases may be differentiated depending on whether we refer to audiovisual presentation, television, or cinema, and in the case of 2D or 3D photography. We will consider that for most shoots in the studio and regardless of the target application, the question does not arise because sensors, lenses, and light are entirely determined. On the other hand, HDR offers new opportunities in outdoor shooting with in the scope of image setting.

There is expected to be exponential development of HDR with 360° images because of immersive vision systems. It is essential to ensure that exposure time and depth are identical in all directions, thereby simplifying the calibration process and continuity between the different images.

In the context of HDR for television, outdoor filming is problematic — for example, when one is shooting sports events on a sunny day: the change from full-sun areas to shaded areas implies that vision engineers located in the control room continuously apply corrections. HDR facilitates the continuity between different camera axes and helps to mitigate differences in exposure and to harmonize depths of field.

In the case of HDR for cinematography and documentaries, the previous remarks remain valid, but HDR application may enable the processing of long shots instead of segmentation in several LDR shots. As an example, HDR enables one to change from a backlight distant shot to a close shot of characters, thus unmasking facial expression without varying the depth of field by opening of the iris. We have seen that HDR is not necessarily flattering for the expressiveness of the image. In this respect, it should be possible to reduce contrast at the beginning of a shot in order for the silhouettes to become legible as the camera is moving forward. One could obtain such an effect in postproduction by keeping a reduced contrast range moving inside and for the duration of the HDR image.

HDR for stereoscopy is probably the most obvious application. It can enable not only a different setting of the image but can also resolve some problems related to visual comfort. Stereoscopic setting frequently contrasts with excessive use of the 2D image. We have already mentioned the case of ghost images induced by fast motion in backlight that can be resolved by HDR or a combination of HDR and HFR. HDR and WCG are obviously closely interlinked. New cinematographic standards should allow the use of a combination of HDR, HFR, WCG, or UHD, suitably chosen according to an artistic project and limited data bandwidth. Digital broadcasting should be flexible, consisting of open standards depending on the directors' choices for specific production.

ACKNOWLEDGMENTS

We thank all the partners and researchers in the NEVEx project for their participation and their support for the shoot in Lectoure and, in particular the following: Catherine Serré from Technicolor as project manager; Yannick Olivier and David Touze from Technicolor, who were responsible for image integration and with whom we had continuous exchanges during the project and shoots; Laurent Chanas and DxO Labs engineers, who calibrated

FIGURE 5.17

(A) Linear underexposed image, (B) linear overexposed image, (C) linear HDR fusion, and (D) color-graded HDR image.

FIGURE 5.18

(A) Linear underexposed image, (B) linear overexposed image, (C) linear HDR fusion, and (D) color-graded HDR image.

all motion picture cameras and ensured images were linearized and fused; Gilles Thery from TF1, who ensured shooting occurred in Lectoure; Jacques Delacoux and Transvideo engineers for their work that allowed perfect display of curves with images to be merged; engineers from Binocle for software, electronic, and mechanical engineering; Forest Finbow (cameraman) and Bérenger Brillante (camera assistant) for the LDR camera; and the Binocle shooting team in Lectoure — Cyril Lèbre (director of photography), Aymeric Manceau (data manager and stereoscopic vision engineer), Thomas Bresard (stereograph technician), Christine Mignard, (camera assistant), and Raphael Aprikian (trainee).

PROCESSING

VIDEO TONE MAPPING

R. Boitard[*,†], **R. Cozot**[*], **K. Bouatouch**[*]

IRISA, Rennes, France[*]
Technicolor, Cesson-Sevigne, France[†]

CHAPTER OUTLINE

Each pixel of a low dynamic range (LDR) image is stored as color components, usually three. The way LDR displays interpret these components to shape color is defined through a display-dependent color space — for example, BT.709 (ITU, 1998) or BT.2020 (ITU, 2012). In contrast, high dynamic range (HDR) pixels represent, in floating point values, the captured physical intensity of light in candelas per square meter. They can also represent relative floating point values. Hence, adapting an HDR image to an LDR display amounts to retargeting physical values, with a virtually unlimited bit depth, to a constrained space (2^{2n} chromaticity values over 2^n tonal level, n being the targeted bit depth). This operation, which ensures backward compatibility between HDR content and LDR displays, is called tone mapping. The bit-depth limitation means that many similar HDR values will be tone-mapped to the same LDR value. Consequently, contrast between neighboring pixels as well as between spatially distant areas will be reduced. Furthermore, LDR displays have a low peak luminance value when compared with the luminance that a real scene can achieve. Consequently, captured color information will have to be reproduced at different luminance levels.

In a nutshell, tone mapping an HDR image amounts to finding a balance between the preservation of details, the spatial coherency of the scene, and the fidelity of reproduction. One usually achieves this balance by taking advantage of the many weaknesses of the human visual system. Furthermore, the reproduction of a scene can sometimes be constrained by an artist or application intent. That is why a lot of tone mapping operators (TMOs) have been designed with different intents, from simulating human vision to achieving the best subjective quality (Reinhard et al., 2010; Myszkowski et al., 2008; Banterle et al., 2011).

In the early 1990s, the main goal of tone mapping was to display computer-generated HDR images on a traditional display. Indeed, use of a simple gamma mapping was not enough to reproduce all the information embedded in HDR images. Although throughout the years TMOs addressed different types of applications, most of them still focused on finding the optimal subjective quality, as the many subjective evaluations attest (Drago et al., 2003; Kuang et al., 2007; Čadík et al., 2008). However, because of the lack of high-quality HDR video content, the temporal aspect of tone mapping has been dismissed for a long time. Thanks to recent developments in the HDR video acquisition field (Tocci et al., 2011; Kronander et al., 2013, 2014), more and more HDR video content is now becoming publicly available (Unger, 2013; Krawczyk, 2006; Digital Multimedia Laboratory, 2014; Lasserre et al., 2013; IRISA, 2015). Soon many applications such as real-time TV broadcasting, cinema movies, and user-generated videos will require video tone mapping.

In this chapter, we propose to evaluate the status of the video tone mapping field when trying to achieve a defined subjective quality level. Indeed, naively applying a TMO to each frame of an HDR video sequence leads to temporal artifacts. That is why we describe, in Section 6.1, different types of temporal artifacts found through experimentation. Then, Section 6.2 introduces state-of-the-art video TMOs — that is to say, TMOs that rely on information from frames other than the frame currently being tone-mapped. In Section 6.3, we present two new types of temporal artifact that are introduced by video TMOs: temporal contrast adaptation (TCA) and ghosting artifacts (GAs). Finally, Section 6.4 presents in more detail two recently published video TMOs.

6.1 TEMPORAL ARTIFACTS

Through experimentation with different HDR video sequences, we encountered several types of temporal artifact. In this section we focus only on those occurring when a TMO is applied naively to each frame of an HDR video sequence. We propose classifying these artifacts into six categories:

1. *Global flickering artifacts* (GFAs; Section 6.1.1),
2. *Local flickering artifacts* (LFAs; Section 6.1.2),
3. *Temporal noise* (Section 6.1.3),
4. *Temporal brightness incoherency* (TBI; Section 6.1.4),
5. *Temporal object incoherency* (TOI; Section 6.1.5),
6. *Temporal hue incoherency* (THI; Section 6.1.6).

This section provides a description of those artifacts along with some examples. Note that all the results are provided with TMOs that do not handle time dependency — namely, TMOs that rely only on statistics of the current frame for tone mapping.

6.1.1 GLOBAL FLICKERING ARTIFACTS

GFAs are well known in the video tone mapping literature and are characterized by abrupt changes, in successive frames, of the overall brightness of a tone-mapped video sequence. These artifacts appear because TMOs adapt their mapping using image statistics that tend to be unstable over time. Analysis of the overall brightness of each frame over time is usually sufficient to detect those artifacts. An overall brightness metric can be, for example, the mean luma value of an image. Note that if it is computed on HDR images, the luminance channel must first be perceptually encoded, with use of, for example, a log transform as proposed in Reinhard et al. (2002), before averaging is done.

To illustrate this type of artifact, we plot in Fig. 6.1 the overall brightness indication for both the HDR sequence and the tone-mapped sequence. Note how the evolution of the overall brightness is stable over time in the HDR sequence, while abrupt peaks occur in the LDR sequence. These artifacts appear because one of the TMO's parameters, which adapts to each frame, varies over time. Fig. 6.2 illustrates such an artifact occurring in two successive frames of a tone-mapped video sequence. The overall brightness has changed because the relative area of the sky in the second frame is smaller, hence reducing the chosen normalization factor (99th percentile).

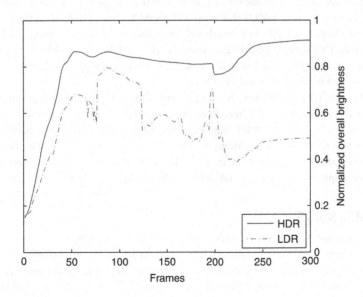

FIGURE 6.1

Example of GFAs. The overall brightness in the HDR sequence is stable over time, while many abrupt variations occur in the LDR sequence. As luminance and luma have different dynamic ranges, they have been scaled to achieve a meaningful comparison.

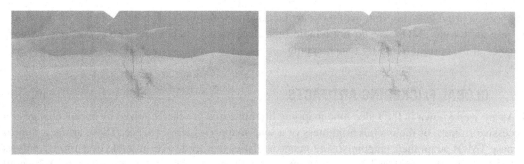

FIGURE 6.2

GFAs due to the use of the 99th percentile on two successive frames of the *Desert* sequence.

To summarize, GFAs mostly occur when one is using TMOs that rely on content-adaptive parameters that are unstable over time. They are usually considered as the most disturbing of the artifacts presented in this section, which is why they have received a lot of attention, as will be seen in Section 6.2.

6.1.2 LOCAL FLICKERING ARTIFACTS

LFAs correspond to the same phenomenon as their global counterpart but on a reduced area. They appear mostly when one is using TMOs that map a pixel on the basis of its neighborhood — namely, local TMOs. Small changes of this neighborhood, in consecutive frames, may result in a different mapping. Edge-aware TMOs are particularly prone to such artifacts as they decompose an HDR image into a base layer and one or more detail layers. As each layer is tone-mapped independently, a difference in the filtering in successive frames results in LFAs.

The top row in Fig. 6.3 represents a zoom on a portion of the computed base layer of three successive frames. Note how the edges are less filtered out in the middle frame compared with the other two frames. Application of the bilateral filter (Durand and Dorsey, 2002) operator results in an LFA in the tone-mapped result (bottom row). Although LFAs are visible in a video sequence, it is tenuous to represent LFAs by means of successive frames. A side effect of LFAs is that they modify the saliency of the tone-mapped sequence as the eye is attracted by these changes of brightness on small areas.

6.1.3 TEMPORAL NOISE

Temporal noise is a common artifact occurring in digital video sequences. Noise in digital imaging is mostly due to the camera and is particularly noticeable in low-light conditions. On images, camera noise has a small impact on the subjective quality; however, for video sequences its variation over time makes it more noticeable. This is why denoising algorithms (Brailean et al., 1995) are commonly applied to video sequences to increase their subjective quality.

As most TMOs aim to reproduce minute details, they struggle to distinguish information from noise. Consequently, most current TMOs increase the noise rather than reducing it. Local TMOs are

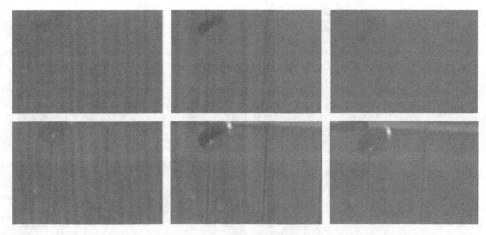

FIGURE 6.3

Example of LFAs when a bilateral filtering TMO (Durand and Dorsey, 2002) is applied to three consecutive frames. Top row: Base layer computed with the bilateral filter. Bottom row: The corresponding tone-mapped result.

particularly prone to such artifacts as they aim to preserve details even in dark areas, which tend to be quite noisy. Furthermore, noise is usually reproduced at a luma level higher than that of a native LDR image, which makes the noise more visible. An example of temporal noise enhanced by the application of local TMOs is illustrated in Fig. 6.4.

6.1.4 TEMPORAL BRIGHTNESS INCOHERENCY

TBI artifacts occur when the relative brightness between two frames of an HDR sequence is not preserved during the tone mapping. As a TMO uses for each frame all its available range, the temporal brightness relationship between frames is not preserved throughout the tone mapping operation. Consequently, a frame perceived as the brightest in the HDR sequence is not necessarily the brightest in the LDR sequence.

For example, TBI artifacts occur when a change of illumination condition in the HDR sequence is not preserved during the tone mapping. Consequently, temporal information (ie, the change of condition) is lost, which changes the perception of the scene (along with its artistic intent). Fig. 6.5 illustrates a TBI artifact, where the overall brightness of both the HDR sequence and the LDR sequence is plotted. Note that although the mean value varies greatly in the HDR sequence, it remains stable in the LDR one. This is because a TMO searches for the best exposure for each frame. As it has no information on temporally close frames, the change of illumination is simply dismissed and the best exposure is defined independently (usually in the middle of the available range). Fig. 6.6 illustrates an example of TBI occurring in consecutive frames of a tone-mapped video sequence. The top row displays the HDR luminance of these frames in false color. The change of illumination conditions occurs when the disco ball light source is turned off. When a TMO is applied, this change of illumination condition is lost (bottom row).

FIGURE 6.4

Example of temporal noise amplification due to the application of a local edge-aware TMO (Gastal and Oliveira, 2011) (left) compared with the global photographic tone reproduction TMO (Reinhard et al., 2002) (right).

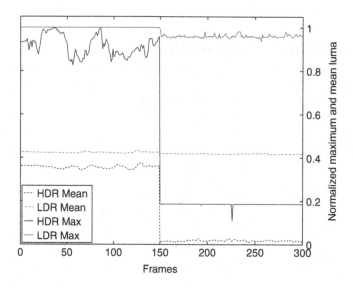

FIGURE 6.5

Example of TBI. The change of the illumination condition (represented by the mean value) in the HDR sequence is not preserved in the tone-mapped result.

| 2463 |
| 276.6 |
| 31.06 |
| 3.489 |
| 0.3918 |
| 0.044 |

FIGURE 6.6

Example of TBI when a change of illumination occurs. False color luminance (top row) and tone-mapped results obtained with a photographic tone reproduction operator (Reinhard et al., 2002) (bottom row). Both frames appear at the same level of brightness although the false color representations indicate otherwise. The color bar indicates the value in candelas per square meter.

TBI artifacts can appear even if no change of illumination condition occurs — that is to say, when the tone mapping adapts to the content. When this adaptation occurs abruptly on successive frames, it gives rise to flickering artifacts as seen previously. However, when this adaptation is smoother — say, over a longer time — the brightness relationship between the HDR and LDR sequences will be slowly disrupted. These artifacts are similar to those that occur when commercial cameras adapt their exposure during a recording (Farbman and Lischinski, 2011). Such an artifact is shown in Fig. 6.7 as the brightest HDR frame (rightmost) is the dimmest one in the LDR sequence. This second cause of TBI artifacts is also a common cause of TOI, which is presented next.

6.1.5 TEMPORAL OBJECT INCOHERENCY

TOI occurs when an object's brightness, stable in the HDR sequence, varies in the LDR sequence. Fig. 6.8 plots the HDR and LDR overall brightness along with the value of a single pixel over several frames. Note that the HDR pixel's value is constant over time, while the overall brightness changes. As the TMO adapts to each frame, the LDR pixel's value changes, resulting in a TOI artifact. Fig. 6.7 illustrates visually such an artifact. When looking at the false color representation of the HDR luminance (Fig. 6.7, top row), one sees the level of brightness of the underside of the bridge to be stable over time. However, after application of a TMO (bottom row), the bridge, which appears relatively

FIGURE 6.7

Example of TBI and TOI artifacts. False color luminance (top row) and tone-mapped result obtained with a photographic tone reproduction operator (Reinhard et al., 2002) (bottom row). The TBI is represented by the overall brightness of each frame that is not coherent between the HDR and LDR frames. The TOI is represented by the brightness of the underside of the bridge, which is similar in the HDR sequence but varies greatly in the LDR sequence. From left to right, frames 50, 100, 150, and 200.

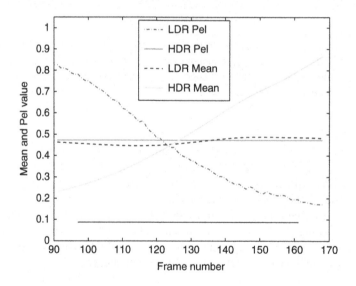

FIGURE 6.8

Illustration of TBI. A pixel's value that is constant in the HDR sequence varies greatly in the LDR sequence. The pixel and mean value have been computed on the UnderBridgeHigh sequence shown in Fig. 6.7.

bright at the beginning of the sequence, is almost dark at the end. The temporal coherency of the bridge in the HDR sequence has not been preserved in the LDR sequence. The adaptation of a TMO to a scene is the source of TBI and TOI artifacts. However, TBI artifacts are of a global nature (difference in overall brightness between frames), while TOI artifacts are of a local nature (difference in brightness between a reduced area over time).

6.1.6 TEMPORAL HUE INCOHERENCY

THI is closely related to TBI as it corresponds to the variation of the color perception of an object rather than its brightness. Such artifacts occur when the balance between tristimulus values in successive frames is not temporally preserved by the tone mapping. The main reason for this imbalance is color clipping. Color clipping corresponds to the saturation of one or more of the tone-mapped color channels (eg, red, green, or blue). Color clipping is a common artifact inherent in tone mapping of still images when one aims to reproduce as well as possible the HDR color (Xu et al., 2011; Pouli et al., 2013). When color clipping is considered as a temporal artifact, it is not the difference between the HDR and LDR reproduction that is important but rather the LDR coherency from frame to frame. Indeed, variations in the tone mapping may saturate one color channel of an area which was not in the previous frame.

To illustrate such an artifact, we generated an HDR sequence with the following characteristics:

- A square area of constant luminance (100 cd/m^2) with two gradients along the CIE u′ and v′ chrominances. The chrominance gradient ranges from -0.25 to 0.25 around the D65 white point.
- A neutral gray border area with a temporally varying luminance ranging from 0.005 to $10,000 \text{ cd/m}^2$.

Fig. 6.9 illustrates a THI due to the clipping of one or more color channels by a TMO. Note the shift in hue illustrated both in Fig. 6.9A (right) and in a zoom on a portion of the tone-mapped frames (Fig. 6.9B).

6.2 VIDEO TMOs

Applying a TMO naively to each frame of a video sequence leads to temporal artifacts. The aim of video TMOs is to prevent or reduce those artifacts. Video TMOs rely on information outside the current frame to perform their mapping. Most current video TMOs extend or postprocess TMOs designed for still images. We have sorted these techniques into three categories depending on the type of filtering:

1. Global temporal filtering (Section 6.2.1),
2. Local temporal filtering (Section 6.2.2),
3. Iterative filtering (Section 6.2.3).

For each category, we provide a description of the general technique along with different state-of-the-art references.

6.2.1 GLOBAL TEMPORAL FILTERING

Global temporal filtering aims to reduce GFAs when global TMOs are used. Indeed, global operators compute a monotonously increasing tone mapping curve that usually adapts to the image statistics of the frame to be tone-mapped. However, abrupt changes of this curve in successive frames result in GFAs. Two main approaches have been formulated so far to reduce those artifacts: filtering temporally either the tone mapping curve or the image statistics.

By application of a temporal filter to successive tone mapping curves, GFAs can be reduced. Such tone mapping curves are usually filtered during a second pass as a first pass is required to compute a

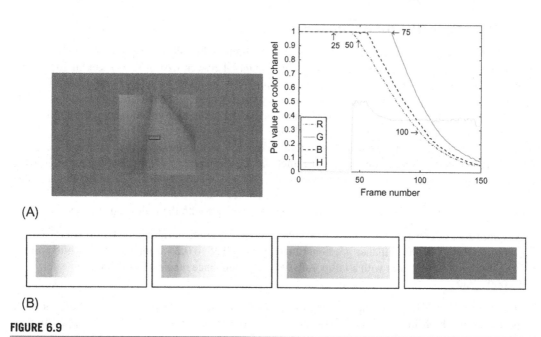

(A)

(B)

FIGURE 6.9

Example of THI due to color clipping. Each color channel desaturates at different temporal positions. (A) Tone-mapped frame 100 of the ColorSquare sequence, the perceptual brightness reproduction operator (Tumblin and Rushmeier, 1993) (left) and the temporal evolution of the central pixel of the square, where RGB denote the three color channels, and H the hue (right). (B) Zoom on the area outlined by the rectangle in frames 25, 50, 75, and 100 (from left to right).

tone mapping curve per frame. The display adaptive operator (Mantiuk et al., 2008) is able to perform such temporal filtering on the nodes of a computed piecewise tone mapping curve. The efficiency of this filtering is illustrated in Fig. 6.10. The top row provides the independently tone-mapped version of three successive frames of an HDR video sequence. The second row displays the corresponding piecewise tone mapping curves on top of their histogram. Note how the tone mapping curve of the middle frame is different from the other two, resulting in a change of overall brightness (GFA) in the tone-mapped result. The third row shows the temporally filtered version of the piecewise tone mapping curves. Finally, the bottom row provides the tone-mapped frames after the GFA has been reduced.

Image statistics can be unstable over time (eg, the 99th percentile, mean value, histogram of the luminance (Ward, 1994), etc.). For example, the photographic tone reproduction operator (Reinhard et al., 2002) relies on the geometric mean of an HDR image to scale it to the best exposure. One temporal extension of this operator filters this statistic along a set of previous frames (Kang et al., 2003). As a consequence, this method smooths abrupt variations of the frame geometric mean throughout the video sequence. This technique is capable of reducing flickering for sequences with slow illumination variations. However, for high variations it fails because it considers a fixed number of previous frames. That is why, Ramsey et al. (2004) proposed a method that adapts this number dynamically. The adaptation process depends on the variation of the current frame key value and that of the previous

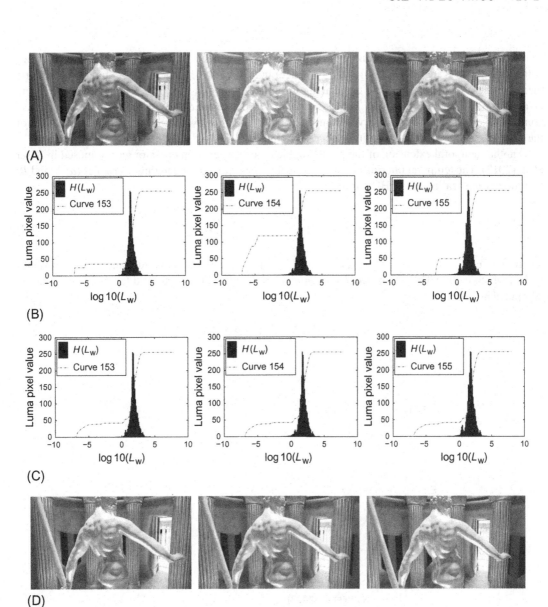

FIGURE 6.10

Reduction of GFAs by temporally filtering the tone mapping curves. The pfstmo implementation of the display adaptive operator (Mantiuk et al., 2008) was used with options -d pd=lcd_office. From left to right, frames 153, 154, and 154 of the Temple sequence. (A) The display adaptive TMO (Mantiuk et al., 2008) without temporal filtering. (B) Histogram and piecewise tone mapping curves without temporal filtering. (C) Histogram and piecewise tone mapping curves with temporal filtering. (D) The display adaptive TMO (Mantiuk et al., 2008) with temporal filtering.

frame. Moreover, the adaptation discards outliers using a min/max threshold. This solution performs better than that of Kang et al. (2003) and for a wider range of video sequences. The computed geometric mean for these techniques and the original algorithm are plotted in Fig. 6.11. The green curve (Kang et al., 2003) smooths every peak but also propagates the resulting smoothed peaks to successive computed key values. The red curve (Ramsey et al., 2004), however, reduces the abrupt changes of the key value without propagating it to successive frames.

Another temporal extension of the photographic tone reproduction operator was proposed in Kiser et al. (2012). The temporal filtering consists of a leaky integrator applied to three variables (a, A, and B) that modify the scaling of the HDR frame:

$$\mathbf{L_s} = \frac{\epsilon \cdot 2^{2(B-A)/(A+B)}}{k}\mathbf{L_w} = \frac{a}{k}\mathbf{L_w}, \tag{6.1}$$

where $A = L_{\max} - k$ and $B = k - L_{\min}$, where $L_{\max}$ and $L_{\min}$ are the maximum and minimum values of $\mathbf{L_w}$, which is the HDR luminance. k corresponds to the geometric mean and the leaky integrator is computed as

$$v_t = (1 - \alpha_v)v_{(t-1)} + \alpha_v v_t, \tag{6.2}$$

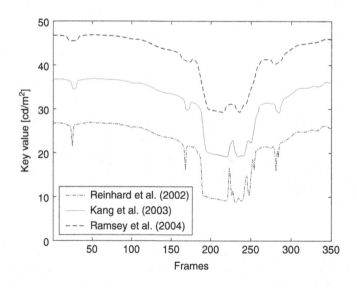

FIGURE 6.11

Evolution of the frame geometric mean (key value) computed for every frame of a video sequence. An offset is added to avoid an overlap between the curves. The smoothing effects of the HDR video operator (Kang et al., 2003) and the adaptive temporal operator (Ramsey et al., 2004) are compared with that of the photographic tone reproduction operator (Reinhard et al., 2002).

where v_t represents any of the three variables a, A, and B at time t and α_v is a time constant giving the strength (leakiness) of the temporal filtering.

Many other TMOs filter their parameters temporally, including those in Pattanaik et al. (2000), Durand and Dorsey (2000), Irawan et al. (2005), and Van Hateren (2006). Most of them aim either to simulate the temporal adaptation of the human visual system or to reduce GFAs.

6.2.2 LOCAL TEMPORAL FILTERING

Local temporal filtering consists in performing a pixelwise temporal filtering with or without motion compensation. Indeed, global temporal filtering cannot apply to local TMOs as such operators rely on a spatially varying mapping function. As outlined previously, local changes in a spatial neighborhood cause LFAs. To prevent these local variations of the mapping along successive frames, video TMOs can rely on pixelwise temporal filtering. For example, the gradient domain compression operator (Fattal et al., 2002) has been extended by Lee and Kim (2007) to cope with videos. This TMO computes an LDR result by finding the output image whose gradient field is the closest to a modified gradient field. Lee and Kim (2007) proposed adding a regularization term which includes a temporal coherency relying on a motion estimation:

$$\sum_{x,y} \|\Delta \mathbf{L_d}(x,y,t) - \mathbf{G}(x,y)\|^2 + \lambda \sum_{x,y} \|\mathbf{L_d}(x,y,t) - \mathbf{L_d}(x + \delta x, y + \delta y, t - 1)\|^2, \tag{6.3}$$

where $\mathbf{L_d}$ is the output LDR luma at the preceding or current frame ($t - 1$ or t) and $\mathbf{G}$ is the modified gradient field. The pairs (x, y) and $(\delta x, \delta y)$ represent, respectively, the pixel location of a considered pixel and its associated motion vectors. The parameter λ balances the distortion to the modified gradient field and to the previous tone-mapped frame.

Another operator (local model of eye adaptation (Ledda et al., 2004)) performs a pixelwise temporal filtering. However, the goal of this operator is to simulate the temporal adaptation of the human eye on a per-pixel basis. Besides increasing the temporal coherency, pixelwise temporal filtering also has denoising properties. Indeed, many denoising operators rely on temporal filtering to reduce noise (Brailean et al., 1995). Performing such a filtering during the tone mapping allows one to keep the noise level relatively low.

6.2.3 ITERATIVE FILTERING

The techniques presented so far in this section focus on preventing temporal artifacts (mostly flickering) when one is tone mapping video sequences. These a priori approaches consist in either preprocessing parameters or modifying the TMO to include a temporal filtering step. Another trend analyzes a posteriori the output of a TMO to detect and reduce temporal artifacts, the reduction consisting in iterative filtering.

One of these techniques (Guthier et al., 2011) aims at reducing GFAs. Such an artifact is detected if the overall brightness difference between two successive frames of a video sequence is greater than a brightness threshold (defined with either Weber's law (Ferwerda, 2001) or Steven's power law (Stevens and Stevens, 1963)). As soon as an artifact is located, it is reduced by an iterative brightness adjustment until the chosen brightness threshold is reached. Note that this technique performs an iterative brightness adjustment on the unquantized luma to avoid loss of signal due to clipping and

FIGURE 6.12

Results of the multiscale operator (Farbman et al., 2008) without (left) and with (right) flicker reduction postprocessing (Guthier et al., 2011). Each image represents two successive frames. Note the flickering artifact on the left image, while it has been removed on the right image after application of flicker reduction postprocessing (Guthier et al., 2011).

quantization. Consequently, the TMO's implementation needs to embed and apply the iterative filter before the quantization step. This technique relies only on the output of a TMO and hence can be applied to any TMO. Fig. 6.12 illustrates the reduction of a GFA when this postprocessing is applied.

6.3 TEMPORAL ARTIFACTS CAUSED BY VIDEO TMOs

In the previous section, we presented solutions to reduce temporal artifacts when one is performing video tone mapping. These techniques target mostly flickering artifacts as they are considered as one of the most disturbing artifacts. However, these techniques can generate two new types of temporal artifact — temporal contrast adaptation (TCA) and ghosting artifacts (GAs) — which we describe in this section.

6.3.1 TEMPORAL CONTRAST ADAPTATION

To reduce GFAs, many TMOs rely on global temporal filtering. Depending on the TMO used, the filter is either applied to the computed tone mapping curve (Mantiuk et al., 2008) or to the parameter that adapts the mapping to the image (Ramsey et al., 2004; Kiser et al., 2012). However, when a change of illumination occurs, as shown in Fig. 6.6, it also undergoes temporal filtering. Consequently, the resulting mapping does not correspond to any of the conditions but corresponds rather to a transition state. We refer to this artifact as temporal contrast adaptation (TCA). Fig. 6.13 illustrates the behavior of the temporal filtering when a change of illumination occurs. Note how the tone mapping curve, plotted on top of the histograms, shifts from the first illumination condition (frame 130) toward the second state of illumination (frame 150; see Fig. 6.6 for the false color luminance). As the tone mapping curve has

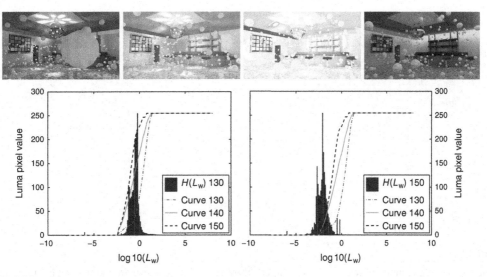

FIGURE 6.13

Example of temporal filtering of tone mapping curves when a change of illumination occurs. Top row: Tone-mapped result (frames 130, 140, 149, and 150) obtained with the display adaptive operator (Mantiuk et al., 2008) with the temporal filtering active (pfsTMO implementation Krawczyk and Mantiuk, 2007). Bottom row: Histograms of frame 130 (left) and frame 150 (right) along with the corresponding tone mapping curves for frames 130, 140, and 150.

anticipated this change of illumination, frames neighboring the change of illumination are tone-mapped incoherently.

These artifacts also occur when one is performing postprocessing to detect and reduce artifacts as in Guthier et al. (2011). Indeed, this technique relies only on the LDR results to detect and reduce artifacts. If one has no information related to the HDR video, then a change of illumination suppressed by a TMO cannot be anticipated or predicted.

6.3.2 GHOSTING ARTIFACTS

Similarly to global temporal filtering, local temporal filtering generates undesired temporal artifacts. Indeed, pixelwise temporal filtering relies on a motion field estimation which is not robust to a change of illumination conditions and object occlusions. When the motion model fails, the temporal filtering is computed along invalid motion trajectories, which results in GAs.

Fig. 6.14 illustrates a GA in two successive frames resulting from the application of the operator of Lee and Kim (2007). This artifact proves that pixelwise temporal filtering is efficient only for accurate motion vectors. A GA occurs when a motion vector associates pixels without a temporal relationship. Those "incoherent" motion vectors should be accounted for to prevent GAs, as these are the most disturbing artifacts (Eilertsen et al., 2013).

FIGURE 6.14

Example of GAs appearing on two successive frames. They are most noticeable around the two forefront columns (red squares). The bottom row shows zooms on the rightmost column.

6.4 RECENT VIDEO TMOs

Recently, two novel contributions have been proposed in the field of video tone mapping. The first one, called zonal brightness coherency (ZBC) (Boitard et al., 2014), aims to reduce TBI and TOI artifacts through a postprocessing operation which relies on a video analysis performed before the tone mapping. The second one (Aydin et al., 2014) performs a ghost-free pixelwise spatiotemporal filtering to achieve high reproduction of contrast while preserving the temporal stability of the video.

6.4.1 ZONAL BRIGHTNESS COHERENCY

The ZBC algorithm (Boitard et al., 2014) aims to preserve the HDR relative brightness coherency between every object over the whole sequence. Effectively, it should reduce TBI, TOI, and TCA artifacts and in some cases THI artifacts. It is an iterative method based on the brightness coherency (Boitard et al., 2012) technique which considered only overall brightness coherency. This method consists of two steps: a video analysis and postprocessing.

The video analysis relies on a histogram-based segmentation as shown in Fig. 6.15. A first segmentation on a per frame basis provides several segments per frame. The geometric mean (called "key value" in the article) of each segment of each HDR frame is computed and used to build a second histogram, which is in turn segmented to compute zone boundaries. The key value $k_z(\mathbf{L_w})$ is computed for each zone of each frame. An anchor is then chosen either automatically or by the user to provide an intent for the rendering.

Once the video analysis has been performed, each frame is tone-mapped with any TMO. Then a scale ratio s_z is applied to each pixel luminance $\mathbf{L_{m,z}}$ of each video zone z to ensure that the brightness ratio between the anchor and the current zone in the HDR sequence is preserved in the LDR sequence (Eq. 6.4):

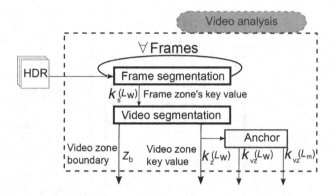

FIGURE 6.15

Details of the video analysis. The *Frame Segmentation* function segments each frame of the sequence and computes each segment's key value. The *Video Segmentation* determines the video zone's boundaries and their corresponding key values. The *Anchor* function determines the anchor zone in the HDR sequence $k_{vz}(\mathbf{L_w})$ and computes its corresponding LDR key values $k_{vz}(\mathbf{L_m})$.

$$\mathbf{L_{zbc,z}} = s_z \mathbf{L_{m,z}},$$
$$s_z = \zeta + (1 - \zeta) \frac{k_{vz}(\mathbf{L_m})}{k_{vz}(\mathbf{L_w})} \frac{k_z(\mathbf{L_w})}{k_z(\mathbf{L_m})}, \tag{6.4}$$

where $\mathbf{L_{zbc,z}}$ is the scaled luminance, ζ is a user-defined parameter, $k_{vz}(\mathbf{L_w})$ is the anchor zone HDR key value, $k_{vz}(\mathbf{L_m})$ is the anchor zone LDR key value, $k_z(\mathbf{L_w})$ is the z zone HDR key value and $k_z(\mathbf{L_m})$ is the z zone LDR key value. Note that the subscript z stands for *zone*.

At the boundaries between two zones, an alpha blending is used to prevent abrupt spatial variations. The whole workflow of this technique is depicted in Fig. 6.16.

Fig. 6.17 presents some results obtained when ZBC postprocessing is used on tone-mapped video sequences where temporal artifacts occurred. The left plot provides results regarding the reduction of the TOI artifact that was illustrated in Fig. 6.8. Thanks to the ZBC technique, the value of the pixel, which was constant in the HDR sequence, is much stabler over time. Note also that the LDR mean value is quite low at the beginning of the sequence, which will most likely result in a loss of spatial contrast in the tone-mapped frames. That is why Boitard et al. (2014) have provided a user-defined parameter which effectively trades off temporal coherency for an increase in spatial reproduction capabilities (see ζ in Eq. 6.4). In the right plot, we show some results regarding the reduction of TBI artifacts. This plot is to be compared with the one in Fig. 6.5. Use of ZBC postprocessing on the Disco sequence allows the change of illumination present in the HDR sequence to be preserved.

More results are available in Boitard et al. (2014) and Boitard (2014), especially regarding the preservation of fade effects and the impact of the different parameters of this technique.

6.4.2 TEMPORALLY COHERENT LOCAL TONE MAPPING

In Section 6.3.2, we explained why pixelwise temporal filtering can cause GAs. However, this type of filtering is the only known solution to prevent LFAs that can arise when local TMOs are used.

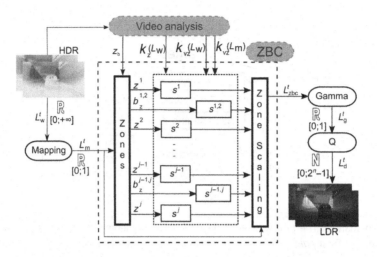

FIGURE 6.16

Complete ZBC workflow with details of the scaling phase. The *Zone's* function determines for each tone-mapped frame and each pixel, the corresponding video zone z^j as well as the video blending zone $b_z^{j,j+1}$. Their respective scaling ratios s^j and $s^{j,j+1}$ are computed. The *Zone Scaling* function applies the scale ratios to the tone-mapped frames. Finally, Q quantizes linearly floating point values in the range $[0;1]$ to integer values on a defined bit depth n (values in the range $[0;2^n - 1]$).

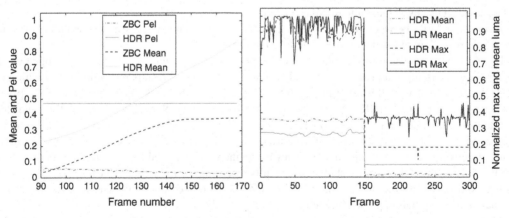

FIGURE 6.17

Reduction of TOI and TBI artifacts using the ZBC post-processing. Left plot is to be compared with Fig. 6.8 while the right one with Fig. 6.5.

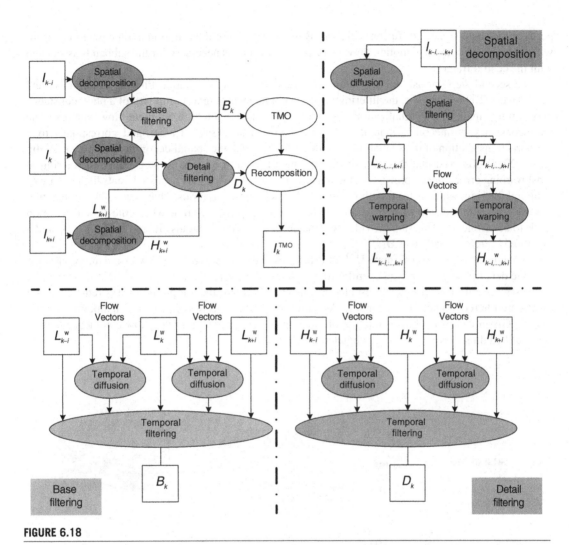

FIGURE 6.18

General workflow of the temporally coherent local operator (Aydin et al., 2014). The spatial decomposition is illustrated in top right corner, while the base and detail filtering are illustrated in the bottom left and right corners, respectively.

Consequently, Aydin et al. (2014) proposed a spatial filtering process to ensure high reproduction of contrast when ghost-free pixelwise temporal filtering is performed. Fig. 6.18 illustrates the workflow of this technique.

This technique considers a temporal neighborhood composed of a center frame $\mathbf{I_k}$ and temporally close frames $\mathbf{I_{k\pm i}}$. In a first step, each frame is decomposed into a base and detail layer by use of a permeability map (spatial diffusion weights). Both subbands are then motion compensated (warped) with a previously computed motion flow. Note that no warping is necessary for the subbands associated with the central frame.

The second step consists of two temporal filters which allow separate filtering of the base and detail layers. To prevent GAs, the filtering relies on confidence weights composed of a photoconstancy permeability map and a penalization on pixels associated with high gradient flow vectors. The photoconstancy permeability map is computed between successive frames and corresponds to a temporal transposition of the spatial diffusion weight on which the spatial decomposition relies. Aydin et al. (2014) observed that this photoconstancy measure can be tuned to stop temporal filtering at most warping errors, hence preventing the appearance of GAs. However, it also defeats the purpose of temporal filtering, which is to smooth medium to low temporal variations. That is why the penalization term has been introduced as it is a good indication of complex motion where the flow estimation tends to be erroneous. This step provides two images, the spatiotemporally filtered base layer $\mathbf{B}_k$ and a temporally filtered detail layer $\mathbf{D}_k$.

To obtain the tone-mapped frame $\mathbf{I}_k^{\mathrm{TMO}}$, the base layer $\mathbf{B}_k$ can be fed to any TMO and then combined with the detail layer $\mathbf{D}_k$, a process similar to that detailed in Durand and Dorsey (2002). This method addresses several of the artifacts presented in this chapter. First, the temporal noise is reduced as the details are filtered temporally. Second, LFAs are minimized thanks to the pixelwise temporal filtering. Finally, GAs are prevented by adaptation of the temporal filtering to a motion flow confidence metric. A simplified illustrative workflow is depicted in Fig. 6.19.

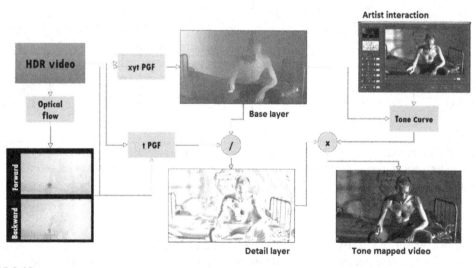

FIGURE 6.19

Illustrative workflow of the temporally coherent local operator (Aydin et al., 2014). PGF, permeability-guided filtering.

Source: Courtesy of Tunç Ozan Aydin.

As this technique is fairly new, extended results with more HDR sequences could help detect new types of artifacts. In particular, it would be interesting to test this method when changes of illumination or cut occur in a sequence. Furthermore, most of the results provided with this technique rely on user interaction to achieve the best trade-off between temporal and spatial contrast. This is not achievable for many applications, such as live broadcasts and tone mapping embedded in set-top boxes.

6.5 **SUMMARY**

In this chapter, we have described known types of temporal artifact that occur when HDR video sequences are tone-mapped. We have categorized video TMOs with respect to how they handle temporal information, and we have shown that although these solutions can deal with certain types of temporal artifact, they can also be a source of new ones. An evaluation of video TMOs (Eilertsen et al., 2013) reported that none of the current solutions can handle a wide range of sequences. However, this study was performed before the publication of the two video TMOs presented in Section 6.4. These two techniques, albeit significantly different, provide solutions to types of artifact not dealt with before.

Table 6.1 gives an overview of the temporal artifacts presented in this chapter, along with possible solutions. From this table, we can see that all of the artifacts described in this chapter have a solution. However, none of the video TMOs presented here encompass all of the tools needed to deal with all different types of artifact. Furthermore, the appearance of more HDR video sequences and applications for video tone mapping will likely result in new types of temporal artifact. Although the two recent contributions have significantly advanced the field of video tone mapping, more work still lies ahead.

Table 6.1 Summary of Temporal Artifacts Along With Their Main Causes and Possible Solutions

Temporal Artifact	Possible Cause	Possible Solutions
Global flicker	Temporal instability of parameters	Global temporal filtering
Local flicker	Different spatial filtering in successive frames	Pixelwise temporal filtering
Temporal noise	Camera noise	Spatial and/or temporal filtering (pixelwise)
TBI (brightness)	Change of illumination adaptation of the TMO	Brightness analysis of each frame
TOI (object)	Adaptation of the TMO	Brightness analysis per zone of frames
THI (hue)	Saturation of color channel (due to clipping)	Hue and brightness analysis per zone of frames
TCA (contrast)	Global temporal filtering	Brightness analysis per zone of frames
Ghosting	Pixelwise temporal filtering	Confidence weighting of pixelwise temporal filtering

REFERENCES

Aydin, T.O., Stefanoski, N., Croci, S., Gross, M., Smolic, A., 2014. Temporally coherent local tone mapping of HDR video. ACM Trans. Graph. 33 (6), 196:1–196:13.

Banterle, F., Artusi, A., Debattista, K., Chalmers, A., 2011. Advanced High Dynamic Range Imaging: Theory and Practice. AK Peters (CRC Press), Natick, MA.

Boitard, R., 2014. Temporal Coherency in Video Tone Mapping. Ph.D. thesis, University of Rennes 1.

Boitard, R., Bouatouch, K., Cozot, R., Thoreau, D., Gruson, A., 2012. Temporal coherency for video tone mapping. In: Proc. SPIE 8499, Applications of Digital Image Processing XXXV, pp. 84990D-1–84990D-10.

Boitard, R., Cozot, R., Thoreau, D., Bouatouch, K., 2014. Zonal brightness coherency for video tone mapping. Signal Process. Image Commun. 29 (2), 229–246.

Brailean, J., Kleihorst, R., Efstratiadis, S., Katsaggelos, A., Lagendijk, R., 1995. Noise reduction filters for dynamic image sequences: a review. Proc. IEEE 83 (9), 1272–1292.

Čadík, M., Wimmer, M., Neumann, L., Artusi, A., 2008. Evaluation of HDR tone mapping methods using essential perceptual attributes. Comput. Graph. 32 (3), 330–349.

Digital Multimedia Laboratory, 2014. DML-HDR Public Database. http://dml.ece.ubc.ca/data/DML-HDR/.

Drago, F., Martens, W.L., Myszkowski, K., Seidel, H.P., 2003. Perceptual evaluation of tone mapping operators. In: Proceedings of the SIGGRAPH 2003 Sketches. ACM Press, New York, NY, p. 1.

Durand, F., Dorsey, J., 2000. Interactive tone mapping. In: Proceedings of the Eurographics Workshop on Rendering. Springer Verlag, Berlin/Heidelberg, pp. 219–230.

Durand, F., Dorsey, J., 2002. Fast bilateral filtering for the display of high-dynamic-range images. ACM Trans. Graph. 21 (3), 257–266.

Eilertsen, G., Wanat, R., Mantiuk, R.K., Unger, J., 2013. Evaluation of tone mapping operators for HDR-video. Comput. Graph. Forum 32 (7), 275–284.

Farbman, Z., Lischinski, D., 2011. Tonal stabilization of video. ACM Trans. Graph. 30 (4), 89:1–89:10.

Farbman, Z., Fattal, R., Lischinski, D., Szeliski, R., 2008. Edge-preserving decompositions for multi-scale tone and detail manipulation. In: ACM SIGGRAPH 2008 papers on, SIGGRAPH '08. ACM Press, New York, NY, pp. 67:1–67:10.

Fattal, R., Lischinski, D., Werman, M., 2002. Gradient domain high dynamic range compression. ACM Trans. Graph. 21 (3), 249–256.

Ferwerda, J., 2001. Elements of early vision for computer graphics. IEEE Comput. Graph. Appl. 21 (4), 22–33.

Gastal, E.S.L., Oliveira, M.M., 2011. Domain transform for edge-aware image and video processing. ACM Trans. Graph. 30 (4), 69:1–69:12.

Guthier, B., Kopf, S., Eble, M., Effelsberg, W., 2011. Flicker reduction in tone mapped high dynamic range video. In: Proc. of IS&T/SPIE Electronic Imaging (EI) on Color Imaging XVI: Displaying, Processing, Hardcopy, and Applications, vol. 7866, pp. 78660C-1–78660C-15.

Irawan, P., Ferwerda, J.A., Marschner, S.R., 2005. Perceptually based tone mapping of high dynamic range image streams. In: Proceedings of the Sixteenth Eurographics Conference on Rendering Techniques, EGSR '05. Eurographics Association, Aire-la-Ville, pp. 231–242.

IRISA, 2015. IRISA Rennes FRVSense Public HDR Database. http://people.irisa.fr/Ronan.Boitard/.

ITU, 1998. Recommendation ITU-R BT.709-3: Parameter values for the HDTV standards for production and international programme exchange. International Telecommunications Union.

ITU, 2012. Recommendation ITU-R BT.2020: Parameter values for ultra-high definition television systems for production and international programme exchange. International Telecommunications Union.

Kang, S.B., Uyttendaele, M., Winder, S., Szeliski, R., 2003. High dynamic range video. ACM Trans. Graph. 22 (3), 319–325.

Kiser, C., Reinhard, E., Tocci, M., Tocci, N., 2012. Real-time automated tone mapping system for HDR video. In: Proceedings of the IEEE International Conference on Image Processing, pp. 2749–2752.

Krawczyk, G., 2006. MPI Public HDR Database. http://www.mpi-inf.mpg.de/resources/hdr/video/.

Krawczyk, G., Mantiuk, R., 2007. Display adaptive pfstmo documentation. http://pfstools.sourceforge.net/man1/pfstmo_mantiuk08.1.html.

Kronander, J., Gustavson, S., Bonnet, G., Unger, J., 2013. Unified HDR reconstruction from raw CFA data. In: IEEE International Conference on Computational Photography (ICCP), pp. 1–9.

Kronander, J., Gustavson, S., Bonnet, G., Ynnerman, A., Unger, J., 2014. A unified framework for multi-sensor HDR video reconstruction. Signal Process. Image Commun. 29 (2), 203–215.

Kuang, J., Yamaguchi, H., Liu, C., Johnson, G.M., Fairchild, M.D., 2007. Evaluating HDR rendering algorithms. ACM Trans. Appl. Percept. 4 (2), 9:1–9:27.

Lasserre, S., Le Léannec, F., Francois, E., 2013. Description of HDR sequences proposed by Technicolor. In: ISO/IEC JTC1/SC29/WG11 JCTVC-P0228. IEEE, San Jose, USA.

Ledda, P., Santos, L.P., Chalmers, A., 2004. A local model of eye adaptation for high dynamic range images. In: Proceedings of the 3rd International Conference on Computer Graphics, Virtual Reality, Visualisation and Interaction in Africa, AFRIGRAPH '04. ACM Press, New York, NY, pp. 151–160.

Lee, C., Kim, C.S., 2007. Gradient domain tone mapping of high dynamic range videos. In: 2007 IEEE International Conference on Image Processing, vol. 3, pp. III-461–III-464.

Mantiuk, R., Daly, S., Kerofsky, L., 2008. Display adaptive tone mapping. ACM Trans. Graph. 27 (3), 68:1–68:10.

Myszkowski, K., Mantiuk, R., Krawczyk, G., 2008. High Dynamic Range Video, vol. 2. Morgan & Claypool, San Rafael, CA, pp. 1–158.

Pattanaik, S.N., Tumblin, J., Yee, H., Greenberg, D.P., 2000. Time-dependent visual adaptation for fast realistic image display. In: Proceedings of the 27th Annual Conference on Computer Graphics and Interactive Techniques, SIGGRAPH '00. ACM Press/Addison-Wesley Publishing Co., New York, NY, pp. 47–54.

Pouli, T., Artusi, A., Banterle, F., 2013. Color correction for tone reproduction. In: Color and Imaging Conference Final Program and Proceedings, pp. 215–220.

Ramsey, S.D., Johnson III, J.T., Hansen, C., 2004. Adaptive temporal tone mapping. In: Computer Graphics and Imaging, 2004, pp. 3–7.

Reinhard, E., Heidrich, W., Debevec, P., Pattanaik, S., Ward, G., Myszkowski, K., 2010. High Dynamic Range Imaging, 2nd Edition: Acquisition, Display, and Image-Based Lighting. Morgan Kaufmann, Los Altos, CA.

Reinhard, E., Stark, M., Shirley, P., Ferwerda, J., 2002. Photographic tone reproduction for digital images. ACM Trans. Graph. 21 (3), 267–276.

Stevens, J.C., Stevens, S.S., 1963. Brightness function: effects of adaptation. J. Opt. Soc. Am. 53 (3), 375–385.

Tocci, M.D., Kiser, C., Tocci, N., Sen, P., 2011. A versatile HDR video production system. ACM Trans. Graph. 30 (4), 41:1–41:10.

Tumblin, J., Rushmeier, H., 1993. Tone reproduction for realistic images. IEEE Comput. Graph. Appl. 13 (6), 42–48.

Unger, J., 2013. HDRv Public Database. http://www.hdrv.org/Resources.php.

Van Hateren, J.H., 2006. Encoding of high dynamic range video with a model of human cones. ACM Trans. Graph. 25 (4), 1380–1399.

Ward, G., 1994. A contrast-based scale factor for luminance display. In: Heckbert, P.S. (Ed.), Graphics Gems IV. Academic Press Professional, Inc., San Diego, CA, pp. 415–421.

Xu, D., Doutre, C., Nasiopoulos, P., 2011. Correction of clipped pixels in color images. IEEE Trans. Vis. Comput. Graph. 17 (3), 333–344.

EVALUATION OF TONE MAPPING OPERATORS FOR HDR VIDEO

7

G. Eilertsen*, J. Unger*, R.K. Mantiuk†

Linköping University, Norrköping, Sweden*
University of Cambridge, Cambridge, United Kingdom†

CHAPTER OUTLINE

7.1 INTRODUCTION

Tone mapping of high dynamic range (HDR) video is a challenging problem. Since a large number of tone mapping operators (TMOs) have been proposed in the literature, it is highly important to develop a framework for evaluation, comparison, and categorization of individual operators so that users can find and assess which is most suitable for a specific problem.

High Dynamic Range Video. http://dx.doi.org/10.1016/B978-0-08-100412-8.00007-3

TMO evaluation has received significant attention for more than a decade, but there is still no standard method for conducting TMO evaluation studies. Since the performance of a TMO is highly subjective, perceptual studies are an important tool in this process. This chapter gives an overview of different approaches on how evaluation of TMOs can be conducted, including experimental setups, the choice of input data, the choice of TMOs, and the importance of parameter adjustment for fair comparisons. The chapter also discusses and categorizes previous evaluations with a focus on the results from the most recent evaluation conducted by Eilertsen et al. (2013). This results in a classification of current video TMOs and an overview of their performance and possible artifacts, as well as guidelines to the problems a video TMO needs to address.

Ideally, a direct, computational comparison between the result of a TMO and a ground truth image would be convenient, leaving out all the possible error sources associated with a subjective comparison experiment. There are also metrics for automatic visibility and quality prediction of HDR images (Mantiuk et al., 2011). However, since the comparisons of TMOs occur after transformation through the human visual system (HVS), the ground truth and tone-mapped image cannot be used directly for numerical comparison. Another method, not involving simulation of the HVS, was suggested by Yeganeh and Wang (2013). The method estimates the structural fidelity and the naturalness of a tone-mapped image, and combines these measures to form a quality index. Although this quality index was shown to correlate with data from subjective evaluations, the heuristics involved in the quality prediction cannot account for all the complex processes involved in a subjective comparison. As of today, there are no models comprehensive enough to entirely replace the human eye in a comparison, taking into account all the different aspects of the HVS. Furthermore, if we seek the most preferred method in terms of some subjective measure, this involves not only the HVS, but also high-level functions of the human brain, functions that are individual and depend on the specific criteria used for comparison.

7.2 SUBJECTIVE QUALITY ASSESSMENT METHOD

Performing a perceptual evaluation study is a complicated task. There are a range of parameters affecting the outcome of the experiment. If we closely define and control the different parts of an evaluation, the outcome is more reliable and consistent. However, it is important to appreciate that the results are fully valid only under the specific conditions used during the evaluation, and that the results may not always generalize well to other situations. This is also reflected in previous studies reported in the literature, where the outcome/result of a perceptual evaluation may vary substantially depending on the experimental setup and evaluation criteria used.

Important aspects in the comparison are, for example, what ground truth, if any, is used (evaluation method), and under which criteria the comparison is made (intent of TMO). It does not make sense to ask what is the best looking tone reproduction if we want to determine which one is closest to how the depicted scene would be perceived in real life, as this criterion is conceptually different (like comparing apples and oranges). However, at the same time it is also important to recognize that a particular TMO intended for a certain purpose can potentially produce results that give a better reproduction than other methods comparing the methods with a different criterion.

Furthermore, even if the actual comparisons and their purpose are closely defined, there are numerous accompanying aspects that may affect the outcome of an experiment. For example, a small difference in how the experiment is explained to the observers can potentially impact their image assessments. Environmental impacts also need to be considered, where the experiment should be designed to be as similar as possible to the typical environment where the methods are supposed to be used. Here it is also important to calibrate the algorithms according to the certain conditions, with, correct adjustment for the particular screen used, for example.

In what follows, we try to categorize some of the many parameters associated with subjective evaluations. These are considered to be among the most influential for a robust and valid evaluation, and should be carefully considered when one is designing a TMO comparison study. Using the categorization, we provide a brief survey in the next section, which includes the most relevant subjective studies of tone mapping methods to date.

It should be noted though that not only is the final outcome of an evaluation sensitive to the implementation design—the actual data produced also need to be examined properly. One crucial aspect is to investigate not just the mean performance of the different methods considered. There should also be a thorough analysis of the variance of the result in order to determine if there are actual statistical proofs for the differences (Mantiuk et al., 2012). Even though a clear distinction can be made, when we consider only the scores or ratings averaged over the participating observers, this does not actually imply statistically significant differences. To reduce the uncertainty, a large number of measurements (observers) may be needed, which in many situations is unfeasible.

7.2.1 **TMO INTENT**

It is important to recognize that different TMOs try to achieve different goals (McCann and Rizzi, 2012); for example, perceptually accurate reproduction, faithful reproduction of colors, or the most preferred reproduction. In an evaluation this particular intent is critical for how the different methods are compared. First, the design of the experiment completely relies on how the algorithms are supposed to be compared, where, for instance, the evaluation method used needs to be in line with the intent of the TMOs. As an example, it does not make sense to compare a TMO with a real-world scene if the evaluation is supposed to be in terms of the subjectively most preferred reproduction, since this could deviate substantially from the real scene. Also, an important aspect of a study is how the experiment is explained to the observer, making sure that the intent of the tone reproduction is clear.

After analyzing the intents of existing operators, we can distinguish three main classes:

Visual system simulators try to simulate the limitations and properties of the HVS. For example, a TMO can add glare, simulate the limitations of human night vision, or reduce colorfulness and contrast in dark scene regions. Another example is the adjustment of images for the difference between the adaptation conditions of real-world scenes and the viewing conditions (including chromatic adaptation).

Scene reproduction operators attempt to preserve the original scene appearance, including contrast, sharpness, and colors, when an image is shown on a device with reduced color gamut, contrast, and peak luminance.

Best subjective quality operators are designed to produce the most preferred images or video in terms of subjective preference. This is a relatively wide term, since the best subjective quality may depend on the situation, and it could be further specified to be in line with a certain artistic goal.

7.2.2 EVALUATION METHOD

One of the central decisions when one is designing a subjective comparison experiment is what is the ground truth we are trying to reproduce. How should this, if possible, be displayed? Should it be displayed at all? The choice depends not only on the application and what is relevant to the study, but also on physical limitations. In Fig. 7.1, the different steps of an HDR capture and reproduction are outlined, highlighting possible locations suited for comparison. Referring to Fig. 7.1, we distinguish the following evaluation methods:

- *Fidelity with reality* methods, where a tone-mapped image is compared with a physical scene. This type of study is challenging to execute, in particular for video because it involves displaying both a tone-mapped image and the corresponding physical scene in the same experimental setup. Furthermore, the task is very difficult for the observers as the displayed scenes differ from real scenes not only in the dynamic range, but they also lack depth cues and have a restricted field of view and limited color gamut. These factors usually cannot be controlled or eliminated. Moreover, this setup does not capture the actual intent when the content needs enhancement.

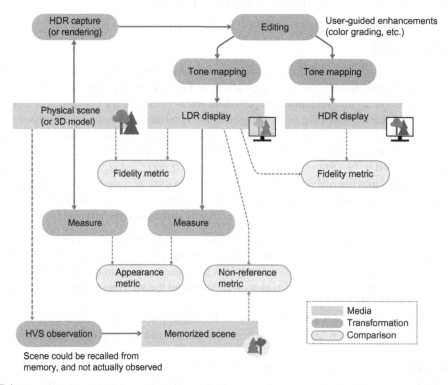

FIGURE 7.1

Evaluation methods in the context of the tone mapping pipeline. The comparisons depicted are made on the basis of subjective metrics and correspond to the evaluation method used. *LDR*, low dynamic range.

Despite these issues, the method has been used in a number of studies (see Table 7.1). It directly tests one of the main intents of tone mapping — namely, its ability to reproduce a convincing picture of a real-world scene as it would be perceived by a human observer.

- *Fidelity with HDR reproduction* methods, where content is matched against a reference shown on an HDR display. Although HDR displays offer a potentially large dynamic range, some form of tone mapping, such as absolute luminance adjustment and clipping, is still required to reproduce the original content. This introduces imperfections in the displayed reference content. For example, an HDR display will not evoke the same sensation of glare in the eye as the actual scene. However, the approach has the advantage that the experiments can be run in a well-controlled environment and, given the reference, the task is easier.

- *Nonreference* methods, where observers are asked to evaluate operators without being shown any reference. In many applications there is no need for fidelity with "perfect" or "reference" reproduction. For example, consumer photography is focused on making images look as good as possible on a device or print alone, as most consumers will rarely judge the images by comparing them with real scenes. Although the method is simple and targets many applications, it carries the risk of running a "beauty contest" (McCann and Rizzi, 2012), where the criteria of evaluation are very subjective. In the nonreference scenario, it is commonly assumed that tone mapping is also responsible for performing color editing and enhancement. But since people differ a lot in their preference for enhancement (Yoshida et al., 2006), such studies lead to very inconsistent results. In some situations though, the purely subjective preference, with no connection to the real scene, is the criterion we want to assess. Many practical situations do not really aim for a result as close to the pictured scene as possible, but aim rather for an aesthetically pleasing outcome given some artistic preferences. In these situations a nonreference method is often needed.
If the factors affecting assessment are well controlled and with a clear description of the evaluation criteria, the method provides a convenient way to test TMO performance against user expectations. Therefore, it has been used in most of the previous studies (see Table 7.1).

- *Appearance match* methods, which compare the appearance of certain attributes in both the original scene and its reproduction (McCann and Rizzi, 2012). For example, the brightness of square patches can be measured in a physical scene and on a display by magnitude estimation methods. Then, the best tone mapping is the one that provides the closest match between the measured perceptual attributes. Even though such direct comparison seems to be a very precise method for evaluation, it poses a number of problems. Firstly, measuring appearance for complex scenes is challenging. While measuring brightness for uniform patches is a tractable task, there is no easy method to measure the appearance of all different image attributes, such as gloss, gradients, textures, and complex materials. Secondly, the match of sparsely measured perceptual attributes does not need to guarantee the overall match of image appearance.

7.2.3 PARAMETER SETTINGS

Most tone mapping algorithms include free parameters that can be tuned to achieve different results. To ensure a fair comparison between different TMOs in an evaluation, it is therefore necessary to perform a thorough adjustment of the parameters. This is, however, often a challenge in itself. The best results are achieved if each algorithm is tweaked independently for each individual input footage to be used in the experiment. However, in this way the operators are tested not as automatic algorithms but rather as editing tools which depend on the skills of the individual performing the adjustment. Therefore, the problem of TMO parameter adjustment is to find a general set of parameters which is optimal for all intended material. Below, we present a summary and categorization of the different adjustment methods used in the evaluation literature:

- Use of *default* parameters is the easiest and most widely used approach. However, even though default parameters are simple to use, they are unlikely to produce the best output for a given TMO; see, for example, (Yoshida et al., 2006; Petit and Mantiuk, 2013; Eilertsen et al., 2013). In many cases, different intents of the tone mapping will benefit from different parameter settings. If an operator is used in a context which differs from the one used during development and tuning, the default parameters may be invalid and may need to be changed. For example, if the parameters have been set to give the best rendering as perceived by the HVS, another calibration may yield a subjectively more preferred result. Finally, another common problem in the literature is that some parameters of certain operators have not been reported with default values.
- *Tuned by experts* could, for example, include parameter adjustment by the experts conducting the experiment or by individual experts in the field. However, this may lead to a bias toward their personal preferences and is not guaranteed to generalize well. Furthermore, since many algorithms are computationally expensive and include a large number of parameters, tweaking by trial and error often means that only a relatively small number of adjustments can be included. That is, only a very sparse sampling of the parameter space is tested, which is unlikely to include the optimal point.
- *Pilot study*, where parameters are adjusted in a separate experiment, designed to be as similar as possible to the following main experiment. While this method has the highest potential, it is also challenging because many TMOs have high-dimensional parameter spaces and may be time consuming to adjust.

  One example of a parameter adjustment experiment was reported by Yoshida et al. (2005). They designed a pilot study where a set of experienced observers were assigned to choose the best image from a selection of tone mappings with different parameter settings. However, similarly to the previous method, this means that a very limited number of possible parameter values were tested, and the optimal set was most likely not included. Another experiment was performed by Eilertsen et al. (2013), where a sparse sampling of the parameter space was also used, but where the values in-between were approximated by interpolation. To find the optimal set of parameters, a set of expert users explored the parameter space using a conjugate gradient method to find a local minimum (Eilertsen et al., 2014). More details on the method are given in Section 7.6.

7.2.4 INPUT MATERIAL

Intuitively, to generalize the results of a subjective study, the selection of the input material used for evaluation should allow a sampling spanning as large a portion of the general population of images as possible. This is to ensure that trends extracted from the evaluation results can be expected to generalize also to other situations. However, if the goal is to find more than general trends — for example, the average performance of a certain operator — the differences between input images make it difficult to draw definite conclusions from the evaluation results. To estimate the expected value of the results, we need to know the individual weights of the images used, drawn from the probability distribution of possible images. For example, if one scene is twice as common as a second scene in the general population of scenes, the results obtained with the first scene should be weighted twice as important in the expected quality value. It is thus difficult to make statistically sound generalizations of the results from an evaluation unless the input footage is selected with care.

The selection of input material is also important to make it possible to distinguish between different operators. For example, a simple TMO in comparison with more advanced algorithms may be able to produce comparable results in basic scenes but fail in more difficult lighting situations (Petit and Mantiuk, 2013).

To categorize, we make a distinction between static and dynamic material:

- *HDR images* are most often produced from a set of differently exposed images (Debevec and Malik, 1997). This imposes limitations on the scene, where, for example, people and other moving objects are difficult to capture, and where common image artifacts such as camera noise are suppressed in the reconstruction.
- *HDR video* has just recently become readily available through the development of HDR video capturing systems (Tocci et al., 2011; Kronander et al., 2013; Froehlich et al., 2014). Previously, HDR video was limited to panning in static HDR panoramas, static scenes with changes in lighting, and computer graphics renderings. The capturing systems provide a set of new challenges, such as complicated transitions in intensity over time, skin colors, and camera noise.

7.2.5 EXPERIMENTAL SETUP

Once the evaluation method and the intent of the TMOs to be evaluated have been specified, another decision is how to present the information to the observers and how they should register their preferences. The decision should be based not only on what information the experiment is supposed to reveal, but also on the human aspects of the design. For example, the difficulty in making the assessments has a direct influence on the accuracy and precision of the result. Also, an experiment that lasts too long for each observer can have implications such as fatigue, which may affect the outcome.

If there is a well-advised and consistent experimental design, different methods for subjective quality assessment are expected to correlate. This has also been demonstrated in different evaluations. Kuang et al. (2007) showed a correlation between pairwise comparisons and rating, but with improved accuracy and precision for pairwise comparisons at the cost of increased experiment length. Čadík et al. (2008) detected no statistical difference between rating and ranking, concluding that for comparison of TMOs, ranking without reference is enough. Mantiuk et al. (2012) showed that four different methods for image quality assessment resulted in the same overall outcome, but differed in precision and experiment length.

Inspecting previous evaluations of TMOs, we find there are three main categories for quality assessment (mentioned in the preceding discussion):

Rating can include assessment of many different attributes at the same time, to more closely measure differences. The measures are either absolute or given relatively as the difference between a pair of displayed images. Since different observers tend to apply different absolute and relative scales, these need to be compensated for. Although the method can provide much information from few displayed conditions, the data tend to have high variance compared with, for example, pairwise comparison. It is a difficult perceptive task to assess the quality level of a certain attribute, compared with selecting one of two images that best fit a certain criterion.

Ranking for TMO evaluation is generally implemented by displaying all TMOs for one scene, and letting the observer sort them according to the evaluation criteria. Although the task is relatively simple, it is difficult to assess all the sequences simultaneously. Practically, the task is most easily performed through a comparison of two images at a time, which in the end is conceptually similar to pairwise comparison.

Pairwise comparison is the perceptually simplest task, where only one of two images should be selected on the basis of closeness to some reference (physical or memory). For N conditions $0.5N(N - 1)$ comparisons are needed to provide all intercondition relations, which would require a lengthy experiment for large N. However, if dynamic algorithms, such as *quicksort* (Silverstein and Farrell, 2001), are used, the number of comparisons can be reduced to about $N \log N$. Also, since the visual task is simple, the comparisons are generally quick, and overall pairwise comparison can be both faster and more precise than rating methods (Mantiuk et al., 2012).

7.3 SURVEY OF TMO EVALUATION STUDIES

There has been a lot of research in quality assessment of TMOs, at least in terms of comparisons on static HDR images. Referring to the categorization in the previous section, Table 7.1 shows a selection of the most relevant studies in the literature, from the first by Drago et al. (2002) to the most recent video TMO evaluation by Eilertsen et al. (2013). From this large body of research, one may be led to believe that it clearly should indicate which particular operators are expected to perform superiorly compared with others. However, as outlined in the previous section, there are many factors affecting the evaluation of TMOs, and one study can lead to a completely different result from a similar study with different experimental conditions. Instead, the information from previous studies has to be viewed in a wider perspective. The evaluations do not provide an unanimous answer to which operator performs best, but rather give insight into their different properties, showing tendencies that could be incorporated in future research and development.

Analyzing the TMO evaluation literature, we find there is a slight tendency to favor global TMOs over local ones. This is the case when we make comparisons in terms of fidelity with both reality (Čadík et al., 2008; Eilertsen et al., 2013) and subjective preference (Akyüz et al., 2007). In fact, in some situations a single exposure directly from the camera, tone-mapped only by the camera response curve, has been shown to outperform sophisticated TMOs (Akyüz et al., 2007). However, in more complicated situations — for example, where important information lies in different intensity regions of the image — a more complex tone curve could be expected to adapt better and render a preferable result (Petit and Mantiuk, 2013). The different results for global and local operators could possibly also be explained by insufficient parameter tuning because global operators are, in general, more robust than local operators. The quality may also be masked by spatial inconsistencies, or artifacts, which are more common when a local tone compression is applied.

In many cases, certain image features have been shown to correlate with the subjective quality, both in perceptual comparisons (Kuang et al., 2007; Čadík et al., 2008) and in comparisons with numerical measures (Delahunt et al., 2005; Yoshida et al., 2006; Petit and Mantiuk, 2013). However, even if

Table 7.1 Categorization of TMO Evaluations

Evaluation Study	Fidelity With Reality	Fidelity With HDR	Nonreference	Rating	Ranking	Pairwise Comparison	Visual System Simulator	Best Subjective Quality	Default	Tuned by Experts	Pilot Study	Image	Video	TMOs	Participants	Scenes
Drago et al. (2002)			×			×	×	×	×[a]			×		7	11	4
Kuang et al. (2004)			×			×		×	×			×		8	33	10
Yoshida et al. (2005)	×		×	×				×			×	×		7	14	2
Delahunt et al. (2005)			×			×		×	×			×		4	20	25
Ledda et al. (2005)		×				×	×		×			×		6	48	23
Ashikhmin and Goyal (2006)	×	×			×		×	×	×	×[b]		×		5	15	4
Yoshida et al. (2006)	×	×	Tuning[c]				×	×		Tuning[c]		×		1[c]	15	25
Kuang et al. (2007)	×	×		×		×	×	×	×			×		7	33	12
Akyüz et al. (2007)			×		×			×	NA			×		3[d]	22	10
Park and Montag (2007)			×			×	×[e]		×			×		9	25	6
Akyüz and Reinhard (2008)	LDR reference[f]					×	×[f]		×			×		7	13	1[f]
Čadík et al. (2008)	×	×		×	×		×				×	×		14	20	3
Villa and Labayrade (2010)	×			×	×		×		NA			×		6	50	5
Kuang et al. (2010)	×	×					×	×	×	×[g]		×		7	23	4
Petit and Mantiuk (2013)			×	×			×	×	×	×[h]			×	4	10	7
Eilertsen et al. (2013)			×	×		×	×				×		×	11	18	5

Many studies involve multiple experiments, using different methods and criteria. For the numbers listed, TMOs refer to the total set of operators considered in the complete study, while participants and scenes are the ones reported for the largest comparative experiment performed in the study.
[a]*TMOs were also matched to have similar overall brightness.*
[b]*Parameters tuned by the TMO authors where default values could not be obtained.*
[c]*Used a generic operator, and the experiment was performed as a tuning of its parameters.*
[d]*Additional to the TMOs, original HDR and single exposures were included.*
[e]*Used a "scientific usefulness" subjective criterion for evaluation.*
[f]*Evaluated the perceived contrast through the Cornsweet-Craik-O'Brien illusion.*
[g]*Two Photoshop TMOs were calibrated by an expert.*
[h]*Five different settings were used, treated as separate operators for evaluation.*
LDR, low dynamic range; NA, not applicable.

single image features can be shown to agree with the overall quality as perceived by an observer, it is difficult to draw any general conclusions about the performance of the evaluated TMO. Hence, the human observer's assessments will likely always be the most reliable method of measuring the quality of TMOs.

Many of the factors defining a subjective study have been shown to correlate strongly. When it comes to the *evaluation method*, some studies demonstrate approximately similar results with reference and nonreference methods (Kuang et al., 2007; Čadík et al., 2008), and using an HDR display and a real-world scene (Kuang et al., 2010). However, Ashikhmin and Goyal (2006) did not find such a relationship, but showed that when no reference was given, there was no evidence of a difference when comparison was made regarding the preferred and most real result (best subjective quality and visual system simulator *intent of TMO*, respectively), which was further supported in Petit and Mantiuk (2013). For different *experimental setups* a correlation in the results is highly expected, which is also demonstrated in TMO evaluations (Kuang et al., 2007; Čadík et al., 2008) as well as in metastudies (Mantiuk et al., 2012).

Despite the many demonstrated correlations, a comparative study is still very sensitive to the choice of the particular experimental setup. It may lead to important differences in the final outcome, where correlations demonstrated in certain situations are not enough to draw definite conclusions. Also, the choice is related not only to the averaged results of a study, but also to important differences and implications for the physical setup and variations in the results. Furthermore, there is not yet any support for correlations in the choice of *input material* (images or video), and the *parameter settings* still remain a problematic and highly influential aspect (Petit and Mantiuk, 2013; Eilertsen et al., 2013). The challenges involved in an evaluation are also reflected in the diverse set of results shown in previous work — it is all about how the experiment is constructed. This further illustrates the importance of clear definitions and restrictions as described in the categorization in Section 7.2.

7.4 EVALUATION STUDIES FOR VIDEO TMOS

Although the area of tone mapping evaluation is quite extensively researched, there are very few studies using HDR video material for the comparisons (see the input material in Table 7.1). This can be explained by the limited availability of HDR video material: HDR video capturing systems first appeared a few years ago (Tocci et al., 2011; Kronander et al., 2013; Froehlich et al., 2014). Now, the question is, how do we evaluate tone mapping in this context using video material? Are there any differences as opposed to making comparisons with single HDR frames in the evaluation of different methods? It would be easiest to infer the results from the many already existing evaluations on static images. However, even though the outcome of such evaluations can be estimated to be approximately the same for many images, it is very unlikely to generalize to video material. This is due to both, differences in the actual tone reproduction with changes in the input signal and the altered subjective impression of material that is moving.

Nontrivial TMOs always rely, at least to some extent, on the input signal to form a tone curve. That is, the mapping f of a pixel value $I_{x,y}$ in the image I is formulated as $f\left(I_{x,y}, I\right)$. The signal dependence could, for example, include the image histogram, the image mean or maximum value, or other statistical measures. Since these statistics can change rapidly from frame to frame in an image sequence, the outcome of the mapping could potentially cause flickering and other temporal incoherences. To temporally stabilize the tone reproduction, a model for adaptation over time is inevitably needed in most situations. This also makes it pointless to compare TMOs for static images without temporal adaptation. The artifacts caused by temporal inconsistencies will likely be the by far most prominent feature, making an evaluation result difficult to interpret.

Another aspect of temporal tone mapping is the visibility/impact of certain image features. Spatial artifacts can, for example, become visually more prominent in the case of extension to the time domain (a temporally inconsistent behavior is often perceptually very easy to perceive). For example, local tone mapping, where the tone curve varies over the image, often gives rise to a spatially varying tone reproduction which is inconsistent with the original HDR input (eg, around edges in the image). Although these artifacts may not be noticeable in a static image, their dependence on the local image statistics can cause fluctuations over time that are visually disturbing.

In comparison with evaluations of TMOs for images, the inclusion of the time domain leads to a significant increase in the information to be assessed in the subjective judgments. The way a human observer perceives a signal that changes over time compared with a static signal is different, in terms of both how the HVS is stimulated and how the higher-level assessments are performed. It is evident that evaluations of TMOs for video have requirements different from those of TMOs for images and that a well-defined experimental setup is even more important (Section 7.2).

7.5 **VIDEO TMO EVALUATION STUDY I**

One of the first evaluations to actually use HDR video sequences for the comparison of tone mapping algorithms was performed by Petit and Mantiuk (2013). This study was not targeted at just providing a ranked list of operators, sorted according to their estimated quality in reproduction. The aim was rather to reveal some of the differences in using a simple S-shaped camera tone curve compared with more sophisticated TMOs. Furthermore, as two different criteria for evaluation were used, and different parameter settings were included, the study also aimed to show what differences this makes.

One motivation for the evaluation was the increasing complexity of the algorithms used for tone mapping. Is this added complexity really necessary in the commonest situations? Most consumer cameras already provide simple tone manipulation, built in to the camera hardware as a postprocessing step. This is usually in the form of a S-shaped mapping function that globally compresses the dynamic range of the input. If such methods were used in the same way in HDR capturing systems, what differences could be expected in terms of subjective quality compared with the more sophisticated solutions?

With reference to the preceding evaluation categorization (see Section 7.2), the setup of the evaluation study was as follows:

- The study was carried out to compare tone mappings with a *nonreference* method, motivated by the difficulties in setup and the fact that the impact of a reference is disputable (Čadík et al., 2008).
- The evaluation used a *best subjective quality* criterion as well as a *visual system simulator* criterion.
- Five different settings of each TMO were used for the experiments, and they were treated separately as different TMOs.
- The operators were compared with use of *HDR video* material, created from panning in static HDR panoramas, and computer-generated scenes.
- For the assessment a *rating* method was used, mainly since pairwise comparison would demand a lengthy procedure owing to the different parameter settings.

As mentioned previously, three of the central questions of the evaluation were (1) to determine if there is a difference between tone mapping algorithms that simulate the HVS and those that try to achieve the subjectively most preferred result, (2) to investigate the differences between default parameters and other possible settings, and (3) to see if sophisticated tone mapping algorithms show measurable improvement over an S-shaped camera tone curve. The study in terms of these questions is described in further detail in the following sections.

7.5.1 TMO INTENT

The evaluation results for different TMOs could potentially depend on what particular criterion is used for the image assessments. To investigate this possibility, and to give a hint regarding the general difference in making comparisons with different criteria, the evaluation used both a preference criterion (*best subjective quality*) and a fidelity with real-world experience criterion (*visual system simulators*). Such comparisons of evaluation criteria have previously been shown to yield approximately the same results when no reference is used (Ashikhmin and Goyal, 2006).

In the end, the experiment showed no statistical difference between the two criteria. The results suggest either that there are no major differences in preferred reproduction and reproduction closest to real life or that the differences are small with no reference provided. That is, without a physical reference for direct comparison, the experienced similarity with the real world is biased toward subjective preferences. Indeed, the memorized image looks nothing like the retinal image (Stone, 2012), and a reference comparison could potentially yield a different result (Ashikhmin and Goyal, 2006).

7.5.2 PARAMETER SETTINGS

If a comparison should reveal differences in subjective quality — for example, between the camera curve and more sophisticated TMOs — another possibility is that these could result from the particular parameter settings of the operators. In an attempt to cover this situation, five different settings of each TMO were provided for the experiments. That is, these settings were treated separately, as different TMOs. Although this subset of different parameters is very small compared with the complete set of possible parameters, it provided a way to distinguish if the default parameters always perform best. The parameters for the different settings were tuned to produce acceptable and visually different results. In the context of sampling the parameter space, this can be explained as an attempt to provide a uniform sampling of the subspace spanned by acceptable values of the altered parameters in a perceptually linear frame of reference.

The outcome showed that fine-tuning the parameters could potentially result in improved quality. Although most cases did not show statistically significant differences in the result with different settings, in some cases it may actually produce an unacceptable or exceptionally good result.

7.5.3 S-SHAPED CURVE PERFORMANCE

In the evaluation, four different TMOs were compared, from Irawan et al. (2005), Kuang et al. (2007), and Mantiuk et al. (2008), as well as a S-shaped camera response curve, similar to those found in consumer-grade cameras.

In the end, there were mostly small differences in performance between the different operators, and no statistical difference between the three best performing ones. This can, at least to some extent, be explained by the altered parameters, where default parameters could have caused some operators to perform significantly worse. Thus, an important conclusion was that one needs to thoroughly analyze the operator parameters when initiating a comparative study. Furthermore, the results obtained from investigation of the different TMOs and parameter settings with use of certain isolated image attributes, also showed a correlation between the quality of the tone reproduction and overall brightness, as well as chroma.

For the camera curve, there were only statistical differences in quality when the scene was more complicated — for example, with higher dynamic range and different properties for different parts of the scene. The camera curve is not as flexible for extension to these situations, where important parts of the image differ significantly in luminance. This highlights the need to carefully select the evaluation input material so the performance of sophisticated algorithms can be distinguished and advantages can be detected.

Although the study revealed a substantial amount of interesting and relevant information, there was no particular emphasis on the differences in tone mapping of static images and video material. Another limitation was that in the material used, some important aspects of video were not captured. Also, the study used a low-sensitivity direct rating method, so the actual differences captured were difficult to prove.

7.6 **VIDEO TMO EVALUATION STUDY II**

The most comprehensive evaluation to date, targeted specifically at HDR video tone mapping, was performed by Eilertsen et al. (2013). The study involved 11 TMOs, all with explicit treatment of the tone mapping over time. This was the first time that material from actual HDR video capturing systems was used in an evaluation, revealing some of the most important problems a video TMO needs to address.

With reference to the categorization in Section 7.2, the setup of the evaluation study was as follows:

- The study applied a *nonreference* method, motivated by the fact that most applications will require the best match to a memorized scene rather than a particular reference.
- The evaluation was targeted at *visual system simulators*, which is one of the commonest types of tone mapping algorithms.
- Parameters were carefully studied, and if they were not defined with default values or were subject to substantial improvement, they were set in a *pilot study*.
- HDR video material was provided from a multisensor HDR camera setup (Kronander et al., 2013), as well as from an RED EPIC camera set to HDR-X mode, and a computer-generated sequence.
- First, a qualitative *rating* experiment was performed, followed by a *pairwise comparison* study.

The survey and evaluation were done as follows: Appropriate TMOs were selected and a parameter calibration experiment was performed. The actual evaluation included a qualitative rating experiment and a pairwise comparison. Finally, from the result a set of important conclusions were drawn. These steps are described in further detail in the following sections.

7.6.1 **TMO SELECTION**

Even though the evaluation was targeted at *visual system simulators*, operators with other intents were also included, since they potentially could produce results matching the performance of operators from the targeted class. Eleven operators were considered:

Visual adaptation TMO (Ferwerda et al., 1996), a global *visual system simulator* operator that uses data from psychophysical experiments to simulate adaptation over time, and effects such as color appearance and visual acuity. The visual response model is based on measurements of threshold visibility as in Ward (1994).

Time-adaptation TMO (Pattanaik et al., 2000), a global *visual system simulator* based on published psychophysical measurements (Hunt, 1995). Static responses are modeled separately for cones and rods, and are complemented with exponential smoothing filters to simulate adaptation in the time domain. A simple appearance model is also included.

Local adaptation TMO (Ledda et al., 2004), a local *visual system simulator*, with a temporal adaptation model based on experimental data. The operator operates on a local level using a bilateral filter.

Maladaptation TMO (Irawan et al., 2005), a global *visual system simulator* based on the work of Ward Larson et al. (1997) for tone mapping and Pattanaik et al. (2000) for adaptation over time. It also extends the threshold visibility concepts to include maladaptation.

Virtual exposures TMO (Bennett and McMillan, 2005), a local *best subjective quality* operator, applying a bilateral filter both spatially for local tone manipulation and separately in the time domain for temporal coherence and noise reduction.

Cone model TMO (Van Hateren, 2006), a global *visual system simulator*, using a dynamic model of cones in the retina. A model of primate cones is used, based on actual retina measurements.

Display adaptive TMO (Mantiuk et al., 2008), a global *scene reproduction* operator, performing a display adaptive tone mapping, where the goal is to preserve the contrasts within the input (HDR) as well as possible given the characteristics of an output display. Temporal variations are handled through a low-pass filtering of the tone curves over time.

Retina model TMO (Benoit et al., 2009), a local *visual system simulator* with a biological retina model, where the time domain is used in a spatiotemporal filtering for local adaptation levels. The spatiotemporal filtering, simulating cellular interactions, yields an output with whitened spectra and temporally smoothed for improved temporal stability and for noise reduction.

Color appearance TMO (Reinhard et al., 2012), a local *visual system simulator* operator, performing a display and environment adapted image appearance calibration, with localized calculations through the median cut algorithm.

Temporal coherence TMO (Boitard et al., 2012), a local *scene reproduction* operator, with a postprocessing algorithm to ensure temporal stability for static TMOs applied to video sequences. Boitard et al. (2012) used mainly the photographic tone reproduction of Reinhard et al. (2002), for which the algorithm is most developed. Therefore, the version used in the survey also used the photographic operator.

Camera TMO, a global *best subjective quality* operator, representing the S-shaped tone curve which is used by most consumer-grade cameras to map the sensor-captured values to the color gamut of a storage format. The curves applied were measured for a Canon 500D DSLR camera, with measurements conducted for each channel separately. To achieve temporal coherence, the exposure settings were anchored to the mean luminance filtered over time with an exponential filter.

7.6.2 **PARAMETER SELECTION EXPERIMENT**

Finding the optimal calibration of an operator, measured by some subjective preference, is an inherently difficult problem because of the many possible parameter combinations to be tested in an experimental setup. To overcome the problem, without resorting to testing only a few selected points of the parameter space, a new method for perceptual optimization (Eilertsen et al., 2014) was used to tune the parameters of each TMO included in the evaluation. The key idea is that the characteristics of an operator can be accurately described by interpolation between a small set of precomputed parameter settings. This allows the user to interactively explore the interpolated parameter space to find the optimal point in a robust way. The method is particularly efficient when dealing with video sequences; since the result of a parameter change needs to be assessed over time, a large number of frames have to be regenerated when a parameter value is changed.

Direct interpolation between a sparse set of points will in many cases generate unacceptable errors due to nonlinear behavior. To prevent this, a linearization of the parameter space was performed, for each operator, before sampling and interpolation. The linearization was formulated by analysis of the image changes $E(p)$ when a certain parameter p was altered. By quantification of the changes in this way, over the range of parameter values, and the requirement that $E(p)$ be constant over all p, the normalized inverse $\hat{f}^{-1}(p)$ of the cumulative sum $f(p)$ of $E(p)$ could subsequently be used as a transformation to a space of linear image changes. Experiments showed that the absolute difference averaged over the image pixels worked well in most situations as a simple measure of image changes. That is, $E(p)$ and $f(p)$ are defined as follows:

$$E(p) = \frac{1}{N} \sum \left| \frac{d\mathbf{I}_p}{dp} \right|, \tag{7.1a}$$

$$f(p) = \int_{p_{\min}}^{p} E(r)\, dr, \tag{7.1b}$$

where $\mathbf{I}_p$ is the input image $\mathbf{I}$ processed with an operator at the parameter value p, and the summation goes over all N pixels in the image. An example linearization, achieved with a simple gamma correction operation for demonstration, is shown in Fig. 7.2. The plots in Fig. 7.2A and B correspond to Eqs. (7.1a) and (7.1b), respectively. Finally, the plot in Fig. 7.2C is of the inverted and normalized cumulative sum $\hat{f}^{-1}(p)$ that is used for transformation to a linear parameter space.

In Fig. 7.3, the linearization of the gamma correction operation is demonstrated on an image. With the transformation it is possible to use only three sample points to compute an interpolated approximation of the parameter changes, with small interpolation errors. Not only does the linearization reduce interpolation errors, but it also makes adjustments easier and more intuitive as it provides a parameter that varies much more closely to what is perceived as linear.

Although the interpolation strategy enables interactive exploration of the parameters of an operator, it is still a difficult problem to actually find the perceptual optimum in a high-dimensional space. The actual search was therefore formulated as a general optimization problem where the objective function was determined by the user's image assessments. The user was given the task to search the parameter space in one direction at a time, choosing directions according to Powell's conjugate gradient method

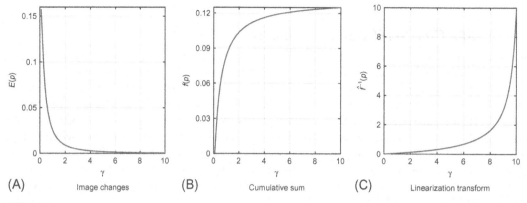

FIGURE 7.2

The mean absolute difference is measured over parameter changes (A), and the cumulative sum (B) is used to create a linearization transform (C). The example uses a simple gamma mapping value as a parameter, $I_p = I^{1/p}$.

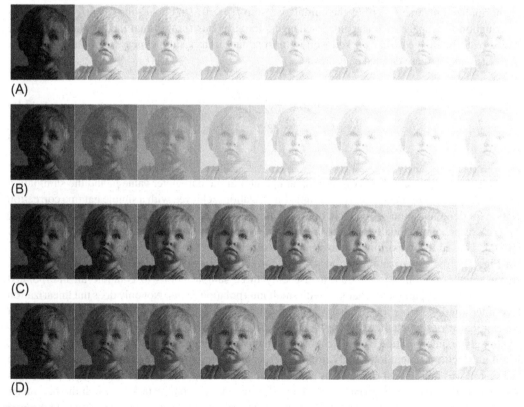

FIGURE 7.3

Example linearization and interpolation, using a simple gamma mapping value as a parameter, $I_p = I^{1/p}$. Without linearization, the interpolation show large errors (B) as compared with interpolation after the linearization has been done (D). (A) Linear parameter changes. (B) Interpolation from three sample points. (C) Linearized changes. (D) Interpolation from three linearized sample points.

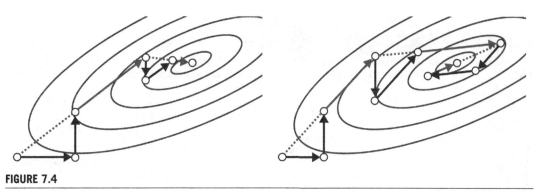

FIGURE 7.4

Left: Example of a search in a two-dimensional space. Right: If errors are introduced in the search, the minimum can still be found with additional iterations.

(Powell, 1964). The method allows one to find a minimum of a nondifferentiable objective function (in this case the user's assessments) in a high-dimensional parameter space. An example of a search in a two-dimensional parameter space using conjugate gradients is illustrated in Fig. 7.4. The iterative nature of the approach makes it robust against errors. If inaccuracies are introduced by the user, the solution will converge more slowly. However, if additional iterations are performed, this will ensure convergence towards a perceptual minimum.

Even though this method provides a robust way for finding a *locally* optimal set of parameters, the *global* perceptual minimum is not guaranteed. Also, performing the calibration procedure for all operators and parameters would mean an expanded pilot study, which would take too long to perform in one go for each observer. To provide the best compromise between default parameters and calibrated ones for the evaluation, they were carefully studied. The calibration experiment was then performed on the parameters where default values either did not provide a reasonable result or were not defined. For each parameter, the linearization function (Eq. 7.1) was evaluated over a range of different sequences, and was then used as a general linearization across all sequences selected for the calibration experiment. The evaluation involved four expert observers, and the final parameters used were calculated as their averaged result.

7.6.3 QUALITATIVE EVALUATION EXPERIMENT

Initial observations on the behavior of the considered TMOs, when exposed to sharp intensity transitions over time, revealed that some have serious temporal coherency problems. In many cases, severe flickering or ghosting artifacts were visible. This is illustrated in Fig. 7.5, where the tone mapping responses of four operators are plotted over time. The outputs are fetched from two different pixel locations; the first (left) is a location where the input is relatively constant over time, and the second (right) is a location with a rapid temporal transition. From the plots, it is clear that the tone-mapped values contain flickering or ghosting artifacts.

From an evaluation standpoint it is problematic to assess the overall quality of a TMO exhibiting severe temporal artifacts. To quantify the different artifacts and tabulate the strengths and weaknesses of each operator included in the evaluation, a qualitative analysis was performed before the actual comparison experiment. This was done as a rating experiment, where five expert observers were

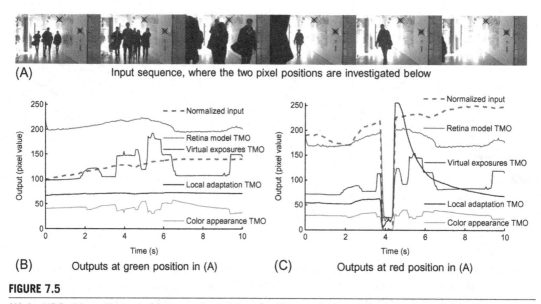

FIGURE 7.5

(A) An HDR video with a set of frames, where the two points indicated are analyzed. (B) and (C) The tone-mapped output of four operators at the positions, plotted over time. (For interpretation of the references to color in this figure legend, the reader is referred to the web version of this article.)

assigned the task of rating different image attributes and artifacts. To measure appearance reproduction, *overall brightness*, *overall contrast*, and *overall color saturation* were included in these ratings. Furthermore, to assess the presence of artifacts, additional measurements were defined: *temporal color consistency* (objects should retain the same hue, chroma, and brightness), *temporal flickering*, *ghosting*, and *excessive noise*.

Fig. 7.6 shows the result from the qualitative evaluation for four TMOs. The figure is divided into ratings of (A) artifacts and (B) color-rendition attributes. From the experiment it was clear that the most salient artifacts were flickering and ghosting. While some inconsistencies in colors and visible noise were accepted as tolerable, even minor flickering or ghosting was deemed unacceptable. It was therefore decided that all operators where either of these artifacts were visible in at least three scenes should be eliminated from further analysis. In the end, four operators were removed with this criterion. They are all shown in Fig. 7.6, and from the plots it is clear that all four TMOs show problems with respect to either flickering or ghosting. This type of analysis was not only helpful in removing operators not suitable for comparison, but also provides insight into many of the challenges involved in tone mapping.

7.6.4 PAIRWISE COMPARISON EXPERIMENT

One of the goals of the evaluation was to reveal which operator could be expected to generate, on average, the best result. In this context, *best* means the tone reproduction closest to what is perceived by the human eye, since the evaluation was performed with visual system simulators (or other TMOs that potentially could deliver a convincing result in line with this criterion).

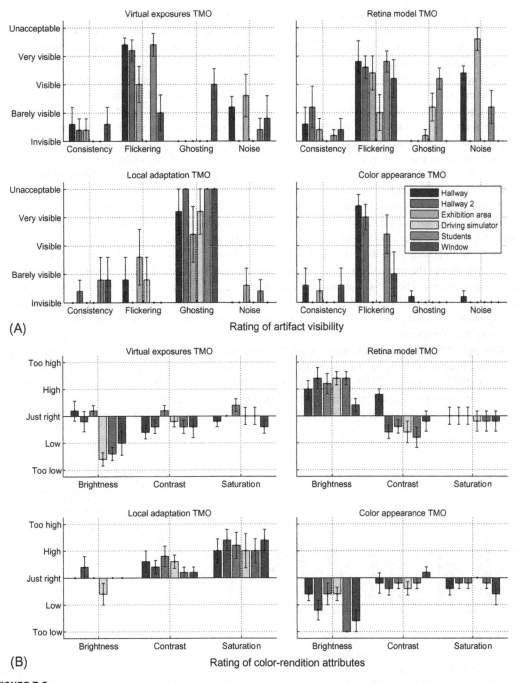

FIGURE 7.6

Example ratings of artifacts (top) and color-rendition problems (bottom). The bars for each attribute represent different sequences.

The evaluation was performed as a direct comparison experiment with the seven operators remaining after the qualitative rating experiment. For the experiment, tone-mapped sequences were shown sequentially in random order, and in each trial the participants were shown two videos of the same HDR scene tone-mapped with two different operators. The observers were then asked to choose the sequence which to them looked closest to how they imagined the real scene would look. In total 18 observers completed the experiment, each evaluating the seven TMOs using five different HDR video sequences. Assuming all pairs are compared, this means $5 \times \frac{1}{2} \times 7 \times 6 = 105$ comparisons, which is time-consuming and potentially exhausting for the observer. To avoid fatigue, the number of comparisons was reduced with the *quicksort* algorithm (Silverstein and Farrell, 2001), adapting the comparisons to be made on the basis of previous comparisons. The algorithm puts more effort into distinguishing closer samples, and with a sorting procedure a complete relative quality estimate for N conditions is provided with about $N \log N$ comparisons.

The outcome of the pairwise comparison experiment is shown in Fig. 7.7. The results are plotted per sequence, and are scaled in units of just noticeable differences (JND). When 75% of the observers select one condition over another, the quality difference between them is defined to be 1 JND. To scale the results in JND units, the Bayesian method of Silverstein and Farrell (2001) was used. In brief, the method maximizes the probability that the collected data explain the experiment under the Thurstone case V assumptions. The optimization procedure finds a quality value for each image that maximizes the probability, which is modeled by the binomial distribution. Unlike standard scaling procedures, the Bayesian approach is robust to unanimous answers, which are common when a large number of conditions are compared.

Although the differences for some sequences are small, and with confidence intervals that do not always indicate that there are any significant differences, two operators seem to be consistently preferred: *display adaptive TMO* and *maladaptation TMO*. Also, the simple camera curve — *camera TMO* — shows good relative performance, and actually came out top for one of the sequences. Another observable pattern is the tendency to reject the *cone model TMO* and the *time-adaptation TMO*.

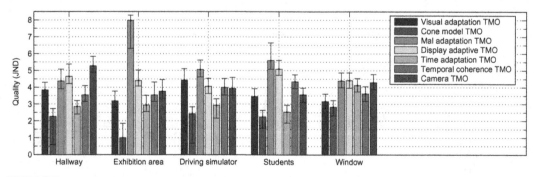

FIGURE 7.7

Results of the pairwise comparison experiment scaled in JND units (the higher, the better) under Thurstone case V assumptions, where 1 JND corresponds to 75% discrimination threshold. Note that absolute JND values are arbitrary and only relative differences are meaningful. The error bars denote 95% confidence intervals computed by bootstrapping.

The scores should not be considered to infer an absolute ranking of the individual TMOs. There are uncertainties in the scores, and the results are not guaranteed to generalize to the entire population of possible sequences. However, they provide an indication of how the operators can be expected to perform in a variety of different situations.

7.6.5 CONCLUSION

One important outcome of the evaluation was due to the use of HDR video camera generated material, with/which contain complicated transitions in intensity both spatially and, most importantly, over time. There are substantial differences in camera-captured HDR video and artificially created material, such as computer-generated scenes, panning in HDR panoramas, and static scenes with changing illumination. This provided genuinely challenging conditions under which the operators under consideration had not yet been, and many TMOs showed artifacts in at least some of the situations.

As a consequence, there was a tendency in the pairwise comparisons to favor the more straight-forward, *global*, tone reproductions. In general, the complicated, *local*, transformations appeared to be more sensitive and prone to flickering and ghosting artifacts. Since these artifacts were established as perceptually most salient, they often give rise to mappings that are unsatisfactory or even unacceptable.

However, the results should not be interpreted as a guideline in favor of global processing as the method of choice for tone transformation. For example, a global strategy is unable to capture important local transitions that need to be preserved in order to maintain an overall level of local contrast corresponding to the original HDR input. Rather, the results show that there still are issues to be addressed in order to achieve high-quality tone reproductions for general HDR video material. To aid in future development of such work, the survey and evaluation concludes with a set of guidelines, pointing out properties a video TMO should possess:

- Temporal model free from artifacts such as flickering, ghosting, and disturbing (too noticeable) temporal color changes.
- Local processing to achieve sufficient dynamic range compression in all circumstances while maintaining a good level of detail and contrast.
- Efficient algorithms, since a large amount of data needs processing, and turnaround times should be kept as short as possible.
- No or very limited requirement of parameter tuning.
- Capability of generating high-quality results for a wide range of video inputs with highly different characteristics.
- Explicit treatment of noise and color.

7.7 SUMMARY

Throughout this chapter, we have emphasized the importance of well-defined and clearly motivated subjective studies. In this context, it is important to appreciate that the outcome of an evaluation, given a certain setup, is only an assessment of the performance of a TMO under the particular conditions chosen. Statistical testing ensures that the results generalize to other observers, but not necessarily to other sequences, intents, etc. If the conditions are chosen with care, however, the outcome can be estimated to generalize reasonably well to an extended selection of conditions. But, all in all, it is not possible to cover all possible scenarios and intents. This is also shown in the diverse set of outcomes from the evaluation studies found in the literature, where the different experimental designs give significantly different results.

During the two last decades we have seen extensive research in the area of tone mapping, and many new TMOs have been developed. However, with the help of thorough evaluations it has been demonstrated that tone mapping is still an open problem and that there are still many problems to be addressed, especially for HDR video.

REFERENCES

Akyüz, A.O., Reinhard, E., 2008. Perceptual evaluation of tone-reproduction operators using the Cornsweet-Craik-O'Brien illusion. ACM Trans. Appl. Percept. 4 (4), 1:1–1:29.

Akyüz, A.O., Fleming, R., Riecke, B.E., Reinhard, E., Bulthoff, H.H., 2007. Do HDR displays support LDR content? A psychophysical evaluation. ACM Trans. Graph. 26 (3).

Ashikhmin, M., Goyal, J., 2006. A reality check for tone mapping operators. ACM Trans. Appl. Percept. 3 (4).

Bennett, E.P., McMillan, L., 2005. Video enhancement using per-pixel virtual exposures. ACM Trans. Graph. 24 (3).

Benoit, A., Alleysson, D., Herault, J., Callet, P., 2009. Spatio-temporal tone mapping operator based on a retina model. In: Computational Color Imaging. Springer-Verlag, Berlin/Heidelberg.

Boitard, R., Bouatouch, K., Cozot, R., Thoreau, D., Gruson, A., 2012. Temporal coherency for video tone mapping. In: Proc. of SPIE 8499, Applications of Digital Image Processing XXXV.

Čadík, M., Wimmer, M., Neumann, L., Artusi, A., 2008. Evaluation of HDR tone mapping methods using essential perceptual attributes. Comput. Graph. 32 (3).

Debevec, P.E., Malik, J., 1997. Recovering high dynamic range radiance maps from photographs. In: Proceedings of the 24th Annual Conference on Computer Graphics and Interactive Techniques, SIGGRAPH '97. ACM Press/Addison-Wesley Publishing Co., New York, NY, pp. 369–378.

Delahunt, P.B., Zhang, X., Brainard, D.H., 2005. Perceptual image quality: effects of tone characteristics. J. Electron. Imaging 14 (2).

Drago, F., Martens, W., Myszkowski, K., Seidel, H.P., 2002. Perceptual evaluation of tone mapping operators with regard to similarity and preference. Research Report MPI-I-2002-4-002, Max-Planck-Institut für Informatik, Stuhlsatzenhausweg, Germany.

Eilertsen, G., Unger, J., Wanat, R., Mantiuk, R., 2014. Perceptually based parameter adjustments for video processing operations. In: ACM SIGGRAPH 2014 Talks, SIGGRAPH '14. ACM, New York, NY, pp. 74:1–74:1.

Eilertsen, G., Wanat, R., Mantiuk, R.K., Unger, J., 2013. Evaluation of tone mapping operators for HDR-video. Comput. Graph. Forum 32 (7), 275–284.

Ferwerda, J.A., Pattanaik, S.N., Shirley, P., Greenberg, D.P., 1996. A model of visual adaptation for realistic image synthesis. In: Proc. of SIGGRAPH '96. ACM, New York, NY.

Froehlich, J., Grandinetti, S., Eberhardt, B., Walter, S., Schilling, A., Brendel, H., 2014. Creating cinematic wide gamut HDR-video for the evaluation of tone mapping operators and HDR-displays. In: Proceedings of SPIE Electronic Imaging.

Hunt, R.W.G., 1995. The Reproduction of Colour. Fountain Press, Tolworth, England.

Irawan, P., Ferwerda, J.A., Marschner, S.R., 2005. Perceptually based tone mapping of high dynamic range image streams. In: Proc. of Eurographics Symposium on Rendering. Eurographics Association, Aire-la-Ville.

Kronander, J., Gustavson, S., Bonnet, G., Unger, J., 2013. Unified HDR reconstruction from raw CFA data. In: Proceedings of the IEEE International Conference on Computational Photography.

Kuang, J., Heckaman, R., Fairchild, M.D., 2010. Evaluation of HDR tone-mapping algorithms using a high-dynamic-range display to emulate real scenes. J. Soc. Inform. Disp. 18 (7).

Kuang, J., Yamaguchi, H., Johnson, G.M., Fairchild, M.D., 2004. Testing HDR image rendering algorithms. In: Proc. IS&T/SID 12th Color Imaging Conference, Scotsdale, Arizona, pp. 315–320.

Kuang, J., Yamaguchi, H., Liu, C., Johnson, G.M., Fairchild, M.D., 2007. Evaluating HDR rendering algorithms. ACM Trans. Appl. Percept. 4 (2).

Ledda, P., Chalmers, A., Troscianko, T., Seetzen, H., 2005. Evaluation of tone mapping operators using a high dynamic range display. ACM Trans. Graph. 24 (3).

Ledda, P., Santos, L.P., Chalmers, A., 2004. A local model of eye adaptation for high dynamic range images. In: Proc. of AFRIGRAPH '04. ACM, New York, NY.

Mantiuk, R., Daly, S., Kerofsky, L., 2008. Display adaptive tone mapping. ACM Trans. Graph. 27 (3).

Mantiuk, R., Kim, K.J., Rempel, A.G., Heidrich, W., 2011. HDR-VDP-2: A calibrated visual metric for visibility and quality predictions in all luminance conditions. ACM Trans. Graph. 30 (4), 40:1–40:14.

Mantiuk, R.K., Tomaszewska, A., Mantiuk, R., 2012. Comparison of four subjective methods for image quality assessment. Comput. Graph. Forum 31 (8).

McCann, J.J., Rizzi, A., 2012. The Art and Science of HDR Imaging. Wiley, Chichester, West Sussex, UK.

Park, S.H., Montag, E.D., 2007. Evaluating tone mapping algorithms for rendering non-pictorial (scientific) high-dynamic-range images. J. Vis. Commun. Image Represent. 18 (5), 415–428.

Pattanaik, S.N., Tumblin, J., Yee, H., Greenberg, D.P., 2000. Time-dependent visual adaptation for fast realistic image display. In: Proc. of SIGGRAPH '00. ACM Press/Addison-Wesley Publishing, New York, NY.

Petit, J., Mantiuk, R.K., 2013. Assessment of video tone-mapping: are cameras' S-shaped tone-curves good enough? J. Vis. Commun. Image Represent. 24.

Powell, M.J.D., 1964. An efficient method for finding the minimum of a function of several variables without calculating derivatives. Comput. J. 7 (2).

Reinhard, E., Pouli, T., Kunkel, T., Long, B., Ballestad, A., Damberg, G., 2012. Calibrated image appearance reproduction. ACM Trans. Graph. 31 (6).

Reinhard, E., Stark, M., Shirley, P., Ferwerda, J., 2002. Photographic tone reproduction for digital images. In: Proc. of SIGGRAPH '02. ACM, New York, NY.

Silverstein, D., Farrell, J., 2001. Efficient method for paired comparison. J. Electron. Imaging 10.

Stone, J.V., 2012. Vision and brain: how we perceive the world. MIT Press, Cambridge, MA.

Tocci, M.D., Kiser, C., Tocci, N., Sen, P., 2011. A versatile HDR video production system. ACM Trans. Graph. 30 (4).

Van Hateren, J.H., 2006. Encoding of high dynamic range video with a model of human cones. ACM Trans. Graph. 25.

Villa, C., Labayrade, R., 2010. Psychovisual assessment of tone-mapping operators for global appearance and colour reproduction. In: Proc. of Colour in Graphics Imaging and Vision 2010, Joensuu, Finland.

Ward, G., 1994. A contrast-based scalefactor for luminance display. In: Heckbert, P.S. (Ed.), Graphics Gems IV. Academic Press Professional, Inc., San Diego, USA.

Ward Larson, G., Rushmeier, H., Piatko, C., 1997. A visibility matching tone reproduction operator for high dynamic range scenes. In: ACM SIGGRAPH 97 Visual Proceedings: The art and Interdisciplinary Programs of SIGGRAPH '97, SIGGRAPH '97. ACM, New York, NY.

Yeganeh, H., Wang, Z., 2013. Objective quality assessment of tone-mapped images. IEEE Trans. Image Process. 22 (2), 657–667.

Yoshida, A., Blanz, V., Myszkowski, K., Seidel, H.P., 2005. Perceptual evaluation of tone mapping operators with real world scenes. In: Proc. of SPIE Human Vision and Electronic Imaging X, San Jose, CA, vol. 5666.

Yoshida, A., Mantiuk, R., Myszkowski, K., Seidel, H.P., 2006. Analysis of reproducing real-world appearance on displays of varying dynamic range. Comput. Graph. Forum 25 (3).

USING SIMULATED VISUAL ILLUSIONS AND PERCEPTUAL ANOMALIES TO CONVEY DYNAMIC RANGE

T. Ritschel[*,†]

*MPI Informatik, Saarbrücken, Germany**

Saarland University, Saarbrücken, Germany[†]

CHAPTER OUTLINE

High Dynamic Range Video. http://dx.doi.org/10.1016/B978-0-08-100412-8.00008-5

8.1 INTRODUCTION

Here we show that introducing visual anomalies can actually be useful to convey dynamic range. Commonly, such anomalies are considered a distracting artifact. In contrast, this chapter argues that they can be used to improve visual realism, in particular high dynamic range, if we are able to predict them by simulation and include them in imagery in an appropriate way.

8.1.1 PRINCIPLE

The principle is always the same and has three main steps: First, if a stimulus is processed by the human visual system (HVS), the result is subject to certain anomalies. Importantly, those anomalies often introduce certain nonlinearities, resulting in a qualitative change, instead of only a quantitative one. This is due to the limits of optics and retinal and cortical processing. Second, because of those nonlinearities, a low dynamic range (LDR) stimulus might not trigger the same anomalies as a high dynamic range (HDR) stimulus or might trigger no anomalies at all. Third and consequently, we introduce the anomalies to the end of inverting cause and effect: a stimulus including anomalies associated with HDR also triggers the sensation of HDR.

8.1.2 OUTLINE

The next four sections introduce example anomalies in detail. We focus on three anomalies: haloes, glare, and afterimages. These are border illusions such as the Cornsweet illusion used in three-dimensional unsharp masking (Ritschel et al., 2008), temporal glare caused by diffraction in the dynamic human eye (Ritschel et al., 2009), and afterimages (Ritschel and Eisemann, 2012), which are due to retinal processing.

8.1.3 ARTISTIC PRACTICE

Further evidence that this concept works is given by works of art such as paintings, where most anomalies discussed in this chapter are used in one way or another.

Use of border illusions as in three-dimensional unsharp masking can be observed, where contrasting shades are juxtaposed to exaggerate local lighting discontinuities. For instance, Salvador Dali extended the apparent dynamic range beyond what is physically possible by countershading along the shadow edges. As in three-dimensional unsharp masking, such countershading effects occur within the context of the depicted scene, not in the context of the two-dimensional painting.

While glare is often an unwanted effect in photography (Raskar et al., 2008), it can be exploited to display HDR content on LDR devices. When a veiling pattern is painted on the image (as a gradient surrounding the light source), the corresponding retinal image is similar to the observation of a real bright object and is thus interpreted by the human brain as brighter (Yoshida et al., 2008). This effect has been used by artists for centuries to improve the apparent dynamic range of their paintings, and it is just as attractive today in a digital imaging context. In traditional animation, dynamic glare effects are used for artistic effect.

8.1.4 TECHNICAL BACKGROUND

The reader is assumed to have some background knowledge of human perception (Palmer, 1999). All techniques explained here have in common that they rely on simplifications of known physiological results describing the cause of an observation. Such simplifications, however, are required to produce results by practical real-time image processing for interactive applications such as image or video display or computer games. It is assumed that the reader is familiar with the basic concepts of massively parallel, fine-grained computation as used by modern graphics hardware (Owens et al., 2007).

The underlying mathematical concepts used for simulation are diverse, ranging from Laplacian mesh smoothing and optical diffraction computed in the frequency domain to solutions of differential equations. While this sounds involved at first, the actual use of those concepts results in code that is simple enough to execute on a graphics chip in a few instructions and can be easily understood by with an applied computer vision or computer graphics background.

8.2 THREE-DIMENSIONAL UNSHARP MASKING

A basic principle behind image understanding is that humans are able to mentally reconstruct an original scenario from visual cues such as shading, occlusions, perspective foreshortening, shadows, and specularities. Local contrast enhancement emphasizes these cues, aiding the interpretation of three-dimensional scenes and complex geometry, which is a common task in applications such as medical diagnostics, computer simulations, geographical navigation, game playing, and film creation. The main problem is in deciding which cues one should emphasize and how to do so with a predictable effect.

The goal of three-dimensional unsharp masking (Ritschel et al., 2008) is to construct a perceptually founded approach for local contrast enhancement of arbitrary interactive three-dimensional scenes. Such an approach should provide easier shape recognition, better visual separation between objects, and a clarification of their spatial arrangement solely by increasing the apparent contrast of specific visual cues. Instead of identifying and modifying cues separately, we look here at their common cause, changes (or gradients) in reflected light. These light gradients include all cues caused by variations in surface geometry, material properties, incoming light properties, and the spatial arrangement of objects; for example, where a surface receives different amounts of incoming light (possibly in shadow), where reflectance properties change, and where specular highlights occur. Three-dimensional unsharp masking strives to simultaneously increase the contrast of all such gradients without breaking coherence with the depicted scene. We call a manipulation coherent if it appears to be a plausible part of the three-dimensional scene instead of something artificially introduced into the two-dimensional image.

To create a scene-coherent enhancement technique for three-dimensional computer graphics, a novel form of unsharp masking is proposed. In generic unsharp masking, an input signal is enhanced by addition of a contrast signal which is a scaled high-pass version of itself. To adapt this principle to three dimensions, the input signal is defined as the outgoing radiance from a surface point to a viewpoint and the contrast signal is defined as the high frequencies in this radiance measured over the mesh surface. These high frequencies are computed by subtraction of a smooth version of the outgoing radiance from the input signal. Scene-coherent smoothing is applied over the mesh surface itself, so the contrast signal can be thought of as lying over the surface. A graphics processing unit (GPU) implementation results in interactive performance and the process fits into the rendering pipeline, allowing simple adjustment of the enhancement strength with an immediately viewable effect. Studies in perception have long analyzed local contrast, especially in relation to the so-called Cornsweet effect, where a specific shape of gradient creates the illusion of contrast. Three-dimensional unsharp masking relates to work by Purves et al. (1999) which shows that the Cornsweet effect is strongest when it is coherent with a three-dimensional scene and its lighting (see Fig. 8.1). Three-dimensional unsharp masking generalizes this observation for all local contrasts to justify a scene-coherent approach.

To ease understanding of the complex shape seen in Fig. 8.2, the shading and occlusion gradients should be made obvious, but enhancements that appear or disappear when the viewpoint is changed

FIGURE 8.1

Illusions from Purves et al. (1999) showing that a Cornsweet profile in a three-dimensional scene produces a much stronger perceptual contrast than simply showing the brightness profile in two dimensions. The top face of the cube appears much darker than the bottom face but in fact they are nearly identical.

Source: Reproduced with permission. © 2008, Association for Computing Machinery.

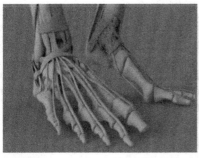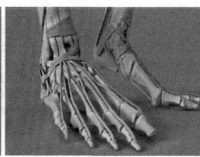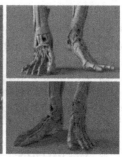

FIGURE 8.2

A naturally illuminated three-dimensional scene (left) and the same scene with three-dimensional unsharp masking enhancement (center). Three-dimensional unsharp masking is coherent with the scene itself, not simply with each rendered frame, permits arbitrary lighting, and is temporally coherent (right).

Source: Reproduced with permission. © 2008, Association for Computing Machinery.

would create inconsistencies that impede understanding. Three-dimensional unsharp masking exaggerates gradients to provide greater visual separation between regions and enhances the shadows that underscore the geometry, doing so coherently over all viewpoints and transitions, and under any type of illumination (Fig. 8.2).

Another instance is the exploration of a scene with details that lie in shadow. The desire is for details in shadow to remain visible, without forcing the shadows to lighten. By enhancing contrast along the shadow edges, the Cornsweet illusion creates a darker appearance without compromising detail visibility, as shown in Fig. 8.5.

8.2.1 THE CORNSWEET ILLUSION

While apparent contrast is most directly related to differences between adjacent regions, it is also impacted by the separating gradient's shape and magnitude (Moulden et al., 1988). If a gradual gradient blends between regions, an enhanced gradient does the opposite by adding distinction, thus increasing perception of contrast. Contrast studies show that a sharp gradient can increase apparent contrast to the point where an equiluminant patch appears to have two contrasting regions purely on account of the edge. Most evocative is the Cornsweet illusion, where neighboring regions are filled with illusory brightness, giving a sense of contrast where there is nearly none (Cornsweet, 2012), as can be seen in Fig. 8.1. This shows that local features can increase the apparent dynamic range far beyond that which is physically present. Recent research in computer graphics has demonstrated the Cornsweet illusion's contrast enhancement abilities in two dimensions: one, for restoration of original HDR contrast to LDR tone-mapped photographs (Krawczyk et al., 2007), and another for restoration of color contrast to grayscale converted images (Smith et al., 2008). Purves et al. (1999) explain that the Cornsweet illusion arises because humans are predisposed to interpreting the stimulus as a result of lighting or reflectance properties. For instance, the gradient could arise from changes in surface reflectance/texture variation, gradients of illumination: penumbra (occlusion), curvature, attenuation with distance, transmittance (partial occlusion). This is why the illusion is weakened if the edge seems to be painted on, or if it becomes discordant with other aspects in the image. It also explains why the illusion is strengthened when it is reinforced by compatible visual cues. In particular, the illusion's salience is increased when it implies depth by incorporating perspective projections. The gradient's orientation should be consistent with lighting, as well as other cues such as texture and background, things that would indicate normal viewing conditions. These findings tell us that although local contrast is interpreted from two-dimensional visual cues, its interpretation is in terms of real-world lighting or reflections. For this reason, the enhancement should be coherent with the entire scene and its viewing.

8.2.2 DEFINITION

Unsharp masking in the two-dimensional image domain proceeds first by smoothing the entire image and then adding the scaled difference between the original and the smooth image, as depicted in Fig. 8.3 (Badamchizadeh and Aghagolzadeh, 2004). More generally, unsharp masking $\mathcal{U}(L)$ of a signal $L(\mathbf{x}) \in \Omega \subset \mathbb{R}^3 \to \mathbb{R}^3$ varying on surface Ω with strength λ and smoothness σ is defined as

$$\mathcal{U}(L) = L + \lambda(L - L_\sigma).$$

In three-dimensional unsharp masking, L is the outgoing radiance from a surface location $\mathbf{x}$ to a viewpoint, and L_σ is that radiance smoothed over the surface Ω of the mesh, defined as

$$L_\sigma(\mathbf{x}) = \int_\Omega L(\mathbf{x})\mathcal{N}(\Delta_\Omega(\mathbf{x}, \mathbf{y}), \sigma)\, d\mathbf{y},$$

where Δ_Ω is measure of the intrinsic distance between two surface locations on the surface Ω, $\mathbf{y}$ is an integration variable that enumerates the shape surface Ω, and $\mathcal{N}$ is a Gaussian function. For simplicity, the difference between the sharp and smooth signals is denoted as the contrast signal $C = L - L_\sigma$. The width of the contrast signals is controlled by a user-chosen smoothness σ.

All three-dimensional unsharp masking operations are performed on radiance, represented in the CIELAB color space because it provides a perceptually uniform lightness channel L that is decorrelated from its a and b chromaticity channels. By its doing so, it is ensured that contrast is measured and modified as apparent lightness and that the hue angle remains unchanged.

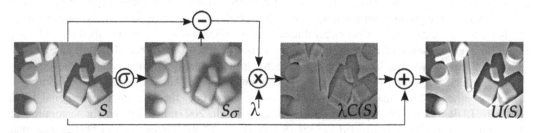

FIGURE 8.3

The process of unsharp masking sharpens a signal S by adding a λ-scaled contrast signal $C(S)$, the difference between the original and a smoothed version S_σ.

8.2.3 **CONTRAST SIGNAL CALCULATION**

We now describe the calculation of the smooth lighting signal L_σ, which gives rise to the contrast signal C. L_σ is the view-dependent reflected light L smoothed "in three dimensions" over the mesh surface. The surface is given as a triangular mesh, consisting of vertices and facets. Laplacian smoothing (Taubin, 1995) is used, which replaces the radiance of each vertex by the average of its neighbors Iterating this filter results in increasingly smooth versions of L_σ. L_σ is rendered for use in the per-pixel calculation of the contrast signal C. However, it is sufficient to calculate and store L_σ per vertex since it is band limited. As opposed to two-dimensional smoothing operations, the smoothing described automatically adapts to the surface orientation and location with respect to the viewpoint, thus undergoing correct perspective foreshortening.

8.2.3.1 *GPU implementation*

The contrast signal calculation above can be implemented with use of a GPU to first perform shading and then smoothing. First, the initial outgoing radiance is computed with an arbitrary, sufficiently fast shading method parallel for all vertices. Instead of triangles being rasterized into the framebuffer from those vertices, the radiance resulting from shading is stored in a GPU array. This GPU buffer encodes L and has as many items as there are vertices. Note that the particular mapping from vertices to this GPU array is arbitrary — in particular, no shape parameterization is required. Second, for smoothing, a parallel program is executed on every item in the GPU buffer (that represents a lit vertex in L), replacing it with the average of all buffer items in its 1-ring (other lit vertices in L). The 1-ring is fixed for one shape and is encoded in two buffers, where the first holds the 1-ring size of each vertex (the valence) and the second holds the indices of each vertex in the 1-ring.

8.2.3.2 *Coherence of the contrast signal*

Let us now consider the origin of high-pass signals contained in C and their temporal coherence. The signals detected need not originate from gradients with specific causes such as geometric discontinuities or depth discontinuities. Instead, they include all gradients in reflected light: surface geometry (variations due to curvature and surfaces facing toward or away from light); reflectance of surface material (variations due to texture and highlights); and incident light (variations due to different amounts of incoming light from different directions and light blocked by occluders creating shadows).

The contrast signal calculation is said to be scene coherent and temporally coherent because it is not affected by view-dependent occlusions (whether a surface point is visible or not in the image), nor is it limited by image space sampling. The contrast signal is best understood through a comparison with a two-dimensional unsharp masking contrast signal. In the three-dimensional unsharp masking case, the signals are continuous along the surface of the mesh and are not interrupted by distant objects that may be acting as occluders, as would occur in a two-dimensional approach (O in Fig. 8.4). This prevents the presence of continuous signals around disconnected objects that change as the viewpoint changes. Three-dimensional unsharp masking contrast signal calculations respect gaps between objects to be consistent with their spatial grouping, which remains temporally consistent. This also prevents brightness changes to a surface on account of its contrast with other distant surfaces appearing in front of or behind it, leading to brightness inconsistency over time (D in Fig. 8.4). In two dimensions, the

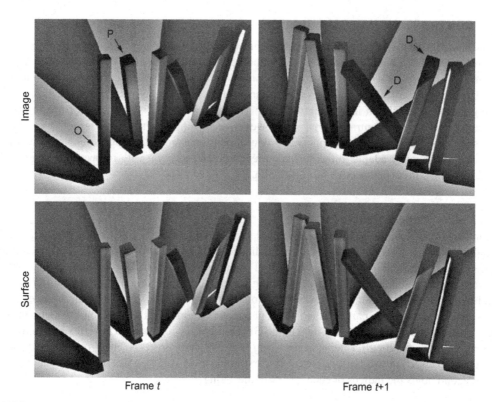

Image

Surface

Frame *t*

Frame *t*+1

FIGURE 8.4

Comparison of two-dimensional (top) and three-dimensional (bottom) unsharp masking over two successive frames (left and right). O, occlusion (compare vertical): the shadow enhancement should not be interrupted by the occluding stick. D, distant (compare horizontal): note the strong change in brightness of surfaces between frames because the sticks are darker or lighter than distant background. P, perspective (compare horizontal): setting a constant filter size that works for shadow edges disregards discontinuities on the sticks.

Source: Reproduced with permission. © 2008, Association for Computing Machinery.

detected high frequencies have the same width in image space, so near and far gradients are treated on the same scale, when in fact they have different sizes. So a wide gradient that is distant may be enhanced as a very narrow gradient (P in Fig. 8.4). Lastly, since three-dimensional unsharp masking contrast is not affected by what is visible or not visible, it does not introduce popping due to visibility changes.

8.2.4 RESULTS

Three-dimensional unsharp masking achieves high-level enhancements similar to those achieved by sharpening of normals (Cignoni et al., 2005), exaggerated shading (Rusinkiewicz et al., 2006), and

  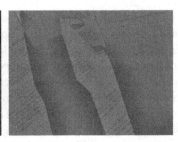

FIGURE 8.5

The apparent dynamic range of the original (left) is increased by three-dimensional unsharp masking (middle), which emphasizes the shadow edges. The contrast signal is shown on the right.

Source: Reproduced with permission. © 2008, Association for Computing Machinery.

depth unsharp masking (Luft et al., 2006). With similar viewpoints and models, it can be seen that the results of three-dimensional unsharp masking are indeed comparable to or better than those of the other methods. Furthermore, it is a comprehensive approach that performs the diverse enhancements with a single simple technique. In addition to making shapes more obvious, depth also appears enhanced owing to the enhanced shading gradients.

Rusinkiewicz et al. (2006) show that varying enhancement strengths according to importance may be desired. They do so over the space of the rendered image with a multiscale approach. The same can be done with three-dimensional unsharp masking within the rendering pipeline, without the need for rendering at multiple scales and recombining. When painting rough gain controls for λ over the three-dimensional object, in this case to indicate importance with both positive and negative values, the effect enhances the face, torch, and hands while deemphasizing the robe and pedestal. Returning to the introductory scenario about detail in shadows, we can see that three-dimensional unsharp masking creates the impression of darker shadows while keeping the text readable in Fig. 8.5. The shadows appear darker because of the Cornsweet illusion, yet do not appear "sharpened."

A main benefit of three-dimensional unsharp masking is its integration in the rendering pipeline and its allowance of interactive parameter adjustment, which is in real time when the original can be rendered in real time. The main bottleneck is the speed of shadow rendering, not the algorithmically more involved Laplacian smoothing step.

8.2.5 RELATED WORK

Three-dimensional unsharp masking falls within three-dimensional computer graphics as a last step in the rendering pipeline. Rendering has succeeded in producing real-to-life imagery, and the next step is to produce more communicative and efficient imagery from that basis.

Given a three-dimensional model, Cignoni et al. (2005) shift object normals, performing normal enhancement to emphasize geometric discontinuities in a single rendered image. In this approach, enhancement strength is controlled by how much the geometry of the mesh is sharpened. Inspired by cartographic illustrations, Rusinkiewicz et al. (2006) introduced a new shading model to expose shape features and surface details by positioning a local light per vertex to achieve maximum contrast. Enhancement strength is controlled by adaptively combining multiple scales of renderings. The enhancement of complex geometry and multiobject scenes addresses the task of communicating the spatial arrangement of various objects and their relationships to each other. Three-dimensional unsharp masking is most closely related to the method of Luft et al. (2006), who focus on enhancing depth perception by unsharp masking the depth buffer. Depth between objects is emphasized by the darkening or lightening of the original image at locations of discontinuities in its accompanying depth map; however, places without depth changes (eg, object and ground intersections) are not enhanced. One important cue for spatial understanding is object and ground plane intersections; however, there is no enhancement because these areas have no depth discontinuity. This work was also presented solely for single frames, and its frame-by-frame application could lead to temporal incoherence when the depth map changes drastically.

Most of the approaches mentioned above do not enhance lighting gradients that occur where geometry is smooth or where there is no depth difference. These are common locations of shadows and specular highlights, which in addition to being necessary cues for proper scene comprehension, can help by underscoring geometry and clarifying spatial relationships (Cavanagh and Leclerc, 1989) and communicating shape and material properties (Fleming et al., 2004). DeCoro et al. (2007) have shown that shadows can benefit from modification.

Existing two-dimensional and 2.5-dimensional approaches applied in a frame-to-frame manner would introduce image space artifacts, such as popping and the effect of enhancements stuck within the image (shower door effect), instead of artifacts within the scene itself.

In the field of expressive rendering, there are techniques for creating many forms of coherent stylized renderings that address the problem of temporal coherence — for example, line drawings of three-dimensional scenes (DeCarlo et al., 2004). Their findings show that the contours depicted (and, by analogy, the contrasts added) cannot simply exist in the two-dimensional projection and slide through image space (creating a shower door effect), but should follow the content and movement of the underlying three-dimensional scene.

8.2.6 DISCUSSION

Without a proper selection of the enhancement parameter λ, adverse percepts may occur for some viewers. The commonest unintended effect is when the enhancement is perceived as a halo (Trentacoste et al., 2012). However, this occurs only for extreme choices for σ. Material characteristics may also be altered. For instance, shiny objects may appear slightly shinier. This can be seen as an advantage that makes cues for material recognition more obvious. The technique may enhance artifacts — for instance, in soft shadows — by bringing below-threshold errors above the visible threshold. To prevent this, one could consider separate rendering passes, with weaker λ for shadows than meshes. However, this is not necessary when high-quality rendering techniques are used. The smooth lighting, described in Section 8.2.2, is sensitive to the mesh tessellation and especially to the uniformity of the length of edges. Therefore, mesh cleaning operations are applied to all surfaces to refine their tessellation. The method does not smooth lighting across disconnected components, so the approach does not properly handle meshes that are not manifold where this is expected. Also, different mesh representations will result in slightly different enhancements.

8.3 **TEMPORAL GLARE**

Glare effects are common in all optical systems that capture an image with bright light sources, caustics, or highlights. In our observation of the everyday world, they are commonly due to light scattering in the human eye. Effectively, instead of having a crisp projected image of bright objects on the retina, surrounding regions are affected by scattered light. This leads, among other effects, to local contrast reduction (also called the *veiling* effect or *disability* glare).

A typical glare pattern for a small light source as perceived by most people with normal vision is depicted in Fig. 8.6, but the actual appearance of the glare varies with viewing conditions and observers.

FIGURE 8.6

Glare effect rendered for a point source with our model. The colorful ring is called the lenticular halo; the fine radiating needles constitute the ciliary corona.

In general, the effects of glare can be divided into *bloom*, a general loss of contrast in the surroundings of the retinal image of the light source (veil), and *flare*, which comprises the *ciliary corona* (the sharp needles) and the *lenticular halo* surrounding the light (Spencer et al., 1995).

Although the glare pattern cannot be reproduced in static images, people usually report that it fluctuates in a fluidlike motion when observed under real-world conditions (Ritschel et al., 2009). Additionally, flickering of the fine needles forming the ciliary corona is readily observable and many people perceive a pulsation of the glare intensity. While these effects are striking, glare has previously only been modeled as a static phenomenon. The dynamics discussed do not occur for cameras, and occur only for eyes.

Here we investigate the temporal aspects of glare appearance and perform simulation of light scattering within the eye based on Fourier optics for high-fidelity glare rendering. Our goal is not only to render glare realistically with real-time performance, but also to show that by better mimicking the real-world experience, we can improve the image brightness impression and overall perceived quality.

8.3.1 A DYNAMIC HUMAN EYE MODEL FOR GLARE

This section describes a time-dependent human eye model that is suitable for real-time computation of plausible dynamic glare. It outlines the anatomical and physiological characteristics of all parts of the eye that contribute to the light-scattering characteristics of glare. Furthermore, it discusses the dynamics shown by these anatomical structures and incorporate their characteristics.

Fig. 8.7 shows a schematic diagram of the human eye. From front to back the optically important parts are the cornea, the aqueous humor, the iris and pupil, the lens, the vitreous humor, and the retina.

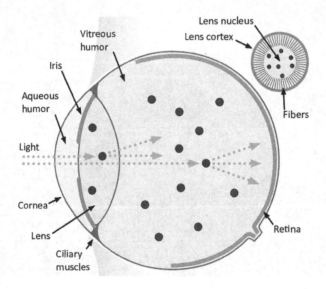

FIGURE 8.7

Anatomy of the human eye. The upper-right inset shows the lens structure.

Source: Reproduced with permission. © 2009, Eurographics Association.

In the following exposition, we skip the specific citations for each aspect of eye physiology, restricting the citations to key articles and computer graphics-related work and refer the interested reader to the original article (Ritschel et al., 2009).

8.3.1.1 Cornea

The main part of the cornea is formed by collagen fibrils, which are densely packed, regularly arranged, cylindrical particles. This special arrangement of the fibrils ensures that they are almost transparent. Between the fibrils there is a transparent ground substance, and a few flat cells are interspersed. These flat cells occupy 35% of the corneal volume and they have a diameter of 15 mm. The flat cells contribute to the glare pattern (25–30%), but the effect is static. This is simulated by having large, sparsely distributed, static particles in the pupil plane.

8.3.1.2 Iris and pupil

The pupil contributes important diffraction to the glare pattern as it is the model's aperture. The iris muscles have the ability to control the size of the pupil. Exposure to a glare source will typically give rise to the pupillary hippus, an involuntary, periodic fluctuation of the pupil size. It is presumably caused by opposing actions of the iris muscles due to the vastly different lighting conditions of the glare source and the background when they are attempting to adjust the pupil. A time-dependent noise function is used to vary the pupil diameter for different glare source intensities.

8.3.1.3 Lens

The lens is an important source of intraocular scattering. Close investigation of the refractive index of lens fiber membranes has shown that they produce significant scattering, and that they are regularly spaced in the lens cortex. This high spatial order of the lens fiber lattice increases the transparency of the lens, but is also a diffraction grating which produces the lenticular halo.

A wave optics model easily accounts for the light diffraction on such a grating pattern. Recently, the ciliary corona has been ascribed to randomly distributed particles in the lens. van den Berg et al. (2005) explain the sharp needles as being the result of a seamless alignment of scaled copies of the same diffraction pattern originating from light of different wavelengths. The motion of the particles in the lens is caused by lens deformations due to accommodative microfluctuations and is quantitatively different from the motion of the particles in the vitreous humor, which is mostly inertia driven.

A simplified two-dimensional space deformation model is used to simulate lens deformations, assuming a radially symmetric lens with heuristic uniform deformation properties (spring stiffness) and a uniform discretization. The accommodation system is known to exhibit temporal variations even during fixation of a stationary stimulus which are due to adaptations of the ciliary muscle. Those variations are introduced as additional forces into the space deformation.

8.3.1.4 Vitreous humor

The vitreous humor contributes relatively little to the glare effect in terms of scattered light energy, and for this reason it is often ignored in glare research. However, the contribution of scattering on particles in the vitreous humor to the ciliary corona is clearly visible in less central (less saturated) glare regions. This visibility is reinforced by temporal effects which are strong attractors of human attention especially in the visual periphery. The vitreous humor is a slightly scattering viscoelastic body. This means that external forces (head movements, saccades) act on the vitreous humor such that it accelerates, rotates, and comes to rest by strong damping. This behavior is modeled as a single rigid body with damped rotation dynamics, whose state is determined by the angle and angular velocity. Damped random forces are generated to mimic saccades and integrate this system by forward Euler integration. The final particles are embedded at random but fixed locations inside this rigid body and do not move relatively to each other. The scatterers are placed following a uniform random distribution throughout the vitreous humor.

8.3.1.5 Retina

The retina contributes to the intraocular scattering. It is unfortunately difficult to model the retinal forward scattering directly with a particle distribution because the particles are intermingled with the receptor cells. A few large particles are used to approximate the retina.

8.3.1.6 Eyelashes and blinking

Eyelashes and blinking can result in long streaks in the glare pattern. Blinking is modeled in the same way as by Kakimoto et al. (2004) by use of bitmaps of eyelids and eyelashes. However, the dynamic framework can produce animated blinking by moving the eyelashes against the pupil. In addition, squinting, which is a normal reaction to a strong, discomforting glare source, is included. It decreases the retinal illumination (and thus the discomfort) since the eyelids cut off the aperture and the eyelashes scatter the incident illumination. We model squinting by keeping the bitmaps closed by a constant amount.

8.3.2 WAVE OPTICS SIMULATION OF LIGHT SCATTERING

To simulate the scattering at obstacles within the different anatomical structures of the eye, an approach based on wave optics similar to that of van den Berg et al. (2005) and Kakimoto et al. (2004) is chosen. Even though the underlying theory is rather complicated, the implementation of scattering even on complex apertures is often simple and generally boils down to taking the Fourier transform (FT) of an aperture function. Considering the human eye as a simplified optical system with an aperture (pupil) and an image plane (retina), we obtain the diffraction pattern for particles and gratings within the lens and the cornea as well as for the eyelashes by computing the incident radiance following the Fresnel approximation of Huygen's principle.

In contrast to Kakimoto et al. (2004), who use Fraunhofer diffraction, we use the more general Fresnel equation. It contains Fraunhofer diffraction as a special case. Fresnel diffraction is considered as more appropriate, owing to the relatively short distance between the pupil and the retina (Hecht, 1998).

8.3.3 IMPLEMENTATION

The afterimages model is implemented to run in real time entirely on recent graphics hardware (GPUs). Despite the involved theory, temporal glare is easily and efficiently implemented: draw a few basic drawing primitives, apply a fast FT (FFT), and perform a special kind of blur. This already constitutes a temporal glare pipeline (see Fig. 8.8). The following paragraphs provide the details.

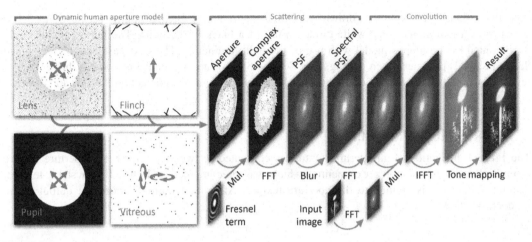

FIGURE 8.8

The temporal glare pipeline. The blue arrows are GPU shaders that transform one or multiple input textures (rectangles) into one or multiple output textures. *IFFT*, inverse FFT; *PSF*, point spread function. (For interpretation of the references to color in this figure legend, the reader is referred to the web version of this article.)

Source: Reproduced with permission. © 2009, Eurographics Association.

8.3.4 HUMAN APERTURE MODEL

We generate the aperture of the human pupil by drawing basic primitives into a texture. All primitives are rendered by 2×2 or higher supersampling. This helps suppress aliasing in time and space, which would be exaggerated by the FFT. Fig. 8.9 shows partial point spread function (PSFs) that include only subsets of the aperture model in order to show how the individual parts contribute to the result. Particles from the lens and the vitreous humor are projected orthogonally onto their aperture plane and are drawn as two-dimensional circles of the same size and color.

The model uses two sets of parameters. Static parameters encode subject-dependent variables: the number of particles, eye size, and other variables. Dynamic parameters are updated for every frame: the blink state, the field luminance, the observer motion, and other variables. First, according to the blink

FIGURE 8.9

Adding scatterers from left to right. Vitreous humor and lens scattering looks similar, but have different dynamics.

Source: Reproduced with permission. © 2009, Eurographics Association.

state, the eyelashes and eyelids are drawn as textured quadrilaterals. Next, the pupil (see Section 8.3.1.2) is drawn as a two-dimensional white circle on top of a black background. The radius of this circle is computed by the hippus model with use of the field luminance. The next pass adds the lens (see Section 8.3.1.3) particles. Then gratings (see Section 8.3.1.4) are drawn as lines with a thickness of a few pixels. Finally, the particles of the vitreous humor are simulated (see Section 8.3.1.4) and drawn.

8.3.5 FRESNEL DIFFRACTION

The Fresnel diffraction of an aperture texture is computed in two steps. First, the aperture texture is multiplied by the complex exponential which was precomputed and stored in a static texture. Second, GPU FFT is applied to the aperture texture. After final normalization, the output is the monochromatic PSF.

8.3.6 CHROMATIC BLUR

The colorful appearance of glare can be modeled in a simple and efficient way (van den Berg et al., 2005). This exploits the fact that a monochromatic PSF F_2 at wavelength λ_2 equals another monochromatic PSF F_1 for wavelength λ_1 whose argument is scaled by λ_2/λ_1. A full-spectrum chromatic PSF is computed by the overlapping of several monochromatic, scaled copies.

8.3.7 CONVOLUTION

The output image is convolved with the colored PSF computed in the previous step. This is different from previous methods, which used only billboards composed onto single bright pixels. Use of the FT convolution theorem allows one to compute the convolution as the multiplication of the FFT of the PSF and an FFT of the output image. Application of a final inverse FFT to all channels yields the HDR input image as seen from the dynamic human eye model. For final display, a gamma mapping is applied to the radiance values to compensate for the monitor's nonlinear response.

Note that such a distinct appearance of the ciliary corona needles as shown in Fig. 8.6 is typical for bright light sources with angular extent below 20 min of arc (the ray-formation angle, Simpson, 1953). Larger light sources superimpose the fine diffraction patterns that constitute the needles of the ciliary corona. This leads to a washing out of their structure. However, the temporal glare effect is still visible because the superimposed needles fluctuate incoherently in time.

8.3.8 RESULTS

The method results in perceptual improvements running at real-time frame rates on a 2.4 GHz CPU with an NVIDIA GeForce 8800 GTX. The performance depends on the aperture resolution. An aperture of 256 pixels squared can be computed in less than 10 ms; a 1024×1024 aperture requires 55 ms. These timings are with 2×2 supersampling of the aperture, 32 samples for the RGB PSF, and no convolution.

8.3.9 **RELATED WORK**

The modeling of glare effects has been used to improve image realism and to convey an impression of high intensity of luminaires in the context of realistic rendering (Spencer et al., 1995), driving simulation (Nakamae et al., 1990), image postprocessing (Rokita, 1993), and tone mapping (Larson et al., 1997). A perceptual study (Yoshida et al., 2008) demonstrates that the impression of displayed image brightness can be increased by more than 20% if high-intensity pixels are convolved in the image with the relatively simple filters used in glare models.

Most of the existing approaches to computer-generated glare, while inspired by knowledge of human eye anatomy and physiology, are based on phenomenological results rather than explicit modeling of the underlying physical mechanisms. A common approach is to design convolution filters which reduce image contrast in the proximity of glare sources up to full image saturation in the glare center. Nakamae et al. (1990) derived such a filter to model the light diffraction on the eye pupil and eyelashes for various wavelengths. Spencer et al. (1995) based their filter on the PSF measured for the optics of the human eye. Glare solutions used in tone mapping (Larson et al., 1997) are mostly based on the approach of Spencer et al. A set of Gaussian filters with different spatial extent, when skillfully applied, may lead to very convincing visual results. This approach is commonly used in computer games and rendering postproduction.

Other glare effects, such as the ciliary corona and the lenticular halo, are often designed offline and placed in the location of the brightest pixel for each glare source as a billboard (image sprite) (Rokita, 1993; Spencer et al., 1995). In the designing of such billboards, seminal ophthalmology references are used such as Simpson (1953). The resulting appearance is very realistic for small pointlike glare sources. However, when billboards are used it is difficult to realistically render glare for glare sources of arbitrary shape and nonnegligible spatial extent.

There have been some successful attempts to model glare based on the principles of wave optics. Kakimoto et al. (2004) proposed a practical model to simulate scattering from a single plane rigid aperture. Three diffraction-causing obstacles — the eyelashes, the eyelids, and the pupil edge — are placed in this plane. The Fraunhofer diffraction formula is then used to determine the diffraction pattern of the obstacle plane on the retina. This pattern is stored as a billboard, placed at high-intensity pixels and blended with the rendered image. Similarly, van den Berg et al. (2005) describe glare as diffraction by particles in the lens. By computing this pattern for multiple wavelengths, they achieved the first physical simulation of the ciliary corona. Temporal glare was first to consider a dynamic eye model that enables us to simulate the temporal fluctuations of glare. The model is also more complete with respect to existing static models by considering all significant contributions to light scattering in the eye.

8.4 **AFTERIMAGES**

8.4.1 **INTRODUCTION**

The HVS adapts to luminance, influencing scene perception. While this process is naturally used by the HVS in the real world, it is necessary to simulate such behavior when one is depicting virtual scenes. Many modern games and simulations rely on simple models and tone corrections to achieve a similar experience, which is important for a realistic look. While previous approaches considered

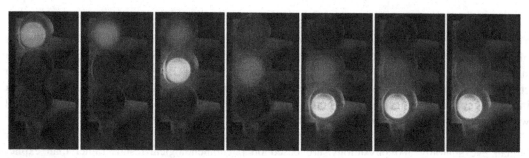

FIGURE 8.10

Afterimage simulation of a traffic light over time. Note the change of colors, blur, and shape over time in the afterimage (For interpretation of the references to color in this figure legend, the reader is referred to the web version of this article.).

Source: Reproduced with permission. © 2012, Eurographics Association.

global manipulations (such as gamma curves), the model of afterimages described here goes further by considering afterimages.

This fatigue-like effect results in an antagonistic color perception which becomes especially visible when one is fixating on a monochrome wall after observing a bright colorful object. The main mechanism behind this process is the bleaching of retinal photoreceptors. In computer graphics, bleaching has previously received attention in the context of tone mapping and color appearance modeling. The computational model of afterimages (Ritschel and Eisemann, 2012) discussed here is based on a localized process that considers temporal, color, and time-frequency behavior (eg, blurred fading over time under varying colors; see Fig. 8.10).

The method results in increased realism concerning the depiction of perceived strong luminance, and can be used to convey HDR content on an LDR medium. While afterimages arise in the observer's eye while looking at an LCD, the brightness of standard screens is insufficient to produce correct afterimages with respect to HDR image content. The model can be used to correct this shortcoming and even leads to an increase in perceived brightness. The approach achieves real-time frame rates for full-high-definition content.

8.4.2 APPROACH

This section explains the model of afterimages and its efficient implementation for real-time performance. The approach gives rise to several applications that will be presented in Section 8.4.9.

8.4.2.1 Model

The different stages of the model are depicted in Fig. 8.11. It assumes photopic conditions (ie, daylight light levels), with color vision dominated by cones, and excludes rods from the simulation. Several processes which are involved in the perception of afterimages are considered. First, world radiance

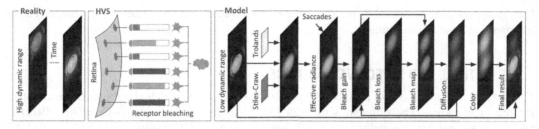

FIGURE 8.11

A stimulus can result in afterimages produced by the HVS, which are simulated by our model. The stimulus (left) is a time-varying HDR radiance map (here, a traffic light). In the HVS (middle), photoreceptors (red circles) of different types (long, medium, and short wavelength) cover the retina (surface patch). The pigment concentration in each receptor is depicted by bars, resulting in differing color perception. Different steps (right box) simulate receptor kinetics computationally (For interpretation of the references to color in this figure legend, the reader is referred to the web version of this article.).

Source: Reproduced with permission. © 2012, Eurographics Association.

is converted into an eye radiance map, with subconscious micro eye movement being taken into account. On the retina, receptor kinetics are used to derive the perceived afterimage stimulus. The simulation includes a model of adaptation state, temporal diffusion, and chromatic influence. The result is combined with the input to derive the complete stimulus. The input to the model is an HDR world radiance map $L_w(\mathbf{x}, \lambda)$ — that is, the display content at position $\mathbf{x}$ for wavelength λ. The model assumes a virtual (ie, "cyclopean") eye, looking forward without saccadic movements. The average over the frame interval s is denoted as L_w.

8.4.3 EYE RADIANCE AND DYNAMICS

To simulate an eye-internal phenomenon such as afterimages, an eye radiance map $L_e(\mathbf{x}, \lambda)$ is produced. It encodes the radiance arriving at $\mathbf{x}$ after compensating for the eye's dynamics. To account for eye dynamics, a fixation function $f : \mathbb{R}^2 \times \mathbb{R}^2 \to \mathbb{R}^2$ maps a location $\mathbf{x}$ at a point in time t to another location $f(\mathbf{x}, t)$. The eye radiance according to this model is then

$$L_e(\mathbf{x}, \lambda) = \int L_w(f(\mathbf{x}, t), \lambda) \, dt.$$

Eye tracking equipment is the optimal solution to acquire f. In the absence of eye tracking, a simple assumption is that the observer fixates on the screen center — that is, $f(\mathbf{x}, t) = \mathbf{x}$ — which is a common assumption for image-appearance modeling (Fairchild, 2013). While this scenario eliminates strong eye movement, small dynamics can still have an important impact on afterimages and need to be considered. The model does not consider saccades or microsaccades and includes tremor (Martinez-Conde et al., 2004) with a spatial convolution using a Gaussian kernel. A viewing distance of approximately 1 m results in a blur kernel of approximately 10 px for a 1280×1024 screen (Martinez-Conde et al., 2004).

Hereby, eye tracking is avoided by our assuming we have a static eye and a blurred stimulus, instead of a dynamic eye and a static stimulus.

8.4.4 RETINAL AND EFFECTIVE RADIANCE

The eye radiance map $L_e(\mathbf{x}, \lambda)$ is converted into retinal radiance (trolands) by the model accounting for the pupil radius. The different sensitivity for different directions (Stiles-Crawford effect) is taken into account by a spatially varying efficiency ratio map. The resulting quantity is called the effective radiance $L(\mathbf{x}, \lambda)$. Finally, the spectral effective radiance is converted into an LMS color space representation $L(\mathbf{x})$ by use of the standard observer LMS profiles. Next, the retinal kinetics of every photoreceptor are simulated.

8.4.5 RETINAL KINETICS

While Baylor et al. (1974) identified a cascade of up to six different chemicals that would need to be accounted for, we found modeling a single chemical sufficient for a qualitative effect. Note that both the six chemicals of Baylor et al. and the single chemical used here are "virtual" in the sense that it cannot be concluded that exactly six individual chemicals are involved in reality. Instead, they should be interpreted as more or less accurate parameters fitted to data (physiological measurements in turtles for Baylor et al.; functionally plausible images for the current purpose). Nonetheless, a more complex multichemical model could be used. Formally, consider the time-space varying ratio of active opsin left in a unit area covered by L, M, and S cones: $r_{\{L,M,S\}} \in (0, 1)$. Furthermore, every receptor's time-space varying change of active opsin concentration over time at position $\mathbf{x}$ and time t is $\dot{r}(\mathbf{x}, t)_{\{L,M,S\}}$. We use here the physics notation $\dot{r}$, where the dot denotes a derivative in time, dr/dt. In the following, the LMS index is dropped and r and $\dot{r}$ are treated as 3-vectors. The change of opsin concentration depends on the effective radiance and the current adaptation (Baylor et al., 1974), following the equation

$$\dot{r}(\mathbf{x}, t) = c_a L(\mathbf{x}, t)(1 - r(\mathbf{x}, t)) - c_d r(\mathbf{x}, t),$$

where c_a and c_d are constants of activation and deactivation, respectively. This equation shows that the change in opsin concentration r follows two mechanisms that are balanced by the concentration itself. When a receptor is exposed to strong effective radiance or is not yet strongly adapted (r is small), the change of adaptation is faster (ie, it activates faster). For a progressed adaptation, the activation is reduced (toward $r = 1.0$). A similar behavior occurs if the effective radiance is low. Interestingly, the opsin deactivation is independent of the effective radiance but is inversely proportional to the current bleaching. Hence, the deactivation is slow when little adaptation has happened, and is fast when much opsin was activated. Choosing $c_a = 0.3$ and $c_d = 1.0c_a$ gave good results for the one-chemical model.

8.4.6 SOLVER

A Newton solver numerically integrates the differential equation (described in Section 8.4.5) forward in time. In each step, the solution is clamped to [0,1]. The adaptation change $\dot{r}$ is maximal for $r = 0$ and proportional to L/n. Hence, a single iteration ($n = 1$) is usually sufficient given the frame rates achieved and changes in L.

8.4.7 DIFFUSION

Afterimages undergo considerable progressive blurring over time. While the underlying principles of this phenomenon are not fully understood, their behavior can be simulated. The time-frequency nature has been measured and enables us to fit a corresponding smoothing process according to these curves. Note that the diffusion blur is not related to the blur due to saccadic eye movements, which is applied to the input only once.

8.4.8 CHROMATIC EFFECT

The bleaching result cannot be simply subtracted from the effective radiance in LMS color space and converted back to RGB for display, as the HVS uses an opponent color processing that needs to be accounted for. Red stimuli lead to green afterimages, orange stimuli lead to blue afterimages, etc. To this end, the bleaching from LMS color space is transformed into an opponent color space (oRGB (Bratkova et al., 2009) is used, and is a linear transform that was conceived according to the mechanism). The approach computes the opponent color of the current bleaching, transforms it back to RGB and composes it with the LDR input for final display.

8.4.8.1 LDR compensation

When HDR content is depicted on an LDR medium, the resulting real afterimage is different (ie, weaker than it would have been had the observer looked at HDR content). To produce a percept closer to the one resulting from the original HDR content, the model can be used to compute the afterimage that the HDR content would have produced, and insert it into the LDR image. If required, the approach can also compensate for the weak afterimage produced by the LDR content in the observer's eye following the same model. In practice, the real LDR afterimage is weak enough to be ignored.

8.4.8.2 Flight of colors

A final component of afterimages is the so-called flight of colors — afterimages that change their hue continuously over time. In practice, this effect is usually only observed when one is closing one's eyes or when one is suddenly in a very dark environment. There are no physiological data available concerning this process and maybe there never will be; the effect is subjective and it seems almost a philosophical question approaching the limits of human introspection. The flight of colors is simulated by the application of a hue rotation in every simulation step.

8.4.8.3 Implementation

An Nvidia GeForce 480 GTX executes the model for m receptors in parallel using OpenGL. It assumes one receptor for every pixel, ignoring their more complex spatial layout and varying density (Deering, 2005). The input to the model is the effective radiance as a GPU vector L of size m and time step s. The current adaptation of every receptor is stored in a GPU vector R, also of size m. All m threads compute a new adaptation state from the current effective radiance and the previous adaptation state, integrating it forward in time in n steps per frame. Using 16-bit floating point precision, the resulting kernel can process 300 megapixels per second with a single iteration. For a multichemical model, each stage requires one GPU buffer.

8.4.9 RESULT

The example applications are discussed next: prediction, tone mapping, and improved brightness appearance.

8.4.10 SIMULATION SCENARIO

The afterimage model discussed allows us to predict visibility in different viewing conditions, which can be critical for many applications (eg, training software for cars, ships, or airplanes) As an example, the effects of afterimages on a traffic light can be simulated under difficult, high-contrast conditions (Fig. 8.10).

8.4.11 HDR PHOTO VIEWER SCENARIO

Depicting high-contrast content as found in many HDR (panorama) images is challenging. To produce a percept closer to the one resulting from the original HDR content, the model computes an afterimage that the HDR content would have produced and inserts it into the LDR image. This is similar to the popular introduction of glare around bright light sources in interactive applications such as computer games. The approach can also be applied to a panoramic HDR image viewer (Fig. 8.12, top).

8.4.12 GAME SCENARIO

The solution can increases the perceived brightness of explosions, fire, and bright light. Further, in a game, the assumptions made (changing content, fixation on screen center) are more likely to hold. In Fig. 8.12, bottom, both afterimages are due to bright daylight shafts in a dark room.

8.4.13 RELATED WORK

This section discusses related work, both from computer graphics and from general vision and physiology.

8.4.14 AFTERIMAGES

Most features of afterimages are explained by the adaptation of the human retina's photoreceptor cells: its receptor kinetics. A receptor is a cell that transforms light in the form of photon energy into an electric signal. This energy transformation is done by a class of proteins called opsins. The chemical processes behind transduction of light to impulses (their kinetics) is complex, but can be predicted with sue of a cascade of coupled equations (eg, for turtles, Baylor et al., 1974). Photoreceptor kinetics, in its simplest form, describes the change of opsin concentration over time (Fig. 8.11) using two basic principles. First, when a receptor is exposed to light, a signal is triggered and the opsin concentration is lowered. The lower the opsin concentration, the less responsive the receptor. Conceptually, opsin flowing out of the receptor leads to a deactivation. When the opsin concentration is zero, the receptor saturates. Nonetheless, opsin is constantly regenerated from deactivated opsin, thereby reactivating the receptor. This process explains the basic principle of afterimages well. They occur when the receptor cell cannot adapt quickly enough to react to illumination changes. The receptor signals are believed to

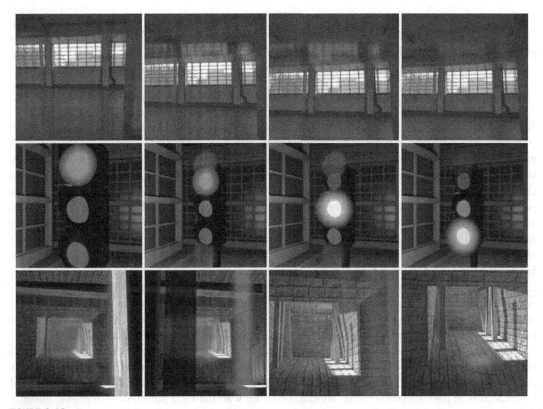

FIGURE 8.12

Video frames of an HDR image viewer (top), a traffic simulation (middle), and a computer game scenario (bottom) running in real time. The strength is exaggerated for better reproduction in print.

Source: Reproduced with permission. © 2012, Eurographics Association.

be combined into the perception of color by opponent processes. When a receptor becomes adapted, the opposite color appears. This agrees with the percept of afterimages (eg, a green stimulus produces a red afterimage). Under normal viewing conditions, the receptor rarely saturates. One reason is saccades (Martinez-Conde et al., 2004): a moving eye receives constantly changing light on the retina. Even when the eye is fixating on a natural stimulus, microsaccades happen, and depending on the constant areas in the stimulus, differently sized areas of adaptation occur, as well as blur that becomes visible in resulting afterimages.

The human retina consists of two types of photoreceptors: cones and rods. Most of the afterimages can be attributed to cones that are active in photopic (daylight) vision and respond to different wavelengths in different ways. Rods are mostly used in dark vision and are saturated in photopic conditions. As these conditions are mostly of interest to us, we will concentrate on cones. Nonetheless, we point out that the perception of afterimages is different between photopic and scotopic conditions, but the kinetics can be similarly described. There are three types of cones, L, M, and S cones, each with a spectral response roughly equivalent to red (long), green (medium), and blue (short) wavelengths. Cones are densely packed in the fovea and are sparser in the periphery. Besides colors, a distinct property of afterimages is their increased blurring over time.

8.4.15 VISUAL COMPUTING

After the first models of photoreceptor kinetics became available (Baylor et al., 1974), hardware to simulate a single receptor was designed (Vallerga et al., 1980) but did not aim at image synthesis. Color-appearance models account for adaptation (Fairchild, 2013). Their extension to image appearance, such as iCAM, considers spatial locality. In all such models, adaptation is time stationary (ie, its change is not considered). Instead, the model predicts how a certain stimulus at a point in time is perceived given previous stimuli over time. Classic computer graphics models of the HVS's adaptation were introduced by Pattanaik et al. (1998) in the context of tone mapping. They model the global adaptation to different lighting conditions in terms of thresholds, but not its localized retinal effect that results in afterimages. The model discussed extends this model by considering localized photoreceptor kinetics and the temporal behavior of afterimage color and sharpness. Adaptation and bleaching are routinely simulated with use of time-varying, global tone adjustments in interactive applications such as computer games. Pajak et al. (2010) proposed a computational model of lightness maladaptation. This model indeed accounts for the local response of the HVS to high contrast. However, their model does not account for afterimages, their color, or fading, and considers only threshold elevation. Recently, Jacobs et al. (2015) described a system for gaze-enabled simulation of phenomena such as glare and afterimages. Use of an eye-tracker correct compensation for the misalignment between afterimages resulting from previous retinal orientations and the current gaze. Their model of afterimages is refined to produce both positive and negative afterimages, both with the right hues and with physiologically motivated hues.

8.5 CONCLUSION

This chapter has discussed and related three visual illusions and visual anomalies that can be used to increase apparent contrast: three-dimensional unsharp masking based on the Cornsweet illusion and a model of temporal glare and afterimages.

Three-dimensional unsharp masking (Ritschel et al., 2008) increases the local contrast of a three-dimensional scene. It is a holistic approach that is perceptually motivated and preserves coherence with the entire depicted scene. Unsharp masking light gradients over the mesh surface avoids temporal artifacts introduced by image-based approaches. Its enhancement better depicts shape details, provides clearer separation between objects, and deepens the impression of a scene's overall dynamic range. Three-dimensional unsharp masking was applied to volume rendering (Tao et al., 2009) to enhance visualizations by smoothing over a lit volume instead of a lit surface. A perceptual study (Ihrke et al., 2009) has shown what values of σ and λ are useful in practice.

Temporal glare simulates the dynamics of the human eye and how they add temporal variation to the perceived glare pattern. The method has a solid basis in eye anatomy and wave optics but remains practical as was demonstrated by its implementation in real time on recent GPU hardware.

Afterimages (Ritschel and Eisemann, 2012) simulated retinal processes to reproduce afterimages based on local bleaching of photoreceptors. Differently from previous work using global bleaching statistics, the work by Ritschel and Eisemann (2012) simulated local bleaching, as well as temporal, color, and time-frequency behavior. Without eye tracking (Jacobs et al., 2015), it has to assume a user fixating on the screen center, which is the main limitation in practice: a subject observing a complex HDR photograph of a city at night on an LDR display would expect afterimages when looking around the picture in an explorative fashion.

8.5.1 COMMON LIMITATIONS

A common limitation of all approaches is that they are difficult to validate. In this chapter we have skipped the perceptual studies which were provided in the original articles. What is common for all these studies is that the success of each method can be measured only indirectly, as no "reference sensation" with which the simulated sensation can be compared is available. We cannot "read out" the retinal image after glare and inclusion of afterimages and compare it with a simulation. Contrast and boundary effects are higher-level effects and are even less accessible to direct comparison. Instead, perceptual experiments (Ritschel et al., 2008; Yoshida et al., 2008; Ihrke et al., 2009; Ritschel et al., 2009; Ritschel and Eisemann, 2012) often allow subjects to adjust parameters of the model until it is perceived to be similar. The resulting parameters give indications on how to tweak the effect best, but do not really validate whether the model is correct or if it has irrelevant or missing parts. How much visual anomalies can really contribute to physiological brightness perception and how much they are a concept of experience or culture is a difficult question to answer. When Jacobs et al. (2015) performed an extensive experiment for brightness matching between images with and without afterimages similarly to that of Ritschel and Eisemann (2012), they found no significant evidence for an effect. Contradictorily, subjects have reported that the addition of effects such as afterimages makes images appear brighter.

8.5.2 EXTENSIONS

Future work on temporal glare could include the conduct of more detailed studies, including more participants to better understand in which conditions temporal glare is important and where not and how individual components contribute to the final result. Future work could also elaborate the glare model (eg, by including real-time subject information such as gaze direction). Afterimages appear only if the content changes (eg, camera movements or animations). Eye motions would also result in changing retinal content in the same way as they would affect the temporal glare dynamics. Use of eye-tracking hardware (Jacobs et al., 2015) is therefore an exciting avenue of future work.

8.5.3 OUTLOOK

The idea of reproducing artifacts to produce the sensation of something that is not there because it cannot be reproduced is more general than luminance contrast. Binocular stereo perception of very bright (Templin et al., 2012) or very dark (Kellnhofer et al., 2014) scenes is different and is subject to important nonlinearities, and considerations similar to those explained above can be applied here: simulating a limitation can strengthen the impression of dynamic range. Other modalities such as sound and haptics are subject to their own anomalies and illusions that could be turned into an advantage as well.

REFERENCES

Badamchizadeh, M., Aghagolzadeh, A., 2004. Comparative study of unsharp masking methods for image enhancement. In: Proc. Multi-Agent Security and Survivability, pp. 27–30.

Baylor, D., Hodgkin, A., Lamb, T., 1974. The electrical response of turtle cones to flashes and steps of light. J. Phys. 242 (3), 685–727.

Bratkova, M., Boulos, S., Shirley, P., 2009. ORGB: a practical opponent color space for computer graphics. IEEE Comput. Graph. Appl. (1), 42–55.

Cavanagh, P., Leclerc, Y.G., 1989. Shape from shadows. J. Exp. Psychol. Hum. Percept. Perform. 15 (1), 3.

Cignoni, P., Scopigno, R., Tarini, M., 2005. A simple normal enhancement technique for interactive non-photorealistic renderings. Comput. Graph. 29 (1), 125–133.

Cornsweet, T., 2012. Visual Perception. Academic Press, New York/London.

DeCarlo, D., Finkelstein, A., Rusinkiewicz, S., 2004. Interactive rendering of suggestive contours with temporal coherence. In: Proc. NPAR, pp. 140–145.

DeCoro, C., Cole, F., Finkelstein, A., Rusinkiewicz, S., 2007. Stylized shadows. In: Proc. NPAR, pp. 77–83.

Deering, M.F., 2005. A photon accurate model of the human eye. ACM Trans. Graphics 24 (3), 649–658.

Fairchild, M.D., 2013. Color Appearance Models. John Wiley & Sons, New York.

Fleming, R.W., Torralba, A., Adelson, E.H., 2004. Specular reflections and the perception of shape. J. Vis. 4 (9), 10.

Hecht, E., 1998. Hecht Optics. Addison-Wesley, Reading, MA.

Ihrke, M., Ritschel, T., Smith, K., Grosch, T., Myszkowski, K., Seidel, H.P., 2009. A perceptual evaluation of 3d unsharp masking. In: Proc. IS&T/SPIE Electronic Imaging, p. 72400R.

Jacobs, D.E., Gallo, O., Cooper, E.A., Pulli, K., Levoy, M., 2015. Simulating the visual experience of very bright and very dark scenes. ACM Trans. Graph. 34 (3), 25:1–25:15.

Kakimoto, M., Matsuoka, K., Nishita, T., Naemura, T., Harashima, H., 2004. Glare generation based on wave optics. In: Proc. Pacific Graphics, pp. 133–140.

Kellnhofer, P., Ritschel, T., Vangorp, P., Myszkowski, K., Seidel, H.P., 2014. Stereo day-for-night: retargeting disparity for scotopic vision. ACM Trans. Appl. Percept. (Proc. SAP) 11 (3), 15.

Krawczyk, G., Myszkowski, K., Seidel, H.P., 2007. Contrast restoration by adaptive countershading. Comput. Graph. Forum 26 (3), 581–590.

Larson, G.W., Rushmeier, H., Piatko, C., 1997. A visibility matching tone reproduction operator for high dynamic range scenes. IEEE Trans. Vis. Comput. Graph. 3 (4), 291–306.

Luft, T., Colditz, C., Deussen, O., 2006. Image enhancement by unsharp masking the depth buffer. ACM Trans. Graph. 25 (3).

Martinez-Conde, S., Macknik, S.L., Hubel, D.H., 2004. The role of fixational eye movements in visual perception. Nat. Rev. Neurol. 5 (3), 229–240.

Moulden, B., et al., 1988. Border effects on brightness: a review of findings, models and issues. Spat. Vis. 3 (4), 225–262.

Nakamae, E., Kaneda, K., Okamoto, T., Nishita, T., 1990. A lighting model aiming at drive simulators. ACM SIGGRAPH Comput. Graph. 24 (4), 395–404.

Owens, J.D., Luebke, D., Govindaraju, N., Harris, M., Krüger, J., Lefohn, A.E., Purcell, T.J., 2007. A survey of general-purpose computation on graphics hardware. Comput. Graph. Forum 26 (1), 80–113.

Pajak, D., Cadík, M., Aydin, T.O., Myszkowski, K., Seidel, H.P., 2010. Visual maladaptation in contrast domain. In: IS&T/SPIE Electronic Imaging, p. 752710.

Palmer, S.E., 1999. Vision Science: Photons to Phenomenology, vol. 1. MIT Press, Cambridge, MA.

Pattanaik, S.N., Ferwerda, J.A., Fairchild, M.D., Greenberg, D.P., 1998. A multiscale model of adaptation and spatial vision for realistic image display. In: Proc. SIGGRAPH, pp. 287–298.

Purves, D., Shimpi, A., Lotto, R.B., 1999. An empirical explanation of the Cornsweet effect. J. Neurosci. 19 (19), 8542–8551.

Raskar, R., Agrawal, A., Wilson, C.A., Veeraraghavan, A., 2008. Glare aware photography: 4D ray sampling for reducing glare effects of camera lenses. ACM Trans. Graph. (Proc. SIGGRAPH) 27 (3), 56.

Ritschel, T., Eisemann, E., 2012. A computational model of afterimages. In: Comp. Graph. Forum (Proc. Eurographics), vol. 31, pp. 529–534.

Ritschel, T., Ihrke, M., Frisvad, J.R., Coppens, J., Myszkowski, K., Seidel, H.P., 2009. Temporal glare: real-time dynamic simulation of the scattering in the human eye. In: Comp. Graph. Forum (Proc. Eurographics), vol. 28, pp. 183–192.

Ritschel, T., Smith, K., Ihrke, M., Grosch, T., Myszkowski, K., Seidel, H.P., 2008. 3d unsharp masking for scene coherent enhancement. In: ACM Trans. Graph. (Proc. SIGGRAPH) 27, 90.

Rokita, P., 1993. A model for rendering high intensity lights. Comput. Graph. 17 (4), 431–437.

Rusinkiewicz, S., Burns, M., DeCarlo, D., 2006. Exaggerated shading for depicting shape and detail. ACM Trans. Graph. (Proc. SIGGRAPH) 25, 1199–1205.

Simpson, G.C., 1953. Ocular haloes and coronas. Br. J. Ophthalmol. 37 (8), 450.

Smith, K., Landes, P.E., Thollot, J., Myszkowski, K., 2008. Apparent greyscale: a simple and fast conversion to perceptually accurate images and video. In: Comp. Graph. Forum (Proc. Eurographics), vol. 27 (2), pp. 193–200.

Spencer, G., Shirley, P., Zimmerman, K., Greenberg, D.P., 1995. Physically-based glare effects for digital images. In: Proc. SIGGRAPH, pp. 325–334.

Tao, Y., Lin, H., Bao, H., Dong, F., Clapworthy, G., 2009. Feature enhancement by volumetric unsharp masking. Vis. Comput. 25 (5–7), 581–588.

Taubin, G., 1995. Curve and surface smoothing without shrinkage. In: Proc. Computer Vision, pp. 852–857.

Templin, K., Didyk, P., Ritschel, T., Myszkowski, K., Seidel, H.P., 2012. Highlight microdisparity for improved gloss depiction. ACM Trans. Graph. (Proc. SIGGRAPH) 31 (4), 92.

Trentacoste, M., Mantiuk, R., Heidrich, W., Dufrot, F., 2012. Unsharp masking, countershading and halos: enhancements or artifacts? Comput. Graph. Forum 31, 555–564.

Vallerga, S., Covacci, R., Pottala, E., 1980. Artificial cone responses: a computer-driven hardware model. Vis. Res. 20 (5), 453–457.

van den Berg, T.J., Hagenouw, M.P., Coppens, J.E., 2005. The ciliary corona: physical model and simulation of the fine needles radiating from point light sources. Invest. Ophthalmol. Vis. Sci. 46 (7), 2627–2632.

Yoshida, A., Ihrke, M., Mantiuk, R., Seidel, H.P., 2008. Brightness of the glare illusion. In: Proc. APGV, pp. 83–90.

COLOR MANAGEMENT IN HDR IMAGING

T. Pouli*, E. Reinhard*, M.-C. Larabi†, M.A. Abebe*

Technicolor, Cesson-Sévigné, France*
University of Poitiers, Poitiers, France†

CHAPTER OUTLINE

High Dynamic Range Video. http://dx.doi.org/10.1016/B978-0-08-100412-8.00009-7

9.1 INTRODUCTION

Much of high dynamic range (HDR) imaging research has its origins in computer graphics. One of the reasons for this is that lighting simulation software naturally produces imagery that can be considered HDR. In fact, it is difficult to design a rendering simulation that outputs pixel values in a constrained range of light values. This has given rise to, for instance, the Radiance lighting simulation system (Ward, 1994). As a consequence, the output of research on HDR imaging has often been presented at computer graphics-related conferences.

On the other hand, almost all research on color systems is presented at conferences and published in journals dedicated to color science. As there was little overlap between computer graphics and color science, for many years HDR imaging was a field of research for the capture, storage, transmission, and display of images with an extended luminance range. Any research that dealt with an extended gamut of colors was published or presented in different places by researchers who operated in a different field.

This can be seen as an understandable albeit somewhat unfortunate situation, because dynamic range and color gamut are consequences of the same imaging and visual processes. The reasons for treating dynamic range and color gamut as separate issues are largely historical, and the fields of HDR imaging and color science are currently rightly converging. In this chapter, we provide a perspective on those aspects of HDR imaging that are important for color science. This gives us an opportunity to outline what has been achieved in this area, and where more research may be required.

To set the stage, we begin by giving a broad overview of how we see the flow of processing from production to display, and we give a flavor of how HDR imaging impacts conventional color management.

9.1.1 FLOW OF PROCESSING

We expect that HDR video will soon originate from several qualitatively different sources, including film production, special effects, and capture by consumer electronics, including mobile devices. Especially concerning consumer electronics, in this chapter we will assume that the capture software will be able to generate true HDR data (ie, excluding footage that is already "fused" into a standard dynamic range). When the software running on the capture device immediately takes the sensor data and merges it to standard dynamic range, any hope of artifact-free postprocessing (such as color management) is directly impaired. Such output is suitable for legacy displays, and is therefore considered outside the scope of this chapter.

The flow of data from a capture device to a display device will in some sense be similar for professional content and for consumer-produced content, albeit that professionally produced content will undergo more steps between the capture device and display device. Focusing on professional content, Fig. 9.1 shows a somewhat simplified flow of processing from capture to final display, specifically focusing on HDR (Reinhard et al., 2014).

Sources of professional HDR footage will include HDR movie cameras, conventional footage scaled up through a process called inverse tone reproduction, and special effects. The footage is color-graded in postproduction houses, where the different sources are brought into unison, and the director's vision is instilled. From the postproduction studio, the content takes an elaborate path with the aim of it

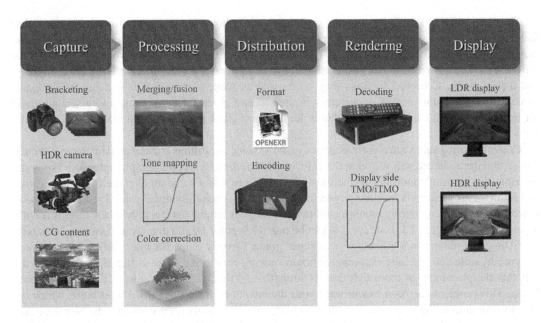

FIGURE 9.1

Stages of processing involved in producing, delivering, and consuming content. *CG*, computer game; *iTMO*, inverse tone mapping operation; *LDR*, low dynamic range; *TMO*, tone mapping operation.

eventually ending up on a screen somewhere. The many and varied transforms that are applied to the content along this pathway enable archiving, facilitate distribution, and render the content on a display with a given specification. The final aim is to reproduce the content on the target display with as high a fidelity as possible. One interpretation of the meaning of "fidelity" is that the consumer sees the content as the director intended it. All steps between the postproduction house and the end consumer therefore need to consider the director's intent as the feature to be optimized. In this context, we see color management as a set of tools to enable the director's intent to be preserved along this pathway from production to end consumer.

The level of control that can be exerted over the footage depends on the destination of the footage. We see two main destinations where content is consumed. These are the cinema and the home. Content that is graded specifically for the cinema undergoes fewer transforms that could affect the director's intent than content aimed at home viewing. Currently, cinemas display two-dimensional movies at a peak luminance of 48 cd/m^2. Three-dimensional movies are displayed with a peak luminance less than half of that. This situation has been the same for decades. However, recent developments suggest that this may soon change. In digital HDR cinema, as in conventional digital cinema, color grading will be performed in a postproduction suite that has the same capabilities as the target cinema. This means that the cinema-goer will have essentially the same experience as the director who joins the colorist in the grading suite to see how his/her movie will look. The distribution chain is set up to not significantly affect the consumers' experience.

The story is not so simple for content that is destined for home viewing. Here, the main intervening factor is the method of delivery. Postproduction houses normally make a second grade for home viewing. The assumed peak luminance is 100 cd/m^2. Different versions of the movie are produced for different markets (often well over 100 versions), including versions with different languages, subtitles, and aspect ratios. Censored versions may be created as well as versions for viewing in airplanes. Nonetheless, the color grading is currently limited to one grade for the cinema and a second grade for home viewing. It can be expected that at least a third HDR grade will be added to this set in the future.

Cinema grading suites are closely matched to the viewing conditions encountered in cinemas, which means that the audience will see what the colorist and the director of photography intended. For home viewing, the viewing conditions in the mastering suite are taken to be an average of home viewing conditions. This includes a mastering monitor calibrated to a peak luminance of 100 cd/m^2, and soon potentially a second mastering monitor calibrated at a higher luminance level (the black level is usually around 0.01 cd/m^2 for LCD displays, but can be much lower for OLED displays). However, even with an HDR grade alongside a 100 cd/m^2 consumer grade, there is scope for many differences between the grading suite and typically encountered home viewing conditions. These differences stem from the fact that the placement of room lighting, its strength, and its color temperature are different in each home. This problem has been exacerbated with the advent of Internet-enabled light bulbs which are able to emit colored light that can be arbitrarily far from commonly used white points. Moreover, the placement, orientation, and size of windows are extremely variable. As such, the viewing conditions in people's homes are relatively unpredictable. A further source of differences stems from the displays on which content is viewed. While in the past the only technology available for watching content was cathode ray tubes, nowadays several more technologies are used, most notably LCD-based televisions, but also plasma televisions and OLED televisions. The trend in display capabilities is toward higher peak luminance, often accompanied by higher black levels in the case of LCD screens. The new OLED displays tend to have comparatively deep blacks as well as a much wider color gamut than other display technologies. Currently, for instance, LCD televisions often reach 300–500 cd/m^2. As an aside, directly displaying a movie that was graded for a peak luminance of 100 cd/m^2 on such a display already begins to destroy the director's intent. Tablet computers routinely reach 700 cd/m^2. In 2014, the first commercial displays entered the market and under favorable circumstances can reach a peak luminance of 1000 cd/m^2. Professional displays can reach higher levels, as they are not bound by energy guidelines.

All this adds up to a wide variability in the signal that is eventually observed by the consumer. The observer is normally adapted according to the room illumination. The display itself is thought to contribute little to the observer's state of adaptation, as the light emitted by the display will be of the same order as the room illumination. However, for very bright displays, especially in dim or dark environments, the display may also contribute to the observer's adaptation. The level and color to which the observer is adapted is one factor determining whether the displayed content is perceived as the director intended. The quality of the display itself as well as its capabilities in terms of peak luminance, black level, color gamut, and bit depth is a second contributing factor. With such an enormous range of display capabilities, and soon at most two different grades available for home viewing (currently only one grade, 100 cd/m^2, is distributed), the onus is increasingly on the consumer display to perform a final processing step on the content to make it suitable for display. As such, display-side color management will become an increasingly involved process, given that each individual display may deviate more or less from the mastering monitor used in the postproduction studio.

9.1.2 COLOR MANAGEMENT

Color management is a catch-all term for a set of processes and algorithms that aim to reproduce content across a range of display devices such that the content is visually matched. For professionally graded content, the aim is to reproduce the content such that it is equivalent to what the colorist and the director of photography observed in the grading suite. In other cases, it may be to reproduce some visual qualities of the scene itself. In essence, color management aims to achieve accurate color reproduction. Several objectives in color reproduction can be defined, such as spectral, colorimetric, exact, equivalent, corresponding, and preferred (Hunt, 1970, 2004). Another classification would distinguish a hierarchy, described as color reproduction, pleasing color reproduction, colorimetric color reproduction, color appearance reproduction, and color preference reproduction (Fairchild, 2013a, 2007). This latter classification has the interesting feature that each subsequent level of color reproduction requires each preceding level to be satisfied. This means, for instance, that to achieve color appearance reproduction, basic color reproduction, pleasing color reproduction, and colorimetric color reproduction also need to be achieved. Current consumer video systems are able to achieve pleasing color reproduction (Fairchild, 2007).

Traditional color management techniques focus on differences between source and target displays and typically account for them with the aid of gamut mapping techniques (Morovič, 2008). This is discussed in Section 9.3. Gamut mapping techniques are particularly important when the target display has a smaller gamut than the source content. In this case, there are colors in the content that cannot be displayed accurately. Gamut mapping would adjust these colors to be inside the target display's gamut, while remaining as close as possible to the original color. To maintain relations between pixels, colors that are already inside the target gamut may also be adjusted slightly. Standardization of this type of color management exists for still images in the form of ICC profiles (ISO 15076-1:2010), and color management within the OpenEXR HDR file format has been proposed (Kainz, 2004). For video there is no such standard.

For HDR content, further adjustments may be necessary. The perception of color is linked to luminance levels, as is, for instance, described by the Hunt effect, which notes that the perceived colorfulness increases with luminance (Fairchild, 2013a). Gamut mapping does not take this into account. When the source and target display have a very different dynamic range, aside from the need for tone reproduction to bring the dynamic range down to the display's dynamic range, there is an additional need to apply color correction. Conversely, if standard dynamic range content is scaled to the capabilities of an HDR display, color correction is also required to match the perception of color. This topic is discussed in more detail in Section 9.4.

Closely related to this set of techniques are those that try to recover information that has already been lost — for instance, due to limitations of the imaging device. Pixels that are out-of-gamut may be corrected with a variety of approximate techniques that try to estimate what the pixel value may have been (Abebe et al., 2015; Guo et al., 2010; Rouf et al., 2012). While it would always be better to avoid the production of out-of-gamut and overexposed pixels, Section 9.5 discusses techniques to partially recover lost information, based on spatial, temporal and cross-channel information.

In the future, we think that source and target viewing environments should also be incorporated into color management, as they have a significant potential to affect how content is perceived. This can be achieved by means of color appearance modeling. Such models require a set of parameters that describe the viewing conditions of the scene depicted in the footage, as well as a matching set of parameters that describe the viewing environment of the observer. These parameters effectively describe the state of adaptation of a hypothetical observer in the scene, as well as the state of adaptation of the observer who is watching the content at home. Color appearance models then apply a transform to the pixels to account for the differential in adaptation. These techniques are described in Section 9.6.

9.2 BACKGROUND

In this section, we briefly review various aspects of color in order to facilitate subsequent discussions. In particular, we define color, color spaces, and color gamuts and discuss appearance correlates as alternative descriptors of color.

9.2.1 COLOR

Color has many possible definitions, which gives a first indication of how difficult it actually is to describe color. It arises from the interaction between light and matter. Such a physical explanation of color would lead to a description of color in terms of spectral radiance (measured in watts per steradian per meter squared per nanometer). This is one definition of color used in radiometry and has applications, for instance, in remote sensing.

The physical measurement of light is a first step toward reproduction of imagery. However, it is no more than a first step. There are many refinements and additional aspects of light to deal with to construct a practical system that can reproduce light. Moreover, light presented to the human visual system is processed in the retina and in the brain, leading to the perception of color. Such color perception cannot be described simply in terms of power as a function of wavelength. The intermediate steps as they are modeled in practical systems are essentially as follows.

The photoreceptors in the retina transduce light into electrical signals that are then transmitted through an optical pathway to the brain. The photoreceptors are all sensitive to a broad range of wavelengths, albeit different types of photoreceptors have a peak sensitivity at different wavelengths. The peak sensitivity is broadly in the blue, green, and red parts of the visible spectrum for the S, M, and L cones. These cones are active at photopic light levels, which are associated with daylight. In dark conditions, the rods are active. The combined light sensitivity of all cone types leads to a broad curve that peaks at 555 nm. This curve is standardized and known as the CIE $V(\lambda)$ luminosity function. A similar curve exists for scotopic lighting conditions, and is known as the CIE $V'(\lambda)$ luminosity function. By weighting spectral radiance according to these curves, we obtain the human visual response to spectra of illumination. The measurement of such weighted quantities is known as photometry.

The three cone types in the human retina allow us to perceive color. As each cone type responds to a broad range of wavelengths with overlapping ranges, the three types together cover the full visible spectrum. A given color spectrum is translated into a triplet of responses, with the strength of the response for each cone type determined by the amount of power in the spectrum at the wavelengths to which the cone type is sensitive. The sensation of color arises from the fact that each cone type has a different peak response.

By and large different spectra result in different responses. However, as a spectrum contains many wavelengths and the visual response of the photoreceptors consists of only three values, inevitably some information is lost. As a consequence, it is possible that some well-chosen spectra produce the same triplet of responses. Such spectra are called metameric with respect to each other. Because humans have three cone types, we are able to build display devices with three color primaries. This also allows us to represent colors with no more than three numbers (sometimes referred to as tristimulus values). As a result, nearly all color spaces are three-dimensional. Moreover, the existence of metamerism allows us to build displays with primaries that have emission spectra different from the spectral response of human cones. Without metamerism, building acceptable color reproduction devices would be nearly impossible.

Cells along the visual pathway are able to analyze the relative response of each cone type and in this way produce the perceptual encoding of the color spectrum that was incident on the retina. This perceptual encoding involves more than just the output of a triplet of photoreceptors. As the human visual system is a complex mechanism with many adaptive processes, it is known that the perception of color is affected not only by the spectrum of the color itself, but also by environmental factors such as the color and strength of the dominant illuminant. Likewise, the amount of illumination surrounding the patch of color of interest affects the perception of the color of the patch itself. This leads to potentially complex spatial interactions that make modeling the appearance of color a difficult endeavor. However, for simple patches shown against a simple background, the higher-level mental representation of color can be described, and this is codefied in a set of color descriptors known as appearance correlates.

9.2.2 COLOR SPACES AND COLOR GAMUTS

A simple model of cone responses is that they output a signal that is a weighted average of the incident spectrum. Having three such weighted averages to describe a color gives rise to a color space, known as the LMS color space. Here L, M, and S stand for "long," "medium," and "short" — the wavelengths at which the different receptor types are maximally sensitive. They correspond roughly to red, green, and blue signals. The digital representation of a color is then a triplet of numbers, which are bounded by some minimum value (often 0 to indicate absence of light) and a maximum value.

As color is represented by three numbers, we could plot the position of a color in a three-dimensional plot. Because each component is bounded by a minimum and a maximum value, the range of representable colors in such a color space is bounded by a volume. This volume is known as the color gamut. If we assume a set of colors is represented in the LMS color space, we can make a three-dimensional point cloud of these colors. If the values of L, M, and S are each bounded between 0 and 1, then this point cloud will be contained within a unit cube.

Some colors representable within such a cube are physically impossible. The spectral response of the L, M, and S cones show significant overlap. This means that when a nonzero value for the M component is specified, the L and S components (or color channels) necessarily have to be nonzero as well. Thus, specifying LMS values of, say, (1, 0, 0) would not represent a physically achievable color, even though it is within the color gamut.

As it is not necessarily feasible to efficiently build capture devices with the same responsivities as the spectral response of the human cones (and the same can be said for the primaries of display devices), there are many color spaces that generally capture a red, a green, and a blue signal. These RGB spaces are typically linked to specific capture and display devices, and are therefore termed "device dependent." They are not, however, connected to absolute physical quantities or even perceptual quantities, although with appropriate measurements they could be transformed into those.

The CIE XYZ color space is a device-independent reference color space that encompasses all colors that humans can perceive. To avoid the problem of requiring negative numbers to represent certain highly saturated colors (or even monochromatic colors), the spectral responses for the three color channels are imaginary (ie, nonphysical). The Y channel in this color space is defined as luminance. The Z channel roughly corresponds to the S-cone response, while the X channel is a linear combination of cone responses such that XYZ tristimulus values are always nonnegative. The XYZ tristimulus values are normalized such that the Y channel is between 0 and 100. This means that X and Z values on occasion may be higher than 100. Transforms between any pair of color spaces can be set up as a conversion between the source color space and XYZ, followed by a conversion from XYZ to the destination color space.

In the CIE XYZ color space, the X and Z channels carry combinations of chromatic and luminance information. It is possible to factor out the luminance information from these two channels, leading to two channels with chromaticity information only. This is achieved by normalization, leading to $x = X/(X+Y+Z)$ and $y = Y/(X+Y+Z)$. A third normalized channel can be computed as $z = 1-x-y$. The combination of Y, x, and y channels leads to a the CIE Yxy color space which contains the same information as the CIE XYZ color space. However, we can directly plot chromatic information in a two-dimensional chromaticity diagram using x and y as the two axes.

Further standard color spaces are CIE L*a*b* and CIE L*u*v*, which are defined through a transform from CIE XYZ. Their input is a pair of XYZ tristimulus values: one for the color to be encoded, and one for an assumed white point. In this case, the white point is the color of the illuminant. As such, these two color spaces offer more than a representation of color. They additionally model the ability of the human visual system to chromatically adapt to an illuminant. As such, they can be seen as rudimentary color appearance models, offering a representation of color as seen under specific illumination. The L channel represents lightness, whereas the a and b channels encode color opponency. Color opponency is a phenomenon that appears in the human visual pathway, where the LMS photoreceptor output is recoded into a different color space where the colors red and green are mutually exclusive, as are the colors yellow and blue. In CIE L*a*b* the a channel represents red-green color opponency (positive values indicate red/magenta; negative values indicate green), while the b channel represents yellow-blue opponency (positive values encode yellow and negative values encode blue). The CIE L*u*v* space functions analogously.

9.2.3 APPEARANCE CORRELATES AND COLOR APPEARANCE MODELS

Human visual perception of color is affected by the observed color itself as well as environmental factors such as the strength and color of illumination. This has given rise to a set of six definitions, each describing a different aspect of color. These appearance correlates and their definitions are as follows (Wyszecki and Stiles, 2000; Fairchild, 2013a):

Brightness. Brightness is the attribute of visual sensation according to which a visual stimulus appears to emit more or less light. It ranges from dim to bright.

Lightness. Lightness is the attribute of visual sensation according to which an area in which a visual stimulus is presented may appear to emit more or less light in proportion to a similarly illuminated area that is perceived as a white stimulus.

Colorfulness. Colorfulness is the attribute of visual sensation according to which the perceived color of an area appears to be more or less chromatic.

Chroma. A visual stimulus may be judged in terms of its difference from an achromatic stimulus with the same brightness. This attribute of visual sensation is called chroma.

Saturation. Saturation is the attribute of visual sensation which allows the difference between a visual stimulus and an achromatic stimulus to be judged regardless of any differences in brightness.

Hue. The attribute of color perception denoted by red, green, blue, yellow, purple, and so on is called hue. A chromatic color is perceived as possessing hue, while an achromatic color is not perceived as possessing hue.

Appearance correlates can be measured through psychophysical experiments. In turn, these experiments have been used to define color appearance models, which take an XYZ tristimulus value, a white point, the dominant illuminant's strength, and other parameters and predict how a patch of color is perceived in terms of brightness, lightness, chroma, colorfulness, saturation, and hue. To uniquely describe a patch of color in this manner, only three of these appearance correlates are required — namely, one of brightness and lightness, one of chroma, colorfulness, and saturation, and finally hue.

9.3 COLOR SPACES FOR HDR AND COLOR WORKFLOWS

In the previous section, we broached the subject of color spaces and discussed the differences between device-dependent and device-independent spaces. Most existing color spaces, however, have been designed to support a limited range of luminance. Because of that, HDR data require different treatment, both when we are modeling and treating color in device-specific scenarios and when we creating perceptual models.

Perceptual color spaces such as CIE L*a*b*, which are by definition device independent, are intended to be approximately uniform. This means that two colors that are spaced by some Euclidean distance in such color spaces have approximately the same perceptual difference, irrespective of where in this color space they are located. In other words, changing the color coordinates by a given amount will always bring about a change in perception of the same magnitude.

The derivation of a perceptually uniform color space requires psychophysical experimentation. This means that uniformity can be claimed only for colors that are within the range of values included in the experiments. As a result, it is not clear whether such color spaces will remain uniform outside their operating ranges. This pertains in particular to HDR data.

The situation is similar with device-dependent color spaces since most existing devices, both for capture and for display, have a limited dynamic range. With the increasing availability of HDR-capable devices, the need for appropriate color spaces and encodings arises. Additionally, new workflows are necessary to deal with such content extensions in the context of production and postproduction. In the following sections we will discuss some of the key efforts in extending color spaces and workflows to the realm of HDR.

9.3.1 PERCEPTUAL COLOR SPACES

If we are interested in evaluating or controlling aspects relating to the perceptual appearance of color, it is crucial that perception plays a role in the construction of the color spaces used for that purpose. Typically, perceptual spaces try to reshape color information so that it better follows perceptual models of color processing. An additional requirement is often that of perceptual uniformity — that is to say, pairs of colors that have the same distance between them should be perceived to be different by a similar amount.

This is important for many applications. For instance, consider a computation of the average of a set of colors. If the color space in which these are encoded is not uniform, then equal distances in the color space might translate to different perceptual changes, therefore leading to a biased measure.

There have been several attempts to construct such color spaces within color science literature, and these have served well for many applications, such as appearance or gamut management. Among those, CIE L*a*b* (CIELAB for short) is the most used and evaluated such space (Wyszecki and Stiles, 2000), and it transforms (X, Y, Z) tristimulus values through a nonlinear transform. It also requires the specification of a nominally white object color stimulus (X_n, Y_n, Z_n) — that is, the white point. If we set $(X_d, Y_d, Z_d) = (X/X_n, Y/Y_n, Z/Z_n)$, then the transform is given by

$$L^* = \begin{cases} 903.3 Y_d & Y_d \le t, \\ 116 Y_d^{1/3} - 16 & Y_d > t, \end{cases} \tag{9.1}$$

$$a^* = 500 \left(f(X_d) - f(Y_d) \right), \tag{9.2}$$

$$b^* = 200 \left(f(Y_d) - f(Z_d) \right), \tag{9.3}$$

where $t = 0.008856$ and $f(x)$ is defined as

$$f(x) = \begin{cases} 7.787x + 16/116 & x \le t, \\ x^{1/3} & x > t. \end{cases} \tag{9.4}$$

Here, the L channel encodes lightness, while a and b encode chromatic information. Specifically, chromatic information is encoded in an opponent manner, akin to the opponent processing stages of the human visual system. Roughly, a encodes chromatic values ranging from red to green, while b encodes yellow to blue. Both channels are achromatic at 0. The relation between the three channels of the CIELAB is illustrated in Fig. 9.2.

A further transform can be applied to CIELAB, recoding the a and b channels into a cylindrical coordinate system, leading to CIE L*C*h* (CIE LCh for short), where C and H represent chroma and hue:

$$L^* = L^*, \tag{9.5}$$

$$C^* = \sqrt{(a^*)^2 + (b^*)^2}, \tag{9.6}$$

$$h^* = \tan^{-1} \left(\frac{b^*}{a^*} \right). \tag{9.7}$$

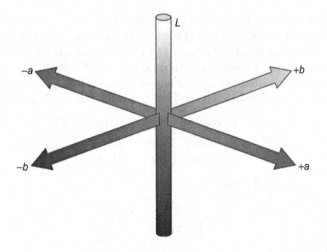

FIGURE 9.2

An illustration of the three channels of the CIELAB color space.

The correlates of lightness, chroma, and hue bring us one step closer to a description of color in terms of human visual perception, although they cannot offer the flexibility and accuracy of a full color appearance model (see Section 9.6). The relative correlate of saturation s may be calculated as

$$s = \frac{C^*}{L^*}.$$ (9.8)

As such, it can be seen that chroma must covary with lightness if the perception of saturation is to remain constant. More recently, a somewhat different form of saturation was proposed (Lübbe, 2008), and has already shown merit in the context of color correction for tone reproduction (Pouli et al., 2013a):

$$s = \frac{C^*}{\sqrt{(C^*)^2 + (L^*)^2}}.$$ (9.9)

The definition of these appearance correlates shows that CIELAB can be seen as the basis of a simple color appearance model. Improved accuracy and predictive power can be obtained by use of more elaborate color appearance models, which are discussed in Section 9.6.

Despite the extensive use of CIELAB, it has several limitations that might restrict its use in the context of extended dynamic range content. The first is that the color space expects relative input. The white point is usually normalized so that $Y_n = 100$ indicates diffuse white. The tristimulus value of the color itself should therefore also be specified in relative units.

Second, CIELAB is not fully perceptually uniform. In particular, blue hues suffer from some nonlinear behavior which could lead to undesired color shifts when one is manipulating colors along lines of constant hue. This is an issue likely to increase in significance since the entertainment industry is moving toward the use of wider color gamuts — for instance, with the adoption of the BT.2020 color space (International Telecommunication Union, 2012).

To reduce this issue and offer better hue linearity, the IPT color space was proposed (Ebner and Fairchild, 1998). While based on similar principles, it offers better hue linearity as well as a simplified transform. The IPT color space is defined relative to CIE XYZ, and involves a linear transform to the LMS color space (thus modeling photoreceptor spectral responses), a nonlinearity applied in this color space, followed by a linear transform to a color opponent space. Under the assumption that the CIE XYZ input is specified for a D65 illuminant, the LMS transform is defined as follows:

$$\begin{bmatrix} L \\ M \\ S \end{bmatrix} = \begin{bmatrix} 0.4002 & 0.7075 & -0.0807 \\ -0.2280 & 1.1500 & 0.0612 \\ 0.0000 & 0.0000 & 0.9184 \end{bmatrix} \begin{bmatrix} X \\ Y \\ Z \end{bmatrix}. \tag{9.10}$$

The nonlinearity for the L channel is then

$$L' = \begin{cases} L^{0.43} & L \geq 0, \\ -(-L)^{0.43} & L < 0. \end{cases} \tag{9.11}$$

The values M' and S' are calculated similarly. The opponent transform is then

$$\begin{bmatrix} I \\ P \\ T \end{bmatrix} = \begin{bmatrix} 0.4000 & 0.4000 & 0.2000 \\ 4.4550 & -4.8510 & 0.3960 \\ 0.8056 & 0.3572 & -1.1628 \end{bmatrix} \begin{bmatrix} L' \\ M' \\ S' \end{bmatrix}. \tag{9.12}$$

After this transform, the I channel represents lightness information, while P encodes red-green color opponency and T encodes yellow-blue color opponency. The range of values is $I \in [0, 1]$ and P, $T \in [-1, 1]$. This color space is known to provide better hue linearity than CIELAB, as shown in Fig. 9.3. However, similarly to CIELAB, the IPT color space is not designed for use with HDR data.

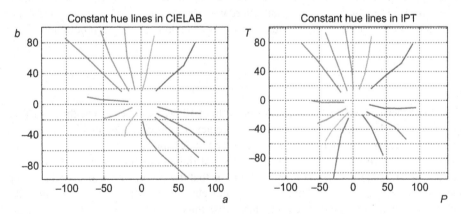

FIGURE 9.3

The lines represent lines of constant perceived hue with increasing chroma in the CIELAB (left) and IPT (right) color spaces. In CIELAB, certain perceived hues show a marked shift, while in IPT, corresponding hues are closer to constant.

Source: Modified from Ebner and Fairchild (1998).

9.3.2 PERCEPTUAL HDR COLOR SPACES

The CIELAB and IPT color spaces were both originally designed to support a dynamic range of approximately 100:1 (Fairchild and Chen, 2011). Recently, however, they were both extended to work with HDR and wide color gamut data. In essence, the nonlinearities in these two color spaces are power functions that have been replaced with physiologically more plausible sigmoidal functions. This allows the resulting hdr-CIELAB and hdr-IPT color spaces to function for colors that are many orders of magnitude below diffuse white as well as many orders of magnitude above diffuse white. These color spaces therefore open the door for HDR colorimetry.

To extend the lightness channel of these perceptual color spaces beyond diffuse white, several perceptual experiments were conducted, following the partition scaling method. In these experiments, three patches were shown. The left and middle patches encoded a predefined lightness difference. The observer was then tasked with adjusting the right patch until its lightness difference with respect to the middle patch appeared equal to that between the left and middle patches. This procedure was followed for both printed and projected patches and for luminance levels above and below diffuse white (Fairchild and Chen, 2011).

An initial effort to extend CIELAB to a higher dynamic range came from Fairchild and Wyble (2010), where the exponential function used to compute the lightness channel in Eq. (9.1) is replaced by the Michaelis-Menten equation, defined as follows:

$$f(Y_d) = 100 \frac{Y_d^\epsilon}{Y_d^\epsilon + 0.184^\epsilon} + 0.02 \tag{9.13}$$

with the exponent ϵ set to 1.5.

On the basis of the findings of the psychophysical studies described above, the constants in Eq. (9.13) can be adjusted, obtaining

$$f(Y_d) = 247 \frac{Y_d^\epsilon}{Y_d^\epsilon + 2^\epsilon} + 0.02 \tag{9.14}$$

with the exponent ϵ set to 0.58. The exponent can be further modified with use of additional knowledge about the scene to account for perceptual phenomena relating to image appearance, such as the Bartleson-Breneman effect and the Stevens effect (Fairchild, 2013b), obtaining

$$\epsilon = 0.58/(\text{sf} \cdot \text{lf}), \tag{9.15}$$

where

$$\text{sf} = 1.25 - 0.25(Y_s/0.184) \text{ for } 0 \leq Y_s \leq 1.0, \tag{9.16}$$

$$\text{lf} = \log 318/\log Y_{abs}, \tag{9.17}$$

where Y_s encodes the relative luminance of the surround and Y_{abs} is the absolute luminance of the diffuse white of the scene, given in candelas per square meter. However, according to Fairchild and Wyble, these adjustments should be treated as suggestions and are not fully validated.

The full formulation of the hdr-CIELAB replaces the exponentiation step of CIELAB with Eq. (9.14) and is given by

$$L_{hdr} = f(Y_d), \tag{9.18}$$

$$a_{hdr} = 5(f(X_d) - f(Y_d)), \tag{9.19}$$

$$b_{hdr} = 2(f(Y_d) - f(Z_d)), \tag{9.20}$$

while the cylindrical coordinates for chroma and hue are given by

$$c_{hdr} = \sqrt{a_{hdr}^2 + b_{hdr}^2}, \tag{9.21}$$

$$h_{hdr} = \tan^{-1}(b_{hdr}/a_{hdr}). \tag{9.22}$$

The formulation of the HDR version of the IPT color space — namely, hdr-IPT — follows a similar procedure. XYZ tristimulus values are first converted to LMS values as for the standard IPT color space, following Eq. (9.10). L', M', and S' values are in turn computed with the Michaelis-Menten–based compressive function given in Eq. (9.14) with the exponent ϵ set to $\epsilon = 0.59/(\text{sf} \cdot \text{lf})$ in this case. Finally, IPT values are obtained with Eq. (9.12).

Although these new color spaces are based on perceptual studies that cover extended dynamic ranges, they have not been sufficiently tested to guarantee their performance. In fact in Fairchild and Chen (2011) they were found to perform no better than their standard counterparts. However, the procedure followed to create them offers some interesting new directions for color space construction, particularly in the context of extended gamut and dynamic range.

9.3.3 DISPLAY-REFERRED COLOR SPACES AND ENCODINGS

The color spaces discussed so far provide a description of color that is independent of any capture or display device. Instead, they represent color in a perceptually oriented way. Since they offer some degree of perceptual uniformity along the different dimensions that they describe (lightness, chroma, hue), they are well suited for color and image processing operations such as gamut mapping. To display images correctly on different devices, this representation needs to be translated to something that is meaningful to the target device itself.

Color spaces that target display applications are known as *output referred* or *display referred* and are typically represented by three color components, encoding red, green, and blue values. They are defined through a set of primaries, given in CIE x, y chromaticity coordinates, as well as a white point, and they cover a subset of the visible color gamut.

Until recently, most display-referred color spaces encoded a color gamut comparable with that of CRT displays. However, with recent developments toward higher dynamic range and wider color gamut displays, the need for extended display-referred spaces has arisen. If we consider specifically television-related standards, this limitation of currently used standards and associated color spaces becomes apparent.

For instance, high-definition televisions currently rely on ITU-R Recommendation BT.709 (ITU, 1998) to define the characteristics for display devices. This standard defines aspects including resolution, frame rate, transfer characteristics, and primary chromaticities that BT.709-mastered

Table 9.1 White Point and Primary Chromaticities for the BT.709 Color Space

	White Point	R	G	B
x	0.3127	0.64	0.30	0.15
y	0.3290	0.33	0.60	0.06

content as well as displays are expected to adhere to. The color gamut defined by this standard covers approximately 36% of the visible gamut, and BT.709 content is typically graded at around 100 cd/m^2.

Although a full definition of BT.709 is beyond the scope of this chapter, the primary chromaticities of this space are given in Table 9.1. For several applications, a luminance-chrominance encoding is used within the context of this standard, instead of the RGB components being used directly. This encoding approximates the opponent-like color spaces discussed in the previous section, separating luma (or luminance) from chromatic channels. This space is known as YC_bC_r or $Y'CbCr$, where Y represents linear luminance and Y' represents nonlinearly compressed luminance (ie, luma).

The YC_bC_r encoding is obtained from RGB as follows, after normalization of the RGB values to a maximum of 1:

$$Y = 0.2126 \times R + 0.7152 \times G + 0.0722 \times B, \tag{9.23}$$

$$C_b = (B - Y)/1.8556, \tag{9.24}$$

$$C_r = (R - Y)/1.5748, \tag{9.25}$$

which can be expressed in matrix form as

$$YC_bC_{r709} = \begin{bmatrix} 0.2126 & 0.7152 & 0.0722 \\ -0.1146 & -0.3854 & 0.5000 \\ 0.5000 & -0.4542 & -0.0458 \end{bmatrix} RGB_{709}^T. \tag{9.26}$$

At the time of writing, consumer televisions currently often exceed a peak luminance output of 400 cd/m^2 and HDR-capable models can exceed a peak luminance output 1000 cd/m^2. For instance, the a high-end Sony model (X950B) achieves 1000 cd/m^2, while similar performance is reported for Samsung's recent offering. Additionally, the introduction of technologies such as OLED and quantum dots in display manufacturing is allowing wider color gamuts to be achieved. Given these improved capabilities, it is clear that BT.709-mastered content is unlikely to match such target displays well, possibly leading to distortions and changing the content reproduction.

To match the extended capabilities of recent and future devices, a new standard has been defined. This is known as BT.2020 (International Telecommunication Union, 2012) and offers extensions in resolution, color gamut, and bit depth. The primary chromaticities for the BT.2020 color space are given in Table 9.2.

Similarly to the previous standards, BT.2020 also offers a luminance-chrominance encoding, which can be obtained as follows:

$$Y = 0.2627 \times R + 0.6780 \times G + 0.0593 \times B, \tag{9.27}$$

Table 9.2 White Point and Primary Chromaticities for the BT.2020 Color Space

	White Point	R	G	B
x	0.3127	0.708	0.170	0.131
y	0.3290	0.292	0.797	0.046

$$C_b = (B - Y)/1.8814, \tag{9.28}$$

$$C_r = (R - Y)/1.4746, \tag{9.29}$$

which can be expressed in matrix form as

$$YC_bC_{r2020} = \begin{bmatrix} 0.2627 & 0.6780 & 0.0593 \\ -0.1396 & -0.3604 & 0.5000 \\ 0.5000 & -0.4598 & -0.0402 \end{bmatrix} RGB_{2020}^T. \tag{9.30}$$

Since YC_bC_r is intended for display applications, a continuous representation is not possible. Consequently, YC_bC_r data can be quantized to a targeted bit depth as follows:

$$\begin{bmatrix} D_Y \\ D_{Cb} \\ D_{Cr} \end{bmatrix} = \text{INT}\left(\begin{bmatrix} 219Y & +16 \\ 224C_b & +128 \\ 224C_r & +128 \end{bmatrix} 2^{n-8} \right). \tag{9.31}$$

In both BT.709 and BT.2020, the YC_bC_r encoding represents chromatic information as a color difference between RGB channels and the luma Y (or luminance Y'). Although this particular color difference encoding is commonly used, other proposals have also been considered by standardization bodies. Specifically, the SMPTE (Society of Motion Picture & Television Engineers, 2014) is considering an encoding known as YD_zD_x, which is constructed on the basis of the CIE 1931 XYZ color space, while a similar encoding based on the CIE Lu'v' color space has also been proposed in the context of HDR encoding specifically (Ward Larson, 1998).

To obtain YD_zD_x values, the following transform is applied on XYZ data:

$$Y = Y, \tag{9.32}$$

$$D_z = \frac{\frac{2763}{2800}Z - Y}{2}, \tag{9.33}$$

$$D_x = \frac{X - \frac{2741}{2763}Y}{2}. \tag{9.34}$$

The YD_uD_v encoding is obtained as follows:

$$Y = L, \tag{9.35}$$

$$D_u = \frac{u'}{0.62} - 0.5, \tag{9.36}$$

$$D_v = \frac{v'}{0.62} - 0.5. \tag{9.37}$$

Although the formulation of these encodings is similar, a question that arises is which is best suited for representing chromatic information given a limited bit depth and how many bits might be necessary. Given increasing demands for extended color gamuts, this question is rapidly becoming more relevant to guide the design (and possibly price) of consumer display devices.

To evaluate the behavior of these color difference encodings in terms of encoding efficiency, Boitard et al. (2015) performed a series of psychophysical studies as well as correlation analysis. To assess the minimum number of bits necessary within a given encoding scheme, a set of patches was shown to observers, encoding both luma and chromatic gradients along different dimensions (see the examples in Fig. 9.4A). Four patches were shown simultaneously, but one was quantized to a given number of bits with a different encoding scheme. The task for observers was to identify which patch was quantized: if the difference from the unquantized patches was visible, that would mean that a higher bit depth was necessary to avoid visible quantization artifacts.

This procedure was repeated at different luminance levels from 0.05 to 50 cd/m^2. The findings from this study are summarized in Fig. 9.4B, showing that the HDR-specific YD_uD_v requires the fewest bits, especially at lower luminance levels. The more commonly used YC_bC_r was found to require nine bits, while YD_zD_x required 10 bits. Although this suggests that YD_uD_v might be a better choice for future encoding schemes, the luminance levels tested in this study were lower than what the vast majority of televisions offer today, indicating that further research is certainly necessary in this direction.

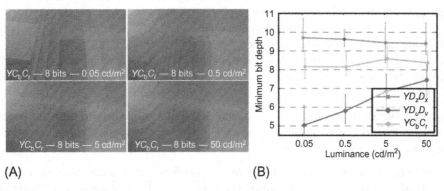

(A) (B)

FIGURE 9.4

Example stimuli and results from psychophysical assessment of color difference encodings. (A) Example stimuli for different luminance levels. (B) Minimum bit depth for different encodings.

Source: Modified from Boitard, R., Mantiuk, R.K., Pouli, T., 2015. Evaluation of color encodings for high dynamic range pixels. In: Human Vision and Electronic Imaging, p. 93941K.

Table 9.3 Primary Chromaticities for the Color Space Defined Within ACES

	R	G	B
x	0.73470	0.00000	0.00010
y	0.26530	1.00000	−0.07700

9.3.4 POSTPRODUCTION WORKFLOWS FOR HDR

The increasing availability of HDR-capable displays also necessitates the development of production and postproduction workflows that can seamlessly address HDR content. The main candidate for this purpose is the solution proposed by the Academy of Motion Picture Arts and Science, which is known as ACES (SMPTE, 2012). ACES stands for "Academy color encoding system" and it describes a color space, image format, and workflow that is intended to be future proof.

The core idea behind ACES is that content mastered within the ACES workflow would theoretically never need to be remastered. To achieve that, the format specified by ACES allows an extremely wide dynamic range and color gamut to be stored and processed, with no loss of information. Within ACES, data are scene referred and linear, with no dependence on the camera or rendering system used to produce the data. The OpenEXR file format is used to store ACES data.

The internal representation used by ACES has several key properties:

- *Wide color gamut:* The gamut encoded within ACES is significantly larger than any output gamut and in fact uses virtual primaries to ensure that it covers the full visible gamut. Table 9.3 lists the primary chromaticities of the ACES color gamut.
- *HDR:* As ACES is intended to be future proof, it allows the encoding and storage of the full dynamic range of film and current digital formats, with more than sufficient room for future expansions. This is also crucial for ensuring that source content is stored and mastered with no loss of quality, irrespective of the output target.
- *High precision:* A floating point representation is used internally to ensure that content is encoded with high precision and without quantization.
- *Scene referred:* To avoid any dependence on devices — both input and output devices — ACES encodes relative scene luminance in a linear manner. A value of 1 is intended to represent the luminance of a 100% reflector.

To translate content from different sources into ACES and from there to different target output devices, several transforms are necessary. Fig. 9.5 illustrates the main components of an ACES workflow, which will be discussed below.

Input device transform. Input content to the ACES workflow may originate from various different sources. Different cameras may have a different response and noise behavior, rendered content may be prepared with a specific target in mind, and so on. The input device transform aims to convert such media from the device- or application-specific space to the scene-referred, linear color space defined within ACES.

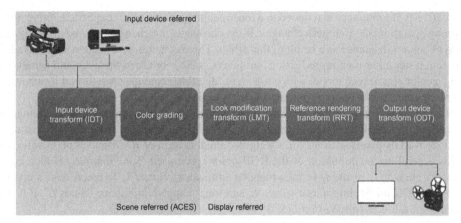

FIGURE 9.5

The main components of the ACES workflow.

As this transform requires knowledge of the specific device, it is intended to be provided by the manufacturer or even included within the device itself. Some camera manufacturers already provide such transforms for their devices (eg, Sony F55/F65 cameras). However, there is not yet widespread support.

Look modification transform. Once content is represented within the ACES color space, look modification transforms can be applied, serving as a starting point for a particular look. Look modification transforms are nondestructive in that they do not directly change image data.

Reference rendering transform. This transform is key to the ACES workflow. Since ACES uses virtual primaries and is scene referred, it is not intended for direct viewing. The reference rendering transform is intended to be universally accepted and it converts ACES to an idealized output color encoding, referred to as the output color encoding specification.

Output device transform. To view ACES-mastered content on a given display, a final transform is necessary, moving from the output color encoding specification to the specified display-referred space, such as BT.709 or BT.2020. Assuming that the given display conforms to particular standards, it should be possible to maintain generalized output device transforms that target different display standards.

9.4 **COLOR CORRECTION**

Typically, tone reproduction and inverse tone reproduction primarily affect a luminance representation of an image. It is not desirable to alter hues. On the other hand, there is a known interaction between lightness and chroma, an effect first described by Hunt, and now known as the Hunt effect

(Fairchild, 2013b). If luminance is reduced in a tone mapping operation, or expanded during an inverse tone mapping operation, the lightness changes. If the chroma is not changed correspondingly, then the perception of color will change as a result of this effect. There is therefore a need to apply an additional color correction step after tone reproduction and inverse tone reproduction. We could achieve this by combining a color appearance model with a tone reproduction technique (Akyüz and Reinhard, 2006), although many applications require computationally simpler solutions.

This leaves several ways for our achieving color correction/preservation in tone mapping. Several of these assume that an HDR image is given in some RGB color space. A luminance channel L can be derived from RGB pixels with use of a weighted combination of R, G, and B pixel values, where the weights depend on the definition of the RGB color space used. Tone mapping is then applied to this luminance channel L, resulting in tone-mapped luminance values L^*. To reconstruct a color image from the HDR RGB image, luminance values L, and tone-mapped luminance values L^*, it is possible to maintain color ratios as follows (Schlick, 1994):

$$R^* = \frac{R}{L}L^*, \quad G^* = \frac{G}{L}L^*, \quad B^* = \frac{B}{L}L^*. \tag{9.38}$$

Maintaining color ratios will, by and large, keep chroma and hue unchanged. While this is desirable for hue, as argued above, it is not desirable for chroma, as the results may look oversaturated. A straightforward solution is to introduce a parameter s, with values between 0 and 1, that is applied as follows (Tumblin and Turk, 1999):

$$R^* = \left(\frac{R}{L}\right)^s L^*, \quad G^* = \left(\frac{G}{L}\right)^s L^*, \quad B^* = \left(\frac{B}{L}\right)^s L^*. \tag{9.39}$$

While for appropriately chosen values of s this method reduces the oversaturation problem, it may affect luminance values for $s \neq 1$. An alternative approach was developed to adjust chroma without affecting luminance values during color image reconstruction (Mantiuk et al., 2009). It achieves this by defining a linear interpolation between chromatic colors and corresponding achromatic colors (Mantiuk et al., 2009):

$$R^* = \left(\left(\frac{R}{L} - 1\right)s + 1\right)L^*, \tag{9.40}$$

$$G^* = \left(\left(\frac{G}{L} - 1\right)s + 1\right)L^*, \tag{9.41}$$

$$B^* = \left(\left(\frac{B}{L} - 1\right)s + 1\right)L^*. \tag{9.42}$$

This approach is better able to preserve luminance values, albeit under certain circumstances undesirable hue shifts may be introduced (Reinhard, 2011). Therefore, the choice between one or the other formulation depends on the targeted application and its tolerance to luminance or hue shifts.

For both methods, setting the parameter s can be done either manually, which requires the specification of a value that is adapted to both content and the tone reproduction operator, or automatically by use of the slope of the tone curve at each luminance level (Mantiuk et al., 2009). For the latter solution, s is a function of the amount of compression or expansion c applied and two constants (k_1, k_2) obtained from the least squares fitting. Automatically determining the value for parameter s is difficult to reconcile with spatially varying tone reproduction operators, as in this case there is no single curve that is amenable to analysis.

To solve this problem and simultaneously minimize both luminance and hue shifts, a correction scheme was developed that was based on recent advances in the design of perceptually uniform color spaces as well as insights into the definition of appearance correlates (Pouli et al., 2013a). In this method, from the IPT color space a cylindrical coordinate system is derived, to give lightness I, chroma C, and hue h for each pixel (Shen, 2009). Saturation is calculated as $s(C, I) = C/\sqrt{C^2 + I^2}$, which is a recent proposal that arguably follows human perception more closely than the traditional formulation for saturation, which is $s(C, I) = C/I$ (Lübbe, 2008). With subscripts i for input HDR, t for tone-mapped, and c for final color-corrected images, the method begins by scaling chroma:

$$C'_t = C_t \frac{I_i}{I_t}.$$

(9.43)

This value is then used to correct chroma on the basis of the ratio between saturations observed in the input HDR and tone-mapped images:

$$C_c = \frac{s(C_i, I_i)}{s(C'_t, I_t)} C'_t.$$

(9.44)

Fig. 9.6 shows results obtained with the three color correction approaches described on images that were tone-mapped by the method of Li et al. (2005).

9.5 RECOVERY OF CLIPPED AND OVEREXPOSED REGIONS

Who has not been disappointed after capturing a sunset or a beautifully illuminated landscape and seeing the brightest parts of the scene appearing clipped and flat in the resulting image. The loss of image details in overexposed and clipped areas can be very annoying, leading to a decrease of the quality of experience. In addition to this subjective aspect, it complicates the objective analysis of the image, potentially leading to the wrong extraction of features.

The problem can be split into two categories: (1) overexposure, where all three channels are clipped and (2) color clipping, where only one or two channels are clipped and therefore some information remains (see Fig. 9.7). Overexposure can occur when the sensor of the camera receives more light than its maximum capacity. Bright areas in that case will not be captured correctly as the sensor saturates. This, for instance, may occur when a scene is captured with incorrect camera settings or when the dynamic range of the scene is higher than that of the camera. On the other hand, color clipping occurs when intensity information for one or two of the color channels is outside the intended device gamut or beyond the capacity of the camera sensor (eg, when a bright and saturated object is captured or near overexposed areas).

Overexposure and color clipping have several consequences for the image. Details in affected areas are lost as the sensor saturates. Additionally, as different color components may saturate, color reproduction may be incorrect in these areas.

Additionally, display capabilities are currently improving at fast rates, with HDR and wide gamut displays starting to become available to consumers. Given the lack of native HDR and wide gamut content, this is creating an unparalleled demand for ways of preparing legacy content so that it takes

FIGURE 9.6

Comparisons between the automatic color correction algorithms of Pouli et al. (2013a) (B and F), Schlick (1994) (C and G), and Mantiuk et al. (2009) (D and H). The images were tone-mapped (A and E) with a spatially varying operator (Li et al., 2005) and then processed with the three correction methods.

Source: Based on results extracted from Pouli et al., 2013a.

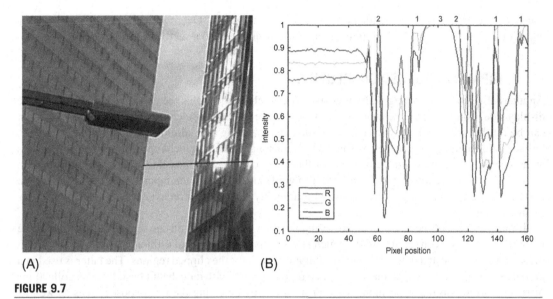

(A)   (B)

FIGURE 9.7

Color clipping: (A) an example image showing clipped and overexposed regions; (B) a plot of the RGB intensities versus the pixel index along the black horizontal line of the image in (A). The numbers at the top in (B) correspond to the number of clipped channels.

advantage of such displays. However, to maximize the visual appeal of legacy content on new displays, it is not sufficient to expand the dynamic range, and it is also desirable to recover the details and color information lost because of overexposure or color clipping.

As an example, consider a standard dynamic range image that is passed through an inverse tone reproduction operator to prepare it for display on an HDR display. If this image contains overexposed areas, inverse tone reproduction will expand the dynamic range, likely placing the overexposed region at the peak luminance of the display. Reproducing a featureless region at such bright luminances will emphasize these defects and thereby reduce the visual experience. In such cases it is highly desirable to fill in such regions with visually plausible detail. The remainder of this section discusses various approaches to mitigate these problems.

9.5.1 CORRECTION DURING ACQUISITION

One solution consists in avoiding the occurrence of overexposed and underexposed pixels directly at the capture stage, where the varying exposures follow a regular pattern. Two methods were proposed for reconstruction of underexposed and overexposed pixels (Nayar and Mitsunaga, 2000). The first is an aggregation approach that involves averaging of the local irradiance values produced by the correctly exposed pixels. The second is based on a bicubic interpolation to simultaneously find the unknown pixels and apply a denoising on the known ones.

More recently, Hirakawa and Simon (2011) proposed performing the HDR reconstruction simultaneously with the demosaicing step by considering that different sensitivities are implied by the different translucencies of the three color filters in a regular Bayer pattern. Overexposed areas are corrected by adoption of the gray-world assumption and are thus projected onto a set of "plausible" colors.

9.5.2 INPAINTING-BASED CORRECTION

An alternative way for dealing with overexposure and clipping locates the corrective processing at the display side. Consequently, the problem of overexposure is then considered as missing information that can be refilled with well-known techniques such as inpainting (Bertalmio et al., 2000). Nevertheless, unlike occluded image regions reconstructed by traditional inpainting techniques, pixels in overexposed areas contain some useful information. On the basis of this, it is possible to take the information recovered from overexposed pixels and use it to guide and constrain the inpainting process (Tan et al., 2003). The inclusion of illumination constraints arguably allows better recovery of shading and textures by inpainting.

If many images of the same scene are available, then this information can be leveraged to direct the inpainting process (Joshi et al., 2010), which starts by predicting global estimates for all layers within the saturation region and then constructs a saturation mask for the clipped regions. The latter is used in a global estimation process such that the masked region is filled with data that fit best. Internet collections of images afford similar opportunities — for instance, by detecting the overexposed areas as well as glare, followed by computing a median color image over the registered images with high photometric consistency (Zhang et al., 2014). Finally, the gradients of the saturated regions are replaced with those of the median image by a method achieving seamless blending.

With the aim to enrich the details of the overexposed and underexposed regions, Wang et al. (2007) proposed an effective HDR image hallucination approach. This approach builds on the assumption that patches of sufficient quality and textures similar to those of the targeted regions exist in the image. Hence, both luma and chroma are corrected in addition to filling the texture of overexposed and underexposed regions. This approach requires user interaction to identify textures that can be used in the filling procedure.

9.5.3 NEIGHBORHOOD-BASED CORRECTION

Instead of using inpainting techniques for filling overexposed areas with content coming from different sources (ie, other parts of the same image, a selected database, or the Internet), many techniques are based on the propagation of information obtained from nearby regions. For instance, Rouf et al. (2012) use the border of the clipped region to estimate a smooth hue distribution and combine it with the gradient information obtained from the unclipped channels. The restoration of clipped areas is achieved by the solving of a Poisson problem. This method generally provides a smooth correction, albeit with the limitation that hue is assumed to be independent of intensity. This assumption may not always hold for all scenes, such as those depicting sunsets.

It can be argued that declipping operations can be done advantageously in CIELAB instead of the more commonly used RGB color spaces (Guo et al., 2010). Processing may proceed by first creating an overexposure map which records how much a pixel is affected by overexposure based on lightness and chroma. Then, tone mapping is used to nonlinearly compress the dynamic range of the well-exposed areas, freeing up range for the newly corrected pixels without requiring additional dynamic range.

9.5.4 **CORRELATION-BASED CORRECTION**

In RGB-like color spaces, the color channels in images tend to exhibit high levels of correlation, meaning that values in one channel are good predictors of the values in other channels (Pouli et al., 2013b). When only one or two channels are clipped, information at the same spatial location within the remaining channels can be used for reconstruction, exploiting this correlation between channels. In the simplest case, this can be done at the scanline level, although in this case only one- and two-channel information can be recovered (Abebe et al., 2014).

In the YC_bC_r color space, strong correlation is present between the saturated and surrounding pixels. Xu et al. (2011) rely on this correlation to restore both luma and chroma of the clipped pixels. Before identification of clipped pixels by means of a threshold, a bilateral filter is applied to remove camera-induced noise. Then, the clipped pixels are grouped depending on their chroma, where the influence of illumination is eliminated by use of only chromatic channels C_b and C_r. For chroma correction, Xu et al. used normalized convolution, associating with each pixel a confidence level with regard to its measurement. Of course, depending on the number of clipped channels, a different strategy is used to recover plausible pixel values.

Fu et al. (2013) designed a YC_bC_r-based algorithm that treats luminance and chrominance separately. Here, the well-known MacAdam ellipse model is mapped to the working color space in order to search for the nearest color. Finally, a nonlinear quantification is adopted, followed by a mapping back to RGB.

Following a different strategy, Zhang and Brainard (2004) proposed an automatic method exploiting the aforementioned correlation between channels at each pixel. This method estimates the clipped color channel(s) using a Bayesian algorithm based on the corresponding unclipped color channel(s) and the prior distribution parameters.

Masood et al. (2009) extended this approach to estimate ratios between RGB channels of clipped pixels by using the relationship between pixels at the same spatial location as well as their neighborhood. Although their use of neighborhood information allows this method to propagate chromatic information into the clipped regions, color bleeding may occur into connected but semantically different regions.

The method proposed by Abebe et al. (2015) relies on cross-channel correlation to correct regions where one or two channels are clipped, similarly to the method of Masood et al. (2009), but applying the correction in an additive rather than a multiplicative manner to avoid overcorrection. To ensure that even disconnected but similar regions are corrected in a coherent manner, a segmentation and region grouping scheme that considers both chromatic differences and distance between regions is used. At the same time, fully overexposed areas are reconstructed with a brightness profile reshaping approach that relies on a combination of inpainting and edge-aware filtering.

9.5.5 **COLOR STATISTICS-BASED CORRECTION**

More complex color statistics have also been used to guide the reconstruction of clipped areas. Mansour et al. (2010) formulated this problem with the assumption that images are nearly sparse in an appropriate transform domain, and consequently the characteristics of clipped pixels can be obtained from the unclipped neighboring pixels. Hence, they applied a hierarchical windowing algorithm, allowing them to select image areas having few clipped pixels. One obtains the correction by solving a constrained ℓ_1 minimization problem for each selected region.

The color lines model has been defined to represent colors in the RGB space, enabling each region of a color image to be represented according to the properties of its associated color line. Exploiting this representation, Elboher and Werman (2011) detect color lines in the color distribution and expand them along a monotonically increasing function. Although this works well for images with distinct, colorful objects with accurate color segmentation, it is more difficult to accurately determine color lines in complex images.

Fig. 9.8 shows a visual comparison of three well-performing declipping methods — namely, those of Guo et al. (2010), Masood et al. (2009), and Abebe et al. (2015). For all images in Fig. 9.8, an HDR reference image was used to derive a standard dynamic range exposure, which was then subjected to each of the declipping algorithms. In addition, Fig. 9.8 shows the output of HDR-VDP-2 version 2.1 (Mantiuk et al., 2011), which provides the probability of observing the distortion: from certain (red) via likely (green) to zero (blue). In this case, the output of the declipping algorithms was compared with the HDR reference image with use of with HDR-VDP-2. Fig. 9.8 shows that the algorithm of Abebe et al., on average, approximates the HDR reference image with the fewest artifacts.

9.6 COLOR APPEARANCE MODELING FOR HDR

As briefly alluded to in the introduction to this chapter, if we were to match a visual experience across different display devices, the first thing to do would be to account for the different capabilities of the displays involved. This would solve a very important part of the color reproduction problem. However, there are still possible discrepancies in the perception of the viewed content, because the respective viewing environments may be mismatched. In a professional production scenario, this can easily occur when the viewing environment in the average household is not the same as the viewing environment in the postproduction studio. Currently, about the only thing that consumers can do is to try to match the viewing environment to the average environment assumed when the 100-nit grade was produced for home viewing.

However, technologies already exist to process footage to match different viewing environments. They go under the general banner of color appearance modeling. Thus, rather than adjust the viewing environment to the content, the content is adjusted to match a given viewing environment. The assumption here is that the viewing environment is what observers adapt to. By matching content to a viewing environment, one implicitly assumes that the perception of color is accurately reproduced across environments. Color appearance models consider only viewing environments, without regard for the displays that are placed within those environments. This means that to achieve a good perceptual match, the previously discussed color management techniques should be applied in addition to color appearance modeling.

The main advantage of using color appearance modeling would be that a more accurate perceptual match can be constructed. The director's intent could therefore be conveyed more accurately. The cost, however, is substantial. The models are comparatively complicated and require measurements that are not generally available. This means that color appearance modeling has yet to be incorporated into standardized and broadly available content delivery methods. They do, however, hold significant promise for the future. It is for this reason that we discuss them in this section in some detail.

(A)   (B)   (C)   (D)   (E)

FIGURE 9.8

Comparison between the HDR reference (A), an extracted low dynamic range clipped image (B), and images produced with the methods of Guo et al. (2010) (C), Masood et al. (2009) (D), and Abebe et al. (2015) (E). Maps illustrating the differences between each image and the HDR reference obtained with HDR-VDP2 (Mantiuk et al., 2011) are shown in the second and fourth row. The first image in these rows shows a map of clipped pixels per channel. All images were tone-mapped for visualization with use of the tone mapping operator of Fattal et al. (2002).

Source: From Abebe, M.A., Pouli, T., Kervec, J., Larabi, M.C., 2015. Color clipping and over-exposure correction. In: Eurographics Symposium on Rendering.

9.6.1 **COLOR APPEARANCE**

The appearance of color can be described by means of appearance correlates, which were discussed in Section 9.2.3. They can be obtained through psychophysical measurements, as discussed in Section 9.6.2. These measurements, described below, are then used to derive computational models that calculate approximate appearance correlates.

In addition, appearance models can be used to transform colors to have appearance correlates that are associated with a different viewing environment. One can do this by applying the color appearance model to calculate appearance correlates. The model is then applied in reverse, with parameters being inserted that describe the target viewing environment. As a result, patches of color viewed in one viewing environment will appear as if they were viewed in a different viewing environment. Color appearance models are discussed in Section 9.6.3.

It would be tempting to view pixels in an image or in a video as a large collection of patches, and thus to apply color appearance models directly to images and video. However, this should be done with caution, as the perception of the color of a given pixel may be influenced by the values of neighboring pixels. This notion has given rise to image appearance models, which aim to predict the perception of pixels in complex images. Image appearance models are usually implemented as spatially varying color appearance models. They are discussed in Section 9.6.4.

Finally, more recently, color and image appearance models were developed to cope with an extended dynamic range. This requires matching models against different psychophysical datasets, notably ones that are obtained while an extended dynamic range is taken into account. HDR color and image appearance models are discussed in Section 9.6.5.

9.6.2 **MEASUREMENTS AND DATASETS**

Color appearance models can be developed on the basis of measurements of the visual system. As natural scenes are often deemed to be too complicated for this purpose, these measurements are performed on a simplified laboratory setup. This setup can be described with a few simple parameters, as follows. The colored patch that is to be assessed (the stimulus) is shown as a disk which subtends a visual angle of 2°. Its color is measured in CIE XYZ with the 2° standard observer. A larger disk of 4° visual angle forms the proximal field. A 10° field forms the background. The background has a relative luminance which can be measured. The proximal field may have the same relative luminance as the background, in which case the background is adjacent to the stimulus. Outside the background is the surround field (or peripheral area), which subtends the remainder of the observer's visual field. The proximal field, background, and surround together form what is known as the adapting field. This configuration is illuminated by a light source with a measured relative color specified as a CIE XYZ tristimulus value and a strength measured in candelas per square meter. The illuminant's strength is the only absolute measurement in this setup.

The surround is normally qualified as either "average," "dim," or "dark." An average surround is assumed when one is observing surface colors in normal illumination conditions (ie, daylight). A dim surround is assumed in conditions that are encountered when one is watching television, while a dark surround corresponds to illumination conditions encountered in the cinema. The qualification of the surround (average, dim, or dark) is one of the variables in the measurements taken to obtain color appearance datasets.

With this description of the viewing environment, it can be appreciated that if either the surround or the illuminant is changed, the appearance of the colored patch will change as well. The appearance of a patch of color can be described by appearance correlates, as discussion in Section 9.2.3. The aim of color appearance models is to capture this relationship. However, color appearance models can be derived only on the basis of psychophysical experiments. There have been several studies that were collected into a single database (the "LUTCHI dataset") (Luo et al., 1991). These studies differ in the way that the setup was built and they were designed for different target applications (emissive materials, reflective materials, textiles, etc.) They were, however, not designed to incorporate a large range of illumination levels.

All studies represented in the LUTCHI dataset followed the same protocol. The experiment consists of a magnitude estimation task, whereby lightness, hue, and colorfulness were estimated for patches of color that were presented under different viewing conditions, as described above. Photometric measurements of each patch and viewing environment were also obtained. Next to each patch, two reference patches were shown. One of these patches is achromatic with a lightness of 40, and the other patch has a reference colorfulness of 40. This allowed observers to anchor their responses. Each patch subtended a visual angle of about 2°. While the LUTCHI dataset covers a large variety of viewing conditions, none exceeded luminance levels of 690 cd/m^2. To understand the appearance of color under extended luminance levels, this experiment was later repeated to obtain similar results over a range of peak luminance values that reached 16,860 cd/m^2 (Kim et al., 2009). A further study determined that the sharpness of the edge of the colored patch can have an effect on the perception of lightness (Kim et al., 2011).

9.6.3 COLOR APPEARANCE MODELS

CIE Technical Committee 1-34 gave the following definition of a color appearance model (Fairchild, 1995, 2013a):

A color appearance model is any model that includes predictors of at least the relative color appearance attributes of lightness, chroma and hue. For a model to include reasonable predictors of these attributes, it must include at least some form of a chromatic adaptation transform. Models must be more complex to include predictors of brightness and colorfulness or to model other luminance dependent effects such as the Stevens effect or the Hunt effect.

With this definition, it should be noted that some simple color spaces such as CIE L*a*b* and CIE L*u*v* (see Section 9.3.1) could be considered rudimentary color appearance models. Most color appearance models contain three distinct steps to calculate appearance correlates:

1. *Chromatic adaptation transform.* This transform changes the tristimulus value of the input patch of color to take into account the color of the dominant illuminant. It is common that such chromatic adaptation transforms are done in a sharpened RGB-like color space. In that color space, each of the three channels of the color patch are scaled according to the corresponding color values of the illuminant. Such independent scaling of three color values is called a *von Kries transform* (Wyszecki and Stiles, 2000).

2. *Nonlinear response compression.* This step effectively models the output of photoreceptors, which constitutes a nonlinearity that can be approximated with relatively straightforward sigmoidal functions when photoreceptors are fully adapted (van Hateren, 2006).

3. *Appearance correlates.* The output of the previous step is then used to compute appearance correlates, which normally proceeds by one first calculating color opponent values, from which chroma and hue are computed. Lightness is computed directly from the output of the nonlinear response compression.

Nearly all existing color appearance models use this high-level approach, and many color appearance models are further refinements and/or elaborations of previous color appearance models. These include the model of Nayatani et al. (1986), the Hunt model (Hunt, 1982), rlab (Fairchild, 1996), zlab (Fairchild, 1999), CIECAM97 (CIE, 1998), and CIECAM02 (Moroney et al., 2002). Among such models, CIECAM02 is the most recent, most standardized, and possibly most straightforward model. One color appearance model has combined the chromatic adaptation and nonlinear response compression into a single step, thereby being a simplified model that has also gained in predictive power relative to CIECAM02 (Kunkel and Reinhard, 2009).

Color appearance models take as input the tristimulus value of a colored patch, as well as a description of a simplified hypothetical viewing environment, to calculate appearance correlates (lightness, chroma, hue, etc., as discussed in Section 9.2.3). In all, color appearance models typically take as input the following parameters:

- (X, Y, Z), the tristimulus value of the colored patch for which appearance correlates will be computed.
- (X_W, Y_W, Z_W), the tristimulus value of the adapting illuminant. The adapting illuminant is the light source that determines the state of adaptation of the observer.
- L_a, the luminance (cd/m^2) of the adapting illuminant.
- Y_b, the relative luminance of the background. This value is on a scale between 0 and 100.
- D, the degree of adaptation, a value between 0 and 1 that indicates to what extent the observer is adapted to the illuminant.

The output consists of the correlates of appearance that were introduced in Section 9.2.3.

There are two different use cases for color appearance models. The first is, as stated, the prediction of the appearance of color under given illumination conditions. For imaging applications, the second use case is conceivably more interesting, as it allows patches of color to be transformed between two viewing environments. This is achieved by one taking the description of the first viewing environment and applying the color appearance model to yield appearance correlates. As color appearance models are designed to be invertible, these correlates can be inserted into the inverse color appearance model, with use of the parameters associated with the destination viewing environment. The result is that an (X, Y, Z) tristimulus value can be viewed under a destination viewing environment, while appearing as if it were seen under the first viewing environment.

9.6.4 IMAGE APPEARANCE MODELS

Color appearance models are useful to predict colors that are presented on a precisely defined background with known illumination. It is tempting to apply such models to images and video, especially to adjust such content for display in various viewing environments. Each pixel would then be treated as an independent patch of color. However, our doing so would result in the implicit assumption that the proximal field, background, and surround of each pixel would be the same, and could be described with the set of parameters described above.

It is recognized that this assumption in general would not be valid. Each pixel is surrounded by a neighborhood of pixels that is highly variable. This means that the viewing environment parameters would have to be adjusted differently for each pixel. This has given rise to spatially varying color appearance models, which go by the name of *image appearance models.* In practice, image appearance models do not simply create a spatially varying map of input parameters. Rather, they incorporate notions of human spatial vision, including contrast sensitivity models.

The spatial filtering that is applied to the image to simulate the local specification of viewing environment parameters differs between models, ranging from simple Gaussian filtering (Zhang and Wandell, 1997; Fairchild and Garret, 2004) to bilateral filtering (Kuang et al., 2007) and median-cut algorithms (Reinhard et al., 2012).

In addition to predicting appearance correlates of lightness, brightness, hue, chroma, colorfulness, and saturation, image appearance models may also predict spatial attributes such as sharpness, contrast, resolution, and graininess (Fairchild and Garret, 2004).

9.6.5 HDR COLOR AND IMAGE APPEARANCE MODELS

Color appearance models were not originally designed to deal with significant mismatches in viewing environment. They are based on the LUTCHI dataset, which included variations in luminance up to 690 cd/m^2. This means that if such models are applied to viewing environments that fall outside these test conditions, extrapolation will occur, and this may impact the quality of the results.

Certain image appearance models, particularly the iCAM and iCAM06 models (Fairchild and Garret, 2004; Kuang et al., 2007), the multiscale observer model (Pattanaik et al., 1998), and more recent HDR image appearance models (Kim et al., 2009; Reinhard et al., 2012) do have the ability to cope with significant mismatches in luminance range between viewing environments.

However, these models tend to be optimized to process content that was captured in conditions brighter than the target viewing environment. As a result, these models have significant commonalities with tone reproduction operators, and can be seen as advanced color-managed versions of such operators.

Most of the image appearance models referred to above follow the example of traditional color appearance models for the processing of images in that they are applied in both forward and reverse mode. In forward mode, per-pixel image appearance correlates are computed with use of spatially varying processing. The reverse step calculates new pixels in a spatially uniform manner. As an example, the iCAM06 image appearance model incorporates the following processing steps (Kuang et al., 2007):

- The input image is converted to the CIE XYZ color space.
- A bilateral filter splits the XYZ image into a base and a detail layer.
- A wide Gaussian convolution kernel is independently applied to the XYZ image to simulate a per-pixel white point.

- Chromatic adaptation and range compression are applied in succession to the base layer. Range compression is guided by the per-pixel white point.
- Details in the detail layer are adjusted and then recombined with the processed base layer to produce a nonlinear output. In other color appearance models, this is equivalent to the nonlinear response compression.
- The result is converted to the IPT color space, where colorfulness and surround adjustments are applied.
- Inverse chromatic adaptation is applied.
- The inverse output model is applied.

It is possible to construct an image appearance model that does not require forward and inverse steps to process an image (Reinhard et al., 2012). This model is based on the observation that what needs to remain constant between viewing conditions is not the input to the photoreceptors, but the output of the photoreceptors. It can be shown that if photoreceptor output is modeled with sigmoidal function

$$V = V_{\max} \frac{I}{I + V_{\max}\, \sigma} \tag{9.45}$$

and the essence of the nonlinear response compression is given by

$$I = L_{\max} \frac{L}{L + L_{\max}\, \tau}, \tag{9.46}$$

then nonlinear response compression does not alter the general behavior of photoreceptor output ($V_{\max}$ is the maximum voltage generated by the photoreceptors, the product of $V_{\max}$ and σ forms the semisaturation constant, and $L_{\max}$ and τ are the color appearance model's equivalents). In other words, whether an image is processed with Eq. (9.46) or not, the receptor output V remains sigmoidal. In this model, $L_{\max}$ and τ can be inferred from the source and target viewing environments involved, leading to precise appearance reproduction. A final refinement of this model is that Eq. (9.46) can be applied independently to each of the three color channels of the LMS color space. Then τ takes values derived from the white point of the dominant illuminant. As a consequence, Eq. (9.46) applies nonlinear response compression as well as chromatic adaptation, so no separate chromatic adaptation transform is necessary.

9.6.6 VISUAL QUALITY

Color appearance models are typically calibrated against the same psychophysical datasets (eg, the LUTCHI dataset, or in the case of HDR applications, also the data provided by Kim et al. (2009)). However, the sheer cost of acquiring such psychophysical data means that the space of viewing conditions can be sampled only sparsely. As a result, there is much room for interpretation of the data, which in turn means that different models that are fitted to such data may produce images that are not visually equivalent to each other. To demonstrate such variability, Fig. 9.9 shows the same scene rendered with different techniques. For each of these renderings, default input parameters were used.

An additional source of variability may be introduced if, for existing images, scene measurements are not available. In those cases it would be desirable to be able to infer plausible scene parameters from the images themselves. However, this is a significantly underconstrained problem, which means that such parameter estimation techniques may not be sufficiently robust.

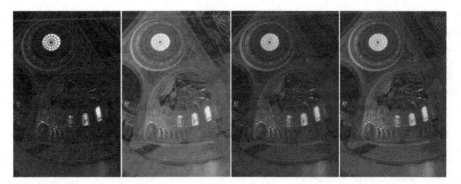

FIGURE 9.9

The Memorial Church image (courtesy of Paul Debevec) rendered with different color and image appearance models. From left to right: iCAM06 (Kuang et al., 2007), the model of Kim et al. (Kim et al., 2009), the model of Kunkel and Reinhard (Kunkel and Reinhard, 2009), and the model of Reinhard et al. (Reinhard et al., 2012).

9.6.7 COLOR APPEARANCE MODELS IN PRACTICE

From a practical perspective, color appearance models offer the advantage that the director's intent may be conveyed across viewing environments, something that is not currently possible. On the other hand, such models require measurement both on-set (or in the postproduction studio) and in the end-user's home. In addition, set/studio-related information needs to be transmitted in the form of metadata, which requires standardization. This is all possible, but will require time, infrastructure, and the political will to make it happen.

9.7 CONCLUSIONS

Color management has traditionally been seen as a problem separate from tone management. However, in terms of mathematical formulation, perceptual underpinnings, and even consumer trends, these two areas are becoming increasingly entangled. In the context of postproduction, where accuracy of color reproduction is of key importance, the introduction of HDR workflows is obviating the need for HDR-specific color management tools. In this chapter we have discussed some of the ongoing efforts for the construction of color-managed workflows in the HDR field, covering both calibrated, scene-referred data and display-referred, device-specific approaches. We have also presented color spaces that enable the encoding, processing, and display of HDR content.

REFERENCES

Abebe, M.A., Pouli, T., Kervec, J., Larabi, M.C., 2014. Correction of over-exposure using color channel corrections. In: IEEE GlobalSIP, Atlanta, GA, pp. 1078–1082.

Abebe, M.A., Pouli, T., Kervec, J., Larabi, M.C., 2015. Color clipping and over-exposure correction. In: Eurographics Symposium on Rendering.

Akyüz, A.O., Reinhard, E., 2006. Color appearance in high-dynamic-range imaging. J. Electron. Imaging 15 (3), 0330001.

Bertalmio, M., Sapiro, G., Caselles, V., Ballester, C., 2000. Image inpainting. In: Conference on Computer Graphics and Interactive Techniques (ACM SIGGRAPH), pp. 417–424.

Boitard, R., Mantiuk, R.K., Pouli, T., 2015. Evaluation of color encodings for high dynamic range pixels. In: Human Vision and Electronic Imaging, p. 93941K.

CIE, 1998. The CIE 1997 interim colour appearance model (simple version), CIECAM97s. Technical Report CIE Pub. 131, Commision Internationale de l'Eclairage.

Ebner, F., Fairchild, M.D., 1998. Development and testing of a color space (IPT) with improved hue uniformity. In: Sixth Color Imaging Conference: Color Science, Systems and Applications, pp. 8–13.

Elboher, E., Werman, M., 2011. Recovering color and details of clipped image regions. Int. J. Comput. Inform. Syst. Indust. Manage. Appl. 3, 812–819.

Fairchild, M.D., 1995. Testing colour appearance models: guidelines for coordinated research. Colour Res. Appl. 20, 262–267.

Fairchild, M.D., 1996. Refinement of the RLAB color space. Colour Res. Appl. 21, 338–346.

Fairchild, M.D., 1999. The ZLAB color appearance model for practical image reproduction applications. Color Res. Appl. 23, 89–94.

Fairchild, M.D., 2007. A color scientist looks at video. In: Proceedings of the 3rd International Workshop on Video Processing and Quality Metrics (VPQM), Scottsdale, AZ.

Fairchild, M.D., 2013a. Color Appearance Models, third ed. John Wiley & Sons, Chichester, UK.

Fairchild, M.D., 2013b. Color Appearance Models. John Wiley & Sons, Chichester, UK.

Fairchild, M.D., Chen, P.H., 2011. Brightness, lightness, and specifying color in high-dynamic-range scenes and images. In: IS&T/SPIE Electronic Imaging, p. 78670O.

Fairchild, M.D., Garret, J., 2004. iCAM framework for image appearance, image differences, and image quality. J. Electron. Imaging 13 (1), 126–138.

Fairchild, M.D., Wyble, D.R., 2010. hdr-CIELAB and hdr-IPT: simple models for describing the color of high-dynamic-range and wide-color-gamut images. In: Color and Imaging Conference, vol. 2010, pp. 322–326.

Fattal, R., Lischinski, D., Werman, M., 2002. Gradient domain high dynamic range compression. ACM Trans. Graph. 21 (3), 249–256.

Fu, J., Peng, H., Chen, X., Mou, X., 2013. Correcting saturated pixels in images based on human visual characteristics. In: SPIE/IS&T Electronic Imaging: Digital Photography IX, vol. 8660.

Guo, D., Cheng, Y., Zhuo, S., Sim, T., 2010. Correcting over-exposure in photographs. In: IEEE International Conference on Computer Vision and Pattern Recognition, pp. 515–521.

Hirakawa, K., Simon, P., 2011. Single-shot high dynamic range imaging with conventional camera hardware. In: IEEE International Conference on Computer Vision, pp. 1339–1346.

Hunt, R.W.G., 1970. Objectives in colour reproduction. J. Photograph. Sci. 18, 205–215.

Hunt, R.W.G., 1982. A model of colour vision for predicting colour appearance. Color Res. Appl. 7, 95–112.

Hunt, R.W.G., 2004. The Reproduction of Colour, sixth ed. Wiley-IS&T Series in Imaging Science and Technology. Wiley, Chichester, UK.

International Telecommunication Union, 2012. BT.2020: Parameter values for ultra-high definition television systems for production and international programme exchange.

ITU, 1998. Recommendation ITU-R BT.709-3: Parameter values for the HDTV standards for production and international programme exchange. International Telecommunications Union.

Joshi, N., Matusik, W., Adelson, E.H., Kriegman, D.J., 2010. Personal photo enhancement using example images. ACM Trans. Graph. 29.

Kainz, F., 2004. A proposal for OpenEXR color management. Available from www.openexr.com/documentation.html.

Kim, M., Weyrich, T., Kautz, J., 2009. Modeling human color perception under extended luminance levels. ACM Trans. Graph. 28 (3), 27:1–27:9.

Kim, M.H., Ritschel, T., Kautz, J., 2011. Edge-aware color appearance. ACM Trans. Graph. 30 (2), 13.

Kuang, J., Johnson, G.M., Fairchild, M.D., 2007. iCAM06: A refined image appearance model for HDR image rendering. J. Vis. Commun. Image Represent. 18 (5), 406–414.

Kunkel, T., Reinhard, E., 2009. A neurophysiology-inspired steady-state color appearance model. J. Opt. Soc. A 26 (4), 776–782.

Li, Y., Sharan, L., Adelson, E., 2005. Compressing and companding high dynamic range images with subband architectures. ACM Trans. Graph. 24 (3), 836–844.

Lübbe, E., 2008. Colours in the Mind — Colour Systems in Reality: A Formula for Colour Saturation. Books on Demand GmbH.

Luo, M.R., Clarke, A.A., Rhodes, P.A., Schappo, A., Scrivener, S.A.R., Tait, C.J., 1991. Quantifying color appearance. Part I. LUTCHI colour appearance data. Color Res. Appl. 16 (3), 166–180.

Mansour, H., Saab, R., Nasiopoulos, P., Ward, R., 2010. Color image desaturation using sparse reconstruction. In: IEEE International Conference on Acoustics Speech and Signal Processing, pp. 778–781.

Mantiuk, R., Kim, K.J., Rempel, A.G., Heidrich, W., 2011. HDR-VDP-2: a calibrated visual metric for visibility and quality predictions in all luminance conditions. ACM Siggraph.

Mantiuk, R., Mantiuk, R., Tomaszweska, A., Heidrich, W., 2009. Color correction for tone mapping. Comput. Graph. Forum 28 (2), 193–202.

Masood, S.Z., Zhu, J., Tappen, M.F., 2009. Automatic correction of saturated regions in photographs using cross-channel correlation. Comput. Graph. Forum 28 (7), 1861–1869.

Moroney, N., Fairchild, M.D., Hunt, R.W.G., Li, C.J., Luo, M.R., Newman, T., 2002. The CIECAM02 color appearance model. In: IS&T 10th Color Imaging Conference, Scottsdale, pp. 23–27.

Morovič, J., 2008. Color Gamut Mapping, first ed. Wiley and Sons, Chichester, UK.

Nayar, S., Mitsunaga, T., 2000. High dynamic range imaging: spatially varying pixel exposures. In: IEEE Conference on Computer Vision and Pattern Recognition (CVPR), pp. 472–479.

Nayatani, Y., Takahama, K., Sobagaki, H., 1986. Prediction of color appearance under various adapting conditions. Color Res. Appl. 11, 62–71.

Pattanaik, S.N., Ferwerda, J.A., Fairchild, M.D., Greenberg, D.P., 1998. A multiscale model of adaptation and spatial vision for realistic image display. In: Proceedings of SIGGRAPH'98, pp. 287–298.

Pouli, T., Artusi, A., Banterle, F., Akyüz, A.O., Seidel, H.P., Reinhard, E., 2013a, Color correction for tone reproduction. In: Proceedings of the 21st IS&T Color Imaging Conference, pp. 215–220.

Pouli, T., Reinhard, E., Cunningham, D.W., 2013b. Image Statistics in Visual Computing. CRC Press, Boca Raton.

Reinhard, E., 2011. Tone reproduction and color appearance modeling: two sides of the same coin? In: Proceedings of the 19th IS&T/SID Color Imaging Conference, pp. 171–176.

Reinhard, E., François, E., Boitard, R., Chamaret, C., Serré, C., Pouli, T., 2014. High dynamic range video chains. In: International Broadcasting Convention (IBC).

Reinhard, E., Pouli, T., Kunkel, T., Long, B., Ballestad, A., Damberg, G., 2012. Calibrated image appearance reproduction. ACM Trans. Graph. 31 (6), 201.

Rouf, M., Lau, C., Heidrich, W., 2012. Gradient domain color restauration of clipped highlights. In: IEEE International Conference on Computer Vision and Pattern Recognition, pp. 7–14.

Schlick, C., 1994. Quantization techniques for visualization of high dynamic range pictures. In: Proceedings of the 5th Eurographics Workshop on Rendering, pp. 7–18.

Shen, S., 2009. Color Difference Formula and Uniform Color Space Modeling and Evaluation. Ph.D. thesis, Rochester Institute of Technology.

SMPTE, 2012. ST 2065-1:2012 — academy color encoding specification (ACES).

Society of Motion Picture & Television Engineers, 2014. High dynamic range electro-optical transfer function of mastering reference displays table. In: SMPTE ST 2084, pp. 1–14.

Tan, P., Lin, S., Quan, L., Shum, H., 2003. Highlight removal by illumination constrained inpainting. In: IEEE International Conference on Computer Vision, Nice, pp. 164–169.

Tumblin, J., Turk, G., 1999. LCIS: a boundary hierarchy for detail-preserving contrast reduction. In: Proceedings of SIGGRAPH' 99, pp. 83–90.

van Hateren, J.H., 2006. Encoding of high dynamic range video with a model of human cones. ACM Trans. Graph. 25 (4), 1380–1399.

Wang, L., Wei, L.Y., Zhou, K., Guo, B., Shum, H.Y., 2007. High dynamic range image hallucination. In: Proceedings of Eurographics Symposium on Rendering.

Ward, G.J., 1994. The radiance lighting simulation and rendering system. In: Proceedings of the 21st Annual Conference on Computer Graphics and Interactive Techniques, pp. 459–472.

Ward Larson, G., 1998. LogLuv encoding for full-gamut, high-dynamic range images. J. Graph. Tools 3 (1), 815–830.

Wyszecki, G., Stiles, W.S., 2000. Color Science: Concepts and Methods, Quantitative Data and Formulae, second ed. John Wiley and Sons, New York.

Xu, D., Doutre, C., Nasiopoulos, P., 2011. Correction of clipped pixels in color images. IEEE Trans. Vis. Comput. Graph. 17 (3), 333–344.

Zhang, X., Brainard, D., 2004. Estimation of saturated pixel values in digital color imaging. J. Opt. Soc. Am. A 21 (12), 2301–2310.

Zhang, X., Wandell, B.A., 1997. A spatial extension of CIELAB for digital color reproduction. J. Soc. Inform. Disp. 5 (1), 61–63.

Zhang, C., Gao, J., Wang, O., Georgel, P., Yang, R., Davis, J., Frahm, J.M., Pollefeys, M., 2014. Personal photograph enhancement using internet photo collections. IEEE Trans. Vis. Comput. Graph. 20 (2), 262–275.

REPRESENTATION AND CODING III

HIGH DYNAMIC RANGE VIDEO COMPRESSION

10

Y. Zhang, D. Agrafiotis, D.R. Bull

University of Bristol, Bristol, United Kingdom

CHAPTER OUTLINE

10.1 INTRODUCTION

The simultaneous dynamic range of the human visual system (HVS) covers a range of approximately four orders of magnitude (Kunkel and Reinhard, 2010). Overall, the HVS can adapt to light conditions with a dynamic range (range of luminance levels) of approximately 14 orders of magnitude (Hood and Finkelstein, 1986; Ferwerda et al., 1996). This range includes the so-called photopic and scotopic vision ranges.

In contrast to the wide range of luminance adaptation exhibited by the HVS, most existing conventional capture and display devices can accommodate a dynamic range of only between two and three orders of magnitude. This is often referred to as "low dynamic range" (LDR). High dynamic range (HDR) imaging has been demonstrated to increase the immersiveness of the viewing experience by capturing a luminance range more compatible with the scene and displaying visual information that covers the full visible (instantaneous) luminance range of the HVS with a larger color gamut (Reinhard et al., 2010). This, however, comes at the cost of much larger storage and transmission bandwidth requirements owing the increased bit depth that is used by most HDR formats to represent this information. Efficient HDR image and video compression algorithms are hence needed that can produce manageable bit rates for a given target perceptual quality. At the time of writing, standardization of compression for wide color gamut and HDR video is still an ongoing process (Sullivan et al., 2007; Winken et al., 2007; Wu et al., 2008; Zhang et al., 2013b; Duenas et al., 2014).

High Dynamic Range Video. http://dx.doi.org/10.1016/B978-0-08-100412-8.00010-3

This chapter offers an introduction to the topic of HDR video compression. We review some of the latest HDR image and video coding methods, looking at both layered (backward-compatible/residual-based) and high-bit-depth (native) HDR compression approaches. In the process, we also examine certain HDR-related aspects of the HVS. This chapter assumes a working knowledge of image and video compression. For further details on this aspect, the reader is refereed to Bull (2014).

10.2 HDR IMAGE STORAGE FORMATS AND COMPRESSION

To preserve the benefits of HDR imaging and avoid taking up excess memory and bandwidth, an efficient HDR format is necessary for storing HDR image and video data. There are three recognized HDR image formats that have been standardized: Radiance RGBE, TIFF, and EXR. At the time of writing, no HDR video format existed.

The Radiance RGBE (.hdr) file format uses run-length coding to represent data with 32 bits per pixel (Ward, 1991, 1994). Eight bits are used for each of the three color channels — red, green, and blue — and an extra eight bits are used for the exponent. The dynamic range covered by this format is more than 76 orders of magnitude. The Radiance format also supports XYZE encoding, which offers full color gamut coverage.

The LogLuv HDR image format (Larson, 1998) comes in two versions that support 24 and 32 bits per pixel, respectively. The LogLuv color space has been adopted as part of the TIFF library (.tiff). The advantage of the LogLuv format is that it stores luminance and chrominance information separately, allowing these values to be directly processed during tone mapping and compression.

The OpenEXR HDR image format (.exr) encodes pixels using 16-bit floating point values for the red, green, and blue channels (Bogart et al., 2003). Each color channel is encoded with use of a half-precision floating point number, where one bit is used for the sign, five bits are used for the exponent, and 10 bits are used for the mantissa. This encoding strategy covers around 10.7 orders of magnitude. The ZIP deflate library and other more efficient lossless wavelet compression tools can be used to achieve, on average, a 60% reduction in storage space compared with uncompressed data (Reinhard et al., 2010).

All of these HDR image formats apply lossless compression to the image data in order to preserve the original HDR information. This makes them unsuitable for applications requiring reduced transmission bit rates and scenarios where storage space is severely limited. JPEG XR (Srinivasan et al., 2007) and JPEG 2000 (Boliek, 2014) are image codecs that offer lossy compression for high-bit-depth (HDR) images (Srinivasan et al., 2007; Richter, 2009; Zhang et al., 2012a). An extension to the JPEG standard called JPEG XT (in the process of standardization at the time of writing), targets backward-compatible, scalable lossy to lossless coding of HDR images (Husak and Ninan, 2012; Richter, 2013b, 2014; Richter et al., 2014). Pinheiro et al. (2014) investigated the performance of three profiles (profiles A, B, and C) of JPEG XT using the signal-to-noise ratio and the feature similarity index as the objective quality metrics. The three different profiles differ in terms of the HDR reconstruction method used in the standard. The results of this performance evaluation indicate that the rate-distortion performances (signal-to-noise ratio) of profiles A and B are similar, showing fast "saturation" at higher bit rates. Profile C exhibits different rate-distortion behavior, with no such "saturation" at high bit rates. Profiles B and C show stronger dependency than profile A on the choice of the tone-mapping operator (TMO) that generates the LDR image. The feature similarity index results suggest that the rate-distortion curves of all three profiles exhibit fast "saturation" at high bit rates.

A backward-compatible JPEG method (JPEG-HDR) was proposed 2005 in Ward and Simmons (2005). JPEG-HDR codes an HDR image in a layered fashion, with an eight-bit tone-mapped image forming the base layer and an image residual in the extension layer. Both layers are coded with standard JPEG. Before the introduction of this method, Ward and Simmons (2004) had proposed another residual-based HDR extension for JPEG. The HDR image is tone-mapped down to an eight-bit LDR image that is coded with JPEG. A ratio image between the original HDR image and the tone-mapped version is downsampled and stored as a tag in the header file. This ratio image can be used by HDR-capable decoders to reconstruct the HDR content, while all other legacy devices would ignore the tag and directly display the tone-mapped LDR image. Spaulding et al. (2003) proposed JPEG backward-compatible layered coding for color gamut extension. In the base layer, an image with a clipped color gamut is encoded. In the enhancement layer, a residual image is formed in a subband. This residual image is defined as the arithmetic difference between an input extended reference input medium metric RGB (ERIMM RGB) color space image and the encoded standard RGB (sRGB) foreground image (limited to eight bits). Korshunov and Ebrahimi (2012, 2013) presented a generic HDR JPEG backward-compatible image compression scheme. They highlight the importance of the TMO in the performance of the codec and its dependence on the content, the device used, and environmental parameters such as backlighting, display type and size, and environment illumination. Korshunov and Ebrahimi (2013) evaluated the performance of their method using three simple TMOs ("*log*," "*gamma*," and "*linear*"). A number of different viewing condition parameters (ie, backlight lighting, display type, display size, environment illumination) were taken into account. The experimental results indicated that the rate-distortion performance of the proposed solutions compared with other methods showed better compression efficiency for the luminance channel on the basis of peak signal-to-noise ratio evaluations.

The method proposed in Richter (2013a) is also JPEG backward compatible and proposes an extended discrete cosine transform process with additional refinement bits. The refinement bits are the result of coding the residual formed after subtraction of the inverse tone-mapped, JPEG-coded LDR layer from the input HDR image. The refinement bits are placed within application markers, thus being hidden from legacy JPEG decoders. The experimental results presented in Richter (2013b) indicate that the high-bit-depth codecs (JPEG 2000 and JPEG XR) offer better HDR rate-distortion performance.

A more detailed discussion of the JPEG XT standard for JPEG-compatible compression can be found in Chapter 12.

10.3 **HDR VIDEO COMPRESSION**

HDR video has significantly higher bit rate requirements than LDR video, not only because of the increased bit depth but also because of increased levels of noise. HDR video compression methods can be grouped into two categories: layered approaches that are backward compatible with standard LDR decoders and high-bit-depth HDR coding methods that code the input HDR signal as close to its native format as possible.

10.3.1 LAYERED (BACKWARD-COMPATIBLE) HDR VIDEO COMPRESSION

Layered HDR compression methods are designed so that legacy decoders, which can cope only with standard dynamic range bit depths and LDR displays (displays that are unable to render HDR content), are still able to decode the base layer and display a tone-mapped version of the HDR image/video. HDR-capable decoders would be able to decode the full stream and deliver the higher-performance HDR content. Fig. 10.1 shows a general block diagram describing the structure of a typical backward-compatible HDR encoder. The base layer encodes a tone-mapped eight-bit LDR representation of the HDR input using a fully compatible legacy encoder/decoder. The enhancement layer contains the difference (residual) between the inverse tone-mapped base layer and the original HDR input. This residual is used for reconstruction of the HDR content to be used with HDR devices.

The TMO plays a significant role in the performance of layered HDR video coding methods. The main goal of tone mapping is to reduce the dynamic range of the input HDR image/video while preserving perceptual aspects in the resulting LDR image/video. Several TMOs have been developed for different purposes, such as photographic tone reproduction (Reinhard et al., 2002), photoreceptor physiology–based modeling (Reinhard and Devlin, 2005), simulation of artistic drawing (Tumblin and Turk, 1999), gradient domain dynamic range compression (Fattal et al., 2002), display adaptive tone mapping (Mantiuk et al., 2008), and user interactive local adjustment tone mapping (Lischinski et al., 2006).

Different visual and coding results are obtained when HDR content is mapped with different TMOs. A number of studies have evaluated, objectively and subjectively, the visual impact of TMOs (Yoshida et al., 2005; Eilertsen et al., 2013; Yeganeh and Wang, 2013; Narwaria et al., 2013a). Narwaria et al. (2012) investigated the relation between TMOs and visual attention. They performed psychovisual experiments that assessed the impact of eight TMOs on visual attention. The results of their experiments suggest that TMOs can modify fixation behavior, with contrast greatly influencing saliency.

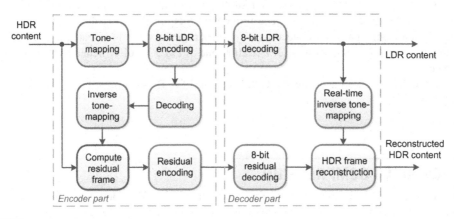

FIGURE 10.1

General structure of a backward-compatible HDR image and video codec.

Source: Zhang et al. (2015).

In most video tone mapping cases a TMO is applied independently to each frame of the HDR video. This produces brightness level variation between consecutive tone-mapped frames, causing flickering artifacts. Hence, when designing a TMO for HDR video, one has to take the temporal dimension into consideration. Lee and Kim (2007) proposed a time-dependent adaptation method for the TMO of Fattal et al. (2002) which exploits the motion information and prevents flickering artifacts. A real-time tone-mapping system for HDR video was proposed in Kiser et al. (2012). This adapts to changes in the scene and reduces the amount of flickering artifacts in the resulting tone-mapped HDR video. The method was implemented in a field programmable gate array for use as a real-time tone-mapping system. Boitard et al. (2014) classified the temporal artifacts generated by TMOs when tone mapping video into different types and surveyed methods designed to remove such artifacts. Aydin et al. (2014) designed a local TMO for avoiding visible artifacts in the spatial and temporal domains. The proposed method decomposes the signal into a base layer and a detail layer. A temporal filter is applied on the detail layer and a spatiotemporal filter is applied on the base layer. An extensive review of TMOs can be found in Reinhard et al. (2010) and Banterle et al. (2011).

Mantiuk et al. (2004) were the first to propose layered coding for backward-compatible HDR video compression. Their method was designed as an extension to the MPEG-4 (Part II) compression standard. The method applies tone mapping (perception-based HDR luminance-to-integer encoding) to form the LDR base layer. The input to the MPEG-4 codec consists of eight-bit RGB data and HDR data in the CIE XYZ color space. The CIE XYZ color space enables encoding of the full visible luminance range and color gamut. The luminance values are quantized with a nonlinear function that distributes the error according to the luminance response curve of the HVS. A color space transformation is introduced that facilitates comparisons between LDR and HDR pixels of different color spaces so that the chrominance data of the residual (enhancement) layer can be determined. Mantiuk et al. (2006a,b) extended this work to include a contrast sensitivity function (CSF)-based prefiltering of the residual stream that removes imperceptible high spatial frequency information, thus reducing the bit rate of the coded sequence.

Mai et al. (2011) proposed a backward-compatible method that aims to find an optimal tone curve for mapping the input HDR video to a backward-compatible eight-bit LDR video format. The optimal tone curve minimizes the quality loss due to tone mapping, encoding/decoding, and inverse tone mapping of the original video. To compute the optimal tone-mapping curve, closed-form and statistical solutions were proposed. The LDR video can be compressed by a conventional video codec such as H.264. The reconstructed video can either be displayed on a conventional LDR display or can be inverse-tone-mapped and augmented by an optional enhancement layer containing an HDR residual signal (also compressed by the codec) for display on an HDR display.

Koz and Dufaux (2014) presented a backward-compatible HDR video compression scheme that uses the tone-mapping method of Mai et al. (2011) and the perceptually uniform mapping of Aydin et al. (2008). Perceptually uniform mapping converts real-world luminance values into perceptually uniform encoded values. The aim of the proposed coding method is to minimize the mean square error between the perceptually uniform-encoded values of the original HDR signal and those of the reconstructed signal. To constrain flickering distortion in the tone-mapped HDR video, their method restrains the average luminance of neighboring frames within the just noticeable differences (JND) interval. Using perceptually uniform mapping, Koz and Dufaux define the perceptually uniform peak-signal to noise ratio that accepts perceptually uniform-mapped values as its input for estimation of the quality of the HDR (or LDR) content when that is displayed on bright displays

(Aydin et al., 2008). The experimental results indicate that this method provides a good trade-off between HDR video compression performance and the flickering distortion introduced by the tone mapping applied in most backward-compatible HDR video compression approaches. Compared with the method of Mai et al. (2011), this method offers better performance in terms of the perceptually uniform peak-signal to noise ratio but achieves a smaller objective score when quality is measured with the HDR-MSE and HDR-VDP metrics.

The method proposed in Le Dauphin et al. (2014) adjusts the TMO depending on the coding type of the video frame (intraframe or interframe) in order to maximize the compression efficiency. The experimental results presented suggest that this adaptation can lead to more than 20% bit rate reduction for the LDR layer and more than 2% reduction for HDR video.

The scalable extension of High Efficiency Video Coding (HEVC) — that is, SHVC, offer both bit-depth and color-gamut scalability. Bit-depth-scalable approaches have been proposed in Winken et al. (2007), Wu et al. (2008), and Liu et al. (2008). Typically, the proposed methods create a backward-compatible eight-bit base layer and an enhancement layer for higher-bit-depth reconstruction (ie, 10-, 12-, or 14-bit). In Bordes et al. (2013) and Duenas et al. (2014), color-gamut scalability was been investigated in SHVC. The proposed method addresses the case where the original enhancement layer uses a color gamut different from that used by the base layer. A prediction tool for color differences between the base layer and the enhancement layer is thus required. This can be useful, for instance, for deployment of HDR or wider color gamut services compatible with legacy LDR devices. Typically, the LDR or high-definition TV uses the ITU-R Rec.709 format, while UHD (maybe HDR) is likely to be based on ITU-R Rec.2020.

10.3.2 HIGH-BIT-DEPTH (NATIVE) HDR VIDEO COMPRESSION

High-bit-depth HDR video compression methods directly code the input high-bit-depth HDR video, as opposed to first separating it into base and enhancement layers through tone mapping. They rely on existing codecs that can handle a high-bit-depth input (eg, JPEG 2000, H.264/AVC, HEVC) with added preprocessing steps and/or modifications/extensions that aim to exploit HDR-specific aspects of the HVS for compression gains.

Motra and Thoma (2010) proposed an adaptive-LogLuv transform which can be used with an existing video encoder such as H.264/AVC. Their approach can represent the luminance channel at any specified bit depth. Eight bits are used for the chrominance channels. This method requires that side information is stored/transmitted with each HDR frame. Round-off quantization noise produced by the logarithmic color space transformation used in this method can have repercussions on visual quality. An extended version of this work was presented in Garbas and Thoma (2011), wherein a temporally coherent adaptive dynamic range mapping was proposed. This generates temporal weights and shift parameters for each frame and makes use of the weighted prediction tool specified by the H.264/AVC codec (Boyce, 2004). An electro-optical transfer function is proposed in Miller et al. (2013) which acts as a perceptual quantizer that maximizes the perceptual quality when one is creating HDR content with bit depths of 10 or 12 bits.

The HVS exhibits nonlinear sensitivity to the distortions introduced by lossy image and video coding. This effect is due to the luminance masking, contrast masking, and spatial and temporal frequency masking characteristics of the HVS. Efficient perception-based compression of HDR imagery requires models that capture accurately the masking effects experienced by the HVS under

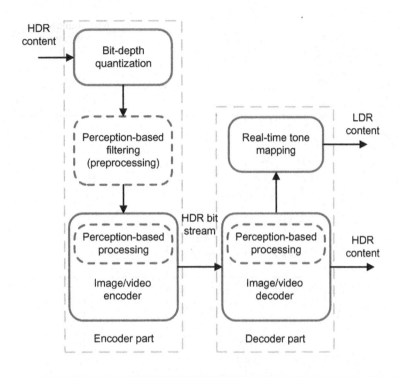

FIGURE 10.2

General structure of perception-based HDR image and video encoding.

Source: Modified from Zhang et al. (2015).

HDR conditions, so that bits are not wasted coding redundant imperceptible information. Fig. 10.2 shows the general structure of perception-based HDR image and video coding.

Zhang et al. (2012a,b) proposed a perception-based HDR image and video compression method for both JPEG 2000 and H.264/AVC. The method uses a discrete wavelet packet transform (DWPT) and applies coefficient weighting factors that are derived from luminance and chrominance CSFs. The aim of the weighting factors is to reduce the amount of imperceptible information that is sent to the encoder. Fig. 10.3 shows the relationship between luminance and chrominance CSFs and a five-level two-dimensional discrete wavelet packet transform decomposition. The perceptual CSF model characterizes the relationship between contrast sensitivity and spatial frequency. In the luminance case (black curve), the HVS is more sensitive to middle spatial frequencies and is less sensitive to lower and higher spatial frequencies. In the chrominance case (blue and red curves), the HVS is more sensitive to lower spatial frequencies and less sensitive to high spatial frequencies. Psychophysical experiments reported in Barten (1999), Mullen (1985), and Mannos and Sakrison (1974) have quantified this phenomenon and describe the ability of the HVS to recognize differences in luminance and chrominance as a function of contrast and spatial frequency. The weighted wavelet coefficients in the case of HDR images are passed to a JPEG 2000 encoder, whereas in the case of HDR video, they are inverse-transformed and the resulting video frames are fed to an H.264/AVC encoder.

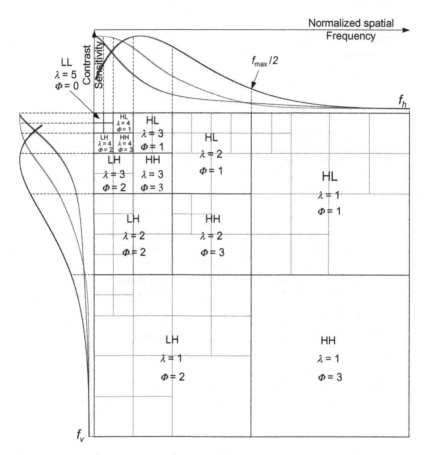

FIGURE 10.3

Relationship between luminance and chrominance CSFs and a five-level two-dimensional discrete wavelet packet transform decomposition.

Source: Zhang et al. (2011).

The experimental results presented in Zhang et al. (2012b) indicate that the method offers visually lossless quality at a significantly reduced bit rate compared with JPEG 2000 (Table 10.1, Fig. 10.4). The HDR images listed in Table 10.1 were adopted from (Debevec and Malik, 1997; Drago and Mantiuk, 2005; Reinhard et al., 2010). Quality was measured with the HDR-VDP (Mantiuk et al., 2005) and HDR-VDP-2 (Mantiuk et al., 2011) quality metrics. The method also outperforms that of Motra and Thoma (2010) for coding HDR video (Fig. 10.5).

The method of Zhang et al. (2012b) assumes that HDR content exhibits the same CSF as standard dynamic range content. However, HDR content and HDR displays can offer a much larger contrast than conventional imaging. Compared with conventional LDR imaging, HDR imaging (HDR images/video displayed on HDR displays) exhibits considerably larger brightness variations. From a psychophysics point of view, this significant contrast extension has the potential to change the noise visibility threshold in HDR content.

Table 10.1 Storage Requirements (kB) of Ten Test HDR Images When Coded With the Method of Zhang et al. (2012b), With JPEG 2000, or With One of the Three HDR Image Formats

HDR Image	Resolution (pixels)	Proposed Method (Visually Lossless)	JPEG 2000 (Visually Lossless)	OpenEXR (.exr)	LogLuv (.tiff) (32 bits)	RGBE (.hdr)
memorial	512 × 768	590	787	1246	1190	1312
AtriumNight	760 × 1016	1151	1308	1983	2412	2547
mpi_atrium_1	1024 × 676	1376	1531	1916	2000	2375
EMPstair	852 × 1136	1211	1411	2076	2667	3040
BristolBridge	2048 × 1536	7332	8195	10,132	7696	9895
BoyScoutFalls	1000 × 1504	2974	3281	4497	4794	5142
BoyScoutTrail5	998 × 1496	2935	3321	4464	5048	5280
BoyScoutTree	998 × 1489	3231	3624	4224	4784	5169
sfmoma1	852 × 1136	1521	1825	1962	2640	2996
WardFlowers	1504 × 1000	2935	3344	3620	4292	4676

Source: Zhang et al. (2012a).

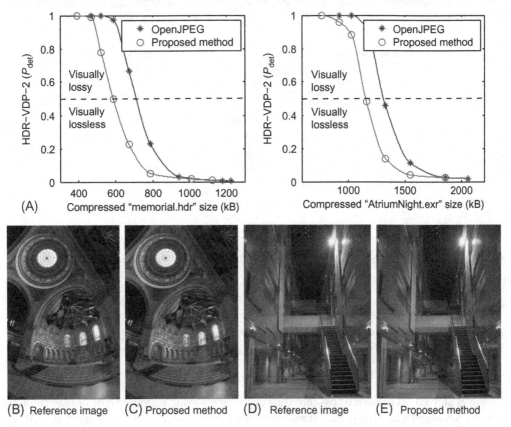

(A) Rate-distortion performance of the method of Zhang et al. (2012b) compared with that of JPEG 2000 (test images, memorial.hdr, EMPstair.hdr). (B)–(E) Visually lossless images obtained with the method of Zhang et al. (2012b) compared with the original HDR image (display adaptive TMO used for tone mapping).

FIGURE 10.4

Source: Zhang et al. (2012a).

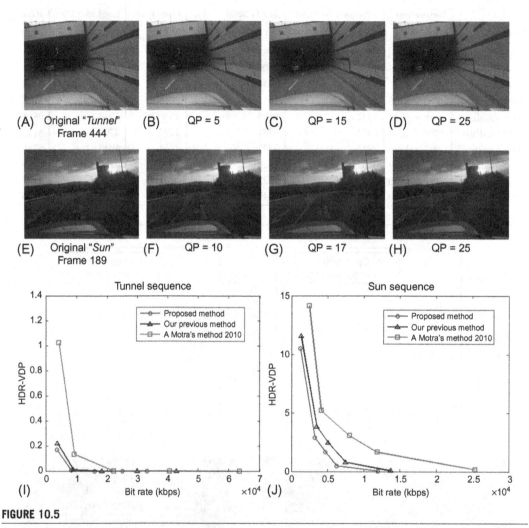

FIGURE 10.5

Original "Tunnel" and "Sun" sequence frames [(A) and (E)] versus frames reconstructed [(B)–(D) and (F)–(H), respectively] with the method of Zhang et al. (2012b) using different QPs. (I) and (J) HDR-VDP (95%) results versus bit rate for the methods of Zhang et al. (2012b), a previous method of Zhang et al. (2011) and the adaptive LogLuv transform of Motra and Thoma (2010) for the "Tunnel" and "Sun" sequences, respectively.

Source: Zhang et al. (2012a).

Zhang et al. (2013a) performed psychovisual experiments using a Dolby HDR display (prototype code name Seymour) and HDR stimuli (luminance range 0.103–3162 cd/m^2) (Zhang et al., 2013a) to examine HDR luminance masking. The experimental results indicated that the noise visibility threshold increases in both low and high luminance background levels, and especially so with a low luminance background. These results suggest that quantization errors could be hidden by a background luminance masking effect in the darker and brighter areas of HDR content.

To exploit this varying distortion sensitivity, a luminance masking profile can be used to map the average luminance intensity to the quantization step of each coded block. Naccari et al. (2012) proposed the intensity-dependent quantization (IDQ) method for the HEVC standard. Their method applies coarser quantization to darker and brighter image areas without introducing coding artifacts that are noticeable to the HVS. The mapping from average pixel intensity to quantization step is performed through an IDQ profile that is fixed for all sequences and is a modified version of the profile proposed in Jia et al. (2006).

An IDQ profile scaling based on a global tone-mapping curve computed for each HDR frame was proposed in Zhang et al. (2013b). A global tone-mapping curve is a function that maps HDR luminance values either to the displayable luminance range (Mantiuk et al., 2008) or directly to LDR pixel values with eight bits per channel (Larson et al., 1997). In Zhang et al. (2013b) the latter case was considered with the histogram adjustment TMO of Larson et al. (1997). Scaling the IDQ profile using a global tone-mapping curve not only enables adaptation to any HDR content, but also offers a more accurate adaptation with respect to the HVS.

With this IDQ profile, the quantization step can be perceptually tuned on a transform unit basis. However, the intensity differential Quantization Parameter (idQP) value depends on the average pixel intensity (μ) of the original content. Given that the original content is not available to the decoder, for the coded bitstream to be properly decoded, μ would have to be communicated to the decoder. This creates significant overhead that increases the resulting bit rate.

To avoid this overhead the method of Zhang et al. (2013b) estimates μ from the predictor used for each coded block, thus avoiding the need for additional signaling. The predictor can be spatial in the case of intraframe coding or temporal for interframe coding. Estimation of μ from the block predictor avoids the introduction of overhead but creates dependencies in the decoding process that can significantly complicate/delay it. More specifically, the inverse quantization process at the decoder is now dependent on the availability of the predictor. For the case of a temporal predictor, the processing required by motion compensation may delay significantly the output of the pixel data since several memory access operations would have to first be performed.

To prevent such data dependency the method makes use of the intensity-dependent spatial quantization (IDSQ) proposed in Naccari et al. (2012). As can be seen in Fig. 10.6, IDSQ does not require availability of the predictor during the inverse quantization step. Instead, perceptual quantization is applied at the point of final reconstruction of the block, when prediction is added to the residual.

With IDSQ, the inverse quantization process is decoupled into two main steps (see Fig. 10.6):

1. *Step 1*: Over the quantized coefficients c, perform inverse quantization as specified in the HEVC standard with quantization parameter (QP). Then perform inverse transform to obtain r'.
2. *Step 2*: Before r' is added to the predictor, perform inverse quantization with $\text{idQP}(\mu)$ to obtain the rescaled residuals $\hat{r}$.

The inverse scaling performed during IDSQ (Fig. 10.6) is performed in the spatial domain, which is the reason for the acronym IDSQ.

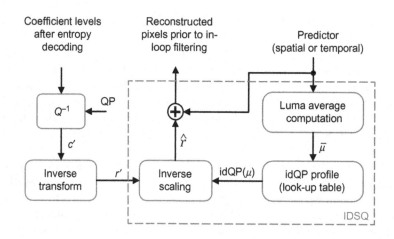

FIGURE 10.6

Block diagram of IDSQ processing.

Source: Zhang et al. (2013b).

IDSQ was integrated into the HM reference codec considered for the HEVC range extensions (HM-10.0-RExt-2.0), and its performance was assessed by measurement of the bit rate reduction relative to the codec without perceptual quantization (Zhang et al., 2013b). The test material used in the performance evaluation consisted of six sequences captured by Zhang et al. (2013b) using a RED EPIC camera using its HDRx function (Fig. 10.7A–F. All the test sequences were then downsampled to 1920 × 1080 and provided as OpenEXR, half float representation files. The input bit depth for the luminance channel was 14 bits, with two eight-bit chrominance channels in LogLuv color space (Zhang et al., 2011). The luma/chroma sampling pattern was 4:4:4. The coding configurations selected for the experiments were the ones described in Flynn and Rosewarne (2013): all intra, random access, and low-delay B. The QP values considered were 12, 17, 22, and 27.

HDR-VDP-2 was used to assess the quality of the reconstructed video. The results obtained (HDR-VDP-2 predicted mean opinion score (MOS)) suggest that the proposed HDR-IDSQ method provides quality perceptually equivalent to that of the anchor but at a significantly lower bit rate. An average bit rate reduction of up to 9.53% was observed when compared with the anchor, with the largest reduction (18.55%) being achieved for "Carnival2."

As with most objective metrics, HDR-VDP-2 has certain limitations in terms of its correlation with subjective results. The experimental results of Narwaria et al. (2013b), for example, suggest that at higher bit rates HDR-VDP-2 scores do not increase in proportion to the subjective scores. To overcome this, recent work by Zhang et al. (2015) focused on conducting objective and subjective quality tests on an extended HDR dataset. The results of these tests support the conclusions in Zhang et al. (2013b), demonstrating a bit rate reduction of up to 42% for the HDR-IDSQ codec compared to an HEVC anchor across various coding conditions.

FIGURE 10.7

Example frames from the test HDR sequences used in the performance evaluation of the method of Zhang et al. (2013b). Nighttime sequences "Carnivalx": (A) "Carnival1," (B) "Carnival2," (C) "Carnival3" and (D) "Carnival4." Daytime sequences: (E) "Library" and (F) "ViceChancelorRoom."

Source: Zhang et al. (2013b).

10.4 SUMMARY

In this chapter, we have presented a review of state-of-the-art HDR image and video coding methods, with an emphasis on the latter. Layered (backward-compatible) methods apply tone mapping before compression to create a base (LDR) layer for legacy decoders and displays. Decoding of the optional enhancement layer enables reconstruction of the HDR output. In this case the choice of TMO can affect the performance of the codec not only in terms of the visual quality but also in terms of the bit rate produced. With layered HDR video compression, attention should also be paid to potential flickering artifacts resulting from tone mapping successive frames. Optimization of the tone-mapping curve can be used to minimize the quality loss due to the tone mapping, encoding/decoding, inverse tone mapping chain followed by layered HDR codecs.

High-bit-depth (native) coding involves standard codecs, such as HEVC, operating on the original HDR input. Rate/quality benefits can be obtained by the introduction of HDR-specific perceptual preprocessing, such as CSF-based prefiltering. Modifications can also be made to the actual codec, as in the case of the IDSQ method of Zhang et al. (2013b, 2015), that exploit HDR-specific redundancies/masking effects of the HVS. This method represents the state of the art in HDR video compression.

REFERENCES

Aydin, T.O., Mantiuk, R., Seidel, H.P., 2008. Extending quality metrics to full luminance range images. In: Human Vision and Electronic Imaging XIII, vol. 6806, p. 68060B.

Aydin, T.O., Stefanoski, N., Croci, S., Gross, M., Smolic, A., 2014. Temporally coherent local tone mapping of HDR video. ACM Trans. Graph. (TOG) 33 (6), 196.

Banterle, F., Artusi, A., Debattista, K., Chalmers, A., 2011. Advanced High Dynamic Range Imaging. AK Peters (now CRC Press), Natick, MA.

Barten, P.G.J., 1999. Contrast Sensitivity of the Human Eye and its Effects on Image Quality. SPIE Press, Bellingham, WA.

Bogart, R., Kainz, F., Hess, D., 2003. OpenEXR image file format. In: Proceedings of ACM SIGGRAPH, Sketches & Applications.

Boitard, R., Cozot, R., Thoreau, D., Bouatouch, K., 2014. Survey of temporal brightness artifacts in video tone mapping. In: HDRi2014—International Conference and SME Workshop on HDR Imaging.

Boliek, M. (Ed.), 2014. Information Technology – The JPEG 2000 Image Coding System: Part 1 (amendment 8), ISO/IEC IS 15444-1/ ITU-TT.800.

Bordes, P., Ye, Y., Alshina, E., Li, X., Kim, S., Duenas, A., Ugur, K., Sato, K., 2013. Description of HEVC scalable extensions core experiment SCE1: color gamut and bit-depth scalability. In: JCTVC-O1101, 15th Meeting: Geneva, CH.

Boyce, J.M., 2004. Weighted prediction in the H.264/MPEG AVC video coding standard. In: Proceedings of International Symposium on Circuits and Systems (ISCAS), vol. 3, p. III-789-92.

Bull, D.R., 2014. Communicating Pictures: A Course in Image and Video Coding. Academic Press, San Diego, CA.

Debevec, P.E., Malik, J., 1997. Recovering high dynamic range radiance maps from photographs. In: Proceedings of Computer Graphics and Interactive Techniques, pp. 369–378.

Drago, F., Mantiuk, R., 2005. HDR Project, Image Gallery, Max-Planck-Institut Informatik, http://resources. mpi-inf.mpg.de/hdr/gallery.html#top.

Duenas, A., Andrivon, P., Ye, Y., Alshina, E., Li, X., Ugur, K., Auyeung, C., Kim, S., 2014. Description of HEVC scalable extensions core experiment SCE1: color gamut scalability. In: JCTVC-Q1101, 17th Meeting: Valencia, ES.

Eilertsen, G., Wanat, R., Mantiuk, R.l., Unger, J., 2013. Evaluation of tone mapping operators for HDR-video. In: Computer Graphics Forum, vol. 32, pp. 275–284.

Fattal, R., Lischinski, D., Werman, M., 2002. Gradient domain high dynamic range compression. In: ACM Transactions on Graphics (TOG), vol. 21. ACM, New York, NY, pp. 249–256.

Ferwerda, J.A., Pattanaik, S.N., Shirley, P., Greenberg, D.P., 1996. A model of visual adaptation for realistic image synthesis. In: Proceedings of Computer Graphics and Interactive Techniques. ACM, New York, NY, pp. 249–258.

Flynn, D., Rosewarne, C., 2013. Common test conditions and software reference configurations for HEVC range extensions. In: JCTVC-L1006, 12th Meeting, Geneva, CH.

Garbas, J.U., Thoma, H., 2011. Temporally coherent luminance-to-luma mapping for high dynamic range video coding with H.264/AVC. In: Proceedings of IEEE International Conference on Acoustics, Speech and Signal Processing (ICASSP), pp. 829–832.

Hood, D., Finkelstein, M., 1986. Chapter 5: Sensitivity to light. In: Boff, K., Kaufman, L., Thomas, J. (Eds.), Handbook of Perception and Human Performance, vol. 1. Wiley, New York, NY.

Husak, W., Ninan, A., 2012. Call for proposals for JPEG HDR. In: ISO /IEC JTC 1/SC 29 /WG 1 N6147.

Jia, Y., Lin, W., Kassim, A.A., 2006. Estimating just-noticeable distortion for video. IEEE Trans. Circ. Syst. Video Technol. 16 (7), 820–829.

Kiser, C., Reinhard, E., Tocci, M., Tocci, N., 2012. Real time automated tone mapping system for HDR video. In: Proceedings of IEEE International Conference Image Processing (ICIP), Orlando, USA.

Korshunov, P., Ebrahimi, T., 2012. A JPEG backward-compatible HDR image compression. In: Proceedings of SPIE Conference on Applications of Digital Image Processing XXXV, p. 84990J.

Korshunov, P., Ebrahimi, T., 2013. Context-dependent JPEG backward-compatible high-dynamic range image compression. Opt. Eng. 52 (10), 102006.1–102006.11.

Koz, A., Dufaux, F., 2014. Methods for improving the tone mapping for backward compatible high dynamic range image and video coding. Signal Process. Image Commun. 29 (2), 274–292.

Kunkel, T., Reinhard, E., 2010. A reassessment of the simultaneous dynamic range of the human visual system. In: Proceedings of the ACM Symposium on Applied Perception in Graphics and Visualization. ACM, Los Angeles, pp. 17–24.

Larson, G.W., 1998. LogLuv encoding for full-gamut, high-dynamic range images. J. Graph. Tools 3 (1), 15–32.

Larson, G.W., Rushmeier, H., Piatko, C., 1997. A visibility matching tone reproduction operator for high dynamic range scenes. IEEE Trans. Vis. Comput. Graph. 3 (4), 291–306.

Le Dauphin, A., Boitard, R., Thoreau, D., Olivier, Y., Francois, E., LeLéannec, F., 2014. Prediction-guided quantization for video tone mapping. In: SPIE 8499, Applications of Digital Image Processing XXXVII, p. 92170B.

Lee, C., Kim, C.S., 2007. Gradient domain tone mapping of high dynamic range videos. In: Proceedings of IEEE International Conference Image Processing (ICIP), pp. III-461–III-464.

Lischinski, D., Farbman, Z., Uyttendaele, M., Szeliski, R., 2006. Interactive local adjustment of tonal values. ACM Trans. Graph. (TOG) 25 (3), 646–653.

Liu, S., Kim, W.S., Vetro, A., 2008. Bit-depth scalable coding for high dynamic range video. In: Society of Photo-Optical Instrumentation Engineers (SPIE) Conference Series, vol. 6822, p. 24.

Mai, Z., Mansour, H., Mantiuk, R., Nasiopoulos, P., Ward, R., Heidrich, W., 2011. Optimizing a tone curve for backward-compatible high dynamic range image and video compression. IEEE Trans. Image Process. 20 (6), 1558–1571.

Mannos, J., Sakrison, D., 1974. The effects of a visual fidelity criterion of the encoding of images. IEEE Trans. Inform. Theory 20 (4), 525–536.

Mantiuk, R., Daly, S., Kerofsky, L., 2008. Display adaptive tone mapping. ACM Trans. Graph. (TOG) 27 (3), 68.

Mantiuk, R., Daly, S.J., Myszkowski, K., Seidel, H.P., 2005. Predicting visible differences in high dynamic range images: model and its calibration. In: SPIE Proceedings, vol. 5666, pp. 204–214.

Mantiuk, R., Efremov, A., Myszkowski, K., Seidel, H.P., 2006a. Backward compatible high dynamic range MPEG video compression. ACM Trans. Graph. (TOG) 25 (3), 713–723.

Mantiuk, R., Kim, K.J., Rempel, A.G., Heidrich, W., 2011. HDR-VDP-2: a calibrated visual metric for visibility and quality predictions in all luminance conditions. ACM Trans. Graph. 30 (4), 40.

Mantiuk, R., Krawczyk, G., Myszkowski, K., Seidel, H.P., 2004. Perception-motivated high dynamic range video encoding. ACM Trans. Graph. (TOG) 23 (3), 733–741.

Mantiuk, R., Myszkowski, K., Seidel, H.P., 2006b, Lossy compression of high dynamic range images and video. In: SPIE Proceedings Vol. 6057: Human Vision and Electronic Imaging XI, p. 60570V.

Miller, S., Nezamabadi, M., Daly, S., 2013. Perceptual signal coding for more efficient usage of bit codes. SMPTE Motion Imaging J. 122 (4), 52–59.

Motra, A., Thoma, H., 2010. An adaptive LogLuv transform for high dynamic range video compression. In: Proceedings of IEEE International Conference Image Processing (ICIP), pp. 2061–2064.

Mullen, K.T., 1985. The contrast sensitivity of human colour vision to red-green and blue-yellow chromatic gratings. J. Physiol. 359 (1), 381–400.

Naccari, M., Mrak, M., Flynn, D., Gabriellini, A., 2012. On intensity dependent quantisation in the HEVC codec. In: JCTVC-I0257, 9th Meeting, Geneva, CH, April.

Narwaria, M., Da Silva, M.P., Le Callet, P., Pepion, R., 2012. Effect of tone mapping operators on visual attention deployment. In: Proceedings of SPIE Conference on Applications of Digital Image Processing XXXV, pp. 1–15.

Narwaria, M., Da Silva, M.P., Le Callet, P., Pepion, R., 2013a. Tone mapping-based high-dynamic-range image compression: study of optimization criterion and perceptual quality. Opt. Eng. 52 (10), 102008.1–102008.15.

Narwaria, M., Perreira Da Silva, M., Le Callet, P., Pepion, R., 2013b. Tone mapping based HDR compression: does it affect visual experience? Signal Proc. Image Commun. 29 (2), 257–273.

Pinheiro, A., Fliegel, K., Korshunov, P., Krasula, L., Bernardo, M., Pereira, M., Ebrahimi, T., 2014. Performance evaluation of the emerging JPEG XT image compression standard. In: IEEE International Workshop on Multimedia Signal Processing (MMSP), pp. 1–6.

Reinhard, E., Devlin, K., 2005. Dynamic range reduction inspired by photoreceptor physiology. IEEE Trans. Vis. Comput. Graph. 11 (1), 13–24.

Reinhard, E., Heidrich, W., Pattanaik, S., Debevec, P., Ward, G., Myszkowski, K., 2010. High dynamic range imaging: acquisition, display, and image-based lighting. Morgan Kaufmann, San Francisco, CA.

Reinhard, E., Stark, M., Shirley, P., Ferwerda, J., 2002. Photographic tone reproduction for digital images. ACM Trans. Graph. 21 (3), 267–276.

Richter, T., 2009. Evaluation of floating point image compression. In: International Workshop on Quality of Multimedia Experience (QoMEx), pp. 222–227.

Richter, T., 2013a, Backwards compatible coding of high dynamic range images with JPEG. In: Data Compression Conference (DCC), 2013, pp. 153–160.

Richter, T., 2013b, On the standardization of the JPEG XT image compression. In: Proceedings of Picture Coding Symposium (PCS), pp. 37–40.

Richter, T., 2014. On the integer coding profile of JPEG XT. In: SPIE Optical Engineering+ Applications, pp. 921719–921719-19.

Richter, T., Artusi, A., Agostinelli, M., 2014. Information technology: scalable compression and coding of continuous-tone still images, HDR Floating-Point Coding, ISO/IEC 18477-7.

Spaulding, K., Woolfe, G., Joshi, R., 2003. Using a residual image to extend the color gamut and dynamic range of an sRGB image. In: Proceeding of IS&T PICS Conference, pp. 307–314.

Srinivasan, S., Tu, C., Zhou, Z., Ray, D., Regunathan, S., Sullivan, G., 2007. An introduction to the HDPhoto technical design. In: JPEG Document WG1N4183.

Sullivan, G.J., Yu, H., Sekiguchi, S., Sun, H., Wedi, T., Wittmann, S., Lee, Y.L., Segall, A., Suzuki, T., 2007. New standardized extensions of MPEG4-AVC/H. 264 for professional-quality video applications. In: IEEE International Conference on Image Processing (ICIP), vol. 1, pp. I-13–I-16.

Tumblin, J., Turk, G., 1999. LCIS: a boundary hierarchy for detail-preserving contrast reduction. In: Proceedings of Computer graphics and interactive techniques. ACM Press/Addison-Wesley Publishing Co., New York, NY, pp. 83–90.

Ward, G., 1991. Real pixels. Graphics Gems II, pp. 80–83.

Ward, G., Simmons, M., 2004. Subband encoding of high dynamic range imagery. In: Proceedings of Symposium on Applied Perception in Graphics and Visualization. ACM, New York, NY, pp. 83–90.

Ward, G., Simmons, M., 2005. JPEG-HDR: a backwards-compatible, high dynamic range extension to JPEG. In: Proceedings of Color Imaging Conference, pp. 283–290.

Ward, G.J., 1994. The radiance lighting simulation and rendering system. In: Proceeding of Computer graphics and interactive techniques. ACM, New York, NY, pp. 459–472.

Winken, M., Marpe, D., Schwarz, H., Wiegand, T., 2007. Bit-depth scalable video coding. In: IEEE International Conference on Image Processing (ICIP), vol. 1, pp. I-5–I-8.

Wu, Y., Gao, Y., Chen, Y., 2008. Bit-depth scalability compatible to H.264/AVC-scalable extension. J. Vis. Commun. Image Represent. 19 (6), 372–381.

Yeganeh, H., Wang, Z., 2013. Objective quality assessment of tone mapped images. IEEE Trans. Image Process. 22 (2), 657–667.

Yoshida, A., Blanz, V., Myszkowski, K., Seidel, H.P., 2005. Perceptual evaluation of tone mapping operators with real-world scenes. In: Electronic Imaging, pp. 192–203.

Zhang, Y., Agrafiotis, D., Naccari, M., Mrak, M., Bull, D., 2013a, Visual masking phenomena with high dynamic range content. In: IEEE International Conference on Image Processing (ICIP), pp. 2284–2288.

Zhang, Y., Naccari, M., Agrafiotis, D., Mrak, M., Bull, D., 2013b, High dynamic range video compression by intensity dependent spatial quantization in HEVC. In: Proceedings of Picture Coding Symposium (PCS), pp. 353–356.

Zhang, Y., Naccari, M., Agrafiotis, D., Mrak, M., Bull, D., 2015. High dynamic range video compression exploiting luminance masking. IEEE Trans. Circuits Syst. Video Technol. 99, 1, DOI: 10.1109/TCSVT.2015.2426552.

Zhang, Y., Reinhard, E., Agrafiotis, D., Bull, D., 2012a, Image and video compression for HDR content. In: Proceedings of SPIE Conference on Applications of Digital Image Processing XXXV, pp. 84990H1–84990H13.

Zhang, Y., Reinhard, E., Bull, D., 2011. Perception-based high dynamic range video compression with optimal bit-depth transformation. In: IEEE International Conference on Image Processing (ICIP), pp. 1321–1324.

Zhang, Y., Reinhard, E., Bull, D., 2012b, Perceptually lossless high dynamic range image compression with JPEG 2000. In: Proceedings of IEEE International Conference Image Processing (ICIP), Orlando, USA, pp. 1057–1060.

HIGH DYNAMIC RANGE AND WIDE COLOR GAMUT VIDEO STANDARDIZATION — STATUS AND PERSPECTIVES

11

E. François, P. Bordes, F. Le Léannec, S. Lasserre, P. Andrivon

Technicolor, Cesson-Sevigne, France

CHAPTER OUTLINE

11.1 INTRODUCTION

The migration from current high-definition (HD) TV (HDTV) technologies to next-generation video services, known as Ultra HD TV (UHDTV), is supported by work in several application standards organizations, including Digital Video Broadcasting (DVB), the Advanced Television Systems Committee (ATSC), and the Blu-ray Disc Association (BDA). The move to UHDTV requires the exploration of video quality dimensions as illustrated in Fig. 11.1. To successfully achieve a successful migration to UHDTV, there are several dimensions that require the attention of standards bodies.

High Dynamic Range Video. **http://dx.doi.org/10.1016/B978-0-08-100412-8.00011-5**

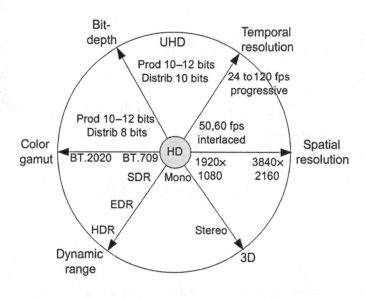

FIGURE 11.1

Six key dimensions toward enhanced video quality.

One of the first areas of focus has been to enhance the picture spatial resolution, which will change from HD (1920 × 1080 pixels) to Quad-HD (3840 × 2160 pixels) or 4K (4096 × 2160 pixels).[1] Another area of improvement revolves around improving the temporal resolution, in which frame rates evolve from 24, 25, and 30 frames per second to 50 and 60 frames per second or even to 100 and 120 frames per second. A third area of focus is the increase of the bit depth used to represent the video signal, which will migrate from eight bits per sample in today's video distribution network to 10 bits per sample. Yet another dimension on which standards bodies must focus is the "relief" — also publicly known as "3D" (stereo video) — which has already been added to HDTV, and is a candidate feature that will enforce immersivity for UHDTV.

Two additional dimensions that are receiving increased attention by standards bodies because of their ability to enhance the video consumption experience revolve around the color gamut and the dynamic range. The color gamut, is going to double from the BT.709 color space (Recommendation ITU-R BT.709-5, 2002) to the so-called BT.2020 color space (Recommendation ITU-R BT.2020-1, 2014). The dynamic range is also expected to increase significantly from the current range (standard dynamic range, or SDR) to extended dynamic range (EDR) or high dynamic range (HDR).

---

[1]The current perspective is to stay with the 4:2:0 chroma format for distribution (chroma component resolution is one quarter of the luma resolution) and in progressive-only scanning (interlace representation would not be considered anymore). For production the chroma format should be 4:2:2 or 4:4:4.

It has turned out that limiting the HDTV-to-UHDTV evolution to an increase in spatial and temporal resolution will not deliver a profound enough impact on the market because the increase in quality is not clearly noticeable by most viewers. As a consequence, the DVB anticipates a UHDTV deployment scenario that will be rolled out in successive phases. Phase 1 is to be completed between the end of 2015 and the middle of 2016. Phase 2, which will have more features, is expected to be available 3 years later (Thomas, 2014).

Phase 1 will introduce increases in resolution (3840×2160 pixels in progressive format) and in wider color gamut (BT.2020), while phase 2 is expected to add other features (mainly high frame rate and HDR). In combination, the addition of phase 1 and phase 2 will deliver improvements that will be clearly visible to consumers. A third phase is envisioned for deployment after 2020, which will extend the resolution to 8K (7680×4320 pixels).

As the industry moves through these phases, one of the major issues that standards bodies must carefully consider revolves around "backward compatibility." This will be critical to avoid unnecessary disruption in the distribution chain, and to ensure a smooth and continuous adaptation to the always-evolving capability of decoding/rendering devices.

The transition to wide color gamut (WCG) and HDR video is also expected to happen in successive waves that will parallel technological improvements for production workflows, as well as for receiving and rendering devices. For this reason, many expect a transition phase that involves an intermediate EDR video technology deployment before adoption and deployment of full HDR.

A first distribution of EDR in the home is expected to use TV sets providing peak brightness of around 1000 cd/m^2 (1 cd/m^2 is commonly known as a "nit") and a dynamic range of 15 stops.

The finalization of High Efficiency Video Coding (HEVC) specification versions 1 and 2, in conjunction with the fast evolution of capture and rendering technologies, offers major opportunities for the video industry to successfully deploy these new services. To avoid the unconstructive development of fragmented standards and to favor interoperable ecosystems, multiple standards organizations have synchronized their efforts to provide the industry with relevant standards that allow the deployment of HDR and WCG video in the short term or mid term. This is work that was ongoing at the time this chapter was written.

This chapter aims to describe these different activities, in terms of organizations, processes, technical solutions, and specifications. Section 11.2 summarizes the main challenges facing HDR and WCG video and discusses the possible evolution that will occur for HDR and WCG video workflows. It also provides an overview of the activities of the different standards organizations addressing HDR and WCG video support. Special focus is placed on the Moving Picture Experts Group (MPEG) and Video Coding Experts Group (VCEG) bodies, as well as their Joint Collaborative Team on Video Coding (JCT-VC), which is dedicated exclusively to the HEVC standard specification.

The two next sections are more technology oriented. Section 11.3 describes key elements that have already been standardized to support the deployment of HDR and WCG video. In particular, a detailed description is provided of the related coding tools adopted in the HEVC standard. Section 11.4 completes the landscape by listing and describing a few promising technologies that are still under investigation.

The concluding section offers several plausible scenarios for follow-on standardization activities that may happen. The impact these potential activities may have on the ecosystem is briefly discussed.

11.2 HDR AND WCG VIDEO WORKFLOWS AND RELATED STANDARDIZATION ACTIVITIES

The dynamic range of a video signal is commonly defined in terms of stops, to measure the ratio between the minimum and maximum luminance of the signal. Basically, a stop is the measure in base 2 logarithm of this ratio. Increasing the dynamic by one stop results in a doubling of the dynamic range. Consequently, 10 stops corresponds to a ratio of about 1000 in linear luminance. The SDR used in current HDTV (Recommendation ITU-R BT.709-5, 2002; EBU Tech 3320, 2014) corresponds to slightly more than 11 stops, with a peak luminance specified at a few hundred candelas per square meter and black levels on the order of 0.1 cd/m^2. Therefore, SDR rendering is far from being able to reproduce the luminosity and the human perception of real-life scenes involving a much higher dynamic range. HDR aims to go beyond the SDR limitation to offer a viewing experience that is much closer to real life.

The same evolution is foreseen for colors. The color gamut specified in BT.709 is a very limited subset of the colors that can be seen by the human eye. As measured in the CIE1931 xy chromaticity diagram (Pedzisz, 2014) the BT.709 gamut covers only 33% of the visible locus. The new BT.2020 will double this coverage, and contains almost all the theoretical Pointer's gamut that encompasses the colors found in the natural world.

Distribution of video signals with wider color gamut and higher dynamic range will have obvious impacts on the complete video chain — from production, to distribution, and ultimately to end-user rendering. Those impacts are discussed in the next subsections in terms of workflow and related standardization activities.

11.2.1 THE HDR AND WCG VIDEO WORKFLOWS

As emphasized in Section 11.1, developments in HDR and WCG capture and rendering devices are expected to soon support the full use of HDR and WCG capabilities. As these new technologies mature, it is important to remember that previous generations of devices will remain in the market. This means that devices with various rendering specifications will coexist throughout the value chain. One of the major challenges facing the industry is to develop workflows — from production to rendering — that are able to deal with this constant evolution and heterogeneity. In particular, backward compatibility with SDR should be properly managed for various applications such as broadcasting and multicasting.

For other applications such as point-to-point video distribution (eg, over-the-top video distribution), this feature is of lower importance. The backward compatibility discussed here relates to two sides: (1) the decoder and (2) the display. Depending on the target applications, different workflows may be considered. Note that there is no compatibility between Advanced Video Coding (AVC; aka H.264) and HEVC, unless scalable HEVC (the so-called SHVC), with its hybrid scalability feature, is used — that is, an AVC base layer (BL) with an HEVC enhancement layer (EL). For UHDTV deployment, the widely preferred track is to rely on the HEVC Main 10 profile configuration as discussed in Section 11.3.

The three workflows discussed below are based on different requirements in terms of backward compatibility (SDR backward compatible versus non-SDR backward compatible) and codec technologies (single-layer nonscalable versus dual-layer scalable). Another aspect is the dual master case, where

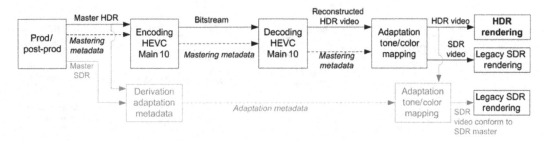

FIGURE 11.2

First HDR workflow: single-layer coding with HDR-to-SDR adaptation on the receiver side.

the production delivers both an SDR version and an HDR version of the content. This leads to another dimension of SDR backward compatibility, which relates to preserving at the rendering side the artistic intent of the SDR version.

The first workflow, illustrated in Fig. 11.2, is built on a single-layer coding design with HDR content, with the potential for additional HDR-to-SDR adaptation on the receiver side. The HDR master is obtained from the production and postproduction in a format adapted to the HEVC Main 10 profile (10 bits per sample, 4:2:0 chroma format). The process generates side mastering metadata that characterize the mastering display and HDR master. The metadata are used for further adaptation of the content to improve its final rendering. The HDR master and mastering metadata are encoded with use of the HEVC Main 10 profile. The bitstream is then decoded by means of a legacy HEVC Main 10 decoder. An adaptation process is then required to adapt the content to the rendering device. This process uses the mastering metadata to properly fit the content to the display capabilities. In particular, rendering on an SDR TV requires a specific HDR-to-SDR adaptation (commonly known as tone mapping) after decoding. For some applications, an SDR-graded version may be provided from the postproduction process in addition to the HDR-graded version. (In some cases for older movies, the HDR grade may even be deduced from already existing SDR-graded version). This leads to a dual-grading scenario. This scenario may need additional adaptation metadata generated from input HDR and SDR content. The data may be used at the rendering stage to convert the decoded HDR video to an SDR version similar to the input SDR master grade, for the purpose of preserving the content creator's artistic intent.

The second workflow, as presented in Fig. 11.3, is also based on single-layer coding. However, the objective of this workflow is to offer direct backward compatibility with SDR displays without the need for an adaptation process after signal decoding. During the first step in this process, the HDR master is preprocessed before encoding so that an SDR (BT.709 conformant) version is generated. Corresponding adaptation metadata are also generated. The SDR version, as well as the mastering metadata and adaptation metadata, is encoded and transmitted. Then, the bitstream is decoded by an HEVC Main 10 decoder to produce an SDR video that can be directly displayed on a legacy SDR display without an additional adaptation process. To revert to the HDR video, an adaptation process (similar to inverse tone mapping) is required. This process may benefit from the mastering and adaptation metadata. As with the first workflow that was described, an SDR master can also be produced in addition to the HDR master. Specific SDR-to-SDR adaptation metadata may be deduced from the input SDR master and

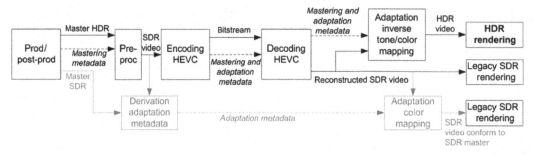

FIGURE 11.3

Second HDR workflow: single-layer coding with SDR-to-HDR adaptation on the receiver side.

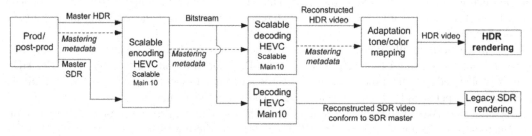

FIGURE 11.4

Third HDR workflow: dual-layer coding.

from the SDR version generated by the preprocessing step. At the rendering stage, those adaptation metadata may be used to convert the reconstructed SDR video to an SDR version close to the input SDR master grade.

The first two workflows that have been described are based on single-layer encoding and decoding, as well as on extra postprocessing steps using metadata to adapt the decoding signal (HDR in the first workflow, SDR in the second workflow) to the capabilities of the display.

The third workflow is based on a scalable dual-layer coding framework, and is shown in Fig. 11.4. Both SDR-graded and HDR-graded versions are supposed to be provided from the production/postproduction process. These two sets of content are then coded by means of a scalable dual-layer codec, with the SDR master as the BL and the HDR master as the EL (residual layer). In this scenario, SDR video that conforms to the artistic intent is directly obtained by decoding of the BL (SDR) bitstream. The decoding of the full bitstream leads directly to the HDR video, which can be rendered, after possible adaptation, on HDR displays.

11.2.2 RELATED STANDARDIZATION ACTIVITIES

To ensure interoperability between deployed HDR and WCG video services, various standards bodies are working to specify the components of the HDR workflows — such as video signal format, metadata,

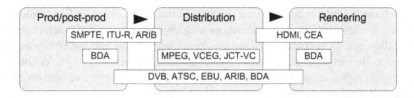

FIGURE 11.5

Simplified video chain and related standardization organizations.

codec, adaptation processes, and interfaces between decoders and displays. A simplified picture of the standardization landscape, indicating where these organizations are mostly active, is shown in Fig. 11.5.

- The Society of Motion Picture and Television Engineers (SMPTE) and the International Telecommunication Union Radiocommunication sector (ITU-R) are working on the definition of the HDR ecosystem and the specification of the video signal format as well as associated metadata for production and distribution. The format includes the following features: spatial resolution, temporal resolution, color gamut and space, and bit depth, as well as nonlinear transfer functions to convert the uncompressed linear signal (scene or displayed light) to or back from the quantized compressed signal (codewords or code values). Obviously, these specifications have an impact on the rendering side because displays should be able to properly interpret the signal format. The components that have already been specified — along with those still under investigation — are discussed in Sections 11.3 and 11.4.
- MPEG (a subgroup of the International Organization for Standardization) and VCEG (a subgroup of the International Telecommunication Union Telecommunication sector) are the two main standards groups focused on video coding and distribution. They are in charge of the HEVC standard specification through the joint team JCT-VC. HEVC already integrates several tools relevant for HDR and WCG video coding, as discussed in Sections 11.3.2 and 11.3.3. MPEG is now working on the performance evaluation of HEVC and on possible complementary solutions, which are discussed in Section 11.4. A call for evidence was issued to evaluate the added value of some extra HDR/WCG dedicated tools. It has resulted in setting up a group studying extensions of HEVC for HDR support.
- DVB, the European Broadcasting Union (EBU), ATSC, and the Association of Radio Industries and Businesses (ARIB) are standards organizations that are addressing the transmission side. These groups are currently discussing system definitions, specifications, and timelines for the deployment of UHDTV. The support of HDR is being considered in this context. For example, the DVB UHDTV timeline is based on a three-phase scenario in which HDR is introduced in the second phase. The associated specification is expected to be issued by 2016, with general deployment targeted for 2018.
- HDR was adopted by BDA. The specification for Ultra HD Blu-ray that includes the HDR and WCG features was finalized in May 2015. Date for deployment is scheduled for early 2016.

- The High-Definition Multimedia Interface (HDMI) Forum and Consumer Electronics Association (CEA) are two bodies working on the interfaces with digital equipment — such as set-top boxes, players, and displays. These two groups are also considering HDR and WCG in their new specifications as they explore the metadata conveyed between receivers or players on one side of the equation and display technologies on the other. Their primary objective is to enable the signal adaptation to the display capabilities.

There are some obvious overlap activities among these different standards groups. As a consequence, it is critical to have official coordination through standards liaisons between the standards bodies. That is why the UHD Alliance — an industrial consortium — has been set up to establish consistency across the landscape of these evolving UHDTV standards.

11.3 HDR AND WCG IN ALREADY EXISTING STANDARDS

Several standards already address certain aspects of HDR and WCG video for the three following points: (1) video signal specification with the definition of a new WCG and the introduction of new transfer functions for HDR signals that better fit the human visual system; (2) specification of metadata; and (3) specification of coding tools specifically adapted to HDR and WCG contents.

11.3.1 HDR AND WCG VIDEO REPRESENTATION

SMPTE and ITU-R are actively working on new specifications adapted to WCG and HDR video. The recent BT.2020 recommendation (ITU-R Rec. 2020, 2014) affects several dimensions — in particular, the frame size (4K, 8K) and the color gamut. As discussed already, and illustrated in Fig. 11.6, the BT.2020 color gamut is twice as wide as the BT.709 color gamut. This provides a bigger palette for content creators and offers higher fidelity in color reproduction.

SMPTE has also issued a new electro-optical transfer function (EOTF) in its recommendation to SMPTE ST 2084 (Miller et al., 2013). The function converts an HDR signal represented in a nonlinear space (codeword space) that is particularly tailored for properties in the human visual system into a linear light domain (display-referred light). The new transfer function, known as perceptual quantization (PQ), is based on the human contrast sensitivity model developed by Barten (2004). The function has been extended to much higher luminance ranges than have been considered so far with SDR signals. The shapes of the PQ EOTF and gamma transfer function used for SDR are shown in Fig. 11.7 for a 10-bit limited-range output signal defined in the codeword interval [64,940]. The PQ is specified up to a luminance of 10,000 cd/m^2, while the gamma function usually applies to signal up to a few hundred candelas per square meter (in Fig. 11.7 the gamma function applies from 0 to 100 cd/m^2 then saturates above 100 cd/m^2).

As was done previously in the AVC standard, some specific optional metadata are used to indicate the format of the video signal in the new HEVC/H.265 standard. These metadata help to properly display the decoded content during the rendering process. They are embedded in a video usability information (VUI) container. The VUI signals the black level and range of the luma and chroma signals

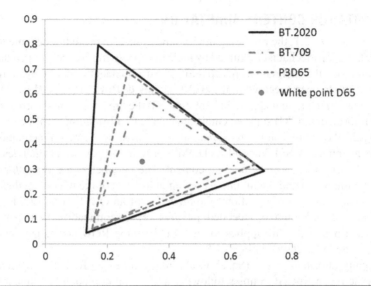

FIGURE 11.6

Comparison of BT.709, P3 (D65 white point) and BT.2020 color gamuts in chromaticity diagram (CIE1931).

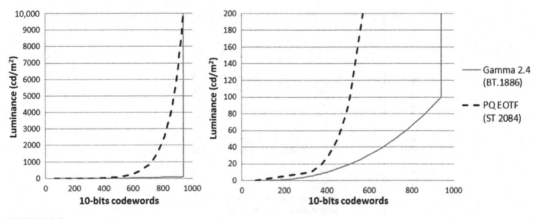

FIGURE 11.7

Comparison of Gamma 2.4 (BT.1886) and PQ EOTF (left) luminance from 0 to 10,000 cd/m². Right: Zoom on 0–200 cd/m².

(ie, limited or full range), the chromaticity coordinates of the source primaries in the CIE1931 xy system, and the optoelectronic transfer function (OETF). HEVC has extended the AVC VUI to insert new entry points corresponding to the signaling of the BT.2020 color primaries and of the inverse EOTF to indicate display luminance levels up to 10,000 cd/m², well beyond the peak of SDR signals.

11.3.2 METADATA FOR CONTENT ADAPTATION

As mentioned in Section 11.2.1, the rendering device landscape is expected to become more and more heterogeneous. Adaptation mechanisms, guided by metadata, are therefore required on the receiver side to properly adapt the signal to the rendering capabilities. Such metadata, optionally conveyed with the main stream, have already been specified in the HEVC standard (Boyce et al., 2014): (1) the mastering display color volume; (2) the tone mapping information; (3) the knee function information; and (4) the color remapping information (CRI) supplemental enhancement information (SEI) messages.

The mastering display color volume metadata, also specified in the recommendation SMPTE ST 2086 and implemented in CEA-861.3 as well as HDMI2.0a, embeds parameters that describe the static characteristics of the display used to grade the content. The parameters are the color primaries, the white point, and the minimum and maximum mastering luminance of the mastering display.

The knee function information SEI allows the modeling of an OETF or an EOTF by a piecewise linear model. Thus, the mapping of decoded pictures for customization to a particular display environment is made possible. This applies to each of the three RGB normalized components. An example of a knee function is shown in Fig. 11.8.

The tone mapping information SEI (Segall et al., 2006), inherited from the previous AVC/H.264 standard, remaps the reconstructed samples following four possible models: (1) linear mapping with clipping; (2) sigmoidal model; (3) user-defined lookup table; or (4) piecewise linear model. Some

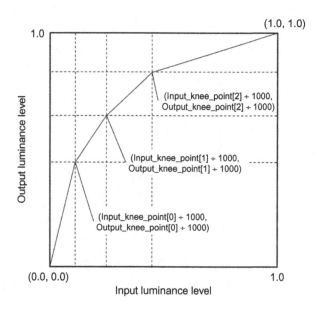

FIGURE 11.8

A knee function with three knee points.

Source: From Boyce, J., Chen, J., Chen, Y., Flynn, D., Hannuksela, M., Naccari, M., Rosewarne, C., Sharman, K., Sole, J., Sullivan, G., Suzuki, T., Tech, G., Wang, Y.K., Wegner, K., Ye, Y., 2014. Edition 2 Draft Text of High Efficiency Video Coding (HEVC), Including Format Range (RExt), Scalability (SHVC), and Multi-View (MV-HEVC) Extensions, Document JCTVC-R1013, 18th JCT-VC Meeting, Sapporo, Japan.

examples are shown in Fig. 11.9. The tone mapping curves are used on the rendering side to adapt the decoded image to the display. The mapping also applies either to the luma or to the RGB components. The knee function SEI previously described is actually a simplified version of the tone mapping SEI. It directly addresses luminance values instead of codewords. However, these models are too simple

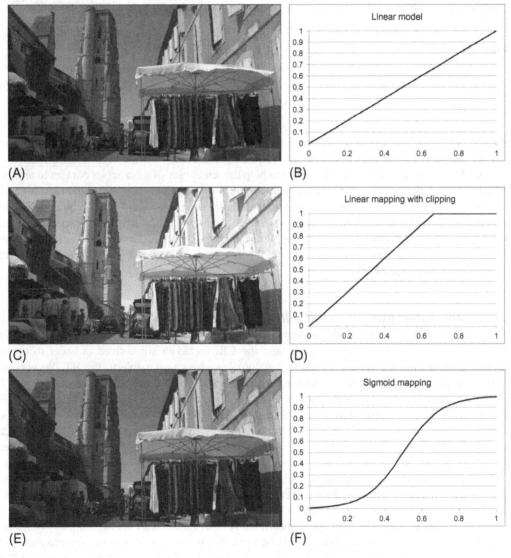

FIGURE 11.9

The tone mapping SEI applied with three different models — right pictures give the shape of the tone mapping function.

Source: From Segall, A., Kerofsky, L., Lei, S., 2006. Tone Mapping SEI Message, Document JVT-T060, 20th JVT Meeting, Klagenfurt, Austria.

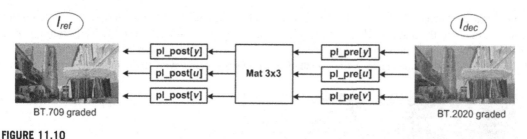

FIGURE 11.10

CRI model.

Source: From Technicolor.

to allow decent color remapping — such as conversion of a color-graded BT.2020 HDR movie into a legacy BT.709 SDR movie that can be displayed on SDR devices.

The CRI SEI message also carries metadata to help the remapping of a particular content to another standardized — or known — color space, with different color gamut, possibly dynamic range, or OETF/EOTF. It may be seen as a generalization of the knee function and tone mapping SEIs by integrating the three color components in the model. It is composed of a first set of piecewise linear functions applied to each color component (pl_pre in Fig. 11.10), a 3 × 3 matrix applied to the three color components, and a second set of piecewise linear functions (pl_post in Fig. 11.10) applied to each color component; see Fig. 11.10.

11.3.2.1 Application example 1: CRI with WCG content

In the use case shown in Fig. 11.11, the content is available in two differently graded versions: BT.2020 and BT.709. The BT.2020 version is coded, and the CRI metadata are derived in order to obtain remapped content from the BT.2020 version that looks as close as possible to the BT.709 version. This is particularly relevant in applications where the artistic intent preservation for both grades is required and the simulcasting of both grades is to be avoided. In Table 11.1, color remapped BT.2020 frames that use CRI are compared with the targeted BT.709 grade, by means of the peak signal to noise ratio (PSNR) computed on the Y′CbCr (commonly known as "YUV") components and the structural similarity (SSIM; Wang et al., 2004) and delta-E76 (Sharma et al., 2005) metrics. The test video sequences grades are the WCG test material provided by JCT-VC (Andrivon and Bordes, 2013). There are 33 piecewise linear curves segments, with 10-bit coded values and the overall CRI size is about four kilobits per frame — which is negligible compared with a typically envisioned UHDTV broadcast bit rate of 15 megabits per second leading to 300 kilobits per frame at 50 frames per second. This is even truer for Ultra HD Blu-ray Disc applications, whose average bit rates are even higher — around 50 megabits per second. A comparison with a basic color space conversion process, as performed by a traditional TV system (Rec. RP 177, 1993), is depicted in Table 11.2 and in Fig. 11.11 with the TableCar material example. Clearly in this example, the CRI allows much better artistic intent preservation than the traditional mapping solution.

FIGURE 11.11

BT.2020 graded (top left), BT.709 graded (top right), BT.2020 color space conversion to BT.709 (bottom left), and BT.2020 color remapped with CRI (bottom right).

Source: From Technicolor.

Table 11.1 Comparison of the BT.2020 Grade Color Remapped to BT.709 Grade With CRI							
Sequence	**PSNR (dB)**			**SSIM**			
1920 × 1080 (10 bits)	**Y**	**U**	**V**	**Y**	**U**	**V**	**Delta-E76 (dB)**
Birthday	40.06	39.22	47.85	0.99	0.95	0.99	44.73
BirthdayFlashPart1	45.77	46.31	48.37	0.99	0.98	0.99	46.26
BirthdayFlashPart2	43.42	44.31	49.28	0.98	0.98	0.99	44.89
Parakeets	45.53	47.52	51.12	0.98	0.98	0.98	46.28
TableCar	33.32	37.39	36.09	0.94	0.93	0.93	41.84
Average	41.62	42.95	46.54	0.976	0.964	0.976	44.80

11.3.2.2 Application example 2: CRI with HDR content

CRI can also be applied to implement HDR-to-SDR conversion. The 2014 Standard Evaluation Material (StEM) has been made available by the Digital Cinema Initiative to industries and

Table 11.2 Comparison of the BT.2020 Grade Color Converted to BT.709 With RP 177

Sequence	PSNR (dB)			SSIM			
1920 × 1080 (10 bits)	Y	U	V	Y	U	V	Delta-E76 (dB)
Birthday	33.26	39.11	34.80	0.99	0.96	0.98	41.62
BirthdayFlashPart1	41.11	42.73	33.99	0.97	0.97	0.99	42.48
BirthdayFlashPart2	36.34	38.78	37.03	0.98	0.98	0.98	41.89
Parakeets	37.57	41.03	37.84	0.98	0.97	0.96	42.41
TableCar	21.88	30.42	26.60	0.93	0.91	0.92	36.32
Average	34.03	38.41	34.05	0.970	0.957	0.966	40.94

Table 11.3 Comparison of the BT.2020 HDR Grade Color Converted to BT.709 SDR Grade With CRI (StEM Content)

StEM	PSNR (dB)			SSIM			
1920 × 1080 (10 bits)	Y	U	V	Y	U[a]	V[a]	Delta-E76 (dB)
Seq_0	42.97	38.02	33.49	0.99	0.86	0.85	41.44
Seq_1	45.90	39.72	40.80	0.98	0.90	0.92	43.52
Seq_2	48.96	38.74	40.03	0.99	0.88	0.91	43.22
Seq_3	49.64	37.50	36.97	0.99	0.89	0.85	42.74
Seq_4	47.82	36.90	36.51	0.99	0.89	0.88	42.21
Average	47.06	38.18	37.56	0.99	0.88	0.88	42.63

[a] Informative only: it is commonly accepted the SSIM is meaningful for luma signal mainly.

organizations that require high-quality program material for test and evaluation purposes (DCI, 2014). It is composed of multiple grades of a unique 12-min-long video segment, including one HDR version (DCI-P3 color space D65 white point, 4000 cd/m^2) and one SDR version (BT.709, 100 cd/m^2). From these versions, two Y'CbCr 10-bit video sequences have been deduced for HDR with use of the PQ inverse EOTF and for SDR with use of the gamma BT.709 OETF. Five scenes each with a duration of 3 s were extracted.

The performance of the CRI for remapping HDR BT.2020 content into SDR BT.709 format is shown in Table 11.3. The remapped HDR version and the targeted SDR version were compared with use of PSNR, SSIM, and deltaE-76 metrics. The objective quality reports were confirmed by means of subjective viewing.

CRI was adopted by BDA in the Ultra HD Blu-ray specification to enable support for a referenced HDR-to-SDR downconversion when the player is connected to an SDR display and plays back an Ultra HD Blu-ray disc that contains one or more HDR video streams (Blu-ray Disc Association, 2015).

11.3.3 HEVC CODING TOOLS — BIT DEPTH AND COLOR GAMUT SCALABILITY

HEVC version 2 issued in July 2014 includes a set of new coding tools in its scalable framework (SHVC) corresponding to workflow 3 presented in Section 11.2.1. Basically, the coexistence of multiple formats for video distribution, coupled with an increased level of network heterogeneity, makes scalable video coding a more promising choice for delivery of content because it offers significant storage and bandwidth efficiencies compared with simulcasting.

SHVC is based on a multilayer framework. Typically a base layer (BL) and an enhancement layer (EL) are both coded by use of interlayer prediction mechanisms that produce an interlayer picture that is used as a reference picture for the prediction. In the use cases and requirements for the scalable coding extensions of HEVC (Luthra et al., 2012a,b), it was identified that changes in resolution in digital video distribution are often associated with a change in color space and/or in bit depth. The main challenge for the bit depth and color gamut scalability (CGS) was to define an efficient interlayer color difference predictor that could anticipate the EL color samples from the collocated reconstructed BL color samples.

The CGS uses a color predictor model based on color lookup tables (CLUTs) as shown in Fig. 11.12. It is based on a meshing of the original three-dimensional input color space, either RGB or YUV. Color parameters are associated with each mesh in order to derive the color values in the targeted color space. The meshing is spatially made up of a set of cubes — called an octant — that can be recursively split into eight smaller octants half sized in all directions. In this way, color space areas where the color transfer functions have somewhat the same characteristics are grouped together into a single octant.

Because of the symmetrical nature of the octree, the grouping operates in the same way in all dimensions of the CLUT. However, the BL and EL video signals may have color disparities that are unequally distributed on the three YUV axes. This may be particularly true for the luma component, which typically carries most of the visual information. That is why the tree may be further split into zero to eight portions in the Y' direction, and possibly nonuniformly partitioned in the chroma directions, as shown in Fig. 11.13.

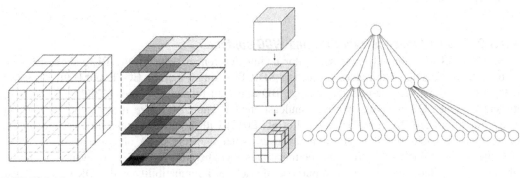

FIGURE 11.12

Example of an RGB three-dimensional CLUT equally partitioning the color space (left) and recursive octant splitting (right).

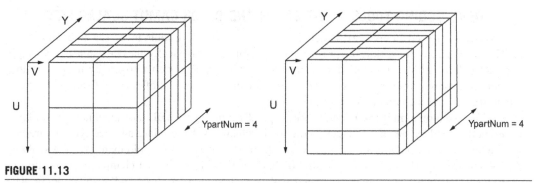

FIGURE 11.13

Three-dimensional CLUT with unequal luma partitioning (left) and nonuniform chroma partitioning (right) of the color space.

The CGS tool can be activated per picture by means of the picture parameters set (PPS) syntax. When activated, the parameters of the CLUT are also coded in the PPS.

11.3.3.1 CGS performance with WCG content

Table 11.4 summarizes the experimental results obtained by application of the SHVC CGS on the WCG test sequences used in JCT-VC (Andrivon and Bordes, 2013), in both all intra (AI) and random access (RA) configurations. The SHVC reference software SHM was used for this experiment and the CGS was configured with default settings (ie, with a CLUT size of $8 \times 2 \times 2$ and 12-bit precision). As shown in Table 11.4, experiments were performed for $1 \times$ and $2 \times$ spatial scalability configurations and quantitative quality is reported as the Bjøntegaard Delta-rate (BD-rate) variation (Bjontegaard, 2008). A negative value indicates an average reduction of the bit rate for a preserved quality. In both cases, the BL resolution is 1920×1080 and the EL bit depth is fixed to 10 bits, while the BL bit depth and EL resolution may vary according to the table entries. The numbers in Table 11.4 show a noticeable BD-rate gain thanks to the CGS.

11.3.3.2 CGS performance with HDR and WCG content

The SHVC CGS tool not only manages color gamut differences between the BL and the EL, but also the transfer function differences — such as SDR BL using a gamma EOTF and HDR EL using the PQ EOTF. Table 11.5 depicts the average BD-rate gain results obtained on the StEM content with and without CGS compared with a simulcast anchor. The CLUT size is set to $8 \times 2 \times 2$ and the test conditions are RA configuration (Li et al., 2013) with no spatial scalability (ie, the spatial ratio is $1 \times$). The achieved BD-rate gains are even better than the gains obtained with the WCG test content (see Table 11.5, left). The demonstrates that the SHVC CGS is relevant for coding SDR and HDR video contents with the purpose of backward compatibility with SDR decoders and displays.

Table 11.4 Performance of the Reference Software SHVC (SHM6.0) Without CGS CLUT Interlayer Prediction

	AI HEVC 1× 10-bit Base			AI HEVC 2× 8-bit Base		
	Y	U	V	Y	U	V
SHM with CGS vs SHM without CGS						
EL only	−49.5%	−51.9%	−57.8%	−25.4%	−25.9%	−31.4%
BL+EL	−15.1%	−20.1%	−31.2%	−13.8%	−14.7%	−20.7%
EL+BL vs simulcast						
CLUT	−35.5%	−39.9%	−50.0%	−26.1%	−26.2%	−31.0%
Reference	−23.9%	−24.7%	−28.3%	−14.5%	−13.3%	−13.4%
	RA HEVC 1× 10-bit Base			RA HEVC 2× 8-bit Base		
	Y	U	V	Y	U	V
SHM with CGS vs SHM without CGS						
EL only	−37.9%	−40.7%	−47.2%	−16.9%	−16.2%	−22.4%
BL+EL	−14.3%	−18.9%	−28.9%	−9.2%	−9.1%	−15.7%
EL+BL vs simulcast						
CLUT	−28.9%	−31.9%	−44.1%	−18.4%	−17.5%	−23.7%
Reference	−17.2%	−16.5%	−22.6%	−10.2%	−9.3%	−9.6%

Table 11.5 Performance of SHVC (SHM7.0) With and Without CGS for Coding SDR (BL) and HDR (EL) (RA, Ratio 1× Configuration)

	StEM 1920 × 1080 at 24 Hz	Y	U	V
SHM with CGS vs SHM	EL only	−57.1%	−62.4%	−70.5%
SHM with CGS vs SHM	Overall BL+EL	−19.4%	−30.4%	−44.1%
SHM with CGS	EL+BL vs simulcast	−41.5%	−48.7%	−57.0%
SHM	EL+BL vs simulcast	−27.9%	−27.8%	−25.2%
	Encoding time		93.4%	
	Decoding time		108.6%	

11.4 OTHER TECHNICAL SOLUTIONS

This section presents other coding technologies that have been proposed to standards bodies and are still under consideration or evaluation. This mostly relates to two fields: (1) signal representation based on alternative coding transfer functions; and (2) coding tools made of new frameworks that combine the HEVC standard with new metadata and postprocessing mechanisms.

11.4.1 ALTERNATIVE OETFS

The PQ inverse EOTF (SMPTE ST.2084) is not the only candidate OETF to map HDR video signals into limited-bit-depth integer values. Other OETFs are currently discussed in SMPTE and ITU-R. An alternative OETF (A-OETF — preliminary work for a hybrid log-gamma OETF) proposed by Borer (2014) is designed from the current BT.709 gamma. It is stretched at a given "shoulder" point, defined by a relative luminance value μ, to support higher dynamic ranges. The design aims to provide some backward compatibility with the BT.709 OETF (and its dual EOTF: ITU-R Rec. BT.1886 – gamma 2.4). Hence, in principle, the same content can be displayed on both SDR and HDR displays without adaptation. Below the luminance value μ, the curve is similar to the BT.709 transfer function. Above this, the curve is stretched to reach a higher peak luminance than the usual BT.709 transfer function, while maintaining its derivability at the luminance value μ. A low value of μ enables higher dynamic range but lower compatibility with the BT.709 transfer function. While the PQ inverse EOTF is designed for an absolute range from 0.005 to 10,000 cd/m^2, the A-OETF is a relative transfer function designed for normalized luminance values of 0 to 1. Practically, the minimum and maximum luminance will therefore depend on the display. To guarantee good compatibility with the gamma function (Borer, 2014), it is suggested that the extension be limited to a factor of 4–8, which would limit the peak luminance to around 1000 cd/m^2 (because BT.709 usually applies up to a few hundred candelas per square meter). The PQ inverse EOTF is designed to go up to much larger values. Fig. 11.14 illustrates the differences between the BT.709 OETF, PQ inverse EOTF, and A-OETF. In this illustration, the BT.709 OETF applies up to 100 cd/m^2, the PQ inverse EOTF applies up to 10,000 cd/m^2 and the A-OETF extends the range of the BT.709 OETF by 4, reaching a peak luminance of 400 cd/m^2.

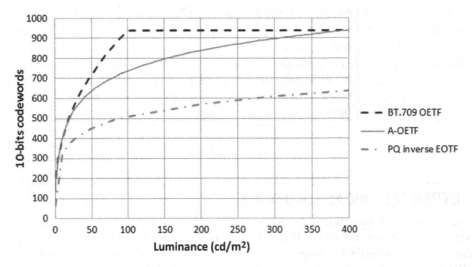

FIGURE 11.14

Comparison of BT.709 OETF, A-OETF, and PQ inverse EOTF.

11.4.2 ALTERNATIVE CODING TECHNOLOGIES

Previous sections presented HDR coding solutions based on a compressive transfer function, followed by a color transformation and a legacy video coding step. HEVC range extensions are designed to support bit depths of more than 10 bits per sample, up to 16 bits per sample. However, HEVC is expected to be deployed with only its 10-bit profile for UHDTV. This may limit the performance of compression, since a 10-bit quantization of the signal may result in important distortion on decoded HDR signals. The needed bit depth and dynamic range are obviously correlated.

Alternative solutions to high-bit-depth HDR signal coding have been designed in order to support a legacy 10-bit HEVC coding — even for HDR distribution. These solutions are generally based on a dual-layer scheme in which each layer uses a limited-bit-depth codec (as shown in Fig. 11.15). In one scenario (François et al., 2013; Kim et al., 2013; Aminlou and Ugur, 2014; Auyeung and Xu, 2014) the signal is split in additive layers that are processed by HEVC configured in eight or 10 bits. The samples are split into most and least significant bits. Different solutions are proposed, with nonoverlapping bits (François et al., 2013) or overlapping bits (Kim et al., 2013; Aminlou and Ugur, 2014; Auyeung and Xu, 2014). In François et al. (2013) the split samples from 4:2:0 high-bit-depth pictures were packed into 4:4:4 low-bit-depth pictures that were encoded by a single-layer encoder, decoded by a single-layer decoder, and re-arranged into 4:2:0 high-bit-depth pictures. In Kim et al. (2013), Aminlou and Ugur (2014), and Auyeung and Xu (2014) the split samples from 4:2:0 high-bit-depth pictures were packed into two layer pictures (BL and EL pictures) and a dual-layer coding was used.

A different approach based on the modulation concept was used by Richter (2013) for still HDR picture coding, and has also been proposed to JCT-VC for HDR video coding (Lasserre et al., 2014; Le Léannec et al., 2014). The approach uses two multiplicative layers and supposes that an HDR signal is of limited dynamic range in a small spatial area. By processing of the signal around its local average, it is possible to adapt to local signal properties, and to get finer quantization compared with a global approach.

A preprocessing step, before the encoding, splits the input signal P_{HDR} into two separate limited dynamic range (LDR) signals, one modulation picture P_{mod} which consists of a low-frequency monochromatic version of the input signal, and one LDR residual picture P_{LDR} corresponding to the

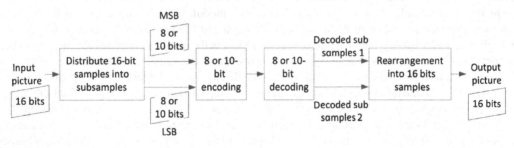

FIGURE 11.15

High-bit-depth coding using limited-bit-depth codecs. *LSB*, least significant bit; *MSB*, most significant bit.

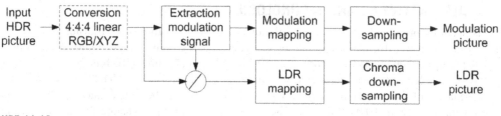

FIGURE 11.16

Simplified block diagram of the dual-modulation preencoding process.

remaining part once the low-frequency part has been extracted. These pictures decompose the original HDR signal as follows:

$$P_{HDR} = P_{mod} {}^\times f^{-1}(P_{LDR}),$$

where f is a nonlinear mapping, conceptually similar to an OETF, used to derive the LDR layer from the multiplicative residual P_{HDR}/P_{mod}. The decomposition process is schematically depicted in Fig. 11.16. The input HDR video is represented by a linear light 4:4:4 chroma format and RGB or XYZ color spaces, otherwise a preconversion linearizes the signal as schematized by the dashed border box signal. The modulation picture is generated by extraction of the low-frequency signal, mapped to an LDR representation ("modulation mapping") and optionally downsampled to profit from the smoothness of the modulation signal as shown in Fig. 11.18. The LDR picture is generated by division of the input HDR video by the modulation picture and by the mapping of this ratio to fit into a limited bit depth. Finally, a color transformation and chroma downsampling are applied according to the chroma format required for the LDR picture.

The LDR derivation process can be achieved according to two scenarios. If SDR backward compatibility is the aim, the process is designed to produce an LDR picture in the Y'CbCr color space. Alternatively, compression performance is preferred over LDR backward compatibility and, in this case, the LDR picture is built in a perceptual color space based on the CIELAB Lab color space (CIE, 2004), with adaptations to optimize the quantization of the chroma components (Lasserre et al., 2014; Le Léannec et al., 2014). In both scenarios, the output of the preprocess is two limited-bit-depth pictures that are separately encoded — for instance, with use of the HEVC Main 10 profile.

On the decoder side — after the decoding of the modulation and the LDR bitstreams — a postprocessing step is applied to the decoded modulation and LDR pictures to generate the original HDR picture. The process is summarized in Fig, 11.17 and is the obvious inverse of the preprocessing. Its inputs are the decoded modulation signal and the decoded LDR signal. Its output is a reconstructed HDR signal. Its format is indicated in an accompanying SEI message. The modulation signal is possibly upsampled (to reach the full resolution of the target output signal) then inverse mapped. The LDR signal is processed through a color resampling and inverse color transform, then an inverse mapping step. The two resulting signals are multiplied and the resulting signal has to be converted to the output format. The implementation of these last two steps can be combined into a single step (dashed black border box in Fig. 11.17) that depends on the target output format.

The preprocessing and postprocessing steps apply per picture. The modulation picture is accompanied by metadata — for instance, a "modulation channel information" SEI message (Le Léannec et al.,

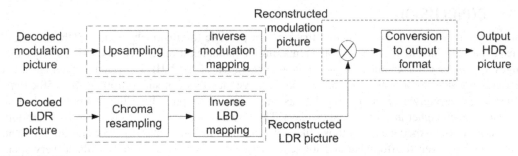

FIGURE 11.17

Postprocessing on the decoder side.

2014) which embeds the parameters required to perform the recomposing process at the decoder — from the decoded modulation picture and the decoded LDR picture. The message also indicates the format of the output signal to be reconstructed. It is possible to enforce the modulation picture to be flat — which results in a single modulation value. In this case, the scheme can use a single-layer coding process and the modulation value can be conveyed as side information (metadata).

Fig. 11.18 illustrates the output of the HDR picture decomposition. The demodulation process can be applied in a closed loop with use of the reconstructed modulation signal after decoding, upsampling, and inverse modulation mapping.

FIGURE 11.18

Source picture (A) and its corresponding modulation picture (B) and LDR picture (C).

Source: From Technicolor.

11.5 CONCLUSION

There is today a convergence of factors that make it possible to effectively deploy HDR and WCG video to the consumer market. Capture and rendering technologies have evolved toward higher dynamic ranges and wider color gamut than the previous generation of SDR HDTV video technologies. On the rendering side, most of the currently available consumer TVs can support up to 400–450 cd/m^2. However, increasingly constraining regulations on power consumption limit the ability to push peak luminance much higher in commercial displays. The visual comfort aspect has also not yet been fully studied. From our experience, this looks like a reasonable limit in terms of power consumption and visual comfort. Even if affordable and low-energy true HDR displays are not available very soon, the current pace of progress means it is reasonable to expect that within the next 10 years displays with higher contrast, intense and detailed darks, and peak luminance up to 4000–5000 cd/m^2 will be widely available. The introduction of HDR TVs to the market should therefore be envisioned as being happening in several phases, with the initial phase supporting EDR with gamut wider than the BT.709 gamut but still far from the BT.2020 gamut. The evolution toward TVs with higher dynamic range and gamut closer to the BT.2020 gamut can be envisioned in the next phases.

The various standardization efforts focused on HDR and WCG video have to take into account this multiphase evolution. As discussed in this chapter, several standards already integrate solutions relevant for HDR and WCG video. Recommendation BT.2020 specifies WCG primaries, and SMPTE ST 2084 specifies an EOTF for HDR signals. HEVC has incorporated these two tools in its specification, and also supports several types of metadata that can be used on the rendering side for the adaptation of signal. HEVC also includes multilayer coding tools that enable color gamut and dynamic range scalability. Other technologies are still under consideration. An important feature that has to be considered is the backward compatibility with SDR decoders and displays. For broadcast applications, this may be a critical point. Backward compatibility is — in principle — achievable by means of scalability, but scalable solutions have noticeable impacts on the distribution workflow that may jeopardize their use. Single-layer solutions offering this backward compatibility are, in this context, of great interest.

In Section 11.2.1, various workflows were discussed. The implementation of these workflows mostly depends on the target applications. An over-the-top HDR and WCG video service, for instance, does not need direct SDR backward compatibility as it is a point-to-point distribution application. The first workflow (Fig. 11.2) can therefore be adapted. In broadcast applications where SDR backward compatibility matters, the second and third workflows (Figs. 11.3 and 11.4) are more relevant. The expectation from the current standardization efforts is to complete the existing standards to deliver the set of tools adapted to these various workflows.

REFERENCES

Aminlou, A., Ugur, K., 2014. On 16-bit Coding, Document JCTVC-P0162, 16th JCT-VC Meeting, San Jose, USA.

Andrivon, P., Bordes, P., 2013. AHG14: Wide Color Gamut Test Material Creation, Document JCTVC-N0163, 14th JCT-VC Meeting, Vienna, Austria.

Auyeung, C., Xu, J., 2014. AhG 5 and 18: Coding of High Bit-Depth Source With Lower Bit-Depth Encoders and a Continuity Mapping, Document JCTVC-P0173, 16th JCT-VC Meeting, San Jose, USA.

Barten, P.G.J., 2004. Formula for the contrast sensitivity of the human eye. In: Proceedings SPIE-IS&T, vol. 5294, pp. 231–238.

Bjontegaard, G., 2008. Improvements of the BD-PSNR Mode, Document ITU-T SG16, VCEG-AI11.

Blu-ray Disc Association, 2015. Bd-rom — audio visual application format specifications — version 3. http://www.blu-raydisc.com/assets/Downloadablefile/BD-ROM_Part3_V3.0_WhitePaper_150724.pdf.

Borer, T., 2014. Non-linear opto-electrical transfer functions for high dynamic range television. In: BBC Research & Development White Paper WHP 283. http://downloads.bbc.co.uk/rd/pubs/whp/whp-pdf-files/WHP283.pdf.

Boyce, J., Chen, J., Chen, Y., Flynn, D., Hannuksela, M., Naccari, M., Rosewarne, C., Sharman, K., Sole, J., Sullivan, G., Suzuki, T., Tech, G., Wang, Y.K., Wegner, K., Ye, Y., 2014. Edition 2 Draft Text of High Efficiency Video Coding (HEVC), Including Format Range (RExt), Scalability (SHVC), and Multi-View (MV-HEVC) Extensions, Document JCTVC-R1013, 18th JCT-VC Meeting, Sapporo, Japan.

CIE, 2004. CIE 15 Technical Report: Colorimetry, International Commission on Illumination, third ed.

EBU Tech 3320, 2014. User requirements for video monitors in television production. Tech 3320, Version 3.0, Geneva. https://tech.ebu.ch/docs/tech/tech3320.pdf.

François, E., Gisquet, C., Laroche, G., Onno, P., 2013. AHG18: On 16-bits Support for Range Extensions, Document JCTVC-N0142, 14th JCT-VC Meeting, Vienna, Austria.

Kim, W.S., Pu, W., Chen, J., Wang, Y.K., Sole, J., Karczewicz, M., 2013. AhG 5 and 18: High Bit-depth Coding Using Auxiliary Picture, Document JCTVC-O0090, 15th JCT-VC Meeting, Geneva, Switzerland.

Lasserre, S., Le Léannec, F., François, E., 2014. High Dynamic Range Video Coding, JCTVC-P0159, 16th JCT-VC Meeting, San Jose, USA.

Le Léannec, F., Lasserre, S., François, E., Touzé, D., Andrivon, P., Bordes, P., Olivier, Y., 2014. Modulation Channel Information SEI Message, Document JCTVC-R0139, 18th JCT-VC Meeting, Sapporo, Japan.

Li, X., Boyce, J., Onno, P., Ye, Y., 2013. Common Test Conditions and Software Reference Configurations for the Scalable Test Model, Document JCTVC-L1009, Geneva, Switzerland.

Luthra, A., Ohm, J.R., Ostermann, J., 2012a. Requirements of Scalable Coding Extensions of HEVC, Document ISO/IEC JTC1/SC29/WG11 N12783, Geneva, Switzerland.

Luthra, A., Ohm, J., Ostermann, J., 2012b. Use Cases of the Scalable Enhancement of HEVC, Document ISO/IEC JTC1/SC29/WG11 N12782, Geneva, Switzerland.

Miller, S., Nezamabadi, M., Daly, S., 2013. Perceptual signal coding for more efficient usage of bit codes. SMPTE Motion Imaging J. 122 (4), 52–59.

Pedzisz, M., 2014. Beyond BT.709. SMPTE Motion Imaging J. 123 (8), 18–25.

Recommendation ITU-R BT.709-5, 2002. Parameter values for the HDTV standards for production and international programme exchange. http://www.itu.int/dms_pubrec/itu-r/rec/bt/R-REC-BT.709-5-200204-I!!PDF-E.pdf, ITU-R.

Recommendation ITU-R BT.2020-1, 2014. Parameter values for ultra-high definition television systems for production and international programme exchange. http://www.itu.int/dms_pubrec/itu-r/rec/bt/R-REC-BT.2020-1-201406-I!!PDF-E.pdf, ITU-R.

Rec. RP 177, 1993. Derivation of Basic Television Color Equations, SMPTE Recommended Practice, RP 177-1993.

Richter, T., 2013. On the standardization of the JPEG XT image compression. In: Proceedings PCS 2013, San Jose, CA, USA, pp. 37–40.

Segall, A., Kerofsky, L., Lei, S., 2006. Tone Mapping SEI Message, Document JVT-T060, 20th JVT Meeting: Klagenfurt, Austria.

Sharma, G., Wu, W., Dalal, E., 2005. The CIEDE2000 color-difference formula: implementation notes, supplementary test data, and mathematical observations. Color Res. Appl. 30 (1), 21–30.

Thomas, Y., 2014. Behind the Scene of UHDTV, Geneva, Switzerland, http://www.hoek.nl/dg/18mrt14/Dutch_Guild_2014_Yvonne_Thomas_handout.pdf.

Wang, Z., Bovik, A., Sheikh, H., Simoncelli, E., 2004. Image quality assessment: from error visibility to structural similarity. IEEE Trans. Image Process. 13 (4), 600–612.

HIGH DYNAMIC RANGE IMAGING WITH JPEG XT

12

T. Richter

University of Stuttgart, Stuttgart, Germany

CHAPTER OUTLINE

12.1 THE JPEG XT STANDARD

The ITU Recommendation T.81 and ISO/IEC Standard 10918-1 for still image coding, commonly known as JPEG (Wallace, 1992; Pennebaker and Mitchell, 1992; ISO/ITU, 1992), is still the dominant codec used for lossy image coding. However, as the quality of sensors has improved and market demands have changed in recent years, JPEG no longer fully addresses all needs of the digital photography market. While JPEG offers a lossy 12 bits per pixel (bpp) mode and also includes a lossless coding mode, both modes are incompatible with the popular eight-bit mode and are rarely implemented for this reason. Most decoders found on the market will not be able to decode such images. Furthermore, it has been seen that standards such as JPEG 2000 (Boliek, 2000; Taubman, 2000) and JPEG XR (Srinivasan et al., 2007), while addressing such needs, were only successful in niche markets and never found acceptance as widely as JPEG. While the list of available features of these newer codecs is certainly long, they require a completely new image processing chain that is hard to establish in the consumer photography market. In short, both standards had limited success in the consumer market because of the lack of backward compatibility.

To modernize the old standard and address the changed needs of the market, the JPEG committee started at its Paris 2012 meeting (ISO SC 29 (WG 1), 2012) a new work item on a fully backward compatible image compression scheme based on 10918-1. The design goals for this compression scheme were to offer a JPEG-compatible, low-complexity codec for high dynamic range (HDR), large color gamut data. In later meetings, it became apparent that extensions to additional directions would be desirable, such as lossless compression or coding of alpha channels, and the definition of the term "high dynamic range" (HDR) itself required clarification; see Section 12.2. Hence, JPEG XT was designed as a multipart standard, where each subsequent part adds additional features, all integrated into a common framework, allowing individual requirements to be addressed by selection of the necessary coding features from the parts as necessary. The current architecture of the JPEG XT standard is outlined in Section 12.4.

Besides all extensions, one leading principle keeps all parts together, and that is backward compatibility with the popular widespread eight-bit Huffman coding mode of JPEG. Backward compatibility here means that legacy decoders are able to reconstruct a lossy low dynamic range (LDR), standard color gamut version of the encoded image, while the complete (full range, lossless, etc.) image data are available only to decoders compliant with the new standard. The common minimum subset of all JPEG XT parts — namely, the eight-bit mode of the legacy JPEG standard plus some widely known extensions defining color-spaces and subsampling, became its own part, ISO/IEC 18477-1. Compatibility with other JPEG modes, such as arithmetic coding or predictive lossless coding, was of no importance as these modes are rarely used in the photography market. Even more, the extensions made within JPEG XT are all derived from known JPEG algorithms such that implementation of JPEG XT on hardware requires little more than two readily available JPEG chips, plus pixelwise postprocessing. Hence, JPEG XT was designed to ease implementation on existing hardware.

12.2 PROBLEM DEFINITION

While the definition of lossless coding or coding of opacity information is certainly obvious, the term "high dynamic range" (HDR) is only vaguely defined, and it is the purpose of this section to introduce the definitions the JPEG committee used to structure its work.

For this, a couple of definitions have to be made. An *f-stop* is commonly known as the logarithm to the base 2 of the quotient of the lightest and the darkest (physical) intensity in a given scene. Hence, the number of f-stops describes the dynamic range.[1]

An *LDR scene* has a dynamic range of at most 10 f-stops, an *intermediate dynamic Range (IDR) scene* has between 10 and 16 f-stops, and everything above is called *high dynamic range* (HDR).

Related to that is the *representation* of a scene as a digital image. An *LDR representation* is defined here, for pure practical purposes, as an image represented by one or three channels of eight-bit integer samples. This is also sometimes called *standard dynamic range* (SDR) and covers the dynamic range

---

[1]Note that this measure is undefined if the intensity should ever be zero (ie, no light at all). This rarely, if ever, happens in nature.

that the popular modes of the legacy JPEG can represent. SDR images are, as in the case of JPEG, also restricted in their color gamut. In digital photography, SDR images are generated by a tone-mapping operation that is applied to the raw sensor signals, compressing the output dynamic range down to 256 possible intensity values.

An *IDR* representation still uses integer samples, but requires more than eight-bit resolution. This is sometimes also called *extended dynamic range*.[2] There is currently no widely accepted standard for IDR image representations for digital photography, although camera vendors advertise proprietary "raw" formats that typically encode integer sensor signals of more than eight-bit precision and thus can be classified as IDR representations. Despite the name, "raw" formats are typically not as "raw" as one might believe: some elementary processing is usually applied to the sensor signals before storage. However, what is common to these proprietary representations is that the final tone-mapping step from the intermediate signal representation to the LDR output is skipped, and has to be performed by the user after the images have been downloaded from the camera, similar to the "printing" of analog negatives. The actual encoding of the preprocessed sensor data is typically lossless. IDR representations are, hence, considered as "digital negatives."

HDR representations are those that require floating point samples because of a dynamic range that is too large to allow representation by integer samples. Since the dynamic range of today's sensors falls well within the range that can be represented by at most 16-bit integers, larger dynamic ranges can be acquired only if several shots with different exposure times are combined into a single digital image.

The classification of scenes and representations is not identical, but related. An LDR scene can surely be represented by a HDR image encoding, but the reverse requires an additional tone-mapping step that alters the content and generates loss. Furthermore, the relation between the physical intensities (ie, *radiance*) and the sample values differs depending on the representation. SDR images are typically gamma-corrected (ie, the sample values are not proportional to the physical intensities, but are instead related to them by a power law). This gamma correction stems originally from analog TV, where it became a necessity because of the nonlinearity of the components used for recording and reproduction of signals (Netravali and Haskell, 1995), but it can also be seen as a very basic tone-mapping procedure (Reinhard et al., 2010). IDR and HDR representations typically use a *linear gamma* encoding (ie, image sample values are proportional to physical intensities). Depending on whether the proportionality constant is known, one also calls these *absolute* or *relative radiance representations*.

12.3 THE HISTORY OF JPEG XT

The original design goal of JPEG XT, as discussed in the initiating 2012 Paris meeting of the JPEG committee, was the backward-compatible representation of HDR images. At that time, the requirements had not been fully settled, and the importance to distinguish between floating point and integer representations and HDR and IDR became evident only several meetings later.

As an answer to this call, the JPEG committee received five proposals at its Shanghai meeting: from the École Polytechnique Fédérale de Lausanne in Switzerland, the Vrije Universiteit Brussel in Belgium, the UK-based company Trellis Management, Dolby in the United States, and the University

---

[2]Confusingly, the MPEG community already denotes this as "HDR."

of Stuttgart in Germany. All proposals expect, as input to the encoder, the IDR or HDR image to be coded, and, sometimes optionally, an additional LDR image which then becomes the SDR image visible to legacy decoders. An additional side stream carries all the information necessary to reconstruct the image to full precision. In principle, this side channel could be coded by any suitable method; however, the committee decided to restrict JPEG XT to entropy coding mechanisms that are as close to JPEG encoding as possible to minimize hardware costs and to reuse as many parts of existing implementations as possible.

The proposals differed insofar as which representations they addressed: the Dolby and Trellis Management proposals covered the HDR case, whereas the University of Stuttgart proposal addressed only IDR use cases and the Vrije Universiteit Brussel proposal addressed only lossless encoding. The École Polytechnique Fédérale de Lausanne proposed a rather generic framework to cover HDR use cases. It took almost one year to fit all proposals into one common architecture and to structure JPEG XT into parts. Part 3 defines a common file format based on a box-type structure very much like JPEG 2000. The box-based file format provides the syntax to express all other parts; it does not define a decoder, but rather the syntactical elements used by the standard as a whole. Part 6 defines IDR coding, part 7 defines coding of HDR sample representations, and part 8 defines lossless coding. Depending on the nature of the proposals, they were integrated into parts 6, 7, or 8, where each part is an extension of the previous part. Part 2 plays a special role by expressing part 7 profile A, while technically equivalent, in a legacy syntax.

The extension for encoding of opacity channels, while already proposed at the 2014 Valencia meeting as part 9, came finally to life at the 2015 Sydney meeting. Future meetings will possibly define additional parts, with such topics as privacy protection of images or regions of images, or encoding of plenoptic images.

More on the encoding algorithms and the joint decoder design of part 6 is discussed in Section 12.4.1, and the HDR extensions are introduced in Section 12.4.3. Lossless coding and coding of opacity data, covered in parts 8 and 9, are beyond the scope of this chapter.

12.4 CODING TECHNOLOGY

While JPEG defines multiple coding tools, including lossless coding, arithmetic coding, and pyramidal coding, only the discrete cosine transform (DCT)-based Huffman coding tools have found widespread use. Even more so, some elements of JPEG as it is understood and used today are surprisingly not part of the original standard at all. The YCbCr color space was first described in JFIF, the "JPEG File Interchange Standard," and was standardized as ISO/IEC 10918-5 as late as 2011 (Brower et al., 2011). Even though the legacy JPEG standard specifies component subsampling, the upsampling procedure and the alignment of components were originally not part of JPEG, but again were described in JFIF. In fact, the original JPEG standard specifies only a codestream that reconstructs samples from data but leaves the interpretation of such samples or their relation to images or color to other standards.

With all the differences between JPEG as it is applied in practical applications and the original ISO standard in mind, the JPEG committee decided to base its new JPEG XT coding technology on the firm grounds of a new standard spelling out all the accepted conventions, and to release this as JPEG XT part 1, ISO/IEC 18477-1. Codestreams following this specification will be reconstructed

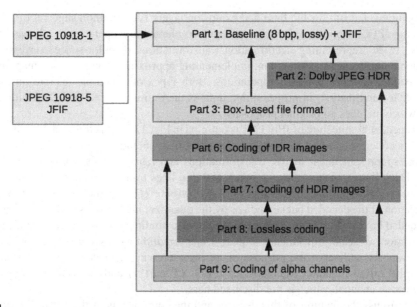

FIGURE 12.1

Overview of the parts of JPEG XT and their relations. Parts 4 and 5 define conformance testing and the reference software and are not included here.

correctly by all existing JPEG decoders, despite dependence on conventions not covered in the original JPEG specifications (Fig. 12.1).

Because legacy JPEG decoders usually support only the eight-bit integer mode described in JPEG XT part 1, JPEG XT part 6 introduces now two orthogonal coding mechanisms to extend the dynamic range: *refinement coding* and *residual coding*. The former method increases the precision of the DCT coefficients, and hence operates in the DCT domain. It is quite similar to the legacy 12-bit mode, although it allows arbitrary bit depths and is, unlike the legacy 12-bit mode, fully backward compatible with the eight-bit mode. The latter method, residual coding, operates entirely in the spatial domain; it extends the bit precision of the *base layer* defined by the legacy codestream elements and the refinement scan by including a second, independently coded *extension layer*. The extension layer coding adopts coding mechanisms that are either directly identical to or closely related to legacy JPEG coding modes.

A backward-compatible signaling mechanism based on the ISO media file format specified in part 3 instructs the decoder how to merge refinement, extension, and base layers into one final image. Metadata and residual and refinement coded data are here embedded into so-called *boxes*, a syntax element JPEG XT has in common with JPEG 2000 and MPEG standards. The box-based syntax allows future extensions of JPEG XT toward applications such as JPIP — an interactive image browsing protocol similar to the popular Google Maps service — that depend on such box structures. The boxes themselves are hidden from legacy decoders by their encapsulation in *application markers*, a generic extension mechanism already defined in the legacy standard. Parts 7 and 8 extend and use these extension mechanisms of part 6 to enable coding of HDR data and lossless coding.

The merging mechanism for the base and extension layers is a common superset of all proposals received by the JPEG committee. It is built from two elementary operations: a linear transformation from $\mathbb{R}^3$ to $\mathbb{R}^3$ representing decorrelation transformations or color-space conversions, and a one-dimensional nonlinear transformation, implementing approximate inverse tone-mapping, inverse gamma correction, or other nonlinear operations. Both types of functions operate on a pixel-per-pixel basis; their combination, in proper order, defines the entire universe of LDR/residual merging operations, which includes all original proposals.

The joint decoder architecture of JPEG XT is outlined in Fig. 12.2. This figure shows all components present in a decoder supporting all profiles and parts 6–8, including integer and lossless coding. Rounded boxes represent linear matrix transformations and regular boxes apply separately to each component. Thick lines transport three channels and thin lines transport a single channel. The dotted boxes are not required for part 6 (ie, IDR coding). *Inverse QNT* is the dequantization procedure (ie, multiplication of the decoded bucket indices by the quantizer bucket sizes in the JPEG quantization matrix). Rounded boxes with the term *NLT* run a pointwise nonlinear transformation, typically inverse gamma correction or scale adjustments. Square *transformation* boxes multiply triples of samples with a linear matrix and serve the purpose of an inverse component decorrelation or color space transformation. Coding of alpha channels (ie, JPEG XT part 9) adds a similar figure for a one-component opacity channel, which is not discussed here.

Before we discuss the features of this decoder and the components of the JPEG XT standard, we will make a couple of observations. First, the legacy JPEG standard covers only the first three top-left boxes in Fig. 12.2, denoted as "T.81 10918-1 Decoder," "FDCT or IDCT," and "Inverse QNT." JPEG, as it is in use today, and as standardized in ISO/IEC 18477-1, adds the two additional boxes to the right: chroma upsampling and transformation from YCbCr to RGB. The lower-left box labeled "Residual Image" represents the decoder for the extension layer, and everything to the right of this box and below the five boxes in the top row implement the merging operations that compute an HDR or IDR image from base layer (top row) and extension layer (bottom row). These operations are all done in the spatial domain, on a pixel-per-pixel basis. The only extension that has been made in the DCT domain is *refinement coding*, indicated by the small boxes that point into the base image and residual image decoder.

The following sections discuss all these extensions, from IDR coding to lossless coding, first introducing the extension mechanisms for IDR coding that are then, in part 7, put to use for HDR coding as well.

12.4.1 IDR CODING

As introduced above, part 6 defines two orthogonal extension mechanisms, one in the DCT domain denoted as *refinement coding* and one in the spatial domain, called *residual coding*. Both possibilities have already been discussed in the video coding community. Residual coding is similar in nature to the HDR extension for MPEG video proposed by Mantiuk et al. (2007, 2006), and refinement coding is not unlike the MPEG extension proposed by Zhang et al. (2011). Both mechanisms can be combined if needed, including the possibility to extend residual coding by refinement coding.

Refinement coding will be discussed first. For that, it is helpful to recapitulate how the *progressive coding mode* of legacy JPEG works (Pennebaker and Mitchell, 1992). Image quality is here progressively improved in two possible directions: the *spectral selection* mechanism allows the encoder to

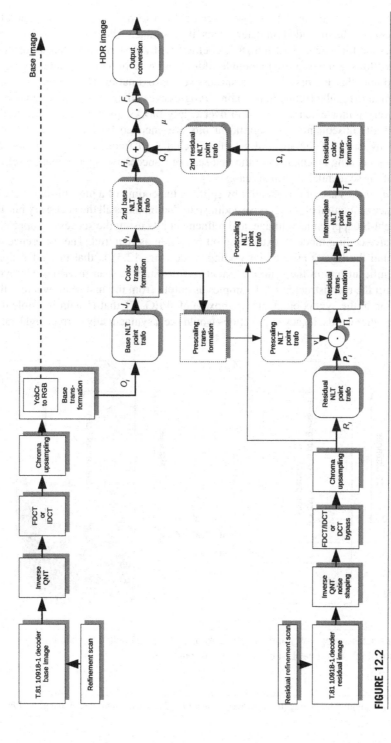

FIGURE 12.2

Decoder design of JPEG XT, merging the functionality of all parts except part 9. For the definition of and motivation for the various boxes, see the text.

select parts of the frequencies in the JPEG zigzag scan pattern to be encoded first, with additional frequency components to be included in later scans if desirable. The *successive approximation* mechanism improves the bit precision of the DCT coefficients by first coding a subset of their most significant bits, then allowing the encoder to include additional lower-order bits in later scans.

It is important to note that the first scan of a successive approximation scan pattern uses a coding mechanism very similar to regular (sequential) coding. Progressive coding uses first a regular sequential scan — with only very minor extensions to skip over empty blocks quickly — to encode the most significant bits, and all subsequent least significant bits are encoded by an alternative *subsequent approximation* entropy coder. If the spectral selection includes all frequencies[3] and block skipping is not used, the entropy coding mechanism of the first scan of a successive approximation scan pattern is *identical* to that of the sequential Huffman scan.

Refinement coding makes use of this identity by splitting the coding of a high-bit-depth DCT block into two parts: a legacy sequential or progressive coding part that includes all the necessary bits required to reconstruct an eight-bit approximate image, and refinement bits using the successive approximation scan of the progressive coding mode to extend the bit precision as required. The difference between refinement coding and the legacy progressive coding mode (Fig. 12.3) is that the latter signals the number of least significant bits in the syntax elements — the start of scan marker — by which the decoder has to upshift the reconstructed DCT samples as output from the first scan, whereas the latter hides this information in the syntax elements (ie, boxes) of JPEG XT that remain invisible to legacy implementations. In other words, legacy applications would see a syntactically correct eight-bit stream,

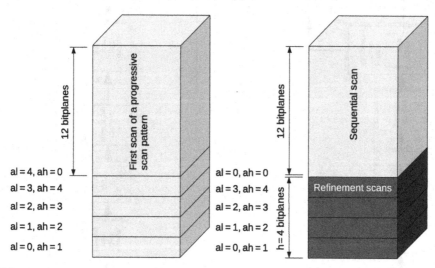

FIGURE 12.3

Relation between the successive approximation mechanism of legacy JPEG (left) and refinement coding (right), including four refinement bits for a total 12-bit spatial sample precision.

---

[3] Actually this is not a valid progressive coding mode because of an artificial restriction introduced in legacy JPEG.

whereas extended applications increase the bit precision by the known successive coding mechanism already specified in the legacy standard.

If one wants to limit the number of bits required for the DCT coefficients to 16 and the number of bits required to perform an integer-to-integer DCT approximation to 32, it is not hard to compute that at most four refinement scans can be included; this constraint is included for purely practical purposes in the current committee draft of the standard, although extensions to more bits would be straightforward.

To introduce *residual coding*, one should first model the merging step of base and extension layer as a similar extension of bit depth through least significant bits, though this time in the spatial domain. In a very simple application of residual coding, the eight most significant bits stem from the legacy JPEG codestream and provide an approximation of the IDR image, whereas as the least significant bits are represented in a side channel making up the residual image IDR images would then be reconstructed by upshifting the samples of the base layer and adding the least significant bits from the residual image. These two operations are done by the addition "$\oplus$" on the right-hand side in Fig. 12.2 and in the box denoted as "Base NLT Point Trafo." The dotted boxes are not used, the box denoted as "Output Conversion" clamps the reconstructed sample values to their valid range, and the box denoted as "Color Transformation" is the identity.

This elementary approach does not, however, withstand a closer analysis of the requirements. IDR images are typically encoded without gamma correction (ie, sample values are proportional to physical intensity), whereas LDR images are gamma-corrected, tone-mapped versions of them. A simple downshift to generate an LDR image from the most significant bits of the IDR image would not work because of the lack of a gamma correction. However, if for image reconstruction purposes the upshift by the LDR image bits is replaced by a simple lookup table that provides a global, simple approximation of the tone mapping applied at the encoder for the generation of the LDR image, both a useful LDR image and an IDR image can be carried in the same file. In such a case, the lookup table (the box denoted as "Base NLT Point Trafo") has 256 entries if the legacy image does not make use of refinement coding and hence has eight-bit precision; any type of error due to the approximation of the tone mapping is then captured by the extension layer. In the case of refinement coding, the lookup table is 2^{h+8} entries long, where h is the number of refinement scans "hidden" from the legacy decoder. Here a finer table approximation allows a more precise approximation of the IDR image, and precise reconstruction with refinement coding alone is possible only if the original tone mapping for generation of the LDR image was a simple global operation. Otherwise, additional errors remain that require correction by a residual scan.

Leaving color issues aside, the reconstruction algorithm for IDR grayscale images defined in part 6 of the JPEG XT standard is hence

$$IDR(x, y) = \Phi(LDR(x, y)) + RES(x, y) - 2^{R_b - 1}, \tag{12.1}$$

where IDR is the reconstructed IDR image, LDR is the base image visible to legacy decoders, RES is the *residual* image hidden in applications markers, and $2^{R_b - 1}$ is an offset creating a residual signal from $RES(x, y)$ that is symmetric around zero. The constant R_b is the bit precision of the IDR image generated. In the presence of refinement coding, the LDR signal will have a bit depth of $8 + h$ instead of 8, where h is the number of refinement scans.

Let us briefly discuss the encoding of the extension layer. Since the extension layer is never seen by a legacy decoder, it is not necessary to constrain entropy coding to the legacy eight-bit Huffman mode of JPEG. If its resolution is not sufficient to reach the desired quality, one can either use the 12-bit mode of JPEG or encode the data with the eight-bit Huffman mode and extend the resolution of the extension layer by refinement scans, as one can for the base layer. In principle, even other entropy coding methods could have been considered, but the committee decided to constrain JPEG XT to algorithms that are as close as possible to existing JPEG technology to ease their implementation.

12.4.2 ENLARGING THE COLOR GAMUT

Even though the original JPEG standard did not define a color space and only specifies a mechanism for how to encode sample values into a codestream, JFIF — later standardized as ISO/IEC 10918-5 — did (Brower et al., 2011). JFIF selected ITU Recommendation BT.601, originally a TV standard, both as a color space and as a decorrelation transformation from an RGB-type description into an opponent chrominance/luminance description. Unfortunately, even this selection is not fully consistent with typical uses of JPEG where images are represented in sRGB rather than Recommendation BT.601. Both color spaces are related but not exactly identical.

For reasons of backward compatibility, part 1 of JPEG XT also specifies the legacy ITU color space. Its color gamut is, however, quite limited, and it is often desirable to encode images expressed in larger color spaces. Simply mapping the primaries of larger color space to the Recommendation BT.601 primaries may be an option in some simple cases, but this will typically not map the white points onto each other; in other words, the colors of the legacy image encoded with wrong primaries when interpreted with Recommendation BT.601 colors might be off, creating an overall undesirable image; such color alterations can be especially irritating in the reproduction of skin colors and should be avoided. They can be corrected, however, by use of the extension layer. Any undesirable color shift can be compensated by the addition of a suitable error signal which will shift the reconstructed sample values to their desired color position. From the perspective of coding efficiency, this approach should, however, be avoided, as the error signal can become large enough to compromise the performance of the code.

To address this predicament, JPEG XT offers three options. First, the LDR primaries can be used consistently for the base image and the full IDR image, then requiring the use of negative color coordinates for out-of-gamut colors. Due to the offset shift in Eqn. 12.1, such sample values are representable by a minimum LDR value and a small, but *positive* residual error.

Second, the legacy transformation between YCbCr and RGB can be replaced by another transformation that maps the legacy samples into a larger gamut color space. This generalized transformation is denoted as "Base Transformation" in Fig. 12.2; it can be considered as the combination of two subsequent linear transformations. The first transforms the Recommendation BT.601 YCbCr opponent sample values into an RGB color space, with color primaries defined as in the mentioned ITU recommendation; the second maps the coordinates of the ITU Recommendation BT.601 color space into the coordinates relative to the target color space. From a color science perspective this transformation is not strictly correct. The transformation is applied in the nonlinear (gamma-corrected) Recommendation BT.601 color space, hence creating errors in the chroma reproduction; the larger the nonlinearity in the inverse tone-mapping process, the larger the errors. Again, such errors can be corrected by an appropriate extension layer, and in the worst case compromise the coding performance, but not the quality of the image.

The third option overcomes this problem by introducing an additional linear transformation after inverse tone mapping, denoted as "Color Transformation" in Fig. 12.2. Here, the decoder first performs the legacy YCbCr to RGB transformation in the nonlinear color space as required by JFIF, and then maps the samples into the (typically) linear target color space of the IDR image by the base nonlinear point transformation, before it finally applies a linear transformation to map the colors to the target space.

While the third approach seems to be ideal from a color science point of view, it comes with problems of its own. None of the transformations can be implemented with infinite precision, and the second color transformation creates additional numerical errors that also require correction. Astonishingly, it is sometimes beneficial to use the second method instead of the third to maximize the coding performance. JPEG XT thus allows the use of all three approaches, even combined, and leaves it to the encoder to signal the desired mechanism.

The color decorrelation in the extension layer is much less critical. At first sight, it seems plausible to transform the residual data also first from YCbCr to RGB with Recommendation BT.601 primaries, and then further transform the data into the target color space. While such transformation are allowed and even present in some profiles, see Section 12.4.3, they not only require one additional matrix transformation, but also add further loss due to numerical inaccuracies. IDR coding as defined in part 6 does not make use of this transformation. Instead, one can take advantage of the fact that the residual image is never displayed directly and does not need to be backward compatible with legacy JPEG applications. Residual coding can be considered as a compressor of three sample values per pixel, without any color space information. While error residuals could be compressed by the residual stream directly without any transformation, it turned out that a decorrelation transformation identical to the Recommendation BT.601 RGB to YCbCr transformation improves compression efficiency as well. This transformation is applied in the box denoted by "Residual Transformation" in Fig. 12.2. The residual nonlinear point transformation is typically only a scalar multiplication or an upshift, necessary to use the full range of eight bits to represent the error signal. It serves no additional purpose, and is even absent for lossless coding.

12.4.3 FROM IDR TO HDR: HDR CODING

The coding mechanisms introduced above are, within the limits the current draft of the standard sets, able to encode images consisting of integer samples of up to 16-bit resolution. Part 7 of the JPEG XT standard is an extension of the coding tools of part 6 to allow reconstruction of images to floating point samples, as required for the representation of HDR images. Similarly to part 6, these extensions are based on two elementary types of operations: a pixel-based nonlinearity, implemented either by a lookup table or by a parameterized curve, and a linear transformation defined by a 3×3 matrix.

These tools, while being part of a common decoder framework, enable several encoder architectures to tackle the problem of HDR coding. As for many other standards, the set of decoding tools is structured into *profiles*, each of which defines a set of allowable decoder tools for codestreams conforming to the corresponding profile. Even though the corresponding encoder architecture is not specified by the standard, it is still helpful to start the discussion with typical encoder designs.

Profiles A and C make use of the approximately logarithmic dependency between the physical luminance (stimulus) and the corresponding response of the human visual system, which has been known for a long time as "Weber's law" (Netravali and Haskell, 1995). Profile C represents images in the logarithmic domain directly, using a piecewise linear approximation of the logarithm that is exactly invertible and hence enables lossless coding, which is, however, standardized in a separate part. The coding algorithm in the logarithmic domain is then identical to part 6 (ie, profile C part 7 is a minimal extension to part 6).

Profile A applies the logarithmic map only to the luminance channel; that is, the HDR image is represented as the product of an LDR base image and a luminance scale factor μ that is encoded logarithmically (see Fig. 12.2). The relation to profile C becomes more apparent if one recalls that this multiplication in the image domain performed by profile A is equivalent to an addition in the logarithmic domain as performed by profile C decoders (ie, the equivalent to the multiplication step by μ in profile A is the addition step immediately before it in profile C; see Fig. 12.2). Profile A also makes use of this addition step, although here to correct small residual errors in the chrominance channels.

Unlike profiles A and C, profile B encoders split the HDR signal along the luminance axis. Samples below a luminance threshold are represented in the base layer, and samples above the threshold are represented in the extension layer. The composition of the base and extension layers in the decoder is performed by a division operation; this division is represented by an addition and logarithmic and exponential preprocessing and postprocessing. Decoders do not need to go through these processing steps, of course, and may apply the division immediately when they detect a profile B codestream. From the perspective of the architecture of the standard, the addition step is here the same addition as applied in part 6, and the step involving multiplication by μ is absent as in profile C.

A somewhat more formal and more precise discussion of the profiles is given now. Profile A, allowing both the prescaling and postscaling nonlinearity but no refinement or secondary nonlinear transformations (Fig. 12.2), reimplements a decoder similar to JPEG-HDR, a backward-compatible JPEG extension originally proposed by Ward and Simmons (2005) and Dolby. The reconstruction algorithm in this specific case is given by

$$\text{HDR}(x,y) = \mu(\text{RES}_0(x,y))(C\Phi(\text{LDR}(x,y)) + \nu(SC\Phi\text{LDR}(x,y)) \cdot \text{RRES}(x,y)). \tag{12.2}$$

In Eq. (12.2), μ is the postscaling factor depending on the luminance signal l of the extension layer. The function $\mu(l)$ relating the luminance of the extension layer to the scaling factor is typically an exponential ramp function of the form $\mu(l) = \exp(al + b)$ with suitable constants a and b (ie, the extension layer luminance l depends logarithmically on the scale factor μ). This nonlinearity is, from the perspective of the standard, implemented by the postscaling nonlinear transformation.

The transformation S is a 3×3 matrix that extracts the luma signal from the RGB base image, and ν adds an offset shift to the base layer luma signal extracted through S. The matrix C finally maps the primaries of the Recommendation BT.601 color space of the base layer onto the primaries of the larger HDR gamut.

The offset shift through the function ν is represented in the standard decoder architecture by the prescaling nonlinear transformation. As said, one typically has $\nu(y) = y + N_0$, where N_0 is the noise floor. It allows for chroma variations due to camera noise even in the absence of a luma signal.

R is an affine (and not linear) transformation that converts the chrominance signal of the extension layer to an additive RGB residual. It does not take the luminance channel of the extension layer into account, which is instead split off before to compute the scale signal μ.

Because of postscaling by μ, the nonlinearity Φ in the base layer, represented by "Base NLT Point Trafo" in Fig. 12.2, does not need to compensate for the effects of tone mapping and is here typically an inverse gamma transformation. It maps the gamma-compensated base image back into a linear radiance space.

Encoding in profile A with prescaling and postscaling transformation enabled is somewhat more complex, and we will next derive the encoder algorithm. The input of the encoder is an LDR-HDR image pair from which a suitable extension layer must be computed. For that, first note that $RRES(x, y)$ is, by design choice, a pure chrominance residual and does not include any luminance information. Thus, if we denote by P_L the projection onto luminance space, we find that

$$P_L(C\Phi(\text{LDR}(x, y)) + \nu(\text{SLDR}(x, y)) \cdot RRES(x, y)) = P_L C\Phi(\text{LDR}(x, y)).$$

This allows the encoder to determine first $\mu(\text{RES}_0(x, y))$ and then, as μ is a known function, $\text{RES}_0(x, y)$ by the quotient of the reconstructed base image luminance and the original image luminance:

$$\mu(\text{RES}_0(x, y)) = \frac{P_L C\Phi(\text{LDR}(x, y))}{P_L \text{HDR}(x, y)}.$$

In the last step, the chroma residuals can be computed:

$$RRES(x, y) = \frac{1}{\nu(\text{SLDR}(x, y))} \left(\frac{\text{HDR}(x, y)}{\mu(\text{RES}_0(x, y))} - C\Phi(\text{LDR}(x, y)) \right).$$

The reader may now verify that the right-hand side lies indeed in the chroma-subspace of the color space — that is, the luma signal is zero (or close to zero, because of numerical inaccuracies). The luminance channel of the residual image contains instead the luminance scale factor.

While profile A uses an explicit multiplication operation to scale the base image luminance to the desired target range, encoding and reconstruction are relatively elaborate, and the standard allows simpler alternatives to achieve similar goals in profiles B and C. Both make use of the functional equation of the logarithm to map the addition in Fig. 12.2 that merges the base layer with the extension layer into a multiplication or division — that is,

$$\text{HDR}(x, y)_i = \exp(\log(C \ \Phi(\text{LDR}(x, y))_i) + \log(\text{RES}(x, y))_i - 2^{R_b - 1})$$

$$= \exp(-2^{R_b - 1}) C \ \Phi(\text{LDR}(x, y))_i \cdot R \ \text{RES}(x, y)_i \quad (i = 0, 1, 2), \quad (12.3)$$

where in Eq. (12.3) exp and log are applied component-wise on the components of the vectorial image data. Similarly, a subtraction in the logarithmic domain becomes a division for the HDR image signal. The reader may want to compare Eq. (12.3) with Eq. (12.1) used for IDR reconstruction in part 6.

This transformation of a multiplication into an addition in the logarithmic domain is the motivating idea for the *secondary base* and *secondary residual nonlinear point transformations* in Fig. 12.2; the logarithm can here also be understood as a transformation into an approximately perceptually uniform domain based on Weber's law.

The Trellis Management XDepth proposal reconstructs the HDR image as the quotient of an inversely gamma-corrected LDR signal reconstructed from the legacy JPEG stream and a residual included in a side channel. This quotient can be rewritten as the difference of two terms in the logarithmic domain, where the log and exp functions necessary for conversion between the linear and the logarithmic domain are expressed in the standard decoder design shown in Fig. 12.1 by the secondary base and secondary residual nonlinear transformations and the output conversion:

$$\text{HDR}(x,y)_i = \sigma \exp(\log(C\Phi(\text{LDR}(x,y))_i) - \log(\Psi(\rho(R\,\text{RES}(x,y))) + \epsilon)_i)$$

$$= \sigma \frac{C\,\Phi(\text{LDR}(x,y))_i}{\Psi(\rho(R\,\text{RES}(x,y)_i)) + \epsilon} \quad (i = 0,1,2). \tag{12.4}$$

The "Output Conversion" in Fig. 12.2 includes here, compared with Eq. (12.3), an additional scale factor σ whose purpose will be described below. As in profile A, the base layer nonlinearity Φ maps the gamma-corrected LDR image back into the linear domain, and Ψ is an additional power map applied in the residual to compress its dynamic range–that is, $\Psi(x) = x^\beta$, where β is a data-dependent exponent selected by the encoder. As in profile A, R is a decorrelation transformation from YCbCr to RGB, although the luma signal is this time not split off but is included in the transformation. Finally, ρ represented in the standard decoder architecture by the "Intermediate NLT Point Trafo" is an affine scaling transformation that maps the RGB signals of the extension layer into a suitable range. It is described next.

Encoding in profile B is much simpler than in profile A as it does neither require prescaling by ν nor postscaling by μ:

$$\rho(R\,\text{RES}(x,y)_i) = \Psi^{-1}\left(\sigma \frac{C\Phi(\text{LDR}(x,y))_i}{\text{HDR}(x,y)}\right).$$

In general, the extension layer computed by the right hand will not be in range of the native JPEG standard. The transformation R^{-1}, however, is there already present as the RGB to YCbCr decorrelation transformation. The purpose of ρ^{-1} is now to perform the necessary scaling and to map the output of Ψ^{-1} into this interval. It is a data-dependent transformation that must be computed by the encoder.

While the profile B algorithm allows for arbitrary LDR images and hence for arbitrary tone mapping of the HDR content, the original XDepth proposal considered only one special choice for the map Φ, the color transformation C, and the LDR image — namely,

$$\text{LDR}(x,y)_i = \min(\Phi^{-1}(\sigma^{-1}\text{HDR}(x,y))_i, 1),$$

where Φ^{-1} is the gamma correction and C is the identity. Hence, the HDR image is first inversely scaled with σ, then gamma-corrected and clamped to the range $[0,1]$. One observes now that this has the following consequences for the residual. As long as the sample value of the HDR image is below σ, no clamping occurs and the quotient defining the extension layer

$$\Psi(\rho(R\,\text{RES}(x,y)_i)) = \sigma \frac{\Phi(\Phi^{-1}(\sigma^{-1}\text{HDR}(x,y)))_i}{\text{HDR}(x,y)_i} = 1$$

becomes the identity. For larger HDR sample values, the LDR image is saturated, and the residual carries all the image information of the overexposed region. The purpose of σ becomes now obvious: It determines which image luminances go into the LDR image and which become part of the extension layer. Using the analog printing process as metaphor, σ controls the "exposure" of the (now digital) negative "HDR(x,y)" into the print "LDR(x,y)"; in this process, the overexposed regions are preserved in the extension layer. Hence the name "exposure value" for the inverse of σ.

While profile A and profile B decoding algorithms — with or without prescaling and postscaling transformation — are clearly invertible as mathematical operations and hence suitable *encoding* algorithms are readily derived, they are *not invertible* without loss as numerical algorithms. This is because both of them depend on floating point representations and use multiplications or transcendental functions, which all introduce (numerical) loss when the (mathematical) inverse is taken. Profile C avoids these problems and extends the decoding algorithm from part 6 (Eq. 12.1) to HDR without depending on floating point arithmetic.

To understand its design, recall the IEEE encoding of floating point numbers (IEEE Computer Society, 2008). In its binary encoding, an IEEE floating point number consists of a sign bit s, a number of k exponent bits e, and a number l of mantissa bits m. The floating point number f is then given by

$$f = (-1)^s \cdot 2^{e-b} \cdot \left(1 + 2^{-l}m\right),$$

where b is an exponent bias that depends on the precision of the floating point number. This number f is now stored in memory as the concatenation of the s, e, and m bits; the very same bit pattern can also be interpreted as integer number i:

$$i = 2^{k+l}s + 2^l e + m.$$

The trick is now that the reinterpretation of f as i, which is a no operation for a computer since it changes only the *interpretation* of the bits rather than their values, is for $s = 0$ an approximation of a logarithm to the base 2, here written as ψ log:

$$i = \psi \log(f) = 2^l(\log_2 f + b) - 2^l \log_2(1 + 2^{-l}m) + m \approx 2^l(\log_2 f + b), \qquad (12.5)$$

where one uses that $\log_2(1 + x) \approx x$ for x between 0 and 1. In fact, the error

$$\epsilon(x) = x - \log_2(1 + x)$$

vanishes at the edges $x = 0$ and $x = 1$.

This observation can now be applied to Eq. (12.3), the multiplicative decoding algorithm, by replacement of the logarithm by ψ log, defined by re-interpretation of bit patterns, and its inverse, denoted by ψ exp. This change implements the reconstruction algorithm as

$$\text{HDR}(x, y)_i = \psi \exp(\psi \log(C\Phi(\text{LDR}(x, y))_i) + \psi \log(\text{RES}(x, y))_i - 2^{R_b-1})$$
$$\approx \psi \exp(-2^{R_b-1})\Phi(C\,\text{LDR}(x, y))_i \cdot R\,\text{RES}(x, y)_i \quad (i = 0, 1, 2), \qquad (12.6)$$

which carries over to lossless compression *and* is a natural extension of the coding algorithm of part 6. Since ψ exp and ψ log require only a reinterpretation of bits, this coding algorithm is also of very low complexity.

Eq. (12.6) can be simplified even further. First of all, if we ignore the color transformation matrix C for a moment, the base layer contributes to the overall HDR image through

$$\psi \log(\Phi(\text{LDR}(x, y))).$$

Now, both operations — the nonlinearity Φ and the pseudologarithm mapping floating points to integers — can be merged into a single operation that consists only of a table lookup process as the combined operation maps integers to integers. Second, since the residual image is never displayed on a monitor for observation, it does not matter whether RES or ψ log RES is encoded. The latter has the advantage that it already consists of integers suitable for JPEG encoding. With these changes, one gets a complete integer-based and numerically invertible reconstruction algorithm:

$$\text{HDR}(x, y)_i = \psi \exp(\hat{\Phi}(\text{LDR}(x, y)) + R \, \text{RES}(x, y) - 2^{R_b - 1}) \quad (i = 0, 1, 2). \tag{12.7}$$

An additional color transformation can now either be applied directly on the LDR image in the gamma-corrected space or be applied outside in the logarithmic space after inverse tone mapping and pseudologarithmic mapping by $\hat{\Phi}$. Since Φ and $\hat{\Phi}$ are nonlinear in general, neither of these operations is identical to the color transformation C in the linear space, and hence additional errors are introduced that can be compensated by the residual image. Hence, coding gain is lost by transformation to an alternative color space. A second strategy, outlined in Section 12.4.2, leaves the color space untouched, but uses negative sample values to represent out-of-gamut colors. Clearly, the original logarithmic and exponential map of Eq. (12.3) cannot express negative values, although an extension of ψ log and ψ exp to the negative axis is straightforward by extension of both functions in an (almost) point-symmetric way around the origin $x = 0$:

$$\psi \log(-x) := -\psi \log(x) - 1 \quad \text{for } x \geq 0. \tag{12.8}$$

The additional subtraction of 1 seems curious, but it allows one to distinguish between the two representations of zero the IEEE floating point format has to offer — namely, $+0$ and -0. Note that one cannot implement this extension of the pseudologarithm by reinterpreting the bit pattern of IEEE floating point numbers, and its implementation requires a conversion from a sign-magnitude representation to a two's complement representation. This conversion, however, is also lossless.

12.5 HARDWARE IMPLEMENTATION

Currently, most JPEG XT implementations are in software, with the exception of the profile stemming from JPEG-HDR, now profile A in part 7; see Section 12.4.3. Even here, the hardware consists of two standard JPEG chips that encode the base layer and the extension layer, plus one embedded processor implementing the operations to compute both layers from the HDR input signal. This setup is a rather typical choice for JPEG XT though. The entire design of the standard enables simple implementations on the basis of existing technology and does not require hardware implementations of algorithms.

All remaining lossy coding parts (ie, part 6 and all other profiles in part 7) can be realized in the very same way as long as refinement coding is not used and eight-bit encoding of the residual image provides sufficient quality.

Refinement coding is an extension based on the progressive scan type (see Section 12.4.1 and Fig. 12.3) and cannot be implemented by postprocessing alone. It will hence not be possible on hardware encoders that support only the sequential Huffman scan type of the legacy JPEG standard. Despite this limitation, the DCT has to be done with higher precision; more details on this are given later.

Lossless coding, as specified in JPEG XT part 8, also specifies the implementation of the DCT and the YCbCr to RGB transformation, and also requires a new entropy coding mode for the extension layer. Hardware encoders may or may not follow these specifications, and are typically not able to support it out of the box.

Finally, we will provide a brief analysis of the required implementation precision and bit depths of the signal paths. The bit precision of the base layer in the spatial domain is at most 12 bits, eight bits from the legacy process and four bits from refinement coding. By analyzing the input/output gain of the DCT, one can see that a 12-bit resolution in the spatial domain requires a 16-bit resolution in the DCT domain. The DCT in part 8 is now specified in fixed point precision such that 32 bits, including integer and fractional bits, are always sufficient for the coefficients and all intermediate results. The part 8 design uses four fractional bits to represent the output of the YCbCr to RGB transformation, and nine fractional bits for the internal representation of the coefficients for the DCT. Clearly, the DCT can at most expand its input by a factor of eight (ie, by three bits). The DCT reaches this maximum expansion gain for the DCT coefficient if the input is constant. In total, this makes $16 + 3 + 4 + 9 = 32$ bits.

The lookup table in the base layer in part 7 profile C and parts 6 and 8 takes the reconstructed sample values of the legacy codestream as input. Tables are therefore 256 entries long for eight-bit images, or 4096 entries long if the range of the legacy image is extended to 12 bits by means of the refinement coding path.

The same considerations also hold for the residual image. If the DCT is enabled for lossy coding, image coefficients in DCT space are at most 16 bits wide, or 12 bits in the spatial domain, again requiring at most 32 bits for intermediate results within the DCT. In the DCT bypass mode used for lossless coding, coefficients are always 16 bits wide and are encoded directly in the spatial domain. No further lookup table is needed in part 8 as the residual coefficients represent the error signal directly.

The remaining decoder logic of the extension layer can be implemented in 16-bit integer logic as the standard requires wraparound (ie, modulo) arithmetic. One minor glitch remains — namely, that the DCT bypass signal of part 8 may create the exceptional amplitude -2^{15}, which is the only DCT amplitude that cannot be represented by the 12-bit Huffman coding mode. Part 8 includes for this exceptional case an "escape code." Similarly, if the residual layer uses the lossless integer-to-integer DCT for part 8 coding, the amplitude of the DCT coefficients may grow as large as $2^{16+4} = 2^{20}$, and an extended Huffman alphabet becomes necessary to allow lossless encoding. For parts 6 and 7, this is never an issue since even the eight-bit sequential Huffman scan type usually provides sufficient quality. Smart encoders may avoid the large amplitude or escape codes of part 8 completely either by using a progressive mode or refinement coding in the extension layer or by ensuring that the quality of the base layer is high enough to avoid error signals overrunning the dynamic range of the legacy coding modes. In the latter case, the extended scan types fall back to the known sequential Huffman scan.

12.6 CODING PERFORMANCE

Evaluation of the performance of JPEG XT and its profiles is unfortunately not straightforward. First, the practical experience in objective and subjective evaluation of IDR and HDR image compression is still limited, and unlike LDR coding, objective quality indices and subjective evaluation procedures are much less tested and understood. More details on IDR image and especially HDR image quality evaluation are given in Chapter 17. Second, the design of JPEG XT itself as a two-layered codec complicates matters because the configuration space of an encoder — base and extension layer quality — is two-dimensional, and the choice of the base image influences the performance of the overall encoder.

12.6.1 IDR CODING PERFORMANCE

IDR image representations are based on integer samples and thus evaluation by established indices such as the peak signal to noise ratio (PSNR) is, at least in principle, possible. Nevertheless, some caution in interpreting such simple indices is certainly required, even if IDR imaging does not cover the same dynamic range as HDR imaging. This is because IDR images are typically not gamma-corrected, unlike LDR images; while gamma correction in the LDR regime creates at least an approximate perceptually uniform space, this is no longer given for IDR and HDR coding. For simplicity, and because of the lack of any better known alternative, only PSNR and multiscale structural similarity (SSIM) (Wang et al., 2003) figures will be presented, despite their known limited correlation to subjective quality. Multiscale SSIM is an objective "top-down" quality index that first separates reference and distorted images into several scales, then segments each scale into small blocks and measures luminance, contrast, and structure deviations within each block before pooling all scores together. It was found that the correlation of SSIM to subjective scores is better than that of PSNR, despite its relatively low complexity. For details, see Wang et al. (2003).

One typically obtains IDR images by demosaicing raw sensor data and either keeping them in the camera color space or transforming them into a scene-referred reference color space. Unlike HDR images, IDR images are typically obtained in a single exposure, although because of improved sensor technology, in bit depths beyond the eight bits per sample. The relation of IDR sample values to physical intensities is here given by the response curve of the camera, which is, however, typically unknown. The test image test in Fig. 12.4 was obtained in such a way (ie, by the demosaicing of "raw" camera streams and their transformation into the 16-bit linear scRGB color space). While this color space defines the same primaries as sRGB, it still allows an extended gamut by including an additive offset in the sample representation. Out-of-gamut colors are hence represented by sample values below the offset, and in-gamut colors are represented by coordinates larger than the offset. Thus, the offset shifts color coordinates which would otherwise be negative into the positive semiaxis.

The images in Fig. 12.4 were then compressed by JPEG XT part 6, then expanded again, and the PSNR and multiscale SSIM (Wang et al., 2003) were measured in the IDR regime and plotted against the total bit rate (Fig. 12.5). Nothing new can be learned about the quality of the base layer alone as this is simply defined by the legacy JPEG standard.

FIGURE 12.4

The images DSC_218 (left) and DSC_532 (right) (both courtesy of Jack Holm) here mapped from scRGB to sRGB to facilitate printing.

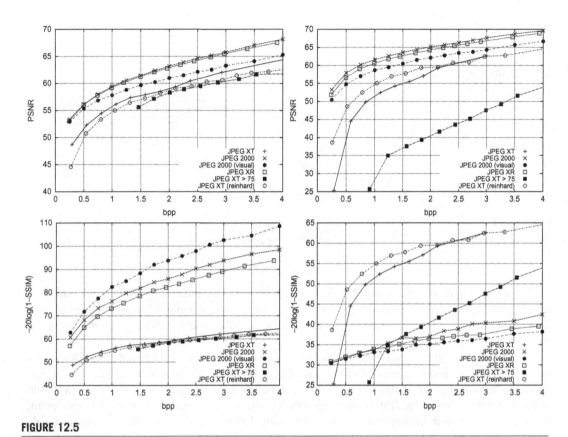

FIGURE 12.5

PSNR (top) and SSIM (bottom) in the IDR domain for DSC_218 (left) and DSC_532 (right) in various configurations.

Two additional codecs have been included in the benchmark — namely, JPEG 2000 and JPEG XR, the former in PSNR optimal and visually optimal mode. JPEG XT part 6 also allows various configurations, three of which have been tested here. For that, recall that JPEG XT requires two images as input, an IDR image and an LDR image. In the first experiment, the corresponding LDR images forming the base layer were obtained by conversion of the scRGB images with a color management system to sRGB. When used in its "perceptual" setting, the color management system used here — namely *TinyCMS* — includes a simple automatic tone mapping to generate perceptually "convincing" results. The base and extension layer quality settings of the encoder were then such that the target quality was maximized for a given bit rate constraint. The resulting curve is denoted as "JPEG XT" in Fig. 12.5.

Leaving the quality parameter of the legacy JPEG codestream unconstrained as done above may, however, result in unpleasing LDR images, an effect that was avoided in the second experiment by the disallowance of quality parameters for the LDR codestream below 75. This shrinks the parameter space for optimization, creates additional overhead, and hence lowers the corresponding PSNR and SSIM scores of the HDR image. This curve is denoted by "JPEG XT > 75" in Fig. 12.5.

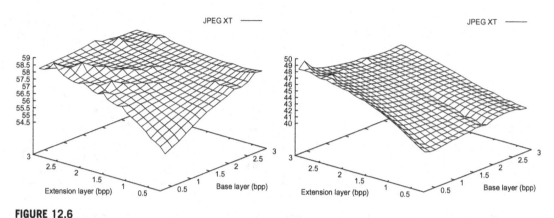

FIGURE 12.6

PSNR in the IDR domain for DSC_218 (left) and DSC_532 (right), parameterized by base and extension layer rate.

For the third experiment, the simple scRGB to sRGB map was replaced by the global Zhang et al. (2011) operator to study the dependency of the quality on the tone-mapping algorithm. Base quality is here again unconstrained. As seen from the curve "JPEG XT (Reinhard)," this has only little impact on the overall performance.

In comparison with more advanced technologies such as JPEG 2000 or JPEG XR, JPEG XT shows, of course, a disadvantage coming from its simpler coding technology and its inclusion of a base channel. Additionally, one should keep in mind that the two former codecs are usually run in PSNR-optimized mode, whereas the JPEG quantization matrix already implements a form of visual weighting by quantizing higher-frequency bands more aggressively than the more visible lower-frequency bands. SSIM scores of JPEG XT are therefore usually closer to those of more advanced technology, although the difference is also image dependent.

The dependency of the overall quality on the base and extension layer quality parameter can be seen much better in a three-dimensional plot (Fig. 12.6, where PSNR is plotted against base layer and extension layer rate). The tone mapper is here again the one defined by *TinyCMS*. The plots show a curious difference between the two images. For the DSC_218 picture, the base and the extension layer contribute approximately equally to the overall PSNR, whereas the bit rate for DSC_532 has to be invested mostly into the extension layer and the overall quality almost does not depend on the base layer rate. The difference lies in the nature of the images. DSC_532 is an almost bitonal image with very light and dark image regions, requiring a very extreme tone mapping to convert it to LDR, whereas DSC_218 is comparably moderate in nature.

12.6.2 CODING PERFORMANCE ON HDR IMAGES

Even when we leave the relevance of such quality indices aside, PSNR and SSIM cannot be directly carried over to the HDR regime as was possible for IDR; this is simply because the definitions of both indices depend on a natural maximum value or maximum brightness which does not exist for HDR. Furthermore, both are not scale independent as required for relative radiance formats (ie, multiplication of all sample values by the same scale alters the value of the index).

A seemingly obvious replacement for the PSNR is the signal-to-noise ratio (SNR), defined as

$$\text{SNR} := 10 \log_{10} \frac{\sum_i x_i^2}{\sum_i (x_i - y_i)^2}, \tag{12.9}$$

because it does not depend on a scale. In Eq. (12.9), x_i is the sample value (grayscale value) of the original image and y_i is the sample value of the reconstructed (distorted) image. Unfortunately, this definition comes with a lot of problems of its own. While the PSNR has some known cases of failure to evaluate image defects correctly according to their visibility, it is even easier to construct defects in HDR images that are misclassified by the SNR. A small relative error in a very light image region will cause an almost invisible defect, but such an error will lower the SNR significantly even though the defect is not obvious to human observers. Similarly, a small defect in a dark image patch may give rise to a significant image defect that is obvious to human observers; the SNR will, however, not differ significantly because the defect is low in amplitude and does not change the denominator much.

A better option to make traditional metrics available for HDR image coding is to first map the absolute luminances, as given by the sample values of the original and the reconstructed image, into a perceptually uniform space, and then apply known metrics in this space instead. If the input scene is described in relative radiance, a suitable scale factor to derive absolute radiance values must be estimated though. A candidate for this is the PU-Map of Aydin et al. (2008); an even simpler map is given by the logarithm, followed by clamping. The motivation for the latter is Weber's law (Netravali and Haskell, 1989), which states that the sensitivity of the human visual system is approximately inversely proportional to the amplitude of the input stimulus.

While this approach seems somewhat ad hoc, one should recall that a similar map is, implicitly, also applied in LDR imaging. Image data here are usually gamma-corrected and hence recorded in a nonlinear space. The gamma correction, even though historically motivated differently, can here be understood as a similar approximation of the map from the physical to an approximately uniform perceptual space.

Despite these attempts to apply known expressions in the HDR regime, a couple of quality indices have been designed explicitly for HDR imaging, and the area of an HDR image and video quality metrics are discussed in much more detail in Chapter 17. For the purpose of this chapter, three quality indices will be used: one of these is Daly's visual difference predictor (Daly, 1992; Mantiuk et al., 2005), extended and enhanced in a later version by Mantiuk et al. (2011), and another is the mean relative square error (Richter, 2009) (MRSE). Unfortunately, currently only limited data exist to assess the performance of these indices themselves — that is, there is only limited evidence that they correlate well to image quality as observed in subjective experiments. The reason is that subjective evaluation of HDR image quality is a relatively young field and not much is known about suitable test conditions and test protocols, partially because equipment to reproduce such images has not been available until recently. Unlike the evaluation of LDR data, the response curve and the properties of the particular monitor used for image reproduction must be taken into consideration as they are another source of nonlinearities. Calibration of the overall signal path from sample values to observable radiance is nontrivial and is beyond the scope of this section.

As in Section 12.6.1, the HDR image quality gained after JPEG XT compression and as estimated by an appropriate index is a function of *two* parameters, the base layer bit rate *and* the extension layer bit rate. The bit rate itself is controlled by a JPEG "quality" parameter which acts as a scaling factor of a suggested default quantization matrix found as an example in the JPEG standard. This choice of the quantization matrix is not necessarily the best for all profiles, although research results for better alternatives are not yet available.

Three objective quality indices have been tried here — namely, PU-PSNR, MRSE, and HDR-VDP2. PU-PSNR first maps the data with the PU2-Map defined in Aydin et al. (2008) into a perceptually uniform space and then measures the PSNR there. The second measures the MRSE in the image domain, and is defined as

$$\text{MRSE} := -10 \log_{10} \sum_i \frac{(x_i - y_i)^2}{x_i^2 + y_i^2}. \tag{12.10}$$

As one can see by a Taylor expansion, this is approximately identical to measurement of the PSNR after application of a logarithmic mapping. The third quality index is HDR-VDP2, defined in Mantiuk et al. (2011), which includes an extensive model of the human visual system, including a map to a perceptual luminance space, separation into perceptually defined frequency bands, masking, and error pooling. More details on HDR-VDP2 are given in Daly (1992) and Mantiuk et al. (2005, 2011).

The HDR images for this study were taken from the HDR Photographic Survey (Fairchild, 2015), which contains 106 images obtained through a multiple-exposure process. Photographs of the same scene were taken multiple times with different exposure times to cover a large range of luminances, and then were computationally merged into a single image per scene. Such computational methods are beyond the scope of this chapter, although the outputs are, in the absence of additional calibration, relative radiance images; that is, the floating point sample values obtained are (approximately) proportional to the luminance of the original scene. For details, see Reinhard et al. (2010). Some (tone-mapped) example images from this test set are shown in Fig. 12.7.

Comparison with subjective quality scores, even though not done in this chapter, requires an additional step — namely, the conversion of relative or absolute radiance data into an absolute radiance display-referred color space that describes the characteristics of the monitor used in the test

FIGURE 12.7

Two images used for the HDR compression performance. Courtesy of Mark Fairchild's HDR Photographic Survey: the BloomingGorse2 image (left) and the 507 image (right). Both images have been tone-mapped for printing.

sessions. Such monitors are already available on the market as professional equipment. In this particular evaluation, a maximum luminance of $4000\,\mathrm{cd/m^2}$ was assumed to define the mapping.

The performance of a JPEG XT encoder depends not only on the HDR input image but also on the selection of the LDR image encoded in the base layer. Here, the base images were taken from the JPEG XT internal "validation test" which was run to check the suitability and correctness of the various proposals. In particular, the tone-mapped LDR images were prepared with Photoshop by a semiautomatic process where parameters were adjusted by a human observer to generate "pleasing" LDR images. While somewhat arbitrary, this procedure mimics the typical HDR workflow for which JPEG XT has been designed. The difference between the test here and the validation test of JPEG XT is that this test uses absolute radiance output-referred images, whereas the latter used the original relative radiance scene-referred images.

Each HDR-LDR image pair was compressed with each of the three profiles of part 7 (profiles A, B, and C), with variation of both base layer and extension layer quality, and measurement of both the extension layer and the base layer bit rate and the image quality, as approximated by PU-PSNR, MRSE, and HDR-VDP2. Fig. 12.8 shows plots of HDR-VDP2 against the base layer and the extension

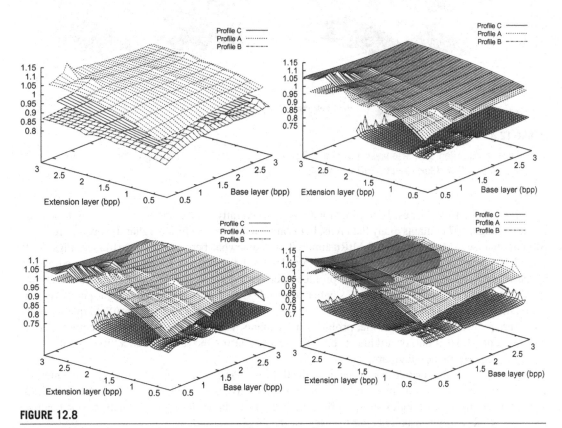

FIGURE 12.8

HDR-VDP2 versus extension and base layer quality for all profiles of part 7. From left to right, top to bottom: BloomingGorse2, 507, McKeesPub, MtRushmore2.

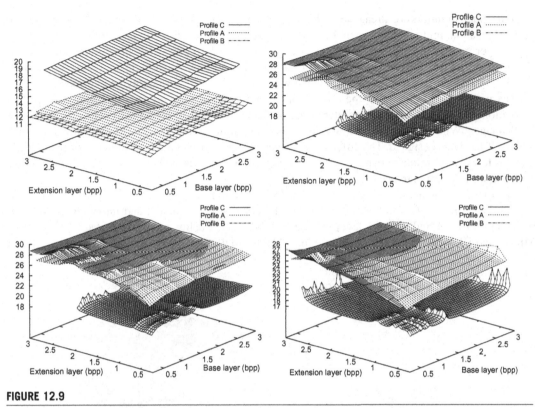

FIGURE 12.9

MRSE versus extension and base layer quality for all profiles of part 7. From left to right, top to bottom: BloomingGorse2, 507, McKeesPub, MtRushmore2.

layer bit rate for four images. BloomingGorse2 is a structurally very complex image with a medium dynamic range, 507 contains many flat areas, but is almost bitonal and has a higher dynamic range than BloomingGorse2. McKeesPub and MtRushmore2 are somewhat between these extremes. Figs. 12.9 and 12.10 show the performance of the same profiles on the very same images, but measured by the MRSE. As can be seen from the figures, profile A performs quite well on complex images and shows an advantage for medium to low bit rates, whereas profile C performs better in the high-quality regime for relatively flat images. What cannot be observed from the plots is that the profile C quality does not saturate toward high bit rates (ie, profile C implements a scalable lossy to lossless compression). Profiles A and B level out toward higher bit rates, and increasing the rate beyond a threshold does not improve the performance anymore.

As can be seen, the relative performance of the three profiles is quite consistent between the metrics except for the complex BloomingGorse2 image, where the advantage is either for profile A or profile C, depending on the quality index used. Subjective tests have shown that the correlation to subjective scores of HDR-VDP2 is usually better than that of the far simpler indices.

A somewhat more conventional one-dimensional plot of quality as measured by HDR-VDP2 versus the bit rate is given in Fig. 12.11. In this case, an extensive search on the two-dimensional

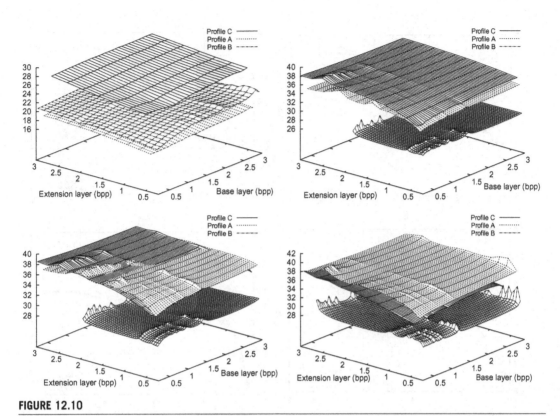

FIGURE 12.10

PU2-PSNR versus extension and base layer quality for all profiles of part 7. From left to right, top to bottom: BloomingGorse2, 507, McKeesPub, MtRushmore2.

base/extension layer rate parameter space has been made to obtain the best possible quality for a given rate constraint. The same trend as in Fig. 12.8 can be observed, showing advantages for profiles A and B for low bit rates and profile C for high bit rates. In this test, the base rate is unconstrained and hence possibly very low; hence, the LDR quality is uncontrolled and may be too low to provide an image of reasonable quality.

Fig. 12.12 shows the results if the base layer quality is constrained to be at least 75, resulting in a good-quality LDR preview. Here, profile C falls even more behind for low bit rates. However, it is the only profile whose quality does not saturate for increasing bit rate. It remains the best choice if scalable lossy to lossless coding is required or really high qualities need to be obtained.

12.7 CONCLUSIONS

JPEG XT is a very flexible and rich image compression framework, allowing lossy and lossless compression of IDR and HDR data, including coding of opacity information. JPEG XT addresses the

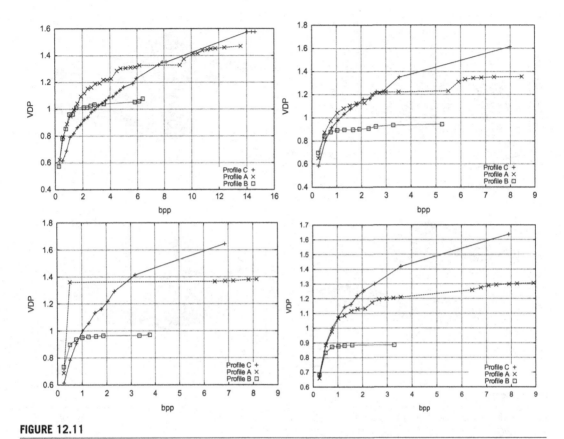

FIGURE 12.11

VDP scores as two-dimensional plots, selecting the best possible base layer-extension layer rate combination to maximize quality. From left to right, top to bottom: BloomingGorse2, 507, McKeesPub, MtRushmore2.

needs of a "modernized JPEG," filling the gaps the legacy JPEG standard left: scalable lossy to lossless compression, support for bit depths between eight and 16 bits, support for floating point samples, and coding of opacity data. Unlike other standardization initiatives, JPEG XT is fully backward compatible with legacy applications and always includes a lossy eight-bit version of the image in its codestream, which is visible to legacy applications. Clearly, carrying the legacy of the more than 20- year-old JPEG coding scheme costs performance, and JPEG XT cannot match more modern compression algorithms such as JPEG 2000 or JPEG XR. However, JPEG XT can be easily implemented on the basis of two readily available JPEG intellectual property cores and a single digital signal processor for postprocessing. Unlike more modern architectures, JPEG XT keeps the legacy JPEG tool chain working, allowing existing software and hardware architectures to access JPEG XT images without the compatibility problems of other image compression formats.

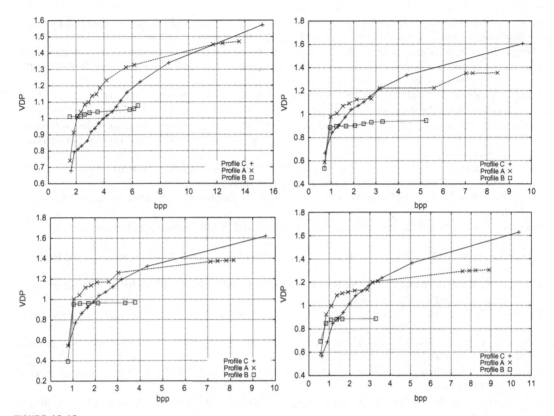

FIGURE 12.12

VDP scores as two-dimensional plots, selecting the best possible base layer-extension rate combination to maximize quality under the constraint of a base image quality of at least 75. From left to right, top to bottom: BloomingGorse2, 507, McKeesPub, MtRushmore2.

REFERENCES

Aydin, T.U., Mantiuk, R., Seidel, H.P., 2008. Extending quality metrics to full luminance range images. In: Proceedings of Human Vision and Electronic Imaging XIII (SPIE 2008).

Boliek, M. (Ed.), 2000. Information Technology — The JPEG2000 Image Coding System: Part 1, ISO/IEC IS 15444-1.

Brower, B., Clark, R., Hinds, A., Lee, D., Sullivan, G. (Eds.), 2011. JPEG File Interchange Format, ISO/IEC 10918-5.

Daly, S., 1992. The visible differences predictor: an algorithm for the assessment of image fidelity. In: SPIE Vol. 1666: Human Vision, Visual Processing, and Digital Display.

Fairchild, M., 2015. The HDR Photographic Survey. Available online at http://rit-mcsl.org/fairchild/HDR.html.

IEEE Computer Society, 2008. IEEE Standard for Floating Point Arithmetic, IEEE Std 754-2008. IEEE, New York.

ISO SC 29 (WG 1), 2012. Proposal for a New Work Item — JPEG Extensions. Available on www.jpeg.org as Document wg1n6164.

ISO/ITU, 1992. Information Technology — Digital Compression and Coding of Continuous-Tone Still Images — Requirements and Guidelines. Published by ISO/IEC as 10918-1 and ITU as T.81.

Mantiuk, R., Daly, S., Myszkowski, K., Seidel, H.P., 2005. Predicting visible differences in high dynamic range images — model and its calibration. In: Proceedings of Human Vision and Electronic Imaging X, IS&T/SPIE's 17th Annual Symposium on Electronic Imaging, 2005, pp. 204–214.

Mantiuk, R., Efremov, A., Myszkowski, K., Seidel, H., 2006. Backward compatible high dynamic range MPEG video compression. In: Proceedings of SIGGRAPH 2006 (Special Issue of ACM Transactions on Graphics).

Mantiuk, R., Krawczyk, G., Myszkowski, K., Seidel, H., 2007. High dynamic range image and video compression — fidelity matching human visual performance. In: Proceedings of 2007th IEEE International Conference on Image Processing (ICIP).

Mantiuk, R., Kim, K.J., Rempel, A.G., Heidrich, W., 2011. HDR-VDP-2: a calibrated visual metric for visibility and quality predictions in all luminance conditions. ACM Trans. Graph. 30 (4).

Netravali, A., Haskell, B., 1995. Digital Pictures: Representation, Compression and Standards, second ed. Springer, Berlin.

Netravali, A.N., Haskell, B.G., 1989. Digital Pictures: Representation and Compression. Plenum Press, New York and London.

Pennebaker, W.B., Mitchell, J.L., 1992. JPEG Still Image Data Compression Standard. Van Nostrand Reinhold, New York.

Reinhard, E., Ward, G., Pattanaik, S., Debevec, P., Heidrich, W., Myszkowski, K., 2010. High Dynamic Range Imaging, second ed. Kaufmann/Elsevier, Boston/Amsterdam.

Richter, T., 2009. Evaluation of floating point image compression. In: Proceedings of International Workshop on Multimedia Experience (QoMEx 2009).

Srinivasan, S., Tu, C., Zhou, Z., Ray, D., Regunathan, S., Sullivan, G., 2007. An Introduction to the HDPhoto Technical Design. JPEG document WG1N4183.

Taubman, D., 2000. High performance scalable image compression with EBCOT. IEEE Trans. Image Process. 9 (7), 1151–1170.

Wallace, G., 1992. The JPEG still picture compression standard. IEEE Trans. Consum. Electron. 38 (1), xviii–xxxiv.

Wang, Z., Simoncelli, E., Bovik, A., 2003. Multi-scale structural similarity for image quality assessment. In: IEEE Asilomar Conference on Signals, Systems and Computers.

Ward, G., Simmons, M., 2005. JPEG-HDR: a backwards compatible, high dynamic range extension to JPEG. In: International Conference on Computer Graphics and Interactive Techniques (ACM SIGGRAPH 2005).

Zhang, Y., Reinhard, E., Bull, D., 2011. Perception-based high dynamic range video compression with optimal bit-depth transformation. In: Proceedings of 2011th IEEE International Conference on Image Processing (ICIP).

DISPLAY

HDR DISPLAY CHARACTERIZATION AND MODELING

13

S. Forchhammer*, J. Korhonen*, C. Mantel*, X. Shu†, X. Wu†

*Technical University of Denmark, Bristol, Denmark**
McMaster University, Hamilton, ON, Canada†

CHAPTER OUTLINE

13.1 INTRODUCTION

The range of the human eye spans from 10^{-6} to 10^8 cd/m^2. Standard consumer devices are capable of displaying only a small fraction of the luminance range that humans observe in nature. Typical consumer television sets and computer screens use an input signal with eight bits per color channel, and operate with a peak luminance in the range of about 80–500 cd/m^2. Thus, these standard consumer devices are capable of displaying only a small fraction of the high luminance range that humans observe in nature. To create more naturalistic representations of digital images and videos, several technologies

High Dynamic Range Video. http://dx.doi.org/10.1016/B978-0-08-100412-8.00013-9

have been developed to provide displays with high dynamic range (HDR) capability. The first HDR displays were intended for professional use, but there are also HDR televisions being launched in the high-end consumer market.

Compared with conventional low dynamic range displays, HDR displays have higher peak luminance, higher contrast ratio, and more accurate representation of colors (10–16 bits per color channel). There is no standard definition of HDR displays, but practical displays marketed with HDR capability usually achieve a peak luminance of at least 2000 cd/m^2 and a contrast ratio of 10,000:1. However, reported contrast ratios should be treated with some caution, because different methods may have been used for contrast measurements.

HDR displays and projectors based on several different technologies have been developed. At the time of writing, most of the practical HDR displays use liquid crystal display (LCD) technology. LCDs can have high peak intensity, because of the bright and power-efficient light-emitting diodes (LEDs) used as backlight and which are available at a reasonable cost. The main disadvantage of LCDs is the limited local contrast: peak intensity can easily be enhanced by use of brighter backlight, but this will also raise the black level locally, because liquid crystals experience backlight leakage. Better local contrast can be achieved by use of plasma technology, but plasma displays have other disadvantages, such as lower power efficiency, lower peak intensity, and image retention (ghost imaging). Because of these disadvantages, plasma technology has not proven to be a competitive alternative to LCDs in the HDR display market segment.

A promising new HDR display technology uses organic LEDs (OLEDs) as pixels (Forrest, 2003). Since each OLED pixel is a light-emitting unit, backlight is not needed, and a high local contrast ratio can be achieved. OLEDs can also reach high peak intensity at a low power. In terms of production cost and lifetime, OLED technology is not yet competitive with LCDs for large displays, but in the long run, OLED displays are expected to dominate the HDR display market.

In this chapter, our main focus is on HDR display technology based on LCDs using LED backlight, the most prominent type at the moment. To improve local contrast in LCDs, local backlight dimming is an essential technique, allowing different backlight segments to be dimmed separately according to the image content to be displayed (Seetzen et al., 2004). First, we introduce the physical characterization of LCDs with LED backlight, including essential concepts for modeling LCDs, such as leakage, clipping, and basic distortion measures. Then, we present the models in more detail and discuss algorithms based on the models for optimization of the image contrast via local backlight dimming. We also extend our discussion to spatiotemporal characteristics observed in video signals and 3D images, and finally, present methods and results on subjective and objective image and video quality modeling and assessment of LED backlight displays.

13.2 HDR IMAGE DISPLAY WITH LED BACKLIGHT

The recent advances in LED technology have brought white high-intensity LEDs to the market at a reasonable price. Compared with traditional cold cathode fluorescent lamps, LEDs have several benefits, including lower power consumption, longer lifetime and smaller size, enabling thinner display panel design. LEDs can also be switched on and off very rapidly, which allows fast dynamic adjustment of backlight intensity (Anandan, 2008). This is an essential feature of LCD architectures capable of

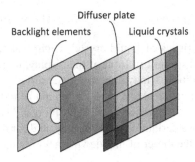

FIGURE 13.1

Structure of an LCD panel including backlight and a diffuser.

local backlight dimming, which is commonly used in HDR displays for professional use, and also in high-end television sets (Seetzen et al., 2004; Cho and Kwon, 2009).

13.2.1 PHYSICAL CHARACTERIZATION

A conceptual illustration of an LCD panel is given in Fig. 13.1. The main components are the backlight, diffuser plate spreading the light smoothly on the display area, and a grid of liquid crystals, forming the pixels of the displayed image (Anandan, 2008). In color displays, each pixel consists of separate subpixels for red, green, and blue color channels, respectively. In conventional LCDs, backlight and the diffuser have been designed to provide uniform distribution of light across the display. In display architectures supporting local backlight dimming, backlight is divided into segments that can be controlled independently. Local dimming allows the use of lower illumination levels in regions where bright content is not present. The basic backlight model is introduced below and more details about local dimming will be given in Section 13.3.

Since liquid crystals are essentially voltage-driven light filters, we can model the relative luminance $l(i,j)$ at pixel position (i,j) as the product of the normalized backlight intensity $b(i,j)$ at the pixel position and the normalized transmittance $t(i,j)$ of the liquid crystal:

$$l(i,j) = b(i,j) \cdot t(i,j). \tag{13.1}$$

In Eq. (13.1), we assume that all the values are normalized to the interval [0,1], where 0 represents the minimum value (backlight is turned off or the liquid crystal blocks all the light) and 1 represents the maximum value (backlight has its full intensity and the liquid crystal is set to its maximum transparency level, ie, as little light is blocked as possible). The maximum value of the backlight may later be chosen for a specific HDR display.

Unfortunately, *light leakage* is usually observed in practical liquid crystals, even if the control signal of the liquid crystal is set to zero. This is why intended black pixels in LCDs do not look entirely black, but look slightly grayish or blueish. To take light leakage into account in the pixel luminance model, we can define a leakage factor ε, expressing the proportion of light that will pass through a pixel that is intended to be black. Including ε, we can redefine the relative pixel luminance as

$$l(i,j) = b(i,j) \cdot t(i,j) + \varepsilon \cdot b(i,j) \cdot (1 - t(i,j)). \tag{13.2}$$

Because of the physical characteristics of liquid crystals, light leakage is dependent on the viewing angle; leakage is less pronounced straight in front of the display, in comparison with a tilted viewing angle (Burini et al., 2013a). In addition, ε may vary slightly at different pixel positions and for different color components. For a fully accurate leakage model, ε should be redefined separately for red, green, and blue subpixels (ε_r, ε_g, and ε_b) as a function of pixel position and viewing angle φ. However, such a model would be highly specific to a certain physical display, and for the sake of simplicity, will not be presented here. Later we discuss the effect of the viewing angle in terms of letting ε be dependent on the viewing angle.

If the backlight is uniform, a constant value for $b(i,j)$ can be used in Eq. (13.2) to derive the observed pixel luminance. If the backlight is at full intensity, $b(i,j) = 1$. However, practical backlights are not accurately uniform, and in local dimming, displays deviating from uniformity are even desired to increase the contrast (eg, for HDR images). For accurate backlight intensity modeling, we need to know the *point spread function* (PSF) of each backlight segment. Basically, the PSF is defined separately for each backlight segment, and it expresses the backlight intensity at different positions when the backlight element in question is turned to maximum intensity. Physical measurements of *luminance*–that is, the luminous intensity of a surface (cd/m^2) — are typically required to define the PSF accurately for any real-life display. However, a Gaussian function, for example, may be used for rough approximation of symmetric backlight segments. In the following, we denote the PSF as $h_k(i,j)$, expressing the resulting backlight intensity at position (i,j) when backlight segment k alone is set to full power.

In practical LCDs, backlight segments are usually not optically separated from each other. This is why backlight segments overlap: at each pixel position, the total backlight intensity is contributed by several backlight elements. Each backlight element contributes to the intensity of a large number of pixels, beyond the backlight segment boundaries. A simple additive model can be used to compute the backlight intensities in different positions as a sum of individual contributions from different backlight segments:

$$b(i,j) = \sum_{k=1,\dots,N} B_k h_k(i,j), \tag{13.3}$$

where B_k denotes the normalized backlight intensity for segment k, and N denotes the total number of backlight segments.

13.2.2 DISPLAY ARCHITECTURES

Dimming of the backlight may be used to increase the dynamic range. This is always the case seen over time, but the effects on the contrast within an image depend on the display architecture.

In small displays, such as those for smartphones or small tablet computers, the backlight typically consists of only one LED, located in one of the corners. A light guide (diffuser) with special characteristics is used to diffuse light uniformly on the screen. Also, large conventional LCDs, especially those in the lower price range without HDR capability, usually have only one backlight segment for the whole display. Adaptive backlight dimming may also be used in displays with only a single backlight segment, referred to as *global backlight dimming* or *0D dimming*. Here dimming will only be effective over time, and there are no local dimming effects.

Displays with several backlight segments can be categorized according to the configuration of the segments. In *direct-lit LCDs*, LEDs are located behind the diffuser, forming a 2D structure of backlight segments. In displays of this type, local backlight dimming is referred to as *2D dimming*. Backlight segments can be allocated in different layouts; for example, SIM2 HDR display uses a hexagonal shape for 2202 controllable backlight segments providing a peak luminance of 4000 cd/m^2 (SIM2, 2014). Another possibility is to allocate LEDs at edges, forming either horizontally or vertically directed backlight segments for *1D dimming*. Edge-lit design is especially popular in flat panel televisions, since edge-lit LCD panels can be built thinner than direct-lit panels. One can arrange edge-lit segments in two columns by allocating separately adjusted LEDs on both sides of the screen. This kind of backlight dimming architecture is called 1.5D dimming as it offers a compromise between 1D and 2D dimming. Different display architectures for backlight dimming are compared in Fig. 13.2. The 2D dimming architecture is recommended for HDR material. In all cases, the PSF will also play a role in the contrast as a function of the distance within the image.

In addition to conventional backlight displays relying on white light sources, it is also possible to use a combination of red, green, and blue LEDs as backlight elements. This kind of architecture allows local backlight dimming to be performed separately for each color channel, termed *3D dimming*. The benefit of 3D dimming is apparent only on pictures where one of the color channels is highly dominating, and 3D dimming has not gained wide popularity in practical backlight-dimming LCDs.

13.2.3 EVALUATING DISTORTION

Since the backlight resolution is lower than the display resolution, it is usually not possible to concurrently eliminate light leakage in dark pixels and provide full backlight for bright pixels when

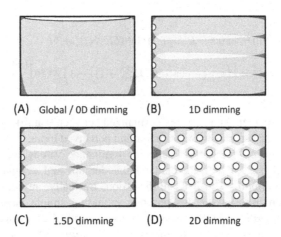

(A)   Global / 0D dimming   (B)   1D dimming

(C)   1.5D dimming   (D)   2D dimming

FIGURE 13.2

Different backlight architectures compared.

they are located close to each other, which leads to nonuniform backlight and *halo* effects (Chen et al., 2012). To achieve the best possible trade-off between leakage in dark pixels and *clipping* the luminance of bright pixels, we first need to find a method to evaluate the distortion caused by leakage and clipping. Assuming that the backlight intensities and PSFs for each segment are known, Eq. (13.3) can be used to compute the backlight luminance at each pixel position.

Since the transmittance levels of liquid crystal cells are set by the image signal fed to the liquid crystal array, the intensities of the red, green, and blue components of each pixel can be solved with Eq. (13.2). In practical displays using adaptive backlight dimming, the original image is often processed to compensate the relative loss of brightness in regions with dark backlight. This process is often referred to as brightness preservation, pixel enhancement, or liquid crystal pixel compensation. Uniformity (or absolute consistency) is best preserved if the backlight model is used to assist in brightness preservation. For the original image we assume that it has been mapped to the range of the display. Thus for an HDR image defined outside the range of the (HDR) display, we assume a tone mapping of the source image has been applied to form the original target image we will try to render on the display.

The computations in the model so far, Eqs. (13.1)–(13.3) are based on relative physical luminance values, and the relationship between physical luminance and perceived lightness (luma) is not linear (Cheng and Chao, 2006). Since images in digital signal processing systems are usually represented in a color space, such as sRGB, where each color component follows a perceptually uniform scale, conversion from the perceptual domain to the physical domain is necessary before transmittance values are used in the model. In displays with peak luminance lower than 100 cd/m^2, a conventional gamma function with $\gamma = 2.2$ can be used to approximate the relationship between physical luminance and perceptual luma. However, for HDR displays with higher peak luminance, a brightness adaptive conversion function would be preferred for more accurate results. Approximating the data from Aydın et al. (2008), we can define conversion functions from physical luminance L to perceptual uniform luma P and vice versa defined as (Korhonen et al., 2011)

$$P(L) = \ln(0.56 \cdot L^{0.88} + 1) / \ln(0.56 \cdot L_{\mathrm{max}}^{0.88} + 1), \tag{13.4}$$

$$L(P) = \left[\left(\exp\left(P \cdot \ln(0.56 \cdot L_{\mathrm{max}}^{0.88} + 1) \right) - 1 \right) / 0.56 \right]^{1.14}. \tag{13.5}$$

In Eqs. (13.4) and (13.5), perceptual luma P is normalized between 0 and 1, so the maximum physical luminance L_{max} (cd/m^2) corresponds with the maximum perceived luma, $P = 1$.

When the relative physical red, green, and blue luminance values are solved by a model, the values can be converted back to the perceptual luma domain and compared against the original target image to evaluate clipping and leakage distortion. Basically, any conventional full-reference image quality metric can be used for this purpose, including simple *mean squared error* (MSE) and *peak signal to noise ratio* (PSNR), as well as more sophisticated models, such as the structural similarity index (Wang et al., 2004) and the *HDR visible difference predictor* (HDR-VDP) (Mantiuk et al., 2005). A block diagram of the objective quality evaluation process is shown in Fig. 13.3.

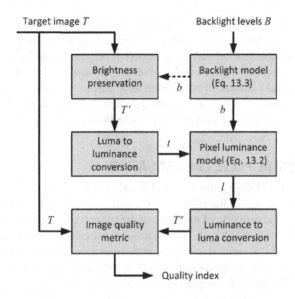

FIGURE 13.3

Objective quality evaluation of backlight dimming artifacts.

13.3 OPTIMIZING LOCAL DIMMING OF LED BACKLIGHT FOR IMAGE DISPLAY

In this section, we consider the modeling of LED backlit displays in more detail, focusing on direct (2D) backlit displays to achieve high contrast within images and as a special case of this edge-lit (1.5D) displays. In both cases a monochrome backlight will be considered. Given a model and related distortion measure, we will further describe how to pose backlight dimming as an optimization problem and how to solve it, including perceptual modeling of the resulting image on the display. We will address the dimming problem as adaptive local dimming reducing the brightness of LED segments depending on the image content (eg, reducing the LED values in dark parts of the image). A target image will be the input. If tone mapping is applied, it is assumed to be performed before the process of displaying the HDR target image. This represents an example of use of the display model.

13.3.1 LOCAL DIMMING OF LED BACKLIGHT

In Section 13.2, the leakage effect due to the liquid crystal elements was introduced. This deficiency may be attenuated by application of local dimming of the backlight segments depending on the image content. Albrecht et al. (2010) approached this by minimizing the leakage (or power consumption) under the constraint of not introducing clipping (ie, a clipper-free reproduction of the image). From a perceptual point of view, we may seek a better visual result with a higher (local) contrast by additionally attenuating the LED values to further reduce the leakage at the expense of introducing clipping.

A number of algorithms for local dimming have been presented and they may be characterized on the basis of what image statistics they use (eg, simple image statistics as the average value of the backlight segment, either directly or taking the square root). This may be generalized to algorithms collecting one or more image histograms and basing the dimming solely on this information. A more complex class of algorithms is based on calculation and evaluation of the displayed pixel values based on modeling the display process. The clipper-free approach (Albrecht et al., 2010) is one example. We will present a class of optimization-based dimming algorithms using a model of the displayed image.

13.3.2 LIQUID CRYSTAL PIXEL COMPENSATION

Given a set of LED values defining the modeled backlight, $b = b(i,j)$ (13.3), there is a choice for adjustment of liquid crystal values for all pixels. We will consider simple pixelwise compensation of liquid crystal values with the property of minimizing the square (or absolute) error. When the backlight is dimmed such that the target value, l_y, is not achieved, even when the liquid crystal element is fully opened, the liquid crystal value is saturated at the maximum value. This is called *hard clipping*. Conversely, when the leakage contribution exceeds that of the target pixel value, the liquid crystal value is set to 0. This may be expressed for target value l_y for a pixel by

$$l = \begin{cases} \varepsilon b, & \text{if } \varepsilon b > l_y, \\ b, & \text{if } b < l_y, \\ l_y, & \text{otherwise.} \end{cases} \tag{13.6}$$

Better visual solutions may be obtained by application of, for example, *soft-clipping*, but we consider the simple solution as a mathematically tractable approach providing a good initial solution (eg, toward determining the backlight values).

13.3.3 DISTORTION OF BACKLIGHT IMAGES AND OPTIMIZING THE BACKLIGHT VALUES

On the basis of the modeled displayed image, l, we can calculate an overall distortion. Having a distortion measure, we can straightforwardly pose an optimization formulation given by minimizing the overall distortion. Besides potentially improving the contrast, another important reason for dimming the backlight is to reduce energy consumption as also reflected in energy norms and labels for television sets today or to limit power dissipation in HDR displays.

The clipper-free solution precisely ensures no clipping but ignores leakage and as such does not involve a distortion measure. A first extension is to consider the distortion of luminance values directly (Shu et al., 2013),

$$\text{minimize } ||l - l_y||, \tag{13.7}$$

subject to Eq. (13.3), Eq. (13.6), and $0 \leq B_k \leq 1$.

As shown in Shu et al. (2013), this can be rewritten as a convex problem. The argument is based on for each pixel our considering each of the cases in Eq. (13.6) (ie, leakage, clipping, and cases with no distortion) and noting that these all provide a linear lower bound on the contribution to distortion of all elements. Thus, for distortion by absolute errors, this optimization problem is linear and for the MSE it is a quadratic problem.

To evaluate and include the power consumption, we note that the variable part of the display power consumed by the backlight may be expressed by a simple sum of LED values. The reason is that the LEDs are time modulated, being turned on in a fraction of the give time slot proportional to the desired value. The LED contribution to the power consumption, p, is thus expressed by

$$p = \frac{1}{N} \sum_{k=1}^{N} B_k.$$ (13.8)

The basic contribution to the power consumption of the platform may be modeled simply as a constant independent of the image content and will not be evaluated or part of the optimization process. Below we will combine distortion and power consumption in one cost function.

Considering the (visual) distortion in Eq. (13.7) directly based on luminance values has the limitation that, as mentioned, the luminance errors do not accurately reflect the visual impression. For this purpose a nonlinear mapping of the errors may be introduced. We consider a nonlinear mapping, f, of luminance, l, into a luma value ideally forming a perceptual uniform domain (13.4)–(13.5). As the gamma function is often used in relation to displays, we choose this as a simple example of a nonlinear mapping — that is, displayed pixel is $x = f(l) = l^{1/\gamma}$. Likewise, we define the target $y = f(l_y) = l_y^{1/\gamma}$ (Burini et al., 2013a, b). The corresponding full-resolution display and target images are denoted $\mathbf{x}$ and $\mathbf{y}$, respectively. The distortion and energy consumption may be combined in one cost function with use of a Lagrangian variable, q, to adjust the relative weighting of the two terms. Furthermore. addition of a weight $\mathbf{w}$ provides the flexibility to adjust the importance of each distorted pixel. This gives the general optimization problem:

$$\text{minimize } \|(\mathbf{y} - \mathbf{x})\mathbf{w}\| + q \times p,$$ (13.9)

where the weight vector $\mathbf{w}$ default is set to unity and the MSE (norm 2) is chosen as image distortion. The power, p, is given by Eq. (13.8).

The displayed image, $\mathbf{x}$, may further include the effects of quantization of LED values (13.3) and liquid crystal values (13.1). This has a marginal effect when one is deciding on LED values, but to calculate representative distortion numbers the quantization effect should be considered.

A suggestion for mapping luminance values into a perceptual uniform presentation was given in Eq. (13.4). We may also incorporate this in the distortion (13.9) either by changing the mapping f or by using the weight function, $\mathbf{w}$, if the absolute values on the display are known. On a similar note, $\mathbf{w}$ may also be used to adjust the relative importance of clipping and leakage or more generally prioritize a selected range of values or spatial regions of interest. The weight function may also be used in connection with tone-mapped HDR images.

Even if the focus is on quality, with little or no concern for the energy, it still makes sense to include the energy term, p, but with a small weight, q. The reason is that this will stabilize the solution to Eq. (13.9) by introducing a regularizing term. There may be (parts of) images with little or no highlight and dark areas and thus a large space of solutions for the LED values will provide (close to) optimal solutions. Even a small power term, q, will reduce the solution space significantly and stabilize the algorithm. Increasing q further will reduce LED values and visually the leakage will be further attenuated at the expense of increasing clipping.

Eq. (13.9) applies equally to direct backlight (2D) and edge-lit (1.5 D or even 1D) displays as the backlight is in these cases described by Eq. (13.3). To solve the minimization (13.9) on, for example, full-high-definition-resolution images, a gradient descent approach was used by Burini et al. (2014). For further speedup, histogram statistics were first collected within blocks of a fixed size in the block-based gradient descent approach.

The formulas above are defined for the luminance and luma or black/white images. We may generalize them to color images given by three color components in a number of ways. We may simply take $\|\mathbf{y} - \mathbf{x}\|$ as the color pixel by pixel distance in the 3D color space of choice. For RGB image representations two alternatives were presented in Burini et al. (2013a, b): a weighted sum of the red, green, and blue contributions or selection of the color component with the minimum value for leakage and the maximum value for clipping for each distorted pixel. Because of the large number of pixels per LED segment, the distortion function used in optimization may be simpler than the ideal distortion function to be used in evaluations. Likewise on the processing side, soft clipping may be applied as part of the pixel compensation, even if hard clipping is applied in the optimization of backlight elements. In Fig. 13.4, the performance for quality versus power is depicted for 48 images from the IEC database (Mantel et al., 2013b). A gradient descent-based version of Eq. (13.9) as described in Burini et al. (2014) was used and compared, among other methods, with the method in Albrecht et al. (2010) as well as with simple methods based on maximum, average, or square root of average pixel values (Mantel et al., 2013b).

Above we focused on global formulations of distortion. This may be extended to more local formulations, truly treating $\mathbf{x}$ and $\mathbf{y}$ as the whole image representations and not just summing contributions pixel by pixel. This extension may include adaptive tone mapping, local contrast,

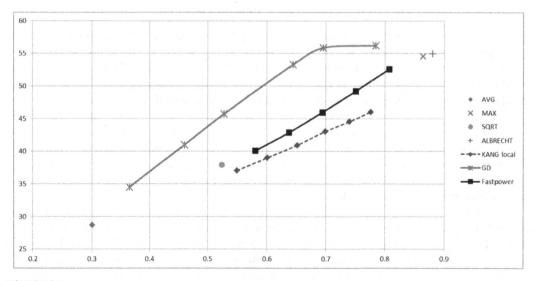

FIGURE 13.4

Quality (PSNR) versus power consumption for 48 images from the IEC database.

backlight artifacts as halos, etc. These issues are treated in other chapters in this book. To encompass this in the optimization in Eq. (13.9), the weight vector may be formulated as a function of the target image, $\mathbf{w}(\mathbf{y})$.

The approach above may be used image by image for image sequences as in video. However, application of full image-by-image optimization may well prove to be too aggressive, when the temporal dimension is included, and lead to flicker. Thus, extension from images by addition of the temporal dimension is nontrivial. The temporal dimension will be addressed in the next section. Restrictions on how much a given LED segment can change from frame to frame provide one approach to control flicker in local backlight dimming (Burini et al., 2014). For quality display of video material, flicker must be attenuated to an acceptable level. Once this has been solved, other issues over time may be to constrain power and variations in quality. A solution for this was presented in Mantel et al. (2013b) to control the quality-power trade-off by adjustment of the weight in Eq. (13.9).

13.4 **LED-BACKLIT 3D VIDEO DISPLAYS**

In addition to 2D image content and video as a sequence of images, LCDs are also capable of displaying stereo 3D images and videos with additional hardware. The model presented in Sections 13.2 and 13.3 will be extended to be continuous in the time domain, as there is a transition in time in the display from one image of a video to the next. The model will also be extended to include the interview crosstalk in stereo display. HDR displays are interesting for stereo display and even more so for multiview display, as these today sacrifice luminance to achieve the 3D effect.

In general, one achieves the illusion of stereo vision by showing two different views of the same scene while making each view visible to only the intended left or right eye. For instance, a thin layer of filter, called a parallax barrier, can be placed in front of the LCD so that each eye sees a different partition (eg, odd or even lines) of the display. If the left and right views are displayed on the two different partitions simultaneously, the viewer will perceive the illusion of three dimensions without wearing any special glasses. These glass-free technologies have the advantage of convenience; however, their 3D effect is inferior comparatively and only visible from certain viewing angles. Another group of technologies requires viewers to wear a pair of passive glasses with polarized filters. Similarly to the previous case, left and right views are displayed simultaneously but on different partitions of the screen; the partitions emit light with polarizations perpendicular to each other and the properly polarized glasses allow each view to reach only the intended eye. Technologies of this type provide good viewing angles and require little extra cost for each additional viewer. However, like the glass-free technologies, they also need to trade off spatial resolution for the 3D illusion, deteriorating the output image quality.

One of the most popular techniques for 3D vision is time-sequential stereoscopic visualization, which alternates the left and right views in sequence on a high refresh rate screen. The high-speed screen and synchronized liquid crystal active shutter glasses collaborate to generate binocular stereo vision, by transmitting the left-eye (right-eye) view while blocking the right-eye (left-eye) view to the left (right) eye. This method has been widely adopted by many stereoscopic displays because of its low cost and ease of implementation. However, the visual quality of liquid crystal stereoscopic displays of this type still leaves much to be desired largely because of the flaw of crosstalk.

Crosstalk, the phenomenon of one eye seeing the image intended for the other eye, which contradicts nature and reality, greatly deteriorates the visual quality of 3D liquid crystal display systems (Wang et al., 2011). The stereoscopic crosstalk is caused by two device limitations of LCDs in combination. First, LCDs refresh each frame from top left to bottom right pixel by pixel sequentially; thus, different pixels start updating at different times depending on their positions on the screen. Second, there is time lag in the transition of the liquid crystal from one gray level to another, and the time lag varies for different gray level changes. As a result, the time interval when a frame is correctly shown on the screen is short or even nonexistent. If the shutter glasses open for longer than this time interval, part of the previous frame becomes visible to the wrong eye, generating the crosstalk; on the other hand, if the shutter glasses do not open for long enough, the displayed images appear too dim through the glasses.

Various techniques have been proposed to suppress crosstalk in time-sequential stereoscopic visualization. One of the methods is called subtractive crosstalk cancelation, which does not directly decrease the light leakage to the wrong eye but cancels its effect by modifying the input signal accordingly (Konrad et al., 2000; Hong, 2012). However, for it to be effective, this method must compress the global dynamic range to create room for the cancellation, which results in undesirable low-contrast output. Although this shortcoming can be partially alleviated by a technique that applies crosstalk cancellation locally in small regions (Doutre and Nasiopoulos, 2011), this technique inevitably brings many other artifacts and difficulties because of the inconsistency of pixel values from region to region.

Another type of crosstalk reduction technique employs the LED local backlight system that provides independent control of the backlight luminance for different pixel blocks. The backlight scanning and strobe methods are in this category (Liou and Tseng, 2009; Liou et al., 2009). In the scanning scheme, the backlight segments are turned on from top to bottom in sequence for a period of time after the pixels in each of the regions have completed their state transitions. In the strobe method, all backlight panels are turned on simultaneously when all the pixels on the screen have stabilized, until the beginning of the next frame.

By selectively reducing light emission, backlight scanning and strobe can effectively reduce crosstalk at the expense of luminance. However, luminance is also a critical element of visual quality; some viewers prefer higher luminance than lower crosstalk. One can obtain a better balance between crosstalk and luminance by optimizing the output luminance while keeping the crosstalk below a given level and the rendition of backlight uniform (Burini et al., 2013b). Similarly to this idea, optimal backlight modulation provides a much simpler closed-form solution to the optimal trade-off between luminance and crosstalk, and hence it can be easily solved by the LCD control hardware (Jiao et al., 2015).

13.4.1 LIQUID CRYSTAL PIXEL COMPENSATION

In 2D LCD systems as described, an image is displayed by attenuation of the light emitted by the backlight source with a thin layer of liquid crystal filter array (13.1). For simplicity, here we will use a single variable i to index pixels in the displayed image, and its intensity value is determined by the transmittance $\tau_i(t)$ of the corresponding unit of the liquid crystal array and the backlight luminance $L_i(t)$ in the time duration T of a frame. Here we emphasize the time-variant properties of both $\tau_i(t)$ and $L_i(t)$. The former is due to transitional behavior of liquid crystal molecules; the latter is required by backlight modulation. To present stereoscopic vision, images for the left eye and the right eye are displayed alternately on a screen, and the liquid crystal shutter glasses (LCG) worn by the viewer select the correct view for each intended eye by rapidly changing their transmittance $g(t)$ accordingly. The (average) luminance of pixel i in stereoscopic display can be modeled as

$$I_i = \frac{1}{T} \int_0^T L_i(t) \cdot \tau_i(t) \cdot g(t) \, dt. \tag{13.10}$$

Without loss of generality, we discuss only one monocular image, the left one, in stereoscopic display in the following discussion. The condition for the image for the other eye can be derived similarly. Without $g(t)$, the model may be used to describe the temporal aspects of backlight dimming for monoview video, extending the time-discrete version in Eq. (13.1).

Design of stereoscopic vision systems usually adopts the following assumptions:

1. The liquid crystal transmittance of pixel i changes to the target level x_i instantly when pixel i is updated, and it stays the same until pixel i receives a new value.
2. The light transmittances of shutter glasses also switch instantly between a constant $\mathcal{G}$ and 0, corresponding to the open and closed states, respectively.
3. The backlight luminance is a positive constant $\mathcal{L}$ when the backlight is on, or 0 when it is off.

In view of the above assumptions, Eq. (13.10) can be approximated as

$$I_i \approx \frac{\Phi(\tilde{L}_i \cap \tilde{g})}{T} \cdot \mathcal{L} \cdot x_i \cdot \mathcal{G}, \tag{13.11}$$

where $\tilde{L}_i$ is the time interval of the backlight at pixel i being on, and $\tilde{g}$ is the LCG open interval. Their intersection, $\tilde{L}_i \cap \tilde{g}$, is the time interval when pixel i is visible to viewer. The function $\Phi(\tilde{t})$, representing the duration of an interval $\tilde{t}$, is defined as

$$\Phi(\tilde{t}) = b - a \text{ if } \tilde{t} = [a, b], \tag{13.12}$$

where a and b are the start and end times of $\tilde{t}$, respectively.

In addition, to ensure that pixel i of one monocular image is displayed correctly without any crosstalk, the backlight panel or shutter glasses can both be on only when the pixel has finished the state transition — that is,

$$\tilde{L}_i \cap \tilde{g} \subseteq \tilde{\tau}_i, \tag{13.13}$$

where $\tilde{\tau}_i$ is the time interval when pixel i is stabilized at the new target level. In a conventional stereoscopic LCD with a constantly-on backlight panel, the visible time interval of a pixel is entirely decided by the LCG open interval $\tilde{g}$; thus, $\tilde{g}$ must be contained within a short interval when every pixel is stable,

$$\tilde{g} \subseteq \bigcap_{i=1}^{N} \tilde{\tau}_i, \tag{13.14}$$

where N is the number of pixels. For some LCDs, however, there is not a moment at which all pixels are stable at the same time; therefore, the LCG open interval $\tilde{g}$ that makes the display crosstalk-free might not even exist.

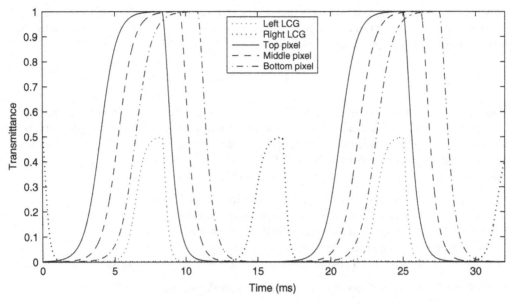

FIGURE 13.5

Transmittance curves of LCG and liquid crystal pixels at various positions. The maximum transmittance of LCG is normalized to 0.5, and the maximum transmittance of liquid crystal pixels is normalized to 1.

For LED-lit LCDs, the accuracy of assumption 3 is acceptable, as the alternation of LED backlight is several orders faster than what is noticeable by the human visual system. However, although the other two assumptions greatly simplify the design and analysis of stereoscopic display systems, they are excessively strong and do not accurately reflect the characteristics of actual hardware. The factors ignored in the approximation, such as the slow response of liquid crystals, are the major causes of crosstalk and low brightness in stereoscopic display. For example, in Fig. 13.5, the more realistic liquid crystal response curves of pixels and LCG show little resemblance to square waves as assumptions 1 and 2 suggest (Tourancheau et al., 2012).

Contrary to assumption 1, which presumes liquid crystal pixels are able to switch to a new transmittance instantly, the actual transition time T_T from a pixel receiving a new value to becoming stable is nonnegligible as shown in Fig. 13.6. Besides the considerable transition time, another practical consideration adding to the complexity of the LCD timing scheme is that pixels at various positions start transmittance transitions at very different times. Normally, pixels receive and start changing to new values line by line from top to bottom, and the offset T_P between the starting times of the first and last pixels can be as long as $T/3$.

Since the transmittance transition time and pixel refreshing time are not insignificant, if the sampling window $\tilde{g}$ is too wide, some pixels are still at the early stage of transition. Thus, the information intended for the other eye may leak into the current view, causing stereoscopic crosstalk. For example, in Fig. 13.5, since the sampling window starts when the bottom pixel is still in transition, crosstalk will be visible at that pixel.

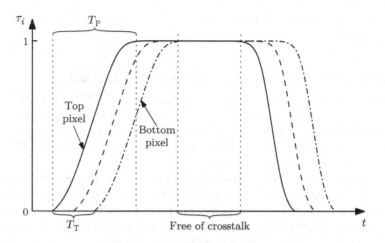

FIGURE 13.6

Timing diagram of liquid crystal pixels. The transmittances of pixels at different positions change from an initial value of 0 to 1 then to 0.

On the other hand, to avoid excessive crosstalk, the sampling window $\tilde{g}$ needs to be narrow enough to ensure that most pixels have reached or are close enough to the target level in the window. By Eq. (13.13), the duration $\Phi(\tilde{g})$ of the sampling window $\tilde{g}$ in Fig. 13.5 must be as short as 2 ms in order to eliminate the appearance of crosstalk in a typical LCD. However, with such a small LCG opening window, only a fraction of the light emitted from the backlight panel could reach to viewer's eyes, and much of the luminance is traded off for crosstalk reduction. To mathematically measure the loss of luminance caused by backlight control and LCG, we can use the image-independent mean luminance, which is defined as

$$E = \frac{1}{N} \sum_{i=1}^{N} \frac{I_i}{x_i}. \tag{13.15}$$

By substituting for I_i using Eq. (13.11), we can approximate E as

$$E \approx \frac{\mathcal{L} \cdot \mathcal{G}}{N \cdot T} \sum_{i=1}^{N} \Phi(\tilde{L}_i \cap \tilde{g}_i), \tag{13.16}$$

because $\tilde{L}_i \cap \tilde{g} \subseteq \tilde{\tau}_i$, E has an upper bound,

$$E \leq \frac{\mathcal{L} \cdot \mathcal{G}}{N \cdot T} \sum_{i=1}^{N} \Phi(\tilde{g}) = \frac{\mathcal{L} \cdot \mathcal{G} \cdot \Phi(\tilde{g})}{T}. \tag{13.17}$$

For a typical LCD, the duration T of a frame is 1/60 s and the transmittance $\mathcal{G}$ of a fully opened shutter glass is less than 0.5 because of the polarizing filter. If the sampling window $\tilde{g}$ is set to only 2 ms to combat crosstalk as shown in Fig. 13.5, then the image-independent mean luminance E is less than

$0.06\,\mathcal{L}$ as introduced in assumption 3. This means that at most 6% of the light from the backlight panel is visible to the viewer.

13.4.2 FORMULATION OF OPTIMAL BACKLIGHT MODULATION

As discussed previously, crosstalk reduction by adjustment of the LCG open interval $\tilde{g}$ might not be feasible in a conventional LCD with a uniform backlight panel. Furthermore, this method also greatly reduces the output brightness, causing a worse problem as viewers might prefer good brightness and contrast. Therefore, trading brightness for crosstalk reduction in conventional LCD systems is not always beneficial to the perceptual visual quality.

However, with LED local backlight systems, crosstalk reduction is possible without sacrificing much of the brightness. Turning on an LED backlight segment makes the pixels within the segment visible to the viewer, while at the same time, a block of pixels can be hidden from the viewer by turning off the corresponding backlight segment. Thus, if we only turn on the backlight of a segment when each pixel within the segment is stable, then the transmittance transitions of liquid crystal pixels are invisible to the viewer, hence eliminating the crosstalk. Because compared with LCG, backlight modulation provides finer control of the visibility of pixels, the same idea of adjusting LCG for crosstalk reduction can also be implemented with only backlight modulation.

Use of backlight modulation instead of LCG does not fundamentally change the idea of crosstalk reduction by hiding pixels from the viewer during their transmittance transitions. Inevitably, the reduced visibility, via the turning off of either LCG or backlight segments, affects the output brightness and deteriorates the perceptual visual quality. To find a better balance between crosstalk and brightness, we formulate crosstalk reduction as an optimization problem maximizing brightness while keeping the crosstalk below a given level as follows:

$$\begin{aligned}
&\text{maximize} \quad E \\
&\text{subject to} \quad \Phi(\tilde{L}_i \cap \tilde{g}_i) = l,\ 1 \le i \le N, \\
&\qquad\qquad\ \ \tilde{L}_i \cap \tilde{g} \subseteq \tilde{\tau}_i,\ 1 \le i \le N.
\end{aligned} \tag{13.18}$$

The objective function of this optimization problem, the image-independent mean luminance E of the display system as defined in Eq. (13.15), measures the perceived output brightness. The first constraint is to guarantee the uniformity of the output. It requires the amount of backlight emitted for each pixel to be identical over the entire display. With this constraint, the approximation of E in Eq. (13.16) is given by

$$E \approx \frac{\mathcal{L} \cdot \mathcal{G}}{N \cdot T} \sum_{i=1}^{N} l = \frac{\mathcal{L} \cdot \mathcal{G}}{T} \cdot l, \tag{13.19}$$

as the backlight luminance $\mathcal{L}$, LCG transmittance $\mathcal{G}$, and frame duration T are constant in the optimization problem, maximizing the objective function E is equivalent to maximizing l.

The second constraint of the optimization problem requires a pixel to be shown only when it is stable as in Eq. (13.13). Before a liquid crystal pixel finishes the transmittance transition to a new level x_i, its transmittance $\tau_i(t)$ is still affected by the previous level y_i intended for the other eye. The amount of crosstalk (ie, distortion caused by the previous level) $d_i(t)$ over time t can be quantified as follows (Tourancheau et al., 2012):

$$d_i(t) = \frac{x_i - \tau_i(t)}{x_i - y_i}. \tag{13.20}$$

Because of the slow response of an LCD, it takes a long time for a pixel to become completely stable at the new level, as shown in Fig. 13.5, making the absolute crosstalk-free time interval very short. Thus, to maintain a reasonable level of display luminance, the crosstalk constraint should be relaxed to allow a pixel to be shown as long as the crosstalk $d_i(t)$ is below a threshold D. This relaxation can be achieved by our redefining the stable time interval $\tilde{\tau}_i$ of a pixel as

$$\tilde{\tau}_i = \{t \mid d_i(t) \leq D\}. \tag{13.21}$$

Because in a typical LCD as in Fig. 13.5 the crosstalk $d_i(t)$ decreases rapidly to a low level at the beginning of the transmittance transition, only a small crosstalk allowance D is necessary to make the duration of the stable interval $\tilde{\tau}_i$ long enough. The LCD response function $\tau_i(t)$ can be acquired from the measurement of a real LCD system or from a mathematical model (Adam et al., 2007), but either way it is constant in the optimization problem; hence, the stable interval $\tilde{\tau}_i$ is also constant with a given crosstalk threshold D.

For the case of an LED local backlight panel, since pixels within a backlight segment are illuminated by the same LED unit, the time interval $\tilde{L}_i$ when the backlight is on is identical for all these pixels. In other words, if pixels i and j are in set S_k of the pixels in backlight segment k, then $\tilde{L}_i = \tilde{L}_j = \tilde{B}_k$, where $\tilde{B}_k$ is the time interval when backlight segment k is on.

Combining all the above points, we finally arrive at a new formulation of the crosstalk reduction problem:

$$
\begin{aligned}
\text{maximize} \quad & l \\
\text{subject to} \quad & \Phi(\tilde{L}_i \cap \tilde{g}) = l, \ 1 \leq i \leq N, \\
& \tilde{L}_i \cap \tilde{g} \subseteq \{t \mid d_i(t) \leq D\}, \ 1 \leq i \leq N, \\
& \tilde{L}_i = \tilde{B}_k, \ 1 \leq i \leq N, \ i \in S_k,
\end{aligned}
\tag{13.22}
$$

where l, $\tilde{g}$, $\tilde{L}_i$, and $\tilde{B}_k$ are variables.

13.5 MODELING AND EVALUATION OF DISPLAY QUALITY

The aim of this section is to present and discuss the various aspects of an LCD HDR display based on local backlight dimming technology and the need to be able to model the display process for quality assessment purposes.

As explained in Sections 13.2 and 13.3, because of the different resolutions of the grid of liquid crystal cells and the backlight segments, for contrasted images no perfect backlight dimming solution exists, but leakage or clipping defects occur. Therefore, the resulting defects need to be modeled and evaluated if one aims to assess the (high) quality of an HDR image displayed on an HDR screen with local dimming capability.

As detailed in Sections 13.2 and 13.3, the complete model of the displayed image, using Eqs. (13.2), (13.4), and (13.6), can be expressed in the following way:

$$l(i,j) = f^{-1}\left(b(x,y) \cdot \left[\frac{f(l_T(i,j))}{b(i,j)} \right]_{\perp \varepsilon}^{T1} \right), \tag{13.23}$$

where f is a function transforming the target perceived luma into physical luminance (both normalized), and $\perp \varepsilon$ and T1 denote the lower and upper bounds on the liquid crystal compensation implying leakage and clipping, respectively. Here the mapping function f is defined such that the perceived luma in general terms represents the desired target.

13.5.1 NECESSITY OF A DISPLAY MODEL FOR EVALUATION OF LOCAL BACKLIGHT DIMMING

There are two ways to model an HDR display: the first one is to characterize the rendering on the display, the way images are reproduced, and the second one is to include it in an image quality assessment chain. Whereas the first indisputably requires a display model, the second has not been thoroughly investigated in the quality assessment literature.

The first way, for example, plays a role when one is displaying HDR images as was done in Mantel et al. (2014) on a SIM2 HDR display, which applies local backlight dimming. For HDR images, tone mapping is generally applied. In our experiment, just a simple scaling was applied when the maximum image value exceeded the peak white value (which was set to 3000 cd/m^2). The rendering was done with the full capacity of the 2202 segments of the display and the MSE-optimal backlight computed with the approach described in Section 13.3 (with the PSF provided by the manufacturer) by means of the gradient descent-based version in Burini et al. (2014). Table 13.1 gives the error of rendering in terms of the signal-to-noise ratio (SNR) and the mean relative squared error (MRSE) (Richter, 2009) due to the display limitations (the SNR is calculated in the gamma-corrected domain) evaluated by the display model expressed in Eq. (13.23) to reflect the actual display of the images. The values given here are for perfect-quality input images and therefore the error stems only from the rendering (ie, clipping, leakage, and quantization). As the rendering is spatially dependent on the image content, the error varies significantly from one image to another. It reflects the rendering complexity specifically for each image and shows that HDR inputs can have very different characteristics: some are easy to render,

Table 13.1 Model-Based Distortion of Six HDR Images Displayed on a SIM2 HDR Screen		
Image Name	**SNR (dB)**	**MRSE (dB)**
Blooming Gorce	49.14	45.45
Canadian Falls	50.15	46.56
McKees Pub	39.65	37.63
MtRushmore2	50.19	43.76
OldFaithful Inn	43.72	40.64
Willy Desk	55.39	35.49

while others "stress" the renderer as also shown by Narwaria et al. (2014) for tone mapping. The SIM2 HDR display used in the experiment can provide locally a peak white value of 4000 cd/m^2, but for the very high values of peak white the actual display may deviate from the model, and hence we chose to operate the display at the lower peak white value of 3000 cd/m^2.

The second way addresses the broader question of the use of display modeling in quality assessment. This was addressed, for example, by Huang et al. (2012), who applied a display model, which includes gamma correction and ambient light reflection, before applying objective quality metrics. They show that use of a display model significantly improved the performance of quality metrics on the database tested.

Each step in the signal chain plays a role. When the quality of the displayed image or video is low, the influence of a display model is less crucial as the error the display introduces can be masked by other artifacts present. However, for settings where the input is an HDR image or just of high quality, the perceived level of errors introduced by the display may become higher than that of other defects. When small-amplitude artifacts or errors are not detected visually, it may remain an open question whether they are not perceivable or whether they are hidden by the rendering errors.

In the case of local backlight dimming, for example, display modeling for quality assessment is crucial as the rendering can be the sole processing applied to the target image. Subjective studies done on images (Mantel et al., 2013a) and highly contrasted videos (Mantel et al., 2015) of original quality displayed on a screen with various local backlight dimming algorithms have shown that observers perceive the difference between the different renderings. Thus, modeling of the dimming algorithms and resulting defects is necessary.

Traditional metrics have been designed to evaluate different types of distortions, such as compression noise and transmission artifacts, and only few attempts have been made to develop metrics tailored for specific display technology (eg, including backlight dimming artifacts). Development of such metrics is restricted by the scarcity of reliable subjective assessment results that could be used as ground truth for assessing the performance of objective metrics. Subjective results on backlight dimming artifacts are specific to the physical display used, and accurate modeling of backlight is a challenging task. This limits the possibilities to create generic annotated quality databases on backlight dimming artifacts.

A few subjective quality assessment studies on backlight dimming artifacts have been reported in the literature, but typically their main purpose is to compare the performance of different backlight dimming algorithms rather than to provide ground truth information for objective quality assessment. Some promising results in objective quality assessment, based on subjective results on a limited set of images, have been achieved with HDR-VDP-2 (Mantel et al., 2013a), but more research is required to confirm the findings on larger sets of both test images and display devices for the HDR domain.

13.5.2 KEY POINTS OF THE BACKLIGHT MODEL

As expressed by Eq. (13.23), besides the PSF measurements necessary to compute the backlight intensity, the key aspects of the display model are the leakage and clipping modeling and the transfer function transforming physical luminance into perceived luminance.

To the best of our knowledge, except for the HDR-VDP-2 metric, all currently published image or video quality metrics take as input the pixel values of the content and not the absolute physically displayed luminance. (The MRSE also takes physical values as input, but uses them only as relative values.) As seen in Section 13.2.3, the mapping function between the two is highly nonlinear and it impacts the objective metrics evaluations. Traditionally, the metrics based on the pixel value assume a gamma relation that is the inverse of that of the display, which often has a value of 2.2 as recommended by the ITU for the electro-optical transfer function (ITU BT.709 or BT.2020). As stated in Section 13.2.3, that might not be accurate for luminance values as high as those HDR displays can provide. Below we briefly touch on the effects of leakage inherent to the combined LCD and backlight technology, which imposes a limitation on the image contrast.

13.5.2.1 Leakage modeling
The leakage defect was defined in Eq. (13.2). It consists of the ratio between the light that goes through a liquid crystal cell when it is shut and the light provided by the backlight.

As leakage produces an increase of the black level, it decreases the contrast of the displayed image and thus impacts the perceived quality.

A subjective study presenting observers with highly contrasted video sequences with no processing other than local backlight dimming on a display with a peak white value of 490 cd/m^2 was presented in Mantel et al. (2015). Objective quality metrics were applied on the modeled stimuli with and without the leakage modeling included in the display model. Without the leakage modeling the metrics are unable to evaluate the quality (they produce negative correlation values, significantly different from the positive ones obtained with leakage modeling) and a quality model designed with partial least squares regression displayed a significant drop in accuracy (R^2 drops from 0.67 to 0.25) without leakage. Therefore, leakage modeling is necessary for objective quality assessment.

13.5.2.2 Leakage dependence on the viewing angle
The leakage highly depends on the viewing angle, meaning that the bigger the viewing angle, the greater the leakage. Fig. 13.7 represents the angular variation of leakage, as measured on a Bang & Olufsen BV7 display, which is an LCD with edge-lit (1.5D) LED backlight. Even though the leaking light starts to stabilize around 40°, the emitted light also decreases and therefore the leakage continues to increase.

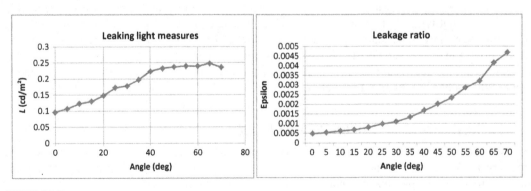

FIGURE 13.7

Variation of leaking light and leakage value as a function of the angle measured on an LED display (Bang & Olufsen BV7).

There are two aspects to this dependency: (1) the location of the observer — whether the observer sits in front of the display or whether the observer is watching it at an angle — and (2) the spatial variation of the leakage value — indeed, even when an observer is sitting in front of the display at a distance of three times the height (as recommended for subjective tests by the ITU), the viewing angle for the screen varies from $-16°$ to $+16°$ from side to side.

In Mantel et al. (2015), the highly contrasted sequences selected were shown to the observers at $0°$ and at an angle of $15°$. The results show that the increased leakage value at $15°$ influences the observers' perception. In this study, two models of the leakage were used as input for the quality calculation: one assuming a constant value over the whole screen and one with values varying spatially. The data show no evidence that inclusion of the spatial angle variation of the leakage over the screen represents an improvement for quality metrics, whereas distinguishing between subjects' viewing angles of $0°$ and $15°$ showed significant improvement.

13.6 CONCLUDING REMARKS

This chapter has presented modeling and characterization of HDR displays based on local backlight dimming of LCDs, with the focus on optimization of image-dependent local dimming. The leakage of liquid crystal elements imposes a bound on the achievable local contrast, an important feature in HDR imaging. Motivated by this fact, the local backlit displays have been modeled to capture this fundamental aspect of the display technology, to describe the effects on displayed images and perceived visual quality, and to optimize the images rendered on the backlit HDR display for high contrast within the individual images.

REFERENCES

Adam, P., Bertolino, P., Lebowsky, F., 2007. Mathematical modeling of the LCD response time. J. Soc. Inform. Disp. 15 (8), 571–577.

Albrecht, M., Karrenbauer, A., Jung, T., Xu, C., 2010. Sorted sector covering combined with image condensation sorted sector covering combined with image condensation: an efficient method for local dimming of direct-lit and edge-lit LCDs. IEICE Trans. Electron. E93-C (11), 1556–1563.

Anandan, M., 2008. Progress of LED backlights for LCDs. J. Soc. Inform. Disp. 16 (2), 287–310.

Aydın, T., Mantiuk, R., Seidel, H.P., 2008. Extending quality metrics to full luminance range images. In: Proc. of SPIE, vol. 6806, San Jose, CA, USA.

Burini, N., Nadernejad, E., Korhonen, J., Forchhammer, S., Wu, X., 2013a. Modeling power-constrained optimal backlight dimming for color displays. J. Display Technol. 8 (9), 656–665.

Burini, N., Shu, X., Jiao, L., Forchhammer, S., Wu, X., 2013b. Optimal backlight scanning for 3D crosstalk reduction in LCD TV. In: Proceedings of the IEEE International Conference on Multimedia and Expo (ICME), pp. 1–6.

Burini, N., Mantel, C., Nadernejad, E., Korhonen, J., Forchhammer, S., Pedersen, J., 2014. Block-based gradient descent for local backlight dimming and flicker reduction. J. Display Technol. 10 (1), 71–79.

Chen, H., Ha, T., Sung, J., Kim, H., Han, B., 2012. Evaluation of LCD local-dimming-backlight system. J. Soc. Inform. Disp. 18 (1), 57–65.

Cheng, W.C., Chao, C.F., 2006. Minimization for LED-backlit TFT-LCDs. In: Proc. of Design Automation Conference, San Francisco, CA, USA, pp. 608–611.

Cho, H., Kwon, O.K., 2009. A backlight dimming algorithm for low power and high image quality LCD applications. IEEE Trans. Consum. Electron. 55 (2), 839–844.

Doutre, C., Nasiopoulos, P., 2011. Crosstalk cancellation in 3D video with local contrast reduction. In: Proc. European Signal Processing Conference, 1884–1888.

Forrest, S.R., 2003. The road to high efficiency organic light emitting devices. Organic Electron. 4 (2–3), 45–48.

Hong, H., 2012. Reduction of spatially non-uniform 3D crosstalk for stereoscopic display using shutter glasses. Displays 33 (3), 136–141.

Huang, T.H., Kao, C.T., Chen, H.H., 2012. Quality assessment of images illuminated by dim LCD backlight. In: Proc. SPIE 8291, Human Vision and Electronic Imaging XVII, pp. 82911Q.

Jiao, L., Shu, X., Cheng, Y., Wu, X., Burini, N., 2015. Optimal backlight modulation with crosstalk control in stereoscopic display. J. Display Technol. 11, 157–164.

Konrad, J., Lacotte, B., Dubois, E., 2000. Cancellation of image crosstalk in time-sequential displays of stereoscopic video. IEEE Trans. Image Process. 9 (5), 897–908.

Korhonen, J., Burini, N., Forchhammer, S., Pedersen, J.M., 2011. Modeling LCD displays with local backlight dimming for image quality assessment. In: Proc. of SPIE, vol. 7866, San Francisco, CA, USA.

Liou, J., Tseng, F., 2009. 120 Hz low cross-talk stereoscopic display with intelligent LED backlight enabled by multi-dimensional controlling IC. Displays 30 (4–5), 148–154.

Liou, J.C., Lee, K., Tseng, F.G., Huang, J., Yen, W.T., Hsu, W.L., 2009. Shutter glasses stereo LCD with a dynamic backlight. In: Proc. of SPIE, vol. 7237, p. 72370X8.

Mantel, C., Burini, N., Korhonen, J., Nadernejad, E., Forchhammer, S., 2013a. Quality assessment of images displayed on LCD screen with local backlight dimming. In: Proceedings of 2013 International Workshop on Quality of Multimedia Experience, Klagenfurt, Austria.

Mantel, C. and Burini, N. and Nadernejad, E. and Korhonen, J. and Forchhammer, S. and Pedersen, J. M., Bech, S., Korhonen, J., Forchhammer, S., Pedersen, J., 2013b. Controlling power consumption for displays with backlight dimming. J. Display Technol. 9 (12), 933–941.

Mantel, C., Ferchiu, S., Forchhammer, S., 2014. Comparing subjective and objective quality assessment of HDR images compressed with JPEG-XT. In: Proceedings of IEEE 16th International Workshop on Multimedia Signal Processing (MMSP), pp. 22–24.

Mantel, C., Bech, S., Korhonen, J., Forchhammer, S., Pedersen, J., 2015. Modeling the subjective quality of highly contrasted videos displayed on LCD with local backlight dimming. IEEE Trans. Image Process. 24 (2), 573–582.

Mantiuk, R., Daly, S., Myszkowski, K., Seidel, H., 2005. Predicting visible differences in high dynamic range images: model and its calibration. In: Proc. of SPIE, vol. 5666, San Jose, CA, USA.

Narwaria, M., Mantel, C., Da Silva, M., Le Callet, P., Forchhammer, S., 2014. An objective method for high dynamic range source content selection. In: Proceedings of the 2014 Sixth International Workshop on Quality of Multimedia Experience, pp. 13–18.

Richter, T., 2009. Evaluation of floating point image compression. In: Proc. International Workshop on Quality of Multimedia Experience, San Diego, CA, USA, pp. 222–227.

Seetzen, H., Heidrich, W., Stuerzlinger, W., Ward, G., Whitehead, L., Trentacoste, M., Ghosh, A., Vorozcovs, A., 2004. High dynamic range display systems. ACM Trans. Graph. 23 (3), 760–768.

Shu, X., Wu, X., Forchhammer, S., 2013. Optimal local dimming for LC image formation with controllable backlighting. IEEE Trans. Image Process. 22 (1), 166–173.

SIM2, 2014. Exclusive High Dynamic Range Display Series. Retrieved 28 January, 2014. http://www.sim2.com/HDR/.

Tourancheau, S., Wang, K., Bulat, J., Cousseau, R., Janowski, L., Brunnström, K., Barkowsky, M., 2012. Reproducibility of crosstalk measurements on active glasses 3D LCD displays based on temporal characterization. In: Proceedings of SPIE, vol. 8288, p. 82880Y-17.

Wang, L., Teunissen, K., Tu, Y., Chen, L., Zhang, P., Zhang, T., Heynderickx, I., 2011. Crosstalk evaluation in stereoscopic displays. J. Display Technol. 7 (4), 208–214.

Wang, Z., Bovik, A., Sheikh, H., Simoncelli, E., 2004. Image quality assessment: from error visibility to structural similarity. IEEE Trans. Image Process. 13 (4), 600–612.

DUAL MODULATION FOR LED-BACKLIT HDR DISPLAYS

14

M. Narwaria, M.P. Da Silva, P. Le Callet

University of Nantes, Nantes, France

CHAPTER OUTLINE

14.1 INTRODUCTION

Liquid crystal displays (LCDs) have more or less completely replaced the traditional cathode ray tube displays in recent years. LCDs offer the advantage of being much thinner and hence more portable. LCDs have also become much cheaper with advances in technology and hence offer an attractive alternative to cathode ray tubes. LCD televisions produce a black and colored image by selectively filtering white light. Millions of individual LCD shutters, arranged in a grid, open and close to allow a metered amount of the white light through. Each shutter is paired with a colored filter to remove all but the red, green, or blue portion of the light from the original white source. Until recently, the source of light in LCDs was cold cathode fluorescent lamps (CCFLs). They have been one of the most popular sources of backlight for LCDs mainly because they are cost-efficient. However, they suffer from several functional disadvantages, such as higher power requirements, low speed switching, and shorter life. Consequently, light-emitting diodes (LEDs) have increasingly replaced CCFLs (Anandan, 2008) as the choice of backlight. An LED is a two-lead semiconductor light source. It is a pn-junction diode, which emits light when activated. When a suitable voltage is applied to the leads, electrons are able to recombine with electron holes within the device, releasing energy in the form of photons. In comparison with fluorescent lights, LEDs have significantly lower power requirements and convert power to light more efficiently, so less is lost as heat, and focus it more precisely, so there is less light leakage, which can cause fuzziness. An LED also lasts much longer than most other lighting technologies (Anandan, 2008). These characteristics coupled with significant advances in semiconductor technologies (thereby

High Dynamic Range Video. http://dx.doi.org/10.1016/B978-0-08-100412-8.00014-0

reducing LED costs) have propelled LEDs to be one of the most preferred sources of light in recent display technologies.

Whatever the source of the backlight (LEDs or CCFLs), the way it is used is the most relevant issue considered in this chapter. Specifically, in most consumer displays, the backlight is mainly uniform and constantly on to power most LCDs and LCD television sets. Hence, the backlight acts merely as a source of constant power and the liquid crystals are modulated to form the picture by adjustment of the amount of light that passes though them (hence they are referred to as uniformly backlit displays). However, such uniform backlight setting can result in low contrast levels and poor clarity especially in darker areas of the picture to be displayed. Moreover, it will also lead to higher power consumption. Thus, nonuniform backlight dimming has increasingly become a popular technology. The LEDs are dimmed locally to create deeper blacks in specific areas of the screen, which can also allow there to be better detail in dark scenes. This allows there to be very high dynamic contrast ratios in comparison with a uniform backlight setting (the ability of LEDs to be switched on and off faster than CCFLs is also an important factor in locally dimmed backlit displays).

The LEDs can be placed directly behind the LCD panel or at one or more of the sides. The former is known as a direct-lit configuration, while the latter is referred to as an edge-lit configuration. Both have their advantages and disadvantages. The edge-lit configuration has the LEDs surrounding the back of the panel and they use guides to light up the whole backplane. The main advantage is cost reduction, because it requires fewer LEDs, and has a reduced halo effect. It also helps to develop thinner display sets. On the downside, the edge-lit configuration can sacrifice the contrast ratio, leading to less sharp pictures. In the direct-lit configuration, the LEDs are placed directly behind the LCD panel and can therefore provide very high contrast ratios. It therefore also allows one to exploit the real benefits of the local backlight dimming. However, direct-lit displays can be much more costly as they require more LEDs, and can also introduce halo effects, especially around bright objects in the scene.

While it can be expected that local backlight dimming will be beneficial both for saving power and for improving picture quality, there are issues that need to be resolved for its full potential to be realized. Specifically, if the backlight is dimmed to an excessive degree, some pixels might not be able to increase their transmittance enough to compensate for the reduced emission of light, causing them to appear darker than intended. This creates an artifact called clipping; the pixels that suffer from it are called clipped pixels. Likewise, when one is viewing a video, the backlight intensity should change in a temporally coherent manner so that artifacts such as flickering are minimized. It should also be kept in mind that the backlight structure is composed of grids (square or hexagonal in many cases) that illuminate a local area. As a consequence, the localization is limited to a region rather than at the pixel level. In other words, each LED can act as a source of power for several pixels and each pixel can receive light from several LEDs. Because of this, bright and dark pixels are often in conflict, particularly when they are in the same region. In such cases, it is therefore nontrivial to obtain a backlight intensity which is optimal for both, because the high luminance required by the former will cause leakage in the latter. The presence of a group or cluster of bright pixels surrounded by dark pixels will also induce a halo artifact (Seetzen et al., 2003), caused by leakage being more evident in the dark pixels close to the bright cluster. Thus, there is a need to dynamically find a trade-off between aspects such as picture quality, power consumption, and the resultant complexity of the backlight dimming algorithm. A survey of literature shows that there are several backlight dimming algorithms for low dynamic range (LDR) displays. Most of these favor one of the aspects in rendering a video mentioned depending on the application considered.

14.2 **DUAL MODULATION FOR BACKLIGHT DIMMING**

Local backlight dimming is motivated by the fact that not all frames require the same amount of light, because some are darker than others. With local backlight, it is possible to vary the light intensity spatially according to the image content; for example, if the image contains both dark and bright areas, the backlight can be dimmed only in the dark ones. As mentioned, most of the existing work in local backlight dimming targets LDR displays and the primary goal is power efficiency (ie, minimization of power consumption). At the same time, contrast of the rendered picture is enhanced by minimization of light leakage. Leakage is caused by imperfections in the liquid crystals preventing the complete obstruction of light when this is required (ie, black pixels) and is particularly visible from wide viewing angles. This raises the black level of the screen, making dark pixels brighter than desired. Local backlight dimming can therefore help to represent the display black level more accurately.

Recall that there is a many-to-many relationship between the backlight segments and pixels in the picture to be rendered. Thus, most existing methods determine the backlight configuration on the basis of the intensities of a local group of pixels. The simplest backlight dimming algorithms use simple global or local image characteristics (eg, the maximum or average pixel value) to determine the backlight. The maximum algorithm sets the intensity of each LED to the maximum pixel value of the corresponding segment, while other algorithms use the mean value (Funamoto et al., 2001). The square root algorithm proposed by Seetzen et al. (2004) uses the square root of the normalized average pixel value (ie, valued between 0 and 1). These algorithms although simple are not reliable because the maximum might not result in any energy saving, especially for large segments, and is very sensitive to noise and prone to flickering, while mean value setting tends to produce excessively dim backlight. To overcome the limitations of methods based on simple statistics, more sophisticated models have been proposed over the years. The method proposed by Cho and Kwon (2009) used a correction term to ofset the average pixel intensity and considers the local difference between maximum and average luminance. A similar method developed by Zhang et al. (2012) computes the said correction term as the ratio of the luminance variance of the segment and the variance when the maximum backlight luminance is equal to the average luminance of the segment. Other methods such as the one introduced by Nam (2011) use both local and global brightness in order to find a better trade-off between enhancing local contrast and preserving the overall appearance of the rendered picture. A local backlight dimming algorithm was developed by Kim et al. (2009), and is based on accounting for the brightness of not just the current segment but the neighboring ones as well. Several other methods, such as those proposed by Kang and Kim (2009), Cho et al. (2011), Kang and Kim (2011), and Nadernejad et al. (2013) are based on image statistics such as histograms. Likewise, Chen et al. (2007) set the initial LED intensities using local weighted histograms. The algorithm presented by Cho et al. (2013) calculated separate RGB histograms for each backlight segment and these were used to reduce the intensity of each LED up to a threshold. The histogram-based method introduced by Lin et al. (2008) computed the cumulative distribution function (from a global histogram) and used its inverse to map a weighted mean of the maximum and average pixel values of each backlight segment to the resulting LED intensity. Another category of backlight modulation methods, such as those proposed by Shu et al. (2013), Albrecht et al. (2010), and Hong et al. (2010), are based on point spread functions (PSFs) to exploit the knowledge of light diffusion and model how lights diffuses from a source. A few methods, such as those introduced by Burini et al. (2013, 2012), focus primarily on power consumption, and aim to achieve a trade-off between clipping and leakage. We do not elaborate on these methods for LDR backlight dimming, but encourage the reader to refer to the respective references. However, most of them share a similar design philosophy. The primary aim is to minimize effects such as leakage and/or clipping on the one hand, and the resultant algorithm complexity on the other. This helps to reduce power consumption and increase contrast of the rendered picture.

As mentioned, the existing studies on backlight dimming target LDR displays. Thus, they do not consider the relatively high resolution backlight LED array. In other words, they do not take into account higher luminance and luminance resolution which is required for HDR picture rendering. As far as HDR is concerned, there are very few studies that address the issue of backlight dimming. The first work on HDR displays with local backlight dimming was reported by Seetzen et al. (2004). It has been used as the HDR baseline method with regard to dual modulation in HDR display systems, and is a simple and general algorithm for HDR dual modulation.

This simple algorithm derives the value of LEDs from a subsampled version of the square root of the luminance of the original HDR image. The value of the LCD panel pixels is determined by division of the original HDR image by the illumination map calculated from the convolution of the value of the backlight LEDs by the PSF of the LEDs. We now describe this algorithm in more detail, and then identify a few issues with it. A schematic diagram of this method is shown in Fig. 14.1.

- The first step is to take the square root of the HDR image I. This may be viewed as a simple strategy to "split" the HDR image into two components, one for LED-based backlight and the other for the LCD front panel.
- Next, the target intensities for every individual LED are computed by downsampling of the square root image to the resolution of the LED array. This is required because there does not exist a one-to-one mapping between pixels and LEDs (ie, we do not have one LED per pixel but have much smaller resolution of the LED array). Also the PSFs of the LEDs will scatter the light from one LED to neighboring ones, resulting in a convoluted source of light for each segment. Therefore, finding the optimal LED configuration for image rendering is nontrivial. Seetzen et al. (2004) approximated the solution with a single Gauss-Seidel iteration over neighboring LED pixels to compute the estimated LED values L. Of course, in practice, one can do this by exploiting convex optimization techniques which seek to minimize the error between the target image and the image obtained after convolution with PSFs. Also note that the said error can be computed either

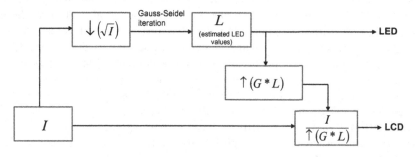

FIGURE 14.1

Baseline method for HDR dual modulation. The method was proposed by Seetzen et al. (2004).

in the physical domain (based on actual luminance) or in perceptual domain (the perceived luminance). The estimated values L will then be used to drive the LED panel.

- Next, the measured PSF of the LED is first approximated with a Gaussian G. Then, the estimated values L are convolved with G and the resulting signal is upscaled to the full resolution.
- Finally, the original HDR image I is divided by the signal in the previous step to compute the signal to drive the LCD panel.

The algorithm described above provides a baseline for the development and comparison of new dual modulation algorithms for HDR content rendering. While this method is simple, it has some issues that deserve further discussion. The first point concerns the use of a square root function as the first step. The original method does not justify this, neither experimentally nor qualitatively. In our opinion, we can view it as a means of equally splitting luminance on LEDs and LCDs although it is not clear if other functions may be used at this step, and what the possible effects (from rendering accuracy and visual quality perspectives) of that are. The second issue concerns the downsampling used in the second step of this method. Downscaling/downsampling essentially leads to averaging of neighboring values. This, in turn, will reduce the maximum luminance and one cannot achieve the target luminance. Obviously, this will affect the quality of the rendered HDR image/video. Also note that the optimization for obtaining backlight corrects the influence of the PSF but does not take into account constraints related to leakage. Thus, there is need for a more customizable optimization. Because videos have temporal information, it is not clear if the same method can be directly applied to render them. Finally, the HDR baseline method does not directly take into account power consumption, which is an important aspect in display systems.

14.3 PROPOSED METHOD FOR DUAL MODULATION

We briefly described the baseline HDR dual modulation method in the previous section, and also identified a few issues with it. In this section, we provide details on the dual modulation method that we developed. It is primarily based on addressing some of the issues identified. The detailed steps in the proposed method for rendering HDR images are now described.

- We begin with the observation that the HDR displays currently available cannot display luminance beyond the specified limit, given the hardware limitations. As a result, the first step in our proposed method aims at saturating the maximum image/video luminance to that of the HDR display to be used. No doubt this will impact the quality of the rendered content because it can lead to loss of details in areas with very high luminance. Unfortunately, this cannot be avoided because display limits. Nevertheless, the HDR display range is still capable of rendering about five orders of magnitude, which is sufficient for the rendering of most natural scenes. Thus, as a first step, we saturate the image luminance to the maximum displayable luminance. This step therefore corresponds to the idea that the HDR content should be graded so that it is display referenced and not scene referenced.
- The next step is to generate a grayscale version of the image. Note that this is not done with a simple mean or with BT.709/BT.601 RGB to Y coefficient. Instead, we used (for each pixel) the maximum of all three color channels. This helps to ensure that the backlight will provide enough light so as to render all color channels correctly and avoid saturation during the final division step

(A) (B)

FIGURE 14.2

Gray level images from two approaches: (A) gray image obtained from weighted combination of red, green, and blue; (B) gray image obtained by taking the maximum value of red, green, or blue. Notice that the image in (B) is brighter.

of the algorithm. An example to illustrate this is shown in Fig. 14.2, in which two grayscale images are shown. The one on the left is obtained by use of a weighted combination of the red, green, and blue components (more specifically, $0.2989R + 0.5870G + 0.1140B$), while the one on the right is obtained if we take the maximum of the three components. The latter image is expectedly brighter and this allows us to properly render all color channels because excess luminance can be clipped, if needed. In contrast, with the image on the left, it may not be always possible to provide the required luminance to each color channel. Lai and Tsai (2008) built local histograms from the maximum RGB value of each segment to develop an LDR dual modulation algorithm and hence this step in the proposed method is conceptually similar to their method but is not the same in terms of implementation.

- The grayscale image generated in the previous step is further used to derive the signals that drive the LED and LCD panels. For this, the image needs to be downsampled to the resolution of the LEDs. However, as mentioned, this can reduce the maximum luminance in the resulting image. Therefore, in our method, we use an additional step of morphological dilation before downsampling. The size and shape of the structuring element depends on the structure of the backlight LED array. Such morphological expansion aims to keep only the maximum value of the illumination map, and so the LEDs will have a lighting value representing not the average of the image area but most of this area. Further, the maximum luminance of the small areas of the image will also be preserved.

- In the next step, the HDR luminance is split into two components on the basis of a power law. This allows flexibility in changing the way the luminance is split depending on the exponent (between 0 and 1). In other words, this allows better control of the distribution between the LEDs and the LCD panel. The default value of 0.5 (this corresponds to the square root as used by Seetzen et al. 2004) allows a balanced distribution. However, it may be advantageous to use a higher power (eg, 0.9) to maximize the use of the dynamics of the LCD panel. This has two important benefits: (1) reduced power backlight, since the LEDs are less stressed; (2) better color reproduction in dark areas. This is illustrated in Fig. 14.3 where Fig. 14.3A shows an HDR image (it is tone-mapped for better display) for which we generated two LED light maps: using exponent (power) values

(A)   (B)   (C)

FIGURE 14.3

Effect of the use of different exponent (power) values on the LED light maps: (A) original HDR image (tone-mapped for better display in print); (B) LED map generated with an exponent value of 0.3; (C) LED map generated with an exponent value of 0.9.

    of 0.3 and 0.9. The resulting LED maps are shown in Fig. 14.3A and B. Notice that the LED map generated with an exponent value of 0.9 is more contrasted, while it is more uniformly lit in the other case, and thus this LED map plays a smaller role in the contrast of the rendered image. In the limiting case of the exponent value being 0, the LED light map will be fully uniform (ie, uniform backlight without any local support). The use of higher exponent, which favors the LCD panel, however, requires a more precise modeling of the response of LEDs. This is because in areas where there are big changes in the local luminance values, use of a high value for the exponent would exaggerate the contrast on both the LCD panel image and the backlight. So any error, for example, in the PSF model will lead to larger reconstruction errors (ie, more artifacts in the rendered HDR image).

- In the next step, the LED map is normalized so that the (target) maximum value of the LCD panel is 255, in order to maximize the use of the dynamic range of the LCD panel. Obviously, this will improve the contrast of the LCD image. An illustration is provided in Fig. 14.4, which shows the images that will drive the LCD panel with and without the said normalization. For the image on the left, we did not normalize the LED map and so the contrast is lower. On the other hand, to obtain

(A)   (B)

FIGURE 14.4

Effect of the normalization of LED light maps on the LCD images: (A) LCD image without LED map normalization; (B) LCD image generated after the LED map is normalized. Because of normalization, the image in (B) uses the entire bit depth (in this case eight bits).

the image on the right, we first normalized the LED maps so that the maximum LCD value is 255, leading to better contrast. We found that such LED map normalization is particularly beneficial in rendering HDR content which is darker (low luminance).

- We then downsample the image to fit the resolution of the LEDs. This is done keeping in mind the position of the LEDs in the display. Note that the LEDs are typically arranged in hexagonal grids rather than as a square grid. The resultant downsampled LED map is further postprocessed in order to round off the values, so values greater than 1 are set to 1, while any zero value is replaced (which could cause division problems during the last steps of the algorithm).

- After we obtain the low-resolution image in the previous step, we generate a low-resolution version of the full-resolution PSF of the LEDs. Then, the LED light map is obtained via optimization. Specifically, the optimization aims to minimize the difference between the theoretical LED light map and the computed light map obtained by considering the PSF of the LEDs. We use a gradient-based optimization. At each iteration, a difference map is computed and a scaled version of this difference map is used to update the computed LED map. The said scaling is based on the idea of applying different penalties for values lower or higher than the target. This helps to avoid underlightening which in most cases cannot be compensated with an LCD panel. Another strategy that we adopted to compute the LED map is error diffusion in order to simulate nonlocal error correction since error at one point will have an impact on neighboring values. Toward this end, we filtered the error (or difference) map with a Gaussian whose parameters can be set to default values or customized. This also helps to account for the smoothness constraint, which is especially relevant to avoid flickering issues with HDR videos. The said optimization is terminated either when the number of iteration exceeds the predefined limit or when a local minimum is reached. In the proposed algorithm the optimization is done at the LED level. This is in contrast to a few algorithms for LDR dual modulation which seek to minimize the error between the finally rendered image and the target image. While the latter approach may be more accurate theoretically in terms of rendering, our approach based on LED optimization is much simpler, faster, and provides acceptable quality of rendering.

- The computed/estimated LED light map values are then rounded off to the nearest integer in the range from 0 to 255 (eight-bit LED panel). Next, a full resolution light map is generated with the computed LED map, the position of the LEDs in the display, and the PSF.

- The LCD values are generated by division of the original (saturated to maximum screen luminance) HDR image by the full resolution LED light map (this is similar to what is done in the baseline method).

- In the last step, the LCD panel values are corrected for both color and gamma. Typically a gamma value of 2.2 is used, but other values can be used (Poynton, 1996). Finally, the resulting values are rounded (any values greater than 255 are clipped to 255) off in the range from 0 to 255 (eight-bit LCD panel).

The above-mentioned steps apply for HDR image rendering only. For videos, we must consider temporal coherency issues and hence some of the steps might need to be either modified or disabled in order to avoid flickering. Such artifacts (flashes) are caused by excessive fluctuations in the value of the LEDs between consecutive frames of a video. Although these illumination differences are theoretically offset by the LCD panel, it appears that the synchronization between the LEDs and the LCD panel is not precise enough. It was therefore necessary to adapt the algorithm to obtain more homogeneous

LED values (but less optimal from the point of view of contrast and power consumption). In other words, the method described for HDR images applies too "aggressive" compensation. Particularly, the morphological expansion used in the third step may result in a large change in the backlight (as mentioned, one of the reasons for this is related to the response times of the LEDs and the LCD, which are not the same). This has the effect of introducing the appearance of flashes (flickering). which are obviously unwanted (a visual example is provided in Fig. 14.9). Thus, this step may need to be disabled for HDR video rendering. Notice that to split the luminance, we used a more general power law instead of the square root. This is beneficial for image rendering because higher powers can be used to reduce power consumption and improve the color rendering (through better use of the dynamics of the LCD panel). However, for video, this can lead to local luminance incoherency between successive frames. Thus, for HDR video, the square root (power 0.5) is more suitable to achieve the initial distribution of the luminance between the LCD panel and the LEDs since it can limit the variation in the distribution between two neighboring frames of a video. The third adaptation for video is to disable the normalization of LCD panel values (to cover the whole dynamic of values). In the case of images, this standardization allows better contrast and better color rendering. However, in rare cases, it can cause global variations in the intensity of the LEDs, leading to the sensation of a flash of light during the transition from one frame to another. Finally, the LED values should be filtered during the optimization in order to limit the variations of the LEDs in successive video frames, thereby avoiding unwanted flashes.

We have described a dual modulation algorithm for HDR images and videos. While there are some restrictions (eg, optimization is based on LED maps), the method is fast and simple to implement, and the rendered HDR content has higher contrast in comparison with that obtained with default display algorithms. We provide some experimental (numerical) evidence to support this, in the next section. First, however, we provide some qualitative results of the proposed method. Fig. 14.5 shows the differences between the two strategies for dual modulation that we adopt in our method for image and video signals. There are three subfigures, for different HDR source contents[1] (content 1, content 2, and content 3). As indicated in Fig. 14.5, the left images correspond to the optimization for images, while the ones on the right correspond to the case of videos. It can be seen that in the case of the images, the LED maps capture the fine and precise luminance variations of the original HDR image better. Further, the corresponding LCD panel has a much better resolution than LEDs. This helps to attain a good trade-off toward capturing local luminance variations in the rendered HDR image. For instance, in Fig. 14.5B one can see when the proposed method is optimized for images, it produces an LED light map which is aggressive and captures brighter areas in the foreground. Likewise, from Fig. 14.5C, we find that the high luminance due to street lights in the background is well represented in the corresponding LED map that has been optimized for images. The case of videos is, however, different, and we find that the proposed method optimized for videos is less aggressive. As mentioned, this is done so that the luminance coherence is maintained and flashes (flickering) are avoided. Thus, in comparison with the case for images, the LED light maps for videos are smoother because the aim is to avoid large local luminance changes. Since we obtain both the LED and the LCD maps, it is possible to combine these two and reconstruct what will actually be displayed on the screen. Thus, we can

---

[1] These HDR video sequences were shot and made available as part of the NEVEx project FUI11, related to HDR video chain study.

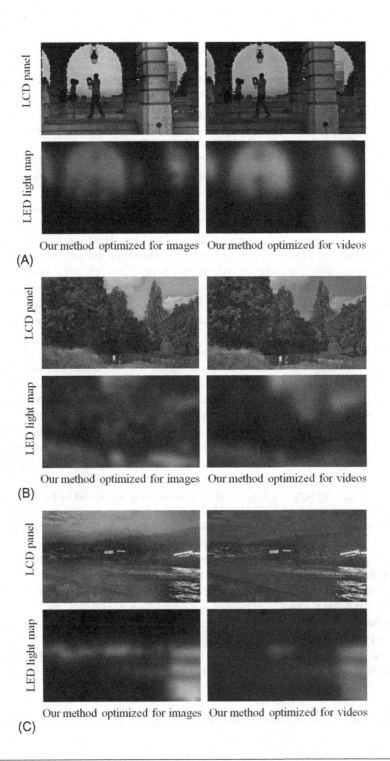

FIGURE 14.5

LCD panel and LED light maps obtained from the proposed method optimized for the two cases: images and videos. (A) Source content 1; (B) Source content 2; (C) Source content 3.

Table 14.1 Error in Rendering Due to Two Strategies in the Proposed Method (One Optimized for Images and the Other Optimized for Videos) and the Corresponding Power Consumption

	MSE (Optimized for Images)	Power Consumption (W)	MSE (Optimized for Videos)	Power Consumption (W)
Content 1	0.74%	235.90	0.34%	330.51
Content 2	0.60%	349.56	1.01%	326.21
Content 3	1.90%	256.86	2.89%	250.74

compute the error in luminance rendering. The said error can be quantified through a mean square error (MSE), which is presented in Table 14.1 for the three HDR contents shown in Fig. 14.5. Table 14.1 shows that the algorithm dedicated to videos usually generates larger errors, mainly due to constraints on the values of the LEDs. We have also presented in Table 14.1 the power consumption (in watts), from which one can easily infer that larger error will lead to lower power consumption and vice versa. The reason is that a more accurate luminance rendering (ie, smaller error) will require the turning on of more LEDs to light up localized bright areas.

Therefore, the proposed dual modulation method optimized for videos is expected to result in bigger errors in rendering high local luminance as compared with the case of images. Finally, we show in Fig. 14.6 the local errors in luminance rendering from the proposed method when optimized for images and videos for content 3, for which the LCD and LED maps are shown in Fig. 14.5C. As can be seen, the method is generally accurate for lower luminance values but results in higher errors especially where there is very high local luminance. It can also be seen that the proposed method optimized for images leads to smaller error (up to 30%). As opposed to this, the method optimized for video can lead to larger local errors (up to 45% error as compared with the actual luminance). There are, of course, two reasons that can, in part, account for such errors. The first one is practical and arises because we

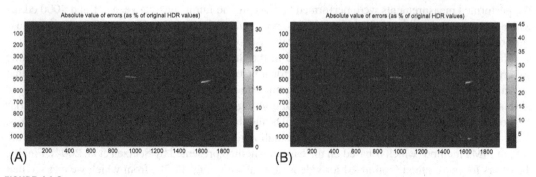

FIGURE 14.6

Error in luminance rendering for the two strategies in the proposed method: (A) error in luminance rendering when the method is optimized for images; (B) error in luminance rendering when the method is optimized for videos. Notice that the error is high in areas that are locally bright. Also observe that optimization for images leads to smaller errors in comparison with those for video.

deliberately allow errors in rendering high luminance over small (localized) areas primarily to maintain global luminance coherence in the rendered HDR video. Thus, we can reduce some of those errors further by more "aggressive" LED compensation at the cost of flickering artifacts. The second reason is theoretical and is due to the PSF of the LEDs. Specifically, the said errors are usually due to a very high brightness requested on a very small area. Because of the (relatively) low power and relatively high spread of the PSF of the LEDs, it is not physically possible to render such high luminance without turning on a large number of LEDs (this in turn will introduce significant errors in darker areas around brighter areas). Thus, a trade-off is what we aim to achieve in the proposed method especially when we are rendering HDR videos.

14.4 ASSESSING THE PERFORMANCE OF A DUAL MODULATION ALGORITHM

Measuring the accuracy of a dual modulation algorithm is nontrivial and requires accurate luminance measurements. This is because the luminance of a point (pixel) on the screen depends both on the value of the pixel on the LCD panel at that location and on the value of the LED at this location. It also depends on the supply of light provided by LEDs that are in the vicinity. To that end, we conducted an experiment in which we measured the error between the actual luminance and the rendered luminance. This provides a scientific way of calibrating and benchmarking the performance of dual modulation methods. To keep the study tractable and possibly eliminate other factors, we chose to display a white square block against a dark background on the HDR display (we used a SIM2 Solar47 HDR display, which has a maximum displayable luminance of 4000 cd/m^2). The idea was to measure the rendered luminance and compare it with the actual luminance of the said block. To account for the fact that HDR content can have local and global dark (or bright) areas, we varied the size of the block such that it fitted only a given percentage of the display height. The width of the block was kept constant. More specifically, the height of the block was varied from 10% to 100% of the display height. The error between the rendered luminance and the actual luminance was measured for each of the covered areas. We performed measurements were performed by varying the luminance between 0.1 and 4000 cd/m^2 with a step factor of $10^{0.2} = 1.584$. However, the practical minimum luminance was 0.4 cd/m^2 because our preliminary tests showed that the probe that we used was not accurate below this value.

The measured errors for the default method in the SIM2 display are shown in Fig. 14.7A, while the errors for the proposed method (optimized for images) are shown in Fig. 14.7B. We can see that our method reduces the error in the rendered luminance especially when the size of the square block is small as compared with the display height and at higher luminance values. This implies that our method allows better compensation between LEDs and the LCD panel at a local level. In other words, it can help to produce very bright areas over a small region, which is more difficult to achieve by the default SIM2 method (which is based on the baseline method presented by Seetzen et al. 2004). Finally, the errors for our method (optimized for videos) are shown in Fig. 14.7C, from which we can see that they are bigger than those for images (the reasons for this have already been discussed). Nevertheless, our method optimized for videos still leads to smaller errors than the default SIM2 method. Thus, our method can lead to a significant reduction in rendering error especially when there are localized high luminance areas in the HDR content. Obviously, this will help to exploit the real benefits of HDR via rendering higher contrast content.

(A)

	10%	20%	30%	40%	50%	60%	70%	80%	90%	100%
0.40		63%	5%	36%	51%	51%	23%	17%	71%	20%
0.63	77%	43%	58%	25%	17%	4%	19%	19%	45%	8%
1.00	72%	28%	31%	44%	39%	5%	20%	11%	21%	17%
1.58	55%	41%	46%	21%	16%	4%	24%	18%	7%	13%
2.51	63%	40%	12%	16%	11%	5%	1%	8%	7%	3%
3.98	64%	35%	18%	10%	9%	6%	4%	2%	12%	2%
6.31	56%	25%	21%	14%	5%	4%	3%	1%	4%	4%
10.00	50%	28%	24%	16%	8%	4%	1%	2%	3%	1%
15.85	44%	30%	24%	16%	12%	9%	6%	2%	2%	1%
25.12	38%	30%	23%	18%	11%	8%	5%	2%	0%	0%
39.81	35%	28%	22%	18%	12%	8%	5%	2%	1%	0%
63.10	34%	27%	23%	17%	12%	8%	4%	3%	1%	0%
100.00	33%	25%	21%	16%	10%	6%	3%	1%	0%	2%
158.49	33%	25%	20%	15%	9%	5%	2%	1%	2%	3%
251.19	33%	23%	18%	14%	8%	4%	1%	1%	2%	3%
398.11	33%	21%	17%	12%	6%	2%	1%	3%	4%	5%
630.96	33%	21%	16%	11%	5%	1%	1%	3%	4%	2%
1000.00	33%	20%	16%	10%	4%	0%	2%	4%	4%	21%
1584.89	35%	22%	16%	11%	6%	2%	0%	1%	9%	24%
2511.89	45%	25%	20%	14%	8%	5%	3%	2%	13%	28%
3981.07	65%	38%	24%	17%	12%	9%	7%	6%	17%	31%

(B)

	10%	20%	30%	40%	50%	60%	70%	80%	90%	100%
0.40	44%	42%	55%	29%	48%	53%	24%	7%	3%	3%
0.63	2%	32%	8%	12%	19%	9%	34%	30%	5%	26%
1.00	13%	8%	1%	5%	9%	7%	17%	4%	20%	11%
1.58	1%	1%	12%	8%	20%	7%	27%	2%	9%	0%
2.51	18%	4%	21%	18%	5%	8%	10%	16%	14%	8%
3.98	18%	14%	19%	20%	13%	15%	12%	9%	8%	14%
6.31	25%	16%	16%	19%	16%	13%	11%	10%	11%	7%
10.00	19%	13%	17%	18%	17%	13%	12%	8%	10%	6%
15.85	19%	16%	18%	18%	16%	12%	8%	10%	6%	8%
25.12	17%	13%	17%	17%	16%	12%	8%	7%	6%	6%
39.81	16%	13%	15%	16%	13%	9%	7%	4%	3%	3%
63.10	14%	11%	14%	13%	11%	6%	3%	2%	0%	2%
100.00	13%	10%	12%	12%	9%	6%	1%	0%	1%	1%
158.49	13%	10%	10%	10%	8%	4%	0%	1%	3%	3%
251.19	12%	8%	9%	9%	7%	3%	2%	3%	4%	5%
398.11	11%	8%	8%	9%	6%	2%	2%	5%	5%	5%
630.96	11%	7%	8%	8%	7%	2%	3%	5%	6%	5%
1000.00	9%	6%	7%	7%	6%	1%	3%	5%	6%	5%
1584.89	7%	5%	5%	6%	4%	0%	4%	6%	6%	5%
2511.89	9%	4%	5%	5%	4%	0%	4%	5%	6%	4%
3981.07	30%	9%	5%	6%	6%	4%	0%	1%	11%	20%

(C)

	10%	20%	30%	40%	50%	60%	70%	80%	90%	100%
1.00	2%	30%	23%	51%	48%	58%	49%	85%	96%	81%
1.58	18%	33%	7%	7%	35%	37%	32%	63%	46%	44%
2.51	21%	31%	7%	3%	0%	22%	19%	27%	26%	25%
3.98	24%	23%	9%	8%	5%	3%	12%	7%	12%	12%
6.31	25%	21%	18%	16%	8%	4%	0%	1%	0%	3%
10.00	27%	25%	22%	19%	20%	10%	10%	6%	7%	8%
15.85	25%	25%	22%	20%	18%	13%	12%	11%	8%	11%
25.12	24%	25%	23%	23%	19%	15%	13%	12%	11%	14%
39.81	24%	24%	22%	21%	19%	18%	14%	12%	12%	13%
63.10	25%	22%	22%	21%	19%	14%	13%	11%	11%	12%
100.00	20%	21%	20%	20%	17%	15%	13%	11%	11%	11%
158.49	19%	20%	19%	19%	17%	14%	11%	10%	9%	12%
251.19	17%	19%	19%	18%	17%	13%	11%	8%	9%	11%
398.11	16%	17%	18%	18%	16%	12%	9%	8%	8%	10%
630.96	15%	16%	17%	17%	15%	11%	9%	7%	7%	9%
1000.00	10%	13%	14%	14%	13%	10%	8%	7%	8%	9%
1584.89	8%	12%	12%	13%	12%	9%	7%	7%	7%	8%
2511.89	13%	10%	12%	12%	11%	9%	7%	6%	15%	25%
3981.07	42%	21%	10%	11%	10%	8%	7%	13%	31%	39%

FIGURE 14.7

Estimated error (%) between the actual luminance and the rendered luminance for the default HDR display mode and the two optimized versions of the proposed method: (A) default SIM2 mode; (B) proposed method (optimized for images); (C) proposed method (optimized for videos). The first row indicates the percentage of the vertical screen height occupied by the square block. The first column lists the approximate targeted luminance values (cd/m^2). Green and reddish colors indicate smaller and bigger errors, respectively (for black and white print version of this paper, please refer to the numbers only).

14.5 SOME PRACTICAL LESSONS FOR HDR CONTENT RENDERING

So far, we have focused on describing and verifying the performance of the proposed dual modulation approach for HDR content rendering. In that context, it will be useful to outline a few practical aspects that could be beneficial for HDR content rendering. We list them in the following in no particular order of importance.

- **HDR content adaptation**: First, in HDR imaging one usually deals with proportional (and not absolute) luminance values. More specifically, unless there is a prior and accurate camera calibration, the luminance values in an HDR video file represent the real-world luminance up to an unknown scale.[2] This, nonetheless, is sufficient for most purposes. Second, the HDR displays currently available cannot display luminance beyond the specified limit, given the hardware limitations. Therefore, the terms low dynamic range (LDR) and high dynamic range (HDR) are also sometimes referred to, respectively, as lower or standard dynamic range and higher dynamic range (to explicitly indicate that the range captured is only relatively higher than LDR but not the entire dynamic range present in a real scene). Although as a matter of convention we do away with such precise distinctions, it does highlight the fact that HDR content needs to be "adapted" according to display luminance limits before the dual modulation algorithm is used.

- **Corrections**: We also found that color and gamma correction played an important role in the quality of the rendered HDR content. Therefore, these should be carefully considered and implemented on the basis of the display characteristics and possibly the HDR content to be rendered. In other words, there is a need to perform calibration measures of the target display in order to apply proper corrections. If these are not done or are wrongly implemented, they can severely affect the rendering. This, in turn, may lead to loss of artistic intention, whose preservation is one of the key goals in HDR imaging (the keen reader is encouraged to refer to Chapter 17 for a more elaborate treatment of this topic).

- **Importance of smooth PSF modeling of LEDs**: In our experiments, we found that the PSF should exhibit smooth behavior and there should not be a discontinuity at the borders. Thus, we approximated it via a Gaussian mixture model, whose parameters we determined by fitting it to the real PSF values. The aim was to obtain a smooth (0 at the borders) model, otherwise the final division of HDR content by the light map can introduce visible artifacts. This is illustrated in Fig. 14.8, which shows the LCD images in two cases: one with a nonsmooth PSF (discontinuity when local bright and dark regions are located close to each other), and the other with a smoother PSF (modeled with a Gaussian mixture model). We have also highlighted the region (marked with red box; dark gray in print versions) where the nonsmooth PSF can cause distortions due to a discontinuity. In contrast, a smoother PSF can avoid such errors.

- **Flickering issues for videos**: Although we have already discussed this issue in detail, we believe it deserves to be reiterated as it is important in the context of HDR video rendering (consumer applications of HDR will typically involve video and not images). The underlying cause of such flickering in video that we observed is due to unequal response times of LEDs and the LCD.

---

[2]Even with calibration, the HDR values represent real physical luminance with certain error. This is because the camera spectral sensitivity functions which relate scene radiance with captured RGB triplets cannot match the luminous efficiency function of the human visual system.

(A)   (B)

FIGURE 14.8

Effect of the PSF on the LCD image: (A) LCD image with a nonsmooth PSF; (B) LCD image with a smooth PSF based on a Gaussian mixture model. Blockiness-like artifacts are visible in the image in (A) since the PSF is not smooth at the borders.

As a consequence, video signal rendering could be from unsynchronized sources, leading to luminance incoherency in successive frames. It follows that we cannot allow big changes in the way we split luminance information (thus we used 0.5 as the power value for equal splitting). As noted before, this can cause larger errors in rendering localized high luminance areas but prevents flashes during viewing of the rendered HDR video. The reader will also recall that we disabled and modified a few steps in the proposed method when optimizing it for videos to prevent flickering. To help the reader further visualize the impact of the unequal response time of LEDs and the LCD, Fig. 14.9 shows the LED light maps obtained for two consecutive video frames (of source content 1). The LED maps shown in Fig. 14.9A and B were obtained by our method optimized for images, while those in Fig. 14.9C and D correspond to the setting when our method was optimized for videos. We have also highlighted (through red boxes; dark gray in print versions) two areas where sharp changes in the LED maps can be noticed even though they correspond to two consecutive video frames (and hence the transition should be smoother). Such localized and sharp changes can introduce flashing when the rendered video is viewed on the HDR display. In contrast, the LED maps shown in Fig. 14.9C and D do not exhibit such incoherent changes between successive frames and hence avoid flickering in the rendered HDR video.

- **Predicting display behavior**: Backlight dimming algorithms can be viewed as a sort of (simple) screen modeling algorithms. For instance, in the proposed method we can reconstruct the rendered signal by using the LED light map and LCD-compensated pixel values. Thus, as demonstrated in this chapter, we can predict local errors or luminance fidelity. This can be helpful for further optimization or modifications that can improve rendering (this will obviously increase the complexity and execution time of the resultant algorithm). Another important advantage of a dual modulation method such as ours is that one can predict the power consumption of the display for a particular content. This can, in turn, allow display optimization from the view point of energy saving, which is a important dimension in today's display systems in the light of recent enforcement of regulations limiting the allowed power consumption for television sets in several countries (European Commission, 2009).

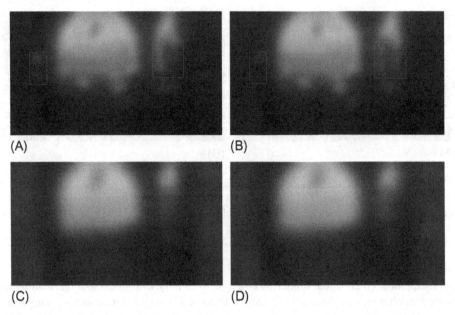

FIGURE 14.9

Illustration of flickering issues in HDR video rendering. LED maps from two consecutive frames for source content 3 are shown. We obtained those in (A) and (B) by optimizing the proposed method for images, while we generated those in (C) and (D) by optimizing the method for videos. The red (dark gray in print versions) boxes in the LED maps in (A) and (B) highlight the areas where sudden changes occur in the backlight (LED) intensity, leading to flickering. The LED maps in (C) and (D) are smoother.

- **Display limitations**: It is worth pointing out that the current LED-based HDR screens are not perfect. This is primarily due to the low resolution of the backlight (LED array). This may limit the capabilities of displays to reproduce very bright or very dark scene details, although Seetzen et al. (2004) noted that is not always a problem. Regarding the LCD panel, full-high-definition resolution may be enough for currently available HDR content, but there is always room for improvement, especially in light of technologies such as ultra high definition. We found that global light leakage can be more important than what could be expected from a single PSF of the LEDs (due to possible interaction of LEDs).
- **Step toward backward compatibility**: The HDR dual modulation algorithm results in two signals: the LED map and the LCD panel pixel information. The LCD image can be considered as a tone-mapped version of the HDR content. Thus, it can be compressed by any legacy coder and the LED map could be transmitted in a lossless fashion. This can provide an alternative approach to HDR video compression. Of course, this approach can be limited by the fact that the LED map and the LCD image are typically display dependent (eg, PSF of the LEDs).

14.6 CONCLUDING REMARKS AND PERSPECTIVES

HDR content visualization is nontrivial (unlike the case of LDR content visualization, which requires minimal or no processing of the content to be displayed). The reason is that the current HDR display technology cannot render the entire luminance range present in the real world. As a result, HDR content needs to be properly adapted before it is rendered. Such rendering requires care since it can affect the level of detail rendered as well as the immersive experience. One approach to HDR content rendering is dual modulation. As the name implies, dual modulation allows the separation of the HDR content into an LED map and an LCD image. We have described one such method that we developed for HDR content rendering. This was based on our addressing some of the issues with the existing HDR baseline method. In particular, we used the generalized power law for luminance splitting, which allows greater flexibility in terms of the adjustment of the luminance split depending on the type of content to be rendered. This lead to two versions of the proposed method: one optimized for images and the other optimized for videos. The latter version is necessary to avoid flickering issues in HDR video rendering. We also compared the rendering error due to the proposed method with that due to the default method in a SIM2 display and found that it reduces the error both for images and for videos. We also outlined a few practical lessons that would be useful for HDR content rendering.

REFERENCES

Albrecht, M., Karrenbauer, A., Jung, T., Xu, C., 2010. Sorted sector covering combined with image condensation: an efficient method for local dimming of direct-lit and edge-lit LCDs. IEICE Trans. Electron. E93-C (11), 1556–1563.

Anandan, M., 2008. Progress of led backlights for LCDs. J. Soc. Inform. Disp. 16 (2), 287–310.

Burini, N., Nadernejad, E., Korhonen, J., Forchhammer, S., Wu, X., 2012. Image dependent energy-constrained local backlight dimming. In: 2012 19th IEEE International Conference on Image Processing (ICIP), pp. 2797–2800.

Burini, N., Nadernejad, E., Korhonen, J., Forchhammer, S., Wu, X., 2013. Modeling power-constrained optimal backlight dimming for color displays. J. Display Technol. 9 (8), 656–665.

Chen, H., Sung, J., Ha, T., Park, Y., 2007. Locally pixel-compensated backlight dimming on led-backlit LCD TV. J. Soc. Inform. Disp. 15 (12), 981–988.

Cho, H., Kwon, O.K., 2009. A backlight dimming algorithm for low power and high image quality LCD applications. IEEE Trans. Consum. Electron. 55 (2), 839–844.

Cho, S.I., Kim, H.S., Kim, Y.H., 2011. Two-step local dimming for image quality preservation in LCD displays. In: 2011 International SoC Design Conference (ISOCC), pp. 274–277.

Cho, H., Cho, B.C., Hong, H.J., Oh, E.Y., Kwon, O.K., 2013. A color local dimming algorithm for liquid crystals displays using color light emitting diode backlight systems. Opt. Laser Technol. 47 (0), 80–87.

European Commission, 2009. Commission regulation (EC) No 642/2009 of 22 July 2009 implementing Directive 2005/32/ec of the European Parliament and of the Council with regard to ecodesign requirements for televisions. Official J. Eur. Union L191, 42–52.

Funamoto, T., Kobayashi, T., Murao, T., 2001. High-Picture-Quality Technique for LCD televisions: LCD-AI. In: Proceedings of the International Display Workshop, pp. 1157–1158.

Hong, J.J., Kim, S.E., Song, W.J., 2010. A clipping reduction algorithm using backlight luminance compensation for local dimming liquid crystal displays. IEEE Trans. Consum. Electron. 56 (1), 240–246.

Kang, S.J., Kim, Y.H., 2009. Image integrity-based gray-level error control for low power liquid crystal displays. IEEE Trans. Consum. Electron. 55 (4), 2401–2406.

Kang, S.J., Kim, Y.H., 2011. Multi-histogram-based backlight dimming for low power liquid crystal displays. J. Display Technol. 7 (10), 544–549.

Kim, S.E., An, J.Y., Hong, J.J., Lee, T.W., Kim, C.G., Song, W.J., 2009. How to reduce light leakage and clipping in local-dimming liquid-crystal displays. J. Soc. Inform. Disp. 17 (12), 1051–1057.

Lai, C.C., Tsai, C.C., 2008. Backlight power reduction and image contrast enhancement using adaptive dimming for global backlight applications. IEEE Trans. Consum. Electron. 54 (2), 669–674.

Lin, F.C., Huang, Y.P., Liao, L.Y., Liao, C.Y., Shieh, H.P.D., Wang, T.M., Yeh, S.C., 2008. Dynamic backlight gamma on high dynamic range LCD TVs. J. Display Technol. 4 (2), 139–146.

Nadernejad, E., Burini, N., Korhonen, J., Forchhammer, S., Mantel, C., 2013. Adaptive local backlight dimming algorithm based on local histogram and image characteristics. In: Proc. SPIE 8652, Color Imaging XVIII: Displaying, Processing, Hardcopy, and Applications, 86520V.

Nam, H., 2011. Low power active dimming liquid crystal display with high resolution backlight. Electron. Lett. 47 (9), 538–540.

Poynton, C.A., 1996. A Technical Introduction to Digital Video. John Wiley & Sons, Inc., New York, NY, USA.

Seetzen, H., Whitehead, L.A., Ward, G., 2003. 54.2: a high dynamic range display using low and high resolution modulators. SID Symp. Dig. Tech. Pap. 34 (1), 1450–1453.

Seetzen, H., Heidrich, W., Stuerzlinger, W., Ward, G., Whitehead, L., Trentacoste, M., Ghosh, A., Vorozcovs, A., 2004. High dynamic range display systems. ACM Trans. Graph. 23 (3), 760–768.

Shu, X., Wu, X., Forchhammer, S., 2013. Optimal local dimming for LC image formation with controllable backlighting. IEEE Trans. Image Process. 22 (1), 166–173.

Zhang, X.B., Wang, R., Dong, D., Han, J.H., Wu, H.X., 2012. Dynamic backlight adaptation based on the details of image for liquid crystal displays. J. Display Technol. 8 (2), 108–111.

PERCEPTION AND QUALITY OF EXPERIENCE

PERCEPTUAL DESIGN FOR HIGH DYNAMIC RANGE SYSTEMS

15

T. Kunkel, S. Daly, S. Miller, J. Froehlich

Dolby Laboratories, Inc., San Francisco, CA, United States

CHAPTER OUTLINE

15.1 INTRODUCTION

Traditional imaging, which includes image capture, processing, storage, transport, manipulation, and display, is limited to a dynamic range lower than the human visual system (HVS) is capable of perceiving. This is due to factors such as the historic evolution of imaging systems and technical and economic limitations (computational power, cost, manufacturing processes, etc.). In addition to the luminance dynamic range limitations, traditional imaging is also overly constrained in its extent of color rendering capabilities compared with what is perceivable.

High Dynamic Range Video. http://dx.doi.org/10.1016/B978-0-08-100412-8.00015-2

Traditional imaging achieved maximum dynamic ranges of three orders of magnitude (OoM), or density ranges of 3.0, or contrast ratios (CRs) of 1000:1, to use terminology from the scientific, hard copy, and direct view display applications, respectively. Ranges between 1.5 and 2.0 OoM were more common. Over time, the limitations of traditional image processing became apparent. In the late 1980s to early 1990s, investigations into image processing beyond the capabilities of the traditional lower dynamic range intensified, beyond the usual incremental improvements of the annual slight increase of film stock density or display contrast. Those approaches and techniques are generally collected under the term "high dynamic range (HDR) imaging." Theoretically, such HDR systems aim to describe any luminance from absolute zero, having no photons, up to infinite luminance levels, which in turn leads to theoretically infinite contrast. For practical reasons, even with HDR systems, the luminance levels are limited, usually by the encoding approach of digital systems. Nevertheless, this still means that HDR imaging systems can be designed to describe extreme luminance levels that are, for example, encountered in astronomy. Even though this sounds enticing, the capture, processing, and especially display of excess dynamic range that cannot be utilized by the HVS is both computationally and economically counterproductive.

This leads to the important question of how much dynamic range is required in HDR imaging systems to neither encode more than can be utilized by human perception nor to deliver insufficient contrast to the HVS, thereby having an impact on image fidelity. Ideally, we want to handle just the right amount of dynamic range to satisfy the viewer's preference, presumably by convincing the HVS that its perceiving contrast levels that are comparable with those in the real world (or a world created by a visual artist).[1]

After an attempt to cover the dynamic range of the HVS, another important element is to model the smallest dynamic range interval the HVS is able distinguish before two levels of gray appear to be identical, or any possible luminance modulation is detectable. This interval is called a "just noticeable difference" (JND). As this interval is not constant from dark to bright light levels, the response curve of the HVS needs to be taken into account as well.

The concept of small dynamic range intervals is crucially important in digital systems, because of quantization. It determines how many unique intervals can be addressed between the darkest and lightest extremes — black and white (eg, 256 unique intervals in an 8-bit integer system). The distribution of those intervals is called the "quantization curve," and is related to the response curve of the HVS.[2] Again, as with the dynamic range described above, the design goal of an imaging system is neither to assign too many quantization steps indistinguishable by the HVS nor to assign too few. A lack leads to quantization steps that are too large, manifesting themselves as contouring or banding artifacts. If the quantization curve does not complement the response curve of the HVS, images may contain both contouring due to few quantization steps in some regions and waste through overquantization in others. Both of these effects can happen in the same image. Therefore, a well-designed imaging system offers quantization intervals and a tone curve that are both well balanced.

Having an accurate complement between the imaging system and HVS along the intensity axis is an integral aspect for image fidelity. Table 15.1 summarizes the fundamental aspects of an HDR system

---

[1] As image capture is discussed in other chapters, this chapter will concentrate on the signal representation and display aspects of HDR ecosystems.

[2] The quantization distribution at the display is usually inversely related to the HVS response curve. For full system design, concepts such as optical to electrical transfer function (OETF), electrical to optical transfer function (EOTF), and optical to optical transfer function (OOTF) are useful, but are not discussed here because of the scope of this chapter.

Table 15.1 The Key Elements of Human Visual Perception in Relation to HDR Imaging

HVS	Imaging System	Description
Perceived black	Minimum luminance	Darkest absolute value possible across all scenes
Perceived highlight	Maximum luminance	Lightest absolute value possible across all scenes
Contrast	Dynamic range	Ratio between minimum and maximum luminance
JND	Quantization step	Interval between two consecutive luminance or color values
Response curve	Quantization curve	Distribution of quantization steps over the full dynamic range (eg, a digital gamma curve)

from an achromatic point of view and brings it into context of the HVS. Nevertheless, imaging is rarely achromatic, and therefore color imaging is an equally important aspect to understand and model accurately. To extend the HDR concepts to color imaging, additional dimensions next to the dynamic range have to be considered, such as color gamut and color volume.

Extending imaging to color adds complexity to a system, especially when one is attempting to efficiently match the capabilities of human color perception. Due to historic reasons, colors visible to the HVS portrayed as 2D color wheels. Ignoring intensity simplified the visualization of the concept of color. The colors an imaging system can represent are defined in a 2D space called a chromaticity or color gamut.

However, our everyday color perception does not fully disentangle intensity from chromaticity. We perceive all hues and levels of saturation in combination with all the shades of intensity visible to the HVS. Therefore, to design a highly performing yet efficient system, it is essential to be able to model human color perception both accurately and efficiently. Thus, it is important to address the intensity dimension in conjunction with chromaticity, which leads to a color volume, of which the intensity dynamic range is a subset. Similarly to finding the "perfect" dynamic range for the HVS, it is desirable to find the right color volume to avoid underrepresenting or overrepresenting colors. Table 15.2 extends Table 15.1 and summarizes the fundamental aspects of an HDR system from a chromatic point of view.

In this chapter, we will identify the aspects of the HVS that determine the dynamic range, quantization, and tone curve requirements. These aspects are then extended into the full color volume, providing guidance for the design and development of both perceptually accurate and at the same time efficient imaging systems.

Table 15.2 The Key Components of Human Visual Perception When Color Is Added to HDR Imaging

HVS	Imaging System	Description
Visible color wheel	Chromaticity or color gamut	2D representation of color, usually ignoring intensity
All shades, hues, and levels of saturation visible to the HVS	Color volume	All colors visible to the HVS under the given environment. This representation is 3D because it includes intensity

15.2 LUMINANCE AND CONTRAST PERCEPTION OF THE HVS

With the advent of HDR capture, transport, and display devices, a range of fundamental questions have appeared regarding the design of those imaging pipeline elements. For instance, viewer preferences regarding overall luminance levels as well as contrast were determined previously (Seetzen et al., 2006), as were design criteria for displays used under low ambient light levels (Mantiuk et al., 2009). Further, recent studies have revealed viewer preferences for HDR imagery displayed on HDR displays (Akyüz et al., 2007), and have assessed video viewing preferences for HDR displays under different ambient illuminations (Rempel et al., 2009).

Especially in the context of display devices, it is important to know how high the dynamic range of a display should be. An implicit goal is that a good display should be able to accommodate a dynamic range that is somewhat larger than the simultaneous dynamic range afforded by human vision, possibly also taking into account the average room illumination as well as short-term temporal adaptive effects. More advanced engineering approaches would try to identify a peak dynamic range so that further physical improvements do not result in diminishing returns.

15.2.1 REAL-WORLD LUMINANCE AND THE HVS

The lower end of luminance levels in the real world is at 0 cd/m^2, which is the absence of any photons (see Fig. 15.1A). The top of the luminance range is open-ended but one of the everyday objects with the highest luminance levels is the sun disk, with approximately 1.6×10^9 cd/m^2 (Halstead, 1993).

A human can perceive approximately 14 $\log_{10}$ units,[3] by converting light incident on the eye into nerve impulses using photoreceptors (see Fig. 15.1B). These photoreceptors can be structurally and functionally divided into two broad categories, which are known as rods and cones, each having a different visual function (Fairchild, 2013; Hubel, 1995). Rod photoreceptors are extremely sensitive to light to facilitate vision in dark environments such as at night. The dynamic range over which the rods can operate ranges from 10^{-6} to 10 cd/m^2. This includes the "scotopic" range when cones are inactive, and the "mesopic" range when rods and cones are both active. The cone photoreceptors are less sensitive than the rods and operate under daylight conditions, forming photopic vision (in these luminance ranges, the rods are effectively saturated). The cones can be stimulated by light ranging from luminance levels of approximately 0.03 to 10^8 cd/m^2 (Ferwerda et al., 1996; Reinhard et al., 2010). The mesopic range extends from about 0.03 to 3 cd/m^2, above which the rods are inactive, and the range above 3 cd/m^2 is referred to as "photopic."

This extensive dynamic range facilitated by the photoreceptors allows us to see objects under faint starlight as well as scenery lit by the midday sun, under which some scene elements can reach luminance levels of millions of candelas per square meter. However, at any moment of time, human vision is able to operate over only a fraction of this enormous range. To reach the full 14 $\log_{10}$ units of dynamic range (approximately 46.5 f-stops), the HVS shifts this dynamic range subset to an appropriate

---

[3]In imaging technology, several logarithmic bases are used. Typically, luminance is provided as base 10 (vision, color, and photographic science), which is often expressed as orders of magnitude (OoM), $\log_{10}$, or simply log. In cinema engineering, f-stops are common, which are log base 2 (or $\log_2$). Hardcopy (print) has traditionally used density, which is $-\log_{10}$ of transmittance or reflectance. A rule of thumb for conversion from $\log_{10}$ to f-stops is to multiply the $\log_{10}$ value by 3.32 (as an approximation).

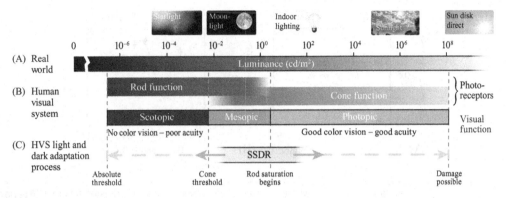

FIGURE 15.1

Real-world luminance levels and the high-level functionality of the HVS.

light sensitivity using various mechanical, photochemical, and neuronal adaptive processes (Ferwerda, 2001), so that under any lighting conditions the effectiveness of human vision is maximized.

Three sensitivity-regulating mechanisms are thought to be present in cones that facilitate this process — namely, response compression, cellular adaptation, and pigment bleaching (Valeton and van Norren, 1983). The first mechanism accounts for nonlinear range compression, therefore leading to the instantaneous nonlinearity, which is called the "simultaneous dynamic range" or "steady-state dynamic range" (SSDR). The latter two mechanisms are considered true adaptation mechanisms (Baylor et al., 1974). These true adaptation mechanisms can be classified into light adaptation and chromatic adaptation (Fairchild, 2013),[4] which enable the HVS to adjust to a wide range of illumination conditions. The concept of light adaptation is shown in Fig. 15.1C.

"Light adaptation" refers to the ability of the HVS to adjust its visual sensitivity to the prevailing level of illumination so that it is capable of creating a meaningful visual response (eg, in a dark room or on a sunny day). It achieves this by changing the sensitivity of the photoreceptors as illustrated in Fig. 15.2A. Although light and dark adaptation seem to belong to the same adaptation mechanism, there are differences reflected by the time course of these two processes, which is on the order of 5 min for complete light adaptation. The time course for dark adaptation is twofold, leveling out for the cones after 10 min and reaching full dark adaptation after 30 min (Boynton and Whitten, 1970; Davson, 1990). Nevertheless, the adaptation to small changes of environment luminance levels occurs relatively fast (hundreds of milliseconds to seconds) and impact the perceived dynamic range in typical viewing environments.

Chromatic adaptation is another of the major adaptation processes (Fairchild, 2013). It uses physiological mechanisms similar to those used by light adaptation, but is capable of adjusting the sensitivity of each cone individually, which is illustrated in Fig. 15.2B (the peak of each of the responses

---

[4]Sometimes "dark adaptation" is also used when the direction of adaptation toward darker levels needs to be distinguished from that toward lighter levels.

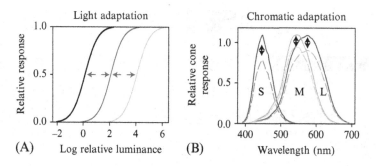

FIGURE 15.2

The concepts of light (and dark) adaptation where the gain of all photoreceptors is adjusted together (A) and chromatic adaptation where the sensitivity of each cone photoreceptor L, M, and S is adjusted individually (B).

can change individually). This process enables the HVS to adapt to various types of illumination so that, for example, white objects retain their white appearance.

Another light-controlling aspect is pupil dilation, which — similarly to an aperture of a camera — controls the amount of light entering the eye (Watson and Yellot, 2012). The effect of pupil dilation is small (approximately 1/16 times) compared with the much more impactful overall adaptation processes (approximately a million times), so it is often ignored in most engineering models of vision.[5] However, secondary effects, such as increased depth of focus and less glare with the decreased pupil size, are relevant for 3D displays and HDR displays, respectively.

15.2.2 STEADY-STATE DYNAMIC RANGE

Vision science has investigated the SSDR of the HVS through flash stimulus experiments, photoreceptor signal-to-noise ratio and bleaching assessment, and detection psychophysics studies (Baylor et al., 1974; Davson, 1990; Kunkel and Reinhard, 2010; Valeton and van Norren, 1983). The term "steady state" defines a very brief temporal interval that is usually much less than 500 ms, because there are some relatively fast components of light adaptation.

The earliest impact on the SSDR is caused by the optical properties of the eye: before light is transduced into nerve impulses by the photoreceptors in the retina and particularly the fovea, it enters the HVS at the cornea and then continues to be transported through several optical elements of the eye, such as the aqueous humor, lens, and vitreous humor. This involves changes of refractive indices and transmissivity, which in turn leads to absorption and scattering processes (Hubel, 1995; Wandell, 1995). On the photoreceptor level, the nonlinear response function of a cone cell can be described by the Naka-Rushton equation (Naka and Rushton, 1966; Peirce, 2007), which was developed from measurements in fish retinas. It was, in turn, modeled on the basis of the Michaelis-Menten equation

---

[5]For example, collected psychophysical data have the effect of pupil dilation already built in to the data (unless an artificial pupil is used).

(Michaelis and Menten, 1913) for enzyme kinetics, giving rise to a sigmoidal response function for the photoreceptors, which can be described by

$$\frac{V}{V_{\max}} = \frac{L^n}{L^n + \sigma^n},$$

where V is the signal response, $V_{\max}$ is the maximum signal response, L is the input luminance, σ is the semisaturation constant, and the exponent n influences the steepness of the slope of the sigmoidal function.[6] It was later found to also describe the behavior of many other animal photoreceptors, including those of monkeys and humans. Later work (Normann and Baxter, 1983) elaborated on the parameter σ in the denominator on the basis of the eye's point spread function and eye movements, and used the model to fit the results of various disk detection psychophysical experiments. One consequence was increased understanding that the state of light adaptation varies locally on the retina, albeit below the resolution of the cone array.

The SSDR has been found to be between 3 and 4 OoM from flicker and flash detection experiments using disk stimuli (Valeton and van Norren, 1983). Using a more rigorous psychophysical study using Gabor gratings, which are considered to be the most detectable stimulus (Watson et al., 1983), and using a $1/f$ noise field for an interstimulus interval to aid in accommodation to the display screen surface as well as matching image statistics, Kunkel and Reinhard (2010) identified an SSDR of 3.6 OoM for stimuli presented for 200 ms (onset to offset). Fig. 15.3 shows the test points used to establish the SSDR, as well as an image of the stimulus and background. The basic assumption is that the contrast of the Gabor grating forms a constant luminance interval. When this interval is shifted away from the semisaturation constant (toward lighter or darker on the x-axis), it remains constant on a log luminance axis ($\Delta L_1 = \Delta L_2$). However, the relative cone response interval (y-axis) decreases because of the sigmoidal nature of the function, leading to $\Delta V_1 > \Delta V_2$. The dynamic range threshold lies at the point where the HVS cannot distinguish the amplitude changes of the Gabor pattern.

We have now established that the SSDR defines the typical lower dynamic range boundary of the HVS, while adaptive processes can extend the perceivable dynamic range from 10^{-6} to 10^8, albeit not at the same time.

However, another outcome of Kunkel and Reinhard's (2010) experiment is that the SSDR also depends on the stimulus duration, which is illustrated in Fig. 15.4. The shortest duration can be understood as determining the physiological SDDR, and the increase in SSDR (ie, the lower luminance goes lower, and the higher luminance goes higher, giving an increased perceptible dynamic range) with longer durations indicates that rapid temporal light adaptation occurs in the visual system.

Sometimes, the SSDR is misinterpreted as being what is needed for the presentation of a single image, but this ignores the fact that the retina can have varying adaptation states (local adaptation) as a function of position, and hence as a function of image region being viewed. This means that the eye angling to focus on a certain image region will cause different parts of the retina to be in different image-dependent light adaptation states. Or, as one uses eye movements to focus on different parts of the image, the adaptation will change despite the image being static. For the case of video, even

---

[6]The Naka-Rushton equation originally did not include an exponent n and is therefore exactly the same as the Michaelis-Menten equation. However, the equation presented is widely used in the field and was adopted as the Naka-Rushton equation (Peirce, 2007; Valeton and van Norren, 1983).

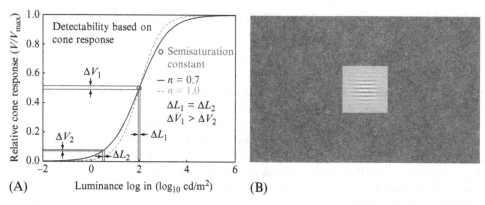

FIGURE 15.3

Useful detectability of image features (here Gabor grating). (A) Detectability based on cone response; (B) Achromatic Gabor Grating with a background $1/f$ noise pattern. Note that the contrast of the grating is amplified for illustration. After Kunkel and Reinhard (2010).

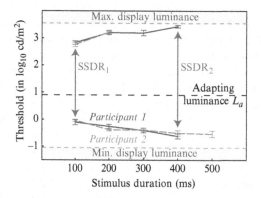

FIGURE 15.4

The dynamic range increases as a function of the display duration of a stimulus. After Kunkel and Reinhard (2010).

more variations of light adaption can occur, as the scenes' mean levels can change substantially, and the retina will adapt accordingly. Therefore, the necessary dynamic range is somewhere in between the SSDR and long-term-adaptation dynamic range.

Our application of interest is in media viewing, as opposed to medical or scientific visualization applications. While this can encompass content ranging from serious news and documentaries to cinematic movies, we refer to this as the "entertainment dynamic range" (EDR). Further, these applications require a certain level of cost-consciousness that is less pressing in scientific display applications. Video and moving picture applications that use EDR are unlikely to allow for full adaptation processes because of cost and most likely do not require them to occur either. Thus, use

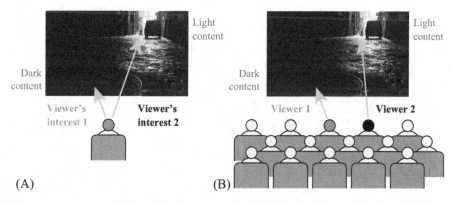

FIGURE 15.5

The SSDR is not enough in real-world applications such as with EDR. (A) Single viewers; (B) Multiple viewers.

of a 14 $\log_{10}$ dynamic range that encompasses complete long-term light adaptation is unrealistic. Nevertheless, they do have strong temporal aspects where the adaptation luminance fluctuates.

To illustrate the necessity of EDR being larger than SSDR, Fig. 15.5 compares common viewing situations (single or multiple viewers) with theater or home viewing settings that address adaptation. In this example, the still image shown can be regarded as a frame in a movie sequence. The dark alley dominates the image area, yet has a bright region in the distance as the alley opens up to a wider street. Depending on where a viewer might look, the overall adaptation level will be different. In Fig. 15.5A, showing a single viewer, the viewer may look from interest area 1 to area 2 in the course of the scene, and that viewer's light adaptation would change accordingly from a lower state to a higher state of light adaptation. Thus, the use of the SSDR would underestimate the need for the dynamic range of this image, because some temporal light adaptation would occur, thereby expanding the needed dynamic range. One could theoretically design a system with an eye tracker to determine where the viewer is looking at any point in time, and estimate the light adaptation level, combining the tracker position with the known image pixel values. The viewer's resulting SSDR could theoretically be determined from this.

However, in the image in Fig. 15.5B, multiple viewers are shown, each looking at different regions. Different light adaptation states will occur for each viewer, and thus a single SSDR cannot be used. Even the use of eye trackers will not allow the use of a single SSDR in this case. Having more viewers would compound this problem, of course, and is a common scenario. Therefore, a usable dynamic range for moving/temporal imaging applications (eg, EDR) lies between the SSDR and the long-term dynamic range that the HVS is able to perceive via adaptive processes.

One approach to determine the upper end of the EDR would be to identify light levels where phototoxicity occurs (the general concepts are well summarized in Youssef et al. (2011). A specific engineering study of phototoxicity (Pattanaik et al., 1998) was made to investigate the blue light hazard of white LEDs. The latter is a problem because the visual system does not have a strong "look away" reflex to the short-wavelength spectral regions of white LEDs (because they consist of a strong blue LED with a yellow phosphor). The results show that effects of phototoxicity occur at 160,000 cd/m^2

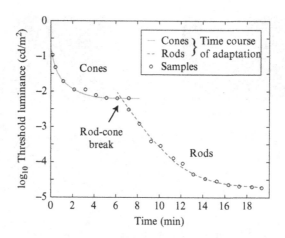

FIGURE 15.6

The time course of dark adaptation. After Riggs (1971).

and higher. Rather than actual damage, another factor to consider for the upper end is discomfort, which is usually understood to begin at 30,000 cd/m² (Halstead, 1993). Well-known snow blindness lies in between these two ranges. Rather than use criteria based on damage or discomfort, another approach is to base the upper luminance level on preferences related to quality factors, which would be lower still than the levels associated with discomfort.[7]

In consideration of the lower luminance end, there are no damage and likely no discomfort issues. However, the lower luminance threshold would be influenced by the noise floor of the cone photoreceptors when leaving the mesopic range due to dark adaptation or if the media content would facilitate adaptation to those lower luminance levels (eg, a longer very low key scene). Time required to adapt to those lower levels would actually occur in the media. The plot in Fig. 15.6 shows that around 6–8 min of dark adaptation is required to engage the rods, from a higher luminance starting point (Riggs, 1971).

15.2.3 STILL IMAGE STUDIES

The vast majority of studies to determine the needed, or appreciated, dynamic range have been for still images. A large body of work has studied viewer preferences for lightness and contrast that seem to be useful for assessment of the HDR parameters. The most recent of these studies (Choi et al., 2008; Shyu et al., 2008) are exemplary as they studied these parameters within a fixed dynamic range window. That is, they studied image-processing changes on fixed displays, not the actual display parameters. Unfortunately, these do not address the question of display range. Further, it is possible to study the dynamic range without invoking image contrast, which can be considered a side effect.

---

[7]If one is planning to conduct such a test, a pretest should be done to avoid one presenting damaging luminance levels to the viewer because of a potential outlier viewer who may prefer uncomfortable light levels.

A key article worth mentioning is that by Kishimoto et al. (2010), who concluded there is lower quality with the increase of brightness, maximizing at around 250 cd/m^2, thus concluding there was no need for HDR displays at all. The study carefully avoided the problems of studying brightness preferences within a fixed display dynamic range by using the then-new technique of global backlight modulation to shift the range further than previously possible. However, the native panel contrast in their study was around 1000:1 to 2000:1. Consequently, with increasing maximum luminance, the black level rose, so the study's results are for neither brightness nor range, but are more for the maximum acceptable black level.

Another key study assessed both dynamic range for capture and display using synthetic psychophysical stimuli as opposed to natural imagery (McCann and Rizzi, 2011). Rather than a digital display, a light table and neutral density filters were used in that study to allow a wider dynamic range than the capability of the then-current displays. It used lightness estimates for wedges of different luminance levels arranged in a circle. These wedge-dissected circles were then presented at different luminance levels, and observer lightness[8] estimates were used to determine the needed range for display, which was concluded to be surprisingly low at approximately 2.0 log$_{10}$ luminance. These low ranges were attributed to the optical flare in the eye, and interesting concepts about the cube root behavior of $L*$ as neural compensation for light flare were developed. However, the low number does not match practitioners' experience, both for still images and especially for video. The most common explanation for their low dynamic range was that the area of the wedges, in particular the brightest ones, was much larger than the brightest regions in HDR display, and thus caused a much higher level of optical flare in the eye, reducing the visibility of dark detail much more than occurs in many types of practical imagery. Another explanation is that their use of lightness estimates did not allow the amplitude resolution needed to assess detectable differences in shadow detail. For video, it was clear their approach did not allow for scene-to-scene light adaptation changes.

In the first study to use an actual HDR display (Seetzen et al., 2006) to study the effects of varying dynamic range intervals (also called "contrast ratio," CR), the interactions between maximum luminance and contrast were studied. This resulted in viewer estimates of image quality (in this case preference, as opposed to naturalness).

These data are important because they show the problem of black level rising with increases in brightness when the contrast is held constant. One can see this in Fig. 15.7 by comparing the curve for the 2500:1 contrast (black curve) with the curve for a 10,000:1 contrast. For the lower CR, quality is lost above 800 cd/m^2, which is due to the black level visibly rising (becoming dark gray) with increasing maximum luminance. The black level rises because of the low CR. On the other hand, the results for the 10,000:1 CR show no loss of quality even up to the maximum luminance studied (6400 cd/m^2). With this high contrast, the black level still rises, but it is one quarter the level of that for the 2500:1 CR, so it essentially remains black enough that it is not a limit to the quality.

Some other key studies for relevant HDR parameters are those focusing on black level (Eda et al., 2008, 2009; Murdoch and Heynderickx, 2012), on local backlight resolution (Lagendijk and Hammer, 2010), and on contrast and brightness preferences as a function of image and viewer preferences (Yoshida and Seidel, 2005). All of these studies have particular problems in determining the dynamic range for moving imagery. Simultaneous contrast has a strong effect on pushing a black level to appear

---

[8]Please keep in mind the well-known differences between luminance, lightness, and brightness. (Fairchild, 2013)

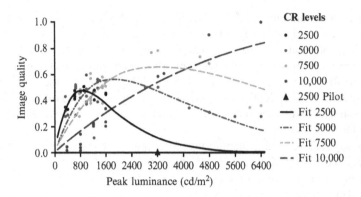

FIGURE 15.7

Preferred image quality based on interactions between maximum luminance and contrast. After Seetzen et al. (2006).

perceptually black, but should not be relied on to occur in all content. For example, dark low-contrast scenes should still be perceived as achieving perceptual black, as opposed to just dark gray. Similar consequences should occur for low-contrast high-luminance scenes. To truly study the needed dynamic range for video, one needs to remove the side effects of image signal-to-noise ratio limits, display dynamic range limits, display and image bit-depth limits, and the perceptual side effects of the Stevens effect and the Hunt effect.

15.2.4 DYNAMIC RANGE STUDY RELEVANT FOR MOVING IMAGERY

Extending still image studies to include motion and temporal components in general adds several new challenges for both experimental design and the analysis of the results. As with the design of any psychophysical study, it is important to reduce the degrees of freedom in the assessment to avoid the results being biased by effects other than the ones specified. Daly et al. (2013) listed design guidelines for psychophysical studies using moving imagery on HDR displays as follows:

- Try to remove all image signal limitations,
- Try to remove all display limitations,
- Try to remove perceptual side effects.

Examples of image signal limits include highlight clipping, black crushing, bit-depth constraints (eg, contouring), and noise. Those problems usually appear as a combination of each other potentially amplifying their impact. For example, the dynamic range reported as being preferred can be biased because of the impact of spatial or temporal noise. Increasing the dynamic range of a signal magnifies its noise, which otherwise would be below the perceptual threshold — for example, in standard dynamic range (SDR) content. Thus, the noise becomes the reason that further dynamic range increases are not preferred. To avoid this, the imagery used as test stimuli were essentially noise-free HDR captures created by the merging of a closely spaced series of exposures resulting in linear luminance, floating-point HDR images.

Examples of display limitations include the actual maximum luminance and minimum luminance of the test display, which should be out of the range of the expected preference. One way to achieve this is to allow the display to reach the levels of discomfort. Another example is the local contrast limits, such as when dual modulation with reduced backlight resolution is used. These were largely removed as explained in the articles.

The test images were specifically designed to assess the viewer preferences without the usual perceptual conflicts of simultaneous contrast, the Stevens effect, the Hunt effect, contrast/sharpness, and contrast/distortion interactions as well as common unintended signal distortions of clipping and tone scale shape changes. The Stevens effect and simultaneous contrast effects were ameliorated by use of low-contrast images. While simultaneous contrast is a useful perceptual effect that can be taken advantage of in the composition of an image/and or scene to create the perception of darker black levels and brighter white levels, you do not want to limit the creative process to require it to use this effect. That is, dark scenes of low contrast should also achieve a perceptual black level on the display, as well as scenes of high contrast, and likewise for bright low contrast appearing perceptually white. The images used were color images, to avoid unique preference possibilities with a black-and-white aesthetic, but the color saturation of the objects and the scene was extremely low, thus eliminating the Hunt effect, which could bias preferences at higher luminances as a result of increased colorfulness (which may or may not be preferred). As a result, the images are testing worst-case conditions for the display. Further, the parameters of black level and maximum diffuse white (sometimes called "reference white") were dissected into separate tests. Contrast was not tested explicitly, as contrast involves content rendering design issues, not display range capability. Note that moving imagery has analogies to audio with its strong dependence on time. In audio, dynamic range is not assessed at a given instant; it is assessed across time (Huber and Runstein, 2009). The same concepts led to study of the maximum and minimum luminances separately, as opposed to contrast in an intraframe contrast framework. The images were dissected into diffuse reflective and highlight regions for the third stage of the experiment.

The dynamic of each tested reflectance image was approximately 1 OoM, as shown in Fig. 15.8. In the perceptual test the images were adjusted by their being shifted linearly on a log luminance axis. The viewer's task was to pick which of two was preferred. The participants were asked for their own preferences, and not for an estimate of naturalness, even though some may have used those personal criteria. To avoid strong light adaptation shifting, the method of adjustment was avoided, which would result in light adaptation toward the values the viewer was adjusting. Instead, the two-alternative forced choice method was applied with interstimulus intervals having luminance levels and timing set to match known media image statistics, such as average scene cut durations.

For the upper bound of the dynamic range, the upper bounds of diffuse reflective objects and highlights were assessed independently. The motivations for this include perceptual issues to be discussed in Section 15.4 as well as typical hardware limitations, where the maximum luminance achievable for large area white is generally less than the maximum achievable for small regions ("large area white" vs "peak white" in display terminology). The connection between these image content aspects and the display limitations is that in most cases highlights are generally small regions, and only the diffuse reflective regions would fill the entire image frame.

Still images were used, but the study was designed to be video cognizant. That is, the timing of the stimuli and interstimuli intervals matches the 2–5 s shot-cut statistics typical of commercial video and movies. This aspect of media acts to put temporal dampening on the longer-term light adaptations as

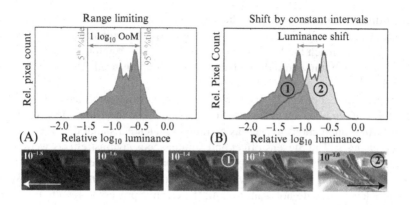

FIGURE 15.8

Illustration of an image having 1.0 OoM (between the 5th and 95th percentile criteria), and being shifted in the log-linear domain in steps of 0.2 OoM. This example is for an image testing for the black level bound of the dynamic range. (A) Range limiting; (B) Shift by constant intervals. After Daly et al. (2013) and Farrell et al. (2014).

described previously. Also, keep in mind that video content may contain scenes that are essentially still images with little or no moving content.

The first stage of this experiment determined preferences for minimum and maximum luminance for the diffuse reflective region of images. The results were not fit well by parametric Gaussian statistics, so they were presented as cumulative distributions. The approximate mean value results of these two aspects of the dynamic range were used to set up the second stage of the experiment, which probed individual highlight luminance preferences. Here, the method of adjustment[9] approach was applied on segmented highlights (with a blending function at the highlight boundary), while the rest of the image luminances were held fixed, as shown in Fig. 15.9. The adjustment increments were 0.33 OoM. Adaptation drift toward the adjustments would be minor because the highlights occupied a small image area.

The segmentation was done by hand, while a blending function was used at the highlight boundaries. The luminance profiles within the highlights were preserved but modulated through luminance domain multiplication, which is also shown in Fig. 15.9. For this stage of the study, images from scenes with full color were used, and the highlights were both specular and emissive. For the specular highlights, the color was generally white, although for metallic reflections, some of the object color is retained in the highlights. For the emissive highlights, various colors occurred in the test data set of approximately 24 images.

The study was done for a direct view display (small screen) and for a cinema display (big screen) (Farrell et al., 2014). In both cases, the dynamic range study was across the full range of spatial frequencies, as opposed to being restricted to the lower frequencies by use of a lower-resolution

---

[9]Because of limitations with the overall length of an experimental run, a two-alternative forced choice was not feasible while allowing for the testing of many images.

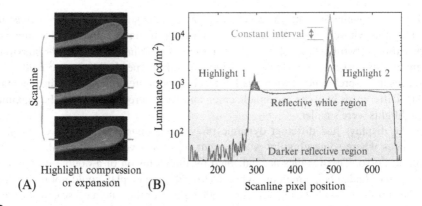

FIGURE 15.9

Highlight preference study via method of adjustment on segmented images. The highlight luminances on the white spoon (A) are remapped in equal intervals while the reflective part of the spoon remains constant. This leads to the scanline plot in (B).

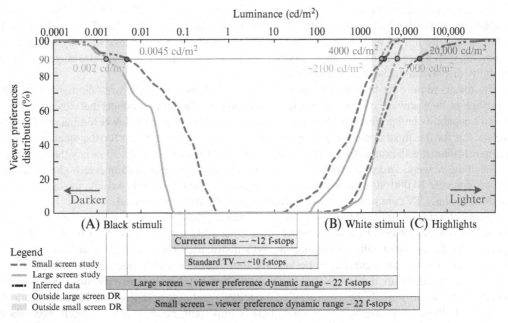

FIGURE 15.10

Preferred dynamic ranges for the smaller direct view display and the cinema. *DR*, dynamic range.

backlight modulation layer (Seetzen et al., 2006). The direct view display had a range of 0.0045–20,000 cd/m^2, while the cinema display had a range of 0.002–2000 cd/m^2.

Fig. 15.10 shows the combined results for black level, maximum diffuse white, and highlight luminance levels for the smaller display and the cinema display. The smaller screen is represented by the dashed curve, while the cinema results are shown as the solid curve. Cumulative curves show how more and more viewers' preferences can be met by a display's range if this range is increased (going darker for the two black level curves on the left, and going lighter for the maximum diffuse white and highlights curves on the right). While both display types had the same field of view, there were differences in the preferences, where darker black levels were preferred for the cinema versus the small display. Differences between the small-screen and large-screen results for the maximum diffuse white and highlights were smaller.

Because the displays had different dynamic ranges, the images were allowed to be pushed to luminance ranges where some clipping would occur (eg, for a very small number of pixels) if the viewer chose to do so. These values are indicated by the inclusion of a double-dash segment in the curves, and via the shaded beige regions. While technically the display could not really display those values (eg, the highlights would be clipped after 2000 cd/m^2 for the cinema case), the viewers adjusted the images to allow such clipping. This is likely because there was a perceived value in having the border regions of the highlights increase even though the central regions of the highlights were indeed clipped. For the most conservative use of these data, the results in the corresponding shaded regions (cinema, small display) should not be used.

Fig. 15.10 can be used to design a system with various criteria. While it is common to set design specifications around the mean viewer response, it is clear from Fig. 15.10 that the choice of such an approach leaves 50% of viewers satisfied with the display's capability, but 50% of the viewers' preferences could not be met. A basic consequence is that the dynamic range of a higher-capability display (ie, with a larger dynamic range) can be adjusted downward, but the dynamic range of a lower-capability display cannot be adjusted upward.

The results require far higher capability than can be achieved with today's commercial displays. Nevertheless, they provide guidance when designing and manufacturing products that exceed the visual system's capability and preferences. Some products can already achieve the lower luminances for black levels, even for the most demanding viewers (approximately 0.005 cd/m^2) for the small screen, for example. Rather, it is the capability at the upper end that is difficult to achieve. To satisfy the preferences of 90% of the viewers, a maximum diffuse white of 4000 cd/m^2 is needed, and an even higher luminance of approximately 20,000 cd/m^2 is required for highlights. At the time of this writing, the higher-quality "HDR" consumer TVs are just exceeding the approximately 1000 cd/m^2 level.

Despite the current limits of today's displays, we can use the data now to design a new HDR signal format. Such a design could be future-proof toward new display advances. That is, if we design a signal format for the most critical viewers at the 90% distribution level, such a signal format would be impervious to new display technologies because the limits of the signal format are determined from the preferences. And those are limited solely by perception, and not display hardware limits. Such a signal format design would initially be overspecified for today's technology, but would have the headroom to allow for display improvement over time. This is important, because it has already been shown that today's signal format limits (SDR, 100 cd/m^2, Rec. 709 gamut, 8 bits) prevents advancement in displays from being realized with existing signals. It is possible to create great demonstrations with new display technologies, but once they are connected to the existing SDR ecosystems, those advantages disappear for the signal formats that consumers can actually purchase. So with those issues in mind, the next section will address the design of a future-proof HDR signal format that takes into account the results in Fig. 15.10.

15.3 **QUANTIZATION AND TONE CURVE REPRODUCTION**

In the imaging technology community, there is often a conflation of dynamic range and bit depth, which are not necessarily linked for the display as elucidated by McCann and Rizzi (2007). For example, it is possible to have an 8-bit 10^6:1 dynamic range display, or a 16-bit 10:1 display. Neither of these would be pragmatic, however. The low bit depth but HDR display would suffer from visible contour artifacts, and the high bit depth but low dynamic range display would not achieve the visibility of its lower-bit modulations. The former display wastes its dynamic range, and the latter display wastes its bits. It is thus clear that there needs to be a balance between dynamic range and bit depth. Increases in dynamic range should be accompanied by increases in bit depth to prevent the visibility of false contours and other quantization distortions. It is generally known that in order to achieve this balance a nonlinear quantization (with respect to luminance) is required.

In addition to identification of luminance extremes and consequently contrast, two additional important aspects for the efficiency of the HVS are quantization (JNDs) and the shape of this quantization (tone curve nonlinearity) over the full dynamic range of the image reproduction system. A simplified example comparing the nonlinearities of a the HVS with gamma encoding is illustrated in Fig. 15.11. The key is that the quantization steps should be matched to the visual system's thresholds. With today's SDR systems, the quantization to a gamma of approximately 2 is approximately matched to the visual system. However, this is only true for dynamic ranges up to 2.0 $\log_{10}$ luminance. When higher dynamic range levels are used, the traditional gamma values no longer match the JND distribution of the HVS. Fig. 15.11 shows uniform quantization in the visual system threshold space (Fig. 15.11A), and what kind of quantization results from the traditional gamma domain quantization (Fig. 15.11B). It has the general tendency to have too few intervals at the dark end and too many at the bright end. As a consequence, contour distortions are often noticed in the dark regions, especially with LCD dynamic ranges exceeding 1000:1 (3.0 $\log_{10}$). Further, in the light regions, interval steps are "wasted" as they cannot be distinguished by the HVS. The result is that the number of distinguishable gray steps (Ward, 2008; Kunkel et al., 2013) was reduced because of the nonlinearity mismatch, as shown in (Fig. 15.11C).

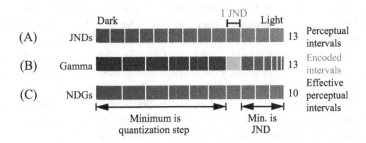

FIGURE 15.11

Illustration of (A) JNDs of the HVS versus (B) quantization of an imaging system (here gamma). For illustrative purposes, the dynamic range is reduced to only 13 intervals (compared with, eg, 256 intervals for 8 bits). The number of distinguishable grays (C) shows that if (A) and (B) do not align, accuracy and efficiency are lost. See also Ward (2008) and Kunkel et al. (2013). *NDGs*, number of distinguishable gray steps.

From these observations of the problems of gamma domain quantization, there is the motivation to derive a new nonlinearity that is more closely tuned to perception. To derive a more suited quantization nonlinearity, there are two methods. At this point it is best to shift the terminology to that used in vision science, and think in terms of sensitivity, which is the inverse of threshold, where threshold is defined in terms of Michelson contrast. Overall, what is needed is an understanding of how sensitivity varies with luminance level. If we assume worst-case engineering design principles, and knowing the crispening effect (Whittle, 1986), which describes that the sensitivity is maximized when the eye is adapted to the luminance level for which the sensitivity is needed, we can refine the question of sensitivity versus luminance level of the system's dynamic range to sensitivity as a function of the light adaptation level, where that level is constrained by the system's dynamic range.

Sensitivity, S, as a function of light adaptation can be determined from the following equation:

$$S = k \times L \frac{dR(L)}{dL} \cong L \times \text{OETF}',$$

where L is luminance, R is the cone response or overall visual system response, and k is a units scaling constant (cg, to go from cone volts to threshold detection). The optical-to-electrical transfer function (the OETF) can be determined from a local cone model (physiology), or from light-adaptive contrast sensitivity function (CSF) behavior (psychophysics) (Miller et al., 2013). Both methods give similar results (Daly and Golestaneh, 2015), attesting to the robustness of the concept. The CSF-based approach will be discussed below in more detail.

Design of new quantization nonlinearities more accurate than gamma was explored by Lubin and Pica (1991) using edge-based psychophysics. In 1993, a new quantization nonlinearity based on the CSF sensitivity (Blume et al., 1993) was developed for medical use and used by the Digital Imaging and Communications in Medicine (DICOM) Standards Committee in the early 21st century (NEMA Standards Publication PS 3.14-2008-2008). This curve, known as the DICOM grayscale standard display function, is based on Barten's model of contrast sensitivity of the human eye (Barten, 1996) and can reproduce an absolute luminance range from 0.05 cd/m² to almost 4000 cd/m² with a 10-bit encoded signal. This standard has been used very effectively for a wide variety of medical imaging devices ever since.

Because a wider range of luminance values was desired for the subset of physical HDR we refer to as "entertainment dynamic range" (EDR), a new nonlinearity was developed based on the same Barten model as the DICOM grayscale standard display function. The model parameters were adjusted for a larger field of view, which was more suitable for entertainment media, and the enhancement of CSF peak tracking (Cowan et al., 2004; Mantiuk et al., 2006) was added to the model calculations. The Barten model calculates an estimate of the CSF on the basis of both luminance and spatial frequency, but the spatial frequency that results in the largest CSF value varies with the luminance level. While DICOM maintained a constant frequency of four cycles per degree for all luminance levels, the new nonlinearity was calculated while allowing the spatial frequency value to always operate at the very peak of contrast sensitivity. The two concepts are illustrated in Figure 15.12.

The sum of the luminance values corresponding to the JNDs from 0 to $L_{\max} = \Sigma\left(1/S(L_a)\right)$ gives an "OETF"[10] which is that of the visual system transducers. But what is needed for display quantization

---

[10]Not to be confused with the OETF of digital cameras, because of the intermediate manipulations that result in an overall optical to optical transfer function OOTF that results in a displayed image that intentionally does not match the scene luminance levels (not even proportionally).

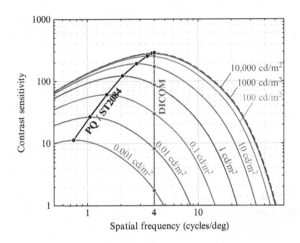

FIGURE 15.12

CSFs for the light adaptation range from 0.001 to 10,000 cd/m^2.

is an EOTF, which can be determined from the inverse of the determined OETF. This result, built by iteration in the OETF space, is described below. First, the luminance range needs to be determined. Here we refer to the results from the viewer preference study of luminance dynamic range, shown in Fig. 15.10. To satisfy 90% of the viewers for the small screen display, a range of approximately 0.005 to 20,000 cd/m^2 is needed as discussed in Section 15.2.4. Those wanting less dynamic range can easily turn down the range (eg, lower the maximum luminance) with no quantization distortion visibility occurring, because the contrast of each code value step is lowered by such an adjustment. On closer examination of the data, there was only one image out of 27 for which the preference was 20,000 cd/m^2, being an outlier (an image with an extremely small specular highlight). There were several images for which the preferences were near 10,000 cd/m^2, however, so this value was chosen as a design parameter for the maximum luminance. For the minimum luminance, the cinema case had a lower value than the small screen case, going down to 0.002 cd/m^2. The signal format should be applicable to both small screen displays expected in the home and large screen displays in the cinema. Because 0.002 cd/m^2 is an extremely low value, there is some rationale in simply selecting zero as the minimum luminance.[11] So the new signal format was designed to range from 0 to 10,000 cd/m^2, and after the design was complete, it was observed that very few code values were between 0 and 0.002 cd/m^2, showing that the choice of zero as a minimum value was worthwhile because the cost was negligible. The curve in Fig. 15.13A was built to cover a range from 0 to 10,000 cd/m^2 with a 12-bit encoded signal by summation of thresholds in the luminance domain.

Because the visual system can operate over very large luminance ranges, it is useful to plot the results in Fig. 15.13A in terms of log luminance (not to imply log luminance is a model of the visual system). This is shown in Fig. 15.13B.

---

[11]While the case of exactly 0 cd/m^2 is physically impossible at practical viewing temperatures, we decided to include that value in the signal format, because the cost is extremely low (just one code value).

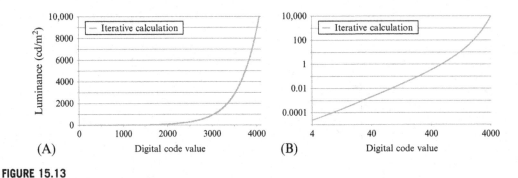

FIGURE 15.13

The curve was designed to cover a range from 0 to 10,000 cd/m² with a 12-bit encoded signal by summation of thresholds in the luminance domain: linear (A) and logarithmic (B).

Because a curve built by iteration is not convenient for compact specification or standardization purposes, it was desirable to create a closed-form model, which was a reasonable match to the iterative calculations. The equation (Nezamabadi et al., 2014) was a very close match to the original curve, and has now been documented by the Society of Motion Picture and Television Engineers (SMPTE) as SMPTE ST 2084, also known as the perceptual quantizer (PQ) curve:

$$L = 10,000 \left(\frac{\max\left[\left(V^{1/m_2} - c_1\right), 0\right]}{c_2 - c_3 V^{1/m_2}} \right)^{1/m_1}, \qquad (15.1)$$

where $m_1 = 0.1593017578125$, $m_2 = 78.84375$, $c_1 = 0.8359375$, $c_2 = 18.8515625$. and $c_3 = 18.6875$.

A comparison of the results from the iterative summing of JNDs and from Eq. (15.1) is shown Fig. 15.14. Note that by virtue of our plotting the data in the log domain (Fig. 15.14A), the differences in the dark region are exaggerated. A better plotting approach would be to plot the vertical axis in

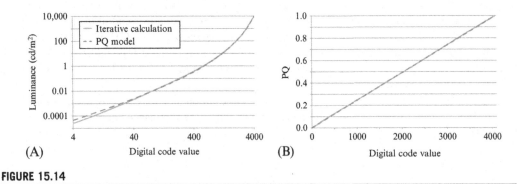

FIGURE 15.14

The results from the iterative summing of JNDs and from Eq. (15.1). (A) Logarithmic axis; (B) PQ axis.

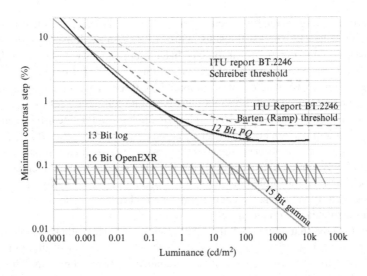

FIGURE 15.15

Efficiency of several common quantization approaches for a range from 0.0001 to 10,000 cd/m².

the units of the modeled perceptual space (ie, a PQ axis) as shown in Fig. 15.14B to better show the visibility of any differences between the equation and the iteration.

15.3.1 PERFORMANCE OF THE PQ EOTF

To assess the performance of this PQ EOTF, the JNDs are compared with other known methods in Fig. 15.15. One standardized threshold model is from Schreiber (1992), and is mentioned in an ITU-R report on ultra-high-definition TV (ITU-R Report BT.2246-4, 2015). It makes the simple assumption that the HVS is logarithmic (following Weber's law) above 1 cd/m², and has a gamma nonlinearity for luminance below that. The report also includes a threshold model based on the Barten CSF (referred to in Fig. 15.15 as "Barten (Ramp)"), which is similar to the iterative PQ curve but uses different Barten model parameters.

The remaining plotted curves are quantization approaches, in a format initially explored by Mantiuk et al. (2004). Shown as the black solid line is the PQ nonlinearity described previously, it is apparent that it lies below the Barten ramp and Schreiber thresholds from the ITU-R report. This means that if either of those is accepted, quantization according to the PQ nonlinearity will result in quantization that is below threshold. However, it might be suggested that the PQ nonlinearity is a better model of HVS thresholds. The vertical positions of the curves depend on the luminance range and the bit depth. For Fig. 15.15, the luminance range is 0–10,000 cd/m². A quantization in the log domain would give a straight line here, and with such an approach, 13 bits are required to keep the quantization better than the 12-bit version of the PQ. For professional workflows, one format is OpenEXR, a floating-point format, which uses 16 bits. It stays below the threshold, but has the higher cost of 16 bits and mantissa management, which is strongly disfavored in application-specific integrated circuit designs.

Another plotted curve is for gamma, for which 15 bits are required to keep the distortions mostly below threshold. For the very dark levels, it still results in quantization distortions above threshold.

As of this writing, the PQ nonlinearity has been independently tested in other laboratories (Hoffman et al., 2015) to see how well it characterizes thresholds for different dynamic ranges. The results are shown in Fig. 15.16, where it was compared for three dynamic ranges against a gamma of 2.2, the DICOM standard, and actual threshold data. Absolute luminance levels were also tested, and were preserved in the plotting (in Fig. 15.16 the plots on the left are for linear luminance and the plots on the right are for log luminance). For the lower dynamic range of 100:1, the DICOM standard and PQ nonlinearity are a good fit to the observer data, while gamma underestimates sensitivity throughout the range. For the middle dynamic range of 1000:1, the DICOM standard now underestimates sensitivity, while the PQ model and the threshold data are still in good agreement. Finally at a 10,000:1 dynamic range, the PQ nonlinearity very slightly underestimates the threshold data, but is a closer fit than the DICOM and gamma models for quantization.

Another recent study that compared the PQ nonlinearity with other visual models used for quantization is shown in Fig. 15.17 (Boitard et al., 2015). In the format of this plot, the best visual model has a slope of zero. This study looked at the luminance range through a small luminance window (0.05–150 cd/m^2) but quantized according to the full range (0–10,000 cd/m^2). While the curve for

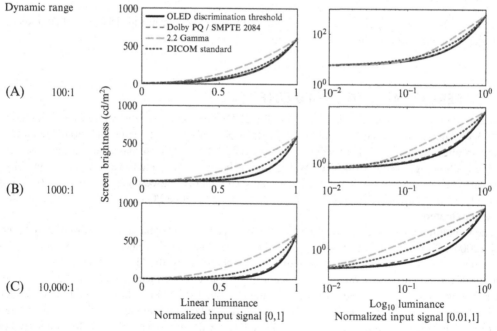

FIGURE 15.16

Comparison of luminance thresholds with three visual models (2.2 gamma, DICOM standard, and the PQ nonlinearity, SMPTE ST 2084) under three different dynamic ranges. After Hoffman et al. (2015). *OLED*, organic LED.

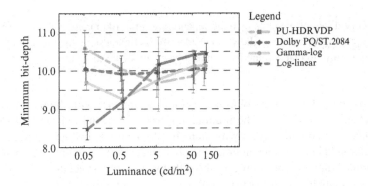

FIGURE 15.17

Minimum bit depth required to encode HDR luminance values with use of several perceptual transfer functions. The error bar locations have been shifted for readability. After Boitard et al. (2015).

gamma-log lies below the PQ curve, and may therefore be considered a better design (resulting in a lower bit depth), it is easy to see that the bit-depth requirement starts to rise at the highest luminances tested, 150 cd/m^2, and would be expected to rise for higher levels, such as needed for any substantial dynamic range, whether 10,000 cd/m^2 in the PQ design or for more short-term maximum luminance levels, such as 1000 cd/m^2.

15.4 PERCEPTION OF REFLECTANCES, DIFFUSE WHITE, AND HIGHLIGHTS

While several key quality dimensions and creative opportunities have been opened up by HDR (eg, shadow detail, handling indoor and outdoor scenes simultaneously, and color volume aspects), one of the key differentiators from SDR is the ability for more accurate rendering of highlights. These can be categorized as two major scene components: specular reflections[12] and emissive regions (also referred to as "self-luminous"). They are best considered relative to the maximum diffuse white luminance in the image.

Most vision research has concentrated on understanding the perception of diffuse reflective objects, and the reasons are summarized well by DeValois and DeValois (1990):

> For the purposes of vision it is important to distinguish between two different types of light entering the eye, self-luminous objects and reflecting objects.... Most objects of visual interest are not self-luminous objects but rather reflective objects.... Self-luminous objects emit an amount of light that is independent of the general illumination. To detect and characterize those objects, an ideal visual system would have the ability to code absolute light level. Reflecting objects, on the other hand,

---

[12]In traditional photography the term "highlights" is sometimes used to refer to any detail near white, such as bridal lace, which may entirely consist of diffuse reflective surfaces. In this chapter we reserve the use of "highlights" to mean the specular or emissive regions in an image.

will transmit very different amounts of light depending on the illumination.... That which remains invariant about an object, and thus allows us to identify its lightness, is the amount of light coming from the object relative to that coming from other objects in the field of view.

In other words, to understand the objects we routinely interact with, humans have evolved very good visual mechanisms to discount the illumination from the actual reflectance of the object. The object's reflectance is important to convey its shape by allowing for shading. Emissive regions, on the other hand, were limited in more ancient times to either objects we did not physically interact with (eg, the Sun, hot magma, lightning, forest fires) or those that were seldom encountered (eg, phosphorescent tides, ignis fatuus,[13] cuttlefish, deep sea fishes). While specular highlights are also reflections, they have distinct properties that are very different from diffuse reflectance. Speculars were often encountered, such as with any wet surface, but their characteristics did not seem as essential to understanding the physical objects as their diffuse reflective components, which dominate most objects. Illustrations of emissive regions and specular highlights are shown in Fig. 15.18.

In traditional imaging on the display side, the range allocated to these highlights is fairly low and most of the image range is allocated to the diffuse reflective regions of objects. For example, in hardcopy print, the highlights will have 1.1 times higher luminance than the maximum diffuse white (Salvaggio, 2008). In traditional video, the highlights are generally set to have luminance no more than 1.25 times the diffuse white. In video terminology, the term "reference white" is often used instead of "diffuse white," taken from the lightest white patch on a test target, and is often set at 80 out of a range of 0–100 (or 235 out of 255). The range 80–100 is reserved for the highlights, the maximum of which is referred to as "peak white," and is set to the maximum capability of the display. Of the various display applications, cinema allocates the highest range to the highlights, up to 2.7 times the diffuse white. These are recommendations, of course, and any artist/craftsperson can set the diffuse white lower in the available range to allocate more range to the highlights if the resulting effect is desired.

However, considering actual measurements shows how much higher the luminance levels of specular regions are compared with those of the underlying diffuse surface. For example, Wolff (1994)

FIGURE 15.18

Emissive light sources, specular highlights, and diffuse white.

---

[13]Luminous swamp gas.

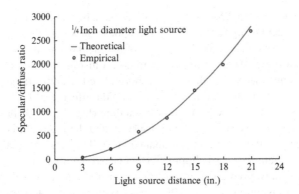

FIGURE 15.19

Measurements showing that the luminance of specular regions can be more than 1000 times higher in comparison with the underlying diffuse surface. After Wolff (1994).

has shown that they can be more than 1000 times higher, as shown in Fig. 15.19. This means the physical dynamic range of the specular reflections vastly exceeds the range occupied by diffuse reflection. If a visual system did not have a way of handling the diffuse regions and illumination with specialized processing as previously described, and saw in proportion to luminance, most objects would look very dark, and most of the visible range would be dominated by the specular reflections.

Likewise, emissive objects and their resulting luminance levels can have magnitudes much higher than the diffuse range in a scene or image. As discussed in Section 15.2.1, the most common emissive object, the disk of the Sun, has a luminance so high (approximately 1.6×10^9 cd/m^2) it is damaging to the eye if it is looked at more than briefly. So, the emissive regions can indeed have ranges exceeding even those of the speculars. A more unique aspect of the emissive regions is that they can also be of very saturated color (sunsets, magma, neon lights, lasers, etc.). This is discussed in further detail in Section 15.3.

With traditional imaging, and its underrepresentation of highlight ranges, the following question arises: What happens to the luminances of highlights? Fig. 15.20 shows example scanlines of various distortions that are commonly encountered. The first scanline (Fig. 15.20A) is the luminance profile of a diffuse reflective object, in particular a curved surface, such as an unglazed porcelain spoon that

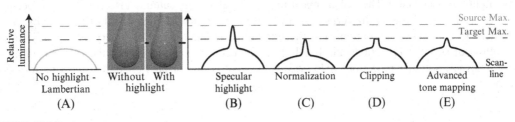

FIGURE 15.20

Effects of highlight rendering, clipping, and compression.

exhibits Lambertian reflection properties (ie, ideally diffuse). In comparison with that, scanline B in Fig. 15.20 illustrates a specular highlight as it would appear on a porcelain spoon that is glazed and therefore glossy. It exceeds the maximum luminance of the display (or the signal), which is indicated as the lower dashed line titled "Target Max." The middle illustration (Fig. 15.20C) shows one type of distortion that is seldom selected — that is, to renormalize the entire range so that the maximum luminance of the highlight just fits within the range. A side effect of this is that the luminances of the diffuse regions are strongly reduced. The fourth scanline (Fig. 15.20D) shows an approach where the diffuse luminances are preserved, or closely preserved, and the highlight is simply truncated, with all of its pixels ending up at the maximum luminance level. This leads to a distortion artifact called "highlight clipping" or sometimes "hard clipping." Details and features within the highlight region are replaced with constant values, giving rise to flat regions in the image, and the resulting look can be quite artificial. The last illustration (Fig. 15.20E) shows typical best practices, and has been referred to as "soft clipping" or "shoulder tonescale." Here the shape and internal details of the highlight are somewhat preserved, and there are no flattened regions in the image. Some of the highlight details are often lost to the quantization of the image or the threshold of the visual system, as the contrast of the details is reduced.

In the 20th century, there were distinct appearances for video and film systems, giving rise to the terms "video look" and "film look." Originally there were numerous characteristics that accompanied each of these systems, and highlight distortions were one of the predominant artifacts: video tended to have hard clipping, whereas film systems had soft clipping, which was generally preferred in professional imaging because it looked less artificial. At this point in time, there are many digital capture systems that include the shoulder effect of film's tonescale, so this particular distinction between digital and film has been largely eliminated. Nowadays, one of the few remaining distinctions between video (digital) and film is the frame rate (Daly et al., 2014).

In the context of HDR displays, the more accurate presentation of specular highlights (assuming the entire video pathway is also HDR) is one of the key distinctions of HDR. In this century, vision science has been less dominated by diffuse reflective surfaces, and a number of perceptual articles have looked closely at specular reflection. More accurate display of speculars reflections can result in better 3D shape perception, even with 2D imagery (Interrante et al., 1997; Blake et al., 1993), especially if it is in motion. Other studies have found specular highlights can aid in better understanding of surface materials (Dror et al., 2004). For stereo imaging, the specular reflections have very unique properties (Muryy et al., 2014). Further, in terms of human preference, there is a long list of objects that are considered to have high value and which have very salient specular reflections (gold, jewelry, diamonds, crystal glass chandeliers, automobile finishes, etc.) as well as low-cost objects that have appeal to the very young (sparkles). The emissive regions have a long history in entertainment (campfires, fireworks, stage lighting, lasers, etc.). There also appear to be physiologically calming effects of certain specular reflections, such as those on the surface of water as vacation destinations with large bodies of water significantly dominate other options. For those who like to entertain evolutionary psychology, all of this allure of specular reflections and highlights in general might derive from the fact that in most natural scenes, specular reflections only arise from what is one of land creatures' most basic need: water.

15.5 ADDING COLOR — COLOR GAMUTS AND COLOR VOLUMES

In the previous sections, we introduced the main concepts of HDR: contrast, quantization, and tone curves. So far, all those concepts were discussed from an achromatic point of view. However, our visual system perceives color, and therefore adding the chromatic component to HDR concepts is a natural consequence.

Of course, HDR imaging offered the capabilities to work with color images from the start. When one is working with real-world luminance levels, the colorimetric accuracy of HDR imaging can be high. However, if tone reproduction and tone mapping as well as efficient compression are taken into consideration, simple color mapping approaches can lead to large perceptual errors. For example, basic tone-mapping approaches (eg, Schlick, 1994; Reinhard et al., 2010) can lead to large variations regarding their output rendition and consequent appearance of a remapped HDR image. This makes the colorimetric and perceptual quantification as well as the consistency of such models challenging.

Therefore, similarly to the treatment of the intensity axis (as discussed in the previous section), it is beneficial to define the right color spaces and mapping approaches in order to alleviate unpredictable variations in rendering results. For this, it is necessary to understand both how the HVS perceives and processes color and how color-imaging systems such as displays can relate to it in the best possible way.

15.5.1 VISIBLE COLORS AND SPECTRAL LOCUS

This section about color spaces is added as context to the discussion of color volumes and therefore is not necessarily exhaustive. See the traditional color science literature (Fairchild, 2013; Wyszecki and Stiles, 2000) for a more thorough discussion of basic colorimetry.

Traditionally, the color components in imaging systems are described as coordinates in a 2D Cartesian coordinate system, which is called a chromaticity diagram. A common type is the CIE xy 1931 chromaticity diagram (Smith and Guild, 1931; Wyszecki and Stiles, 2000) shown in Fig. 15.21A. There, all chromaticity values visible to the HVS appear inside the horseshoe-shaped spectral locus, while colors outside this area are not visible to the HVS and can be called "imaginary colors" (nevertheless, they can be useful for color space transforms and computations).

Colors that lie on the boundary of the spectral locus are monochromatic (approximating a single wavelength as indicated). Typical representatives of such monochromatic colors are laser light sources, but they can also be created, for example, by diffraction or iridescence. Examples appearing in nature are the colors of rainbows, marine mollusks such as the pāua (*Haliotis iris*), and the Giant Blue Morpho butterfly (*Morpho didius*). The straight line at the lower end of the "horseshoe," connecting deep violet and deep red (at approximately 380 and 700 nm, respectively) is called the "line of purples" and is formed by a mixture of two colors, making it a special case as no single monochromatic light source is able to generate a purple color.

15.5.2 CHROMATICITY AND COLOR SPACES

A color space describes how physical responses relate to the HVS. It further allows the translation from one physical response to another (eg, from a camera via an intermediate color space and ultimately

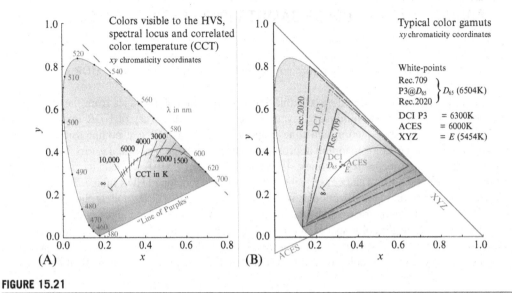

FIGURE 15.21

Chromaticity diagrams in CIE *xy* showing the fundamental components of color imaging and color spaces.

to the physical properties of a display system). Most color spaces used with the aim of additive[14] image display have three color primaries.[15] Those three primaries will map to three points in a 2D chromaticity diagram, thereby, when connected, spanning a triangle (as illustrated for several color spaces in Fig. 15.21B). This triangle contains all the colors that can be represented (or realized) by this color space and is called the 2D "color gamut"[16] Further, colors that lie outside the triangle and thus are not representable are called "out-of-gamut colors" (Reinhard et al., 2010).

Typical additive display systems such as CRTs, LCDs, organic LED displays, and video projectors use phosphors, color filters, or lasers with a distinct wavelength combination, which define their primaries. However, these primaries differ from device to device, resulting in each of them having a different color gamut. Nevertheless, for accurate and perceptually correct rendering of imagery it is important to have color imagery displayed as intended (eg, a particular color shown on a display matches its chromaticity values sent to the display).

The conventional approach to alleviate this problem is to design an imaging system that closely matches an industry standard color space. For consumer devices, this industry standard is usually ITU-R

---

[14]On the basis of the principles of trichromacy and additive color mixing, it is possible to approximate the perceptual spectrum of colors with the minimum of three color primaries (Wyszecki and Stiles, 2000). Color processes can also be subtractive (eg, with print).

[15]There are display systems that use more than three primaries (such as addition of white or yellow subpixels) (Roth and Weiss, 2007; Ueki and Nakamura, 2007).

[16]Sometimes, the term "color gamut" can refer to a 3D color volume, but in the context of this chapter it is intended to be 2D.

Recommendation BT.709-6, 2015, whose primaries form a reasonable approximation to the capabilities of TVs and computer displays. The triangular area "spanned" by the Rec. 709 primaries is shown in Fig. 15.21B. Nevertheless, because of economic and design considerations as well as effects of aging, a match to one of the industry standards is not guaranteed, leading to an incorrect color rendition.[17]

To resolve this problem of inevitable differing primaries, new technologies have been developed that can match, for example, an HDR and wide gamut color signal to the actual physical properties of a particular display device (independently of the above-mentioned industry standards) (Brooks, 2015).

For professional use, where economic decisions are less restrictive and calibration can be tighter, the Digital Cinema Initiative (DCI) defined the P3 color gamut that is mainly intended for display systems such as reference monitors and cinema projectors (SMPTE RP 431-1:2006; SMPTE RP 431-2:2011) (also shown in Fig. 15.21B).

Recently, the ITU-R released a new recommendation with a wider color gamut whose color primaries lie on the spectral locus (ITU-R Recommendation BT.2020-1, 2014). Because of its large gamut, the latter color space is now considered for use in consumer and professional fields. Nevertheless, because of the triangular nature, those three color gamuts still do not completely cover the color gamut of the HVS.

The largest color spaces shown in Fig. 15.21B are ACES (SMPTE ST 2065-1:2012) and CIE XYZ (from which the actual *xy* chromaticity diagram is derived). Those two color gamuts are designed to cover all visible colors in the "horseshoe." However, to achieve this with three primaries, many chromaticity values lie outside the visible gamut of the HVS.

15.5.3 COLOR TEMPERATURE AND CORRELATED COLOR TEMPERATURE

Another important attribute to describe color is the white point, which can be precisely described by its tristimulus value (eg, in XYZ). However, a more usual way to do so is to provide its "correlated color temperature" (CCT). The CCT is related to the concept of color temperature, which is derived from a theoretical light source known as a blackbody radiator, whose spectral power distribution is a function of its temperature only (Planck, 1901). Thus, "color temperature" refers to the color associated with the temperature of such a blackbody radiator, which can be measured in Kelvins (K) (Fairchild, 2013; Hunt, 1995). For example, the lower the color temperature, the "redder" the appearance of the radiator (while still being in the visible wavelength range). Similarly, higher temperatures lead to a more bluish appearance. The colors of a blackbody radiator can be plotted as a curve (known as the "Planckian locus") on a chromaticity diagram, which is shown in Fig. 15.21A. However, real light sources only approximate blackbody radiators. Therefore, of more practical use is the CCT. The CCT of a light source resembles the color temperature of a blackbody radiator that has approximately the same color as the light source.

White points are provided wither as a Kelvin value (eg, 9300 K, 12,000 K[18]) or as standardized representations of typical light sources. The CIE defined several standard illuminants (Fairchild, 2013;

---

[17]Display calibration to one of the industry standard color spaces can often alleviate this problem as long as the display is capable of reaching that standard.

[18]Such "bluish" high CCT white points are common with consumer TVs in "standard," "vivid," "dynamic," or similar nonreference modes.

Reinhard, 2008) that are used in imaging applications.[19] Commonly used white points in this field are the daylight illuminants, or D-illuminants (D_{50}, D_{55}, D_{65}), of which D_{65} (CCT of 6504 K) is used by a large number of current color spaces (eg, in Rec. 709 and Rec. 2020). Another important white point is Illuminant E. It is the white point of XYZ and exhibits equal-energy emission throughout the visible spectrum (equal-energy spectrum) (Wyszecki and Stiles, 2000). Being a mononumeric value, the CCT does not describe white-point shifts that are perpendicular to the Planckian locus. This mathematical simplification is usually not apparent when one is working with traditional white points, as they are commonly lying on or close to the Planckian locus. However, if they lie further away from the Planckian locus, this will lead to green or magenta casts in images. The HVS can also adjust to white-point and illumination changes (Fairchild, 2013). This is facilitated by the chromatic adaptation process introduced earlier (see Fig. 15.2).

Now that we have discussed the properties of white points, note that a common omission is to ignore the color of the black point. Nevertheless, it should be taken into account to avoid potential colorcasts in the dark regions.

15.5.4 COLORIMETRIC COLOR VOLUME

The chromaticity diagram shown in Fig. 15.22A discards luminance and shows only the maximum extent of the chromaticity of a particular 2D color space. In reality, color is 3D as presented in Fig. 15.22B, and is referred to as a color volume. When luminance is plotted on the vertical axis, it becomes apparent that, for a three primary color space, the chromaticity of colors that show a higher luminance level than the primaries (R, G, or B in Fig. 15.22) will collapse toward the white point (W in Fig. 15.22). A consequence is that although certain colors can be situated inside the 2D chromaticity diagram such as the example emissive color from the fire-breather in Fig. 15.22D, they can actually be too bright to still fit inside the smaller volume (Fig. 15.22C). However, it is possible to render the same color with a larger color volume (eg, with 1000 cd/m^2), which is also illustrated in Fig. 15.22C. Therefore, a sufficiently large color volume is an important aspect when one is dealing with higher dynamic range and wide gamut imaging pipelines.

So far in this chapter we have defined the minimum and maximum luminance as well as their color (CCT or white point). With these and the primaries, we have described the extrema of a color volume. Now, quantization and nonlinear behavior inside a color volume need to be addressed in a similar way as introduced earlier with the achromatic tone curve.

An example is given by comparison of the plots in Fig. 15.23A and B.[20] Both plots show the horseshoe of chromaticities visible to the HVS as well as the previously discussed industry standard color spaces. However, the two plots use different reference systems to compute the chromaticity values. Instead of CIE xy 1931 in Fig. 15.23A, the chromaticity coordinate system of Fig. 15.23B is CIE $u'v'$ 1976 (which is based on CIE LUV) (CIE Colorimetry, 2004).

---

[19]There are several other CIE illuminants that are less common with imaging applications, such as Illuminant A, which approximates a blackbody radiator with a color temperature of 2856 K or the group of F-illuminants, describing the white points of fluorescent light sources (CIE Colorimetry, 2004).

[20]For the simplicity of explanation, this is first illustrated as a 2D chromaticity representation again before the concepts are extended to the full 3D color volume.

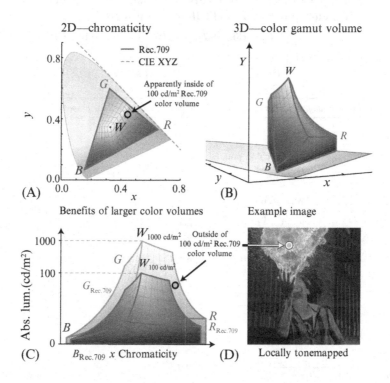

FIGURE 15.22

Assessing if a color such as the example color in the fire (D) can be displayed in a given color space is misleading in a 2D chromaticity diagram (A). A 3D color volume (B) reveals this information (C).

The latter is widely used with movie and video engineering as CIE $u'v'$ is considered to be a more perceptually uniform chromaticity diagram. Perceptual uniformity is an important aspect of a color space for encoding efficiency and if any remapping operations have to be performed. It is also important if an assessment of the size of a color gamut or volume is desired. In the display community, color spaces are often compared by their coverage of all visible chromaticities in the xy or $u'v'$ diagram. Even though $u'v'$ is considered to be more perceptually uniform, it still has limitations in this regard. Therefore, the area of a color space computed in CIE xy or $u'v'$ does not exactly correlate with a perceptual measure. Table 15.3 provides a comparison of the chromaticity area of three industry standard color spaces. It is apparent that the values for CIE xy and $u'v'$ do not match. However, it is to be expected that the results from a more perceptually uniform color space are more meaningful.

Similarly to luminance, perceptual uniformity in chromaticity is not automatically given by the sampling of physical quantities of light, even if they are weighted by the spectral response curve of the rods and cones. In the early 1940s Wright (1941) and MacAdam (1942) quantified the perceptual nonuniformity by JND ellipses as shown in Fig. 15.23A and B. Ultimately, their work led to more perceptually uniform chromaticity representations such as the one provided by CIE $u'v'$ 1976 as shown in Fig. 15.23B. As can be seen the perceptual uniformity for $u'v'$ performs better compared to CIE xy 1931, but the MacAdam JND ellipses are still not rendered perfectly circular, as would be the case

Table 15.3 Relative Area of Three Common Color Gamuts When Plotted in CIE *xy* and CIE *u′v′* Coordinate Systems

	CIE *xy* (Fig. 15.21A and B)	CIE *u′v′* (Fig. 15.21C)
Rec. 709/sRGB	33.2%	33.3%
DCI P3	45.5%	41.7%
Rec. 2020	63.4%	57.2%

The percentage describes the coverage of the spectral locus (the "horseshoe").

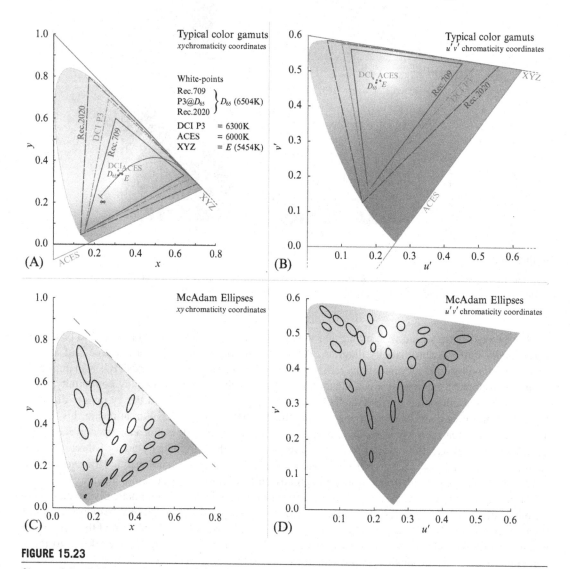

FIGURE 15.23

Chromaticity diagrams in CIE *xy* 1931 and CIE *u′v′* 1976 showing the common industry color spaces (A, B) as well as the MacAdam JND ellipses for observer PGN from MacAdam (1942), size amplified 10 times (C, D).

for a perfect uniform color space. A further drawback when quantizing chromaticity in $u'v'$ is the low resolution for yellows and skin tone areas.

In addition to the remaining perceptual nonuniformity in the chromatic domain, chromatic JNDs are also dependent on adapting luminance (which is based on the state of light adaptation and with that on the level of ambient illumination as discussed in Section 15.2.1). Thus, when aiming for perceptual uniformity we cannot deal with luminance and chromaticity independently.

15.5.5 THE CONCEPT OF COLOR APPEARANCE MODELING

Further improvement of perceptual accuracy including uniformity requires that environment factors be taken into account. Besides the already introduced adapting luminance, there are several additional aspects that have an impact on how colors appear to the HVS, such as the background behind a color patch or directly neighboring colors or pixels (eg, causing the effect of simultaneous contrast; Fairchild, 2013). Further, the dominant color of the illumination influences our chromatic adaptation (see Section 15.2.1). Therefore, the impact of those aspects on colors, especially in the context of image display, needs to be taken into account when one is designing an advanced imaging system. Modeling and describing this are the domain of color appearance models (CAMs) or image appearance models (Fairchild, 2013).

CAMs are designed to be perceptually uniform throughout the whole color volume while taking into account aspects of the viewing environment to provide the perceptual correlates of lightness, chroma, and hue.[21] The concept of those appearance correlates is illustrated in Fig. 15.24. There, very bright and very dark stimuli appear desaturated and thus having smaller radii — for example, because the limit of the SSDR has been reached (Section 15.2.2) with the reaching of the level of cone pigment bleaching and their noise floor, respectively. Maximum saturation can be reached for stimuli with average lightness.

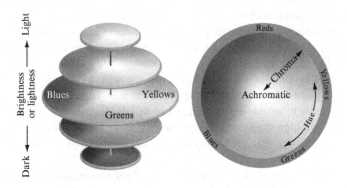

FIGURE 15.24

The concept of the general appearance correlates lightness, chroma, and hue.

[21]"Brightness" and "colorfulness" are additional appearance correlates that are less commonly used with image pipeline and display applications. Further, "saturation" can be calculated from the other appearance correlates.

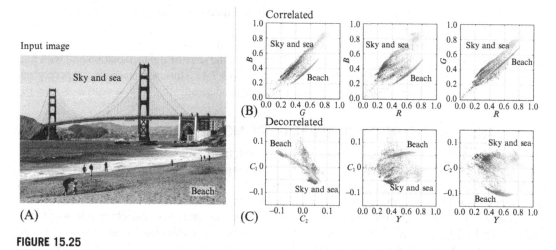

FIGURE 15.25

Comparison of correlated (B) and decorrelated (C) image dimensions from the source image (A).

To use perceptually uniform and appearance-based color manipulation based on appearance correlates, input color values provided, for example, as RGB, XYZ, or LMS[22] tristimulus values need to be decorrelated. The reasoning for this step is illustrated in Fig. 15.25. When the individual channels (here R, G, and B) from Fig. 15.25A are plotted against each other as shown in Fig. 15.25B, it is apparent that all three channels show a correlation between them (each channel contains information about both intensity and chromatic information). In Fig. 15.25C, the color channels Y, C_1, and C_2 are decorrelated from each other, (eg, by use of a principal component analysis, PCA). Now, the intensity information is mostly independent of the chromatic information.

Decorrelation with perceptually accurate principal component axes followed by a conversion of C_1 and C_2 from Cartesian to polar coordinates leads to correlates such as lightness, chroma, and hue as illustrated in Fig. 15.25.

The decorrelation of tristimulus values is also common with video compression approaches. In contrast to improving appearance accuracy, the foremost motivation with this use case is image compression efficiency such as chroma subsampling, which exploits the lower contrast sensitivity bandwidth of the HVS for high-frequency chroma details. One example is the YC_bC_r (Rec. 709/Rec. 2020) color difference model that is widely used with high-definition and ultra-high-definition broadcasts.

In addition to the decorrelation, a component of nonlinear compression is also added simulating that behavior of the HVS (eg, the sigmoidal behavior of the cone photoreceptors as introduced in Section 15.2.2).

---

[22]LMS describes the response signals from the cone photoreceptors. It can be derived from XYZ.

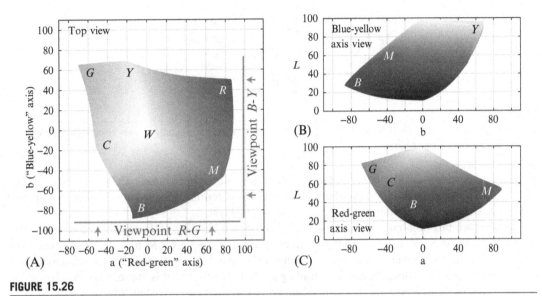

FIGURE 15.26

sRGB rendered in CIECAM02. Here, the adapting luminance L_a is fixed to 20 cd/m^2 and the white point is D$_{65}$ as specified in (ISO, 2004)

CIE L*a*b* and CIECAM02 provide two commonly used CAMs[23] and color appearance spaces that aim for perceptual uniformity. As an example, Fig. 15.26 shows the sRGB color volume projected into the CIECAM02 appearance space. Note the expanded areas for yellow (light gray in print versions) and brown (dark gray in print versions) tones (upper right corner of B) compared with $u'v'$.[24]

Although the design of CAMs is an improvement over non-appearance-based or perceptually uniform color spaces, these CAMs are both based on reflective colors, making them suitable only for SDR levels. Therefore, these existing CAMs still have limitations, especially when it comes to processing large color volumes as required with higher dynamic range and wide gamut imaging systems and displays.

The research community is actively working on improving CAMs. For example, Kunkel and Reinhard (2009) have presented progress with matching color appearance modeling closer to the expectations of the HVS. Further, Kim et al. (2009) studied perceptual color appearance under extended luminance levels.

Also, deriving the optimal color quantization needed for HDR and wide color gamut imaging systems is a challenge because input parameters for CAMs such as the adaptation field cannot be assumed as fixed for most HDR viewing scenarios as explained by the requirements of the EDR introduced in Section 15.2.2 and illustrated in Fig. 15.5. Another example is the impact of

---

[23]CIE LAB was not originally designed as a CAM, but as it fulfills the minimum definition of modeled parameters, such as taking the chromatic adaptation into account, and it can be considered as one (Fairchild, 2013).

[24]Note that the color volume of the sRGB color space collapses toward black because color discrimination is reduced for dark colors compared with bright colors with the same chromaticity. sRGB black does not reach $L = 0$ because of the 1 cd/m^2 black value (ISO, 2004).

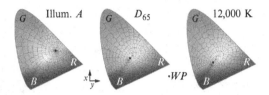

FIGURE 15.27

The change of distribution of JND "voxels" as a function of the white point. Each cell approximates 50 JND voxels. After Kunkel et al. (2013).

chromatic adaptation and the white point on both chromatic and luminance JND steps. Identifying those parameters can be useful to avoid or at least reduce contouring effects that otherwise can appear with changes in ambient illumination (eg, with mobile displays).

To approach this problem, Kunkel et al. (2013) proposed building a 3D color space based on the same principles as SMPTE ST 2084/PQ by freely parameterizing a CAM to find the smallest JND ellipsoids for any viewer adaptation in any viewing environment on any display. This approach extends the 1D JND approach to three dimensions, leading to JND "voxels." As this approach is based on color appearance modeling, aspects such as the white point can easily be adjusted, which is illustrated in Fig. 15.27.

With this approach, it is further possible to count all those voxels leading to the number of distinguishable colors. This can potentially lead to more accurate assessments when one is quantifying the color volume of an imaging system or display compared with solely stating the image area in CIE xy or CIE $u'v'$ as shown in Table 15.3.

Recently, Froehlich et al. (2015) introduced a color space that they designed by finding the optimal parameters for a given color space model by minimizing cost functions such as the variance in size for a given JND-ellipsoid set for quantization visibility, hue linearity, and decorrelation to further increase perceptual accuracy. Further, Pytlarz et al. (2016) have introduced a new color space named ICtCp that improves perceptual uniformity for extended dynamic range and wide color gamut imaging pipelines while maintaining the simplicity of YCbCr. This approach can help facilitate the adoption of more accurate color processing in both existing and new imaging workflows.

More in-depth information regarding color appearance modeling is presented in Chapter 9. Further, a good reference is the book by Fairchild (2013).

15.5.6 LIMITATIONS OF PERCEPTUALLY BASED MANIPULATION OF IMAGERY

It is important to keep in mind that image processing based on perceptual principles is not universally the ideal approach. There are image manipulations that are best performed in different spaces. For example, image processing operations that simulate physical processes should be performed in linear light (eg, in RGB or a spectral representation). A good example for an image manipulation operation that simulates physical processes is accurate motion blur and defocus, which are frequently used in postproduction as well as in frame rate upconversion algorithms in consumer displays. An illustration of the differences when blur is applied is provided in Fig. 15.28. Here, the source image (Fig. 15.28A) is convolved both in linear light (Fig. 15.28B) and in SMPTE 2084/PQ (Fig. 15.28C), with only the linear light option rendering the bokeh of a lens as physically expected.

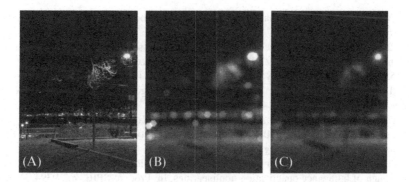

FIGURE 15.28

Defocus rendered in the perceptual domain (PQ) versus linear light. The original image (A) is defocused with a hexagonal convolution kernel (lens blur) in linear RGB (B) and in PQ RGB (C). After Brinkmann (2008).

15.6 SUMMARY

This chapter has covered key perceptual issues relating to the design of a color HDR signal format that also encompasses a wide color gamut. Experiments using specialized laboratory displays to determine the range were described, with full consideration of multiple viewers, wider field of view presentations, and creative intent for sequential imagery (video and movies). By means of psychophysical models, the design then explored how to best quantize that range, and how color gamut can impact that design. The range of the signal format was designed around perceptual needs, and avoided short-term constraints such as today's display capabilities at consumer price levels. Thus, we have shown how a future-proof signal format can be designed, achieving the best possible quality with today's displays, but also for displays that will have much higher capabilities. HDR imagery rendered and stored in this signal format will be robust against the legacy issues that have been plaguing imagery since the earliest days of film.

REFERENCES

Akyüz, A., Fleming, R., Riecke, B., Bülthoff, H., 2007. Do HDR displays support LDR content? A psychophysical evaluation. ACM Trans. Graph. 26 (3), 38.

Barten, P.G.J., 1996. Contrast Sensitivity of the Human Eye and Its Effects on Image Quality. SPIE Optical Engineering Press, Bellingham, WA.

Baylor, D.A., Hodgkin, A.L., Lamb, T.D., 1974. The electrical responses of turtle cones to flashes and steps of light. J. Physiol. 242, 685–727.

Blake, A., Bülthoff, H., Sheinberg, D., 1993. Shape from texture: ideal observers and human psychophysics. Vis. Res. 33 (12), 1723–1737.

Blume, H., Daly, S., Muka, E., 1993. Presentation of medical images on CRT displays — a renewed proposal for a display function standard. In: Proc. SPIE V. 1897 Image Capture, Formatting, and, Display, 2150-231.

Boitard, R., Mantiuk, R., Pouli, T., 2015. Evaluation of color encodings for high dynamic range. In: Proceedings of the SPIE, vol. 9394.

Boynton, R.M., Whitten, D.N., 1970. Visual adaptation in monkey cones: recordings of late receptor potentials. Science 170 (3965), 1423–1426.

Brinkmann, R., 2008. The Art and Science of Digital Compositing, second ed.: Techniques for Visual Effects, Animation and Motion Graphics. Morgan Kaufmann, San Francisco, CA, USA.

Brooks, D., 2015. The art of better pixels. SMPTE Motion Imaging J. 124 (4), 42–48.

Choi, S., Luo, M., Pointer, M., Rhodes, P., 2008. Investigation of large display color image appearance I and II. J. Imaging Sci. Technol. 52 (4), 040904–040904-11.

CIE Colorimetry, 2004. Publication No. 15:2004, third ed. CIE, Vienna.

Cowan, M., Kennel, G., Maier, T., Walker, B., 2004. Contrast sensitivity experiment to determine the bit-depth for digital cinema. SMPTE Motion Imaging J. 113, 281–292.

Daly, S., Golestaneh, S.A., 2015. Use of a local cone model to predict essential CSF light adaptation behavior used in the design of luminance quantization nonlinearities. In: Proc. of SPIE, vol. 9394, Human Vision and Electronic Imaging, San Francisco, CA.

Daly, S., Kunkel, T., Sun, X., Farrell, S., Crum, P., 2013. Viewer preferences for shadow, diffuse, specular, and emissive luminance limits of high dynamic range displays. In: SID Display Week, Paper 41.1, Vancouver, Canada.

Daly, S., Xu, N., Crenshaw, J., Zunjarrao, V., 2014. A psychophysical study exploring judder using fundamental signals and complex imagery. In: SMPTE Annual Technical Conference & Exhibition, Hollywood, CA.

Davson, H., 1990. Physiology of the Eye, fifth ed. Pergamon Press, New York.

DeValois, R., DeValois, K., 1990. Spatial Vision. Oxford University Press, Oxford, p. 26.

Dror, R., Willsky, A., Adelson, E., 2004. Statistical Characterization of Real World Illumination. J. Vis. 4, 821–837.

Eda, T., Koika, Y., Matsushima, S., Ayama, M., 2008. Effect of blackness level on visual impression of color images. In: Proceedings of SPIE, vol. 6806, p. 68061B-1-8.

Eda, T., Koika, Y., Matsushima, S., Ozaki, K., Ayama, M., 2009. Influence of surround luminance upon perceived blackness. In: Proceedings of SPIE, vol. 7240.

Fairchild, M.D., 2013. Color Appearance Models, third ed. Wiley-IS&T Series in Imaging Science and Technology, New York.

Farrell, S., Kunkel, T., Daly, S., 2014. A cinema luminance range by the people, for the people: viewer preferences on luminance limits for a large screen environment. In: SMPTE Annual Technical Conference & Exhibition, Hollywood, CA.

Ferwerda, J.A., 2001. Elements of early vision for computer graphics. IEEE Comput. Graph. Appl. 21 (5), 22–33.

Ferwerda, J.A., Pattanaik, S.N., Shirley, P., Greenberg, D.P., 1996. A model of visual adaptation for realistic image synthesis. In: Proceedings of the 23rd Annual Conference on Computer Graphics and Interactive Techniques, New York, NY, USA, pp. 249–258.

Froehlich, J., Kunkel, T., Atkins, R., Pytlarz, J., Daly, S., Schilling, A., Eberhardt, B., 2015. Encoding color difference signals for high dynamic range and wide gamut images. In: Proceedings of the 23rd Color Imaging Conference, Darmstadt, Germany, pp. 240–247.

Halstead, C., 1993. Brightness, luminance and confusion. Inform. Display 9 (3), 21–24.

Hoffman, D., Johnson, P.V., Kim, J.S., Vargas, A.D., Banks, M.S., 2015. 240 Hz OLED technology properties that can enable improved image quality. J. Soc. Inform. Disp. 22 (7), 346–356.

Hubel, D.H., 1995. Eye, Brain and Vision. W. H. Freemann and Co. Reprint Edition. See also http://hubel.med. harvard.edu.

Huber, D.M., Runstein, R.E., 2009. Modern Recording Techniques, seventh ed., p. 513. Focal Press, Amsterdam.

Hunt, R.W.G., 1995. The Reproduction of Colour, fifth ed. Fountain Press Ltd., Tolworth, England.

Interrante, V., Fuchs, H., Pizer, S.M., 1997. Conveying the 3D shape of smoothly curving transparent surfaces via texture. IEEE Trans. Vis. Comput. Graph. 3 (2), 98–117.

ISO, 2004. ISO 22028-1:2004: Photography and Graphic Technology—Extended Colour Encodings for Digital Image Storage, Manipulation and Interchange—Part 1: Architecture and Requirements. ISO, Geneva.

ISO (2004) ISO 22028-1:2004: Photography and Graphic Technology—Extended Colour Encodings for Digital Image Storage, Manipulation and Interchange—Part 1: Architecture and Requirements. ISO, Geneva

ITU-R Recommendation BT.709-6, 2015. Parameter values for the HDTV standards for production and international programme exchange. Technical report. ITU, Geneva. Formerly CCIR Rec. 709.

ITU-R Recommendation BT.2020-1, 2014. Parameter values for ultra-high definition television systems for production and international programme exchange. Technical report. ITU, Geneva.

ITU-R Report BT.2246-4, 2015. The present state of ultra-high definition television. Technical report. ITU, Geneva.

Kim, M.H., Weyrich, T., Kautz, J., 2009. Modeling human color perception under extended luminance levels. ACM Trans. Graph. 28 (3), 1–9, article 27.

Kishimoto, K., Kubota, S., Suzuki, M., Kuboto, Y., Misawa, Y., Yamane, Y., Goshi, S., Imai, S., Igarashi, Y., Matsumoto, T., Haga, S., Nakatsue, T., 2010. Appropriate luminance of LCD television screens under real viewing conditions at home. J. Inst. Image Inform. Television Eng. 64 (6), 881–890.

Kunkel, T., Reinhard., E., 2009. A neurophysiology-inspired steady-state color appearance model. J. Opt. Soc. Am. A 26 (4), 776–782, http://dx.doi.org/10.1145/344779.344810.

Kunkel, T., Reinhard, E., 2010. A reassessment of the simultaneous dynamic range of the human visual system. In: Symposium on Applied Perception in Graphics and Vision, pp. 17–24.

Kunkel, T., Ward, G., Lee, B., Daly, S., 2013. HDR and wide gamut appearance-based color encoding and its quantification. In: 30th IEEE Picture Coding Symposium.

Lagendijk, E., Hammer, M., 2010. Contrast requirements for OLEDs and LCDs based on human eye glare. In: SID Display Week Tech Digest 14.4.

Lubin, J., Pica, A., 1991. A non-human quantizer matched to human visual performance. SID Digest 30 (3), 619–622.

MacAdam, D.L., 1942. Visual sensitivities to color differences in daylight. J. Opt. Soc. Am. 32 (5), 247–273.

Mantiuk, R., Krawczyk, G., Myszkowski, K., Seidel, H.P., 2004. Perception-motivated high dynamic range video encoding. ACM Trans. Graph. (Proc. SIGGRAPH) 23 (3), 733.

Mantiuk, R., Myszkowski, K., Seidel, H., Informatik, M.P.I., 2006. Lossy compression of high dynamic range images and video. In: Human Vision and Electronic Imaging, p. 60570V.

Mantiuk, R., Rempel, A.G., Heidrich, W., 2009. Display considerations for night and low-illumination viewing. In: Proceedings of the 6th Symposium on Applied Perception in Graphics and Visualization. ACM, New York, NY, pp. 53–58.

McCann, J., Rizzi, A., 2007. Camera and visual veiling glare in HDR images. J. Soc. Inform. Disp. 15 (9), 721–730.

McCann, J., Rizzi, A., 2011. Limits of HDR in humans. The Art and Science of HDR Imaging. Wiley SPSE Press, New York (Chapter 12).

Michaelis, L., Menten, M.L., 1913. Die kinetik der invertinwerkung. Biochem. Z. 49.

Miller, S., Daly, S., Nezamabadi, M., 2013. Perceptual signal coding for more efficient usage of bit codes. SMPTE Motion Imaging J. (May/June).

Murdoch, M.J., Heynderickx, I.E.J., 2012. Veiling glare and perceived black in high dynamic range displays. J. Opt. Soc. Am. A 29 (4).

Muryy, A., Fleming, R., Welchman, A., 2014. Key characteristics of specular stereo. J. Vis. 14 (14), 1–26.

Naka, K.I., Rushton, W.A.H., 1966. S-potentials from luminosity units in the retina of fish (Cyprinidae). J. Physiol. 185, 587–599.

NEMA Standards Publication PS 3.14-2008, 2008. Digital Imaging and Communications in Medicine (DICOM), Part 14: Grayscale Standard Display Function. National Electrical Manufacturers Association, Rosslyn, VA.

Nezamabadi, M., Miller, S., Daly, S., Atkins, R., 2014. Color signal encoding for high dynamic range and wide color gamut based on human perception. In: Color Imaging at SPIE's Electronic Imaging. Proc. SPIE 9015, Color Imaging XIX: Displaying, Processing, Hardcopy, and Applications, 90150C.

Normann, R., Baxter, B., 1983. Photoreceptor contribution to contrast sensitivity: applications in radiological diagnosis. IEEE Trans. Syst. Man Cybernetic SMC-13 (5), 944–953.

Pattanaik, S.N., Ferwerda, J.A., Fairchild, M.D., Greenberg, D.P., 1998. A multi- scale model of adaptation and spatial vision for realistic image display. In: SIGGRAPH '98: Proceedings of the 25th Annual Conference on Computer Graphics and Interactive Techniques. ACM, New York, NY, pp. 287–298.

Peirce, J., 2007. The potential importance of saturating and supersaturating contrast response functions in visual cortex. J. Vis. 7 (6), 1–10, 4.

Planck, M., 1901. Über das Gesetz der Energieverteilung im Normalspectrum. Ann. Phys. 3, 553–563.

Reinhard, E., 2008. Color Imaging: Fundamentals and Applications. http://dl.acm.org/citation.cfm?id=1386684.

Reinhard, E., Ward, G., Pattanaik, S., Debevec, P., Heidrich, W., Myszkowski, K., 2010. High Dynamic Range Imaging: Acquisition, Display, and Image-Based Lighting, second ed., The Morgan Kaufmann Series in Computer Graphics. Morgan Kaufmann, San Francisco, CA, USA.

Rempel, A., Heidrich, W., Li, H., Mantiuk, R., 2009. Video viewing preferences for HDR displays under varying ambient illumination. In: Proceedings of the 6th Symposium on Applied Perception in Graphics and Visualization. ACM, New York, NY, pp. 45–52.

Riggs, L., 1971. Vision. In: Kling, J.W., Riggs, L.A. (Eds.), Woodworth and Schlosberg's Experimental Psychology, third ed. Holt, Rinehart, and Winston, New York.

Roth, S., Weiss, N., 2007. Multi-Primary LCD for TV Applications, Genoa Color Technologies. In: Society for Information Display (SID) Symposium, Long Beach, CA, USA.

Salvaggio, N., 2008. Basic Photographic Materials and Processes. Focal Press, New York.

Schlick, C., 1994. Quantization techniques for the visualization of high dynamic range pictures. In: Shirley, P., Sakas, G., Müller, S. (Eds.), Photorealistic Rendering Techniques. Springer-Verlag, Berlin/Heidelberg/New York.

Schreiber, W.F., 1992. Fundamentals of Electronic Imaging Systems, third ed. Springer-Verlag, Berlin.

Seetzen, H., Li, H., Ye, L., Heidrich, W., Whitehead, L., Ward, G., 2006. Observations of luminance, contrast and amplitude resolution of displays. Proc. Soc. Inform. Display 37 (1), 1229–1233.

Shyu, M., Tang, D., Pong, B., Hsu, B., 2008. A study of preferred image quality with the manipulation on image attributes for LCD display. 3rd International Workshop on Image Media Quality and Its Applications, IMQA.

Smith, T., Guild, J., 1931. The C.I.E. colorimetric standards and their use. Trans. Opt. Soc. 33 (3), 73–134.

SMPTE RP 431-1, 2006. D-Cinema Quality — Screen Luminance Level, Chromaticity and Uniformity. The Society of Motion Picture and Television Engineers Family: D-Cinema Quality.

SMPTE RP 431-2, 2011. D-Cinema Quality — Reference Projector and Environment. The Society of Motion Picture and Television Engineers Family.

SMPTE ST 2065-1, 2012. Academy Color Encoding Specification (ACES).

Ueki, S., Nakamura, K., 2007. Five-primary-color 60-inch LCD with novel wide color gamut and wide viewing angle. In: Society for Information Display (SID) Symposium, San Antonio, TX, USA.

Valeton, J.M., van Norren, D., 1983. Light adaptation of primate cones: an analysis based on extracellular data. Vis. Res. 23 (12), 1539–1547.

Wandell, B.A., 1995. Foundations of Vision. Sinauer Associates, Sunderland, MA.

Ward, G., 2008. 59.2: defining dynamic range. SID Symp. Digest Tech. Pap. 39 (1), 2168-0159.

Watson, A.B., Yellot, J.I., 2012. A unified formula for light-adapted pupil size, J. Vis. 12 (10), 12.

Watson, A., Barlow, H., Robson, J., 1983. What does the eye see best? Nature 302, 419–422.

Whittle, P., 1986. Increments and decrements: luminance discrimination. Vis. Res. 26 (10), 1677–1691.

Wolff, L., 1994. On the relative brightness of specular and diffuse reflection. In: CVPR.

Wright, W.D., 1941. The sensitivity of the eye to small colour differences. Proc. Phys. Soc. 53 (2), 93.

Wyszecki, G., Stiles, W.S., 2000. Color Science: Concepts and Methods, Quantitative Data and Formulae (Wiley Series in Pure and Applied Optics). Wiley-Interscience, New York.

Yoshida, A., Blanz, V., Myszkowski, K., Seidel, H., 2005. Perceptual evaluation of tone mapping operators with real-world scenes. In: Proceedings of SPIE, vol. 5666, pp. 192–203.

Youssef, P.N., Sheibani, N., Albert, D.M., 2011. Retinal light toxicity. Eye (London) 25 (1), 1–14.

QUALITY OF EXPERIENCE AND HDR: CONCEPTS AND HOW TO MEASURE IT

16

M. Narwaria*, M.P. Da Silva*, P. Le Callet*, G. Valenzise†, F. De Simone†, F. Dufaux†

*University of Nantes, Nantes, France**
Telecom ParisTech, CNRS LTCI, Paris, France†

CHAPTER OUTLINE

16.1 INTRODUCTION

Humans perceive the outside visual world through the interaction between light energy (usually measured in candelas per square meter) and the eyes. The luminance creates a sensation which helps us to recognize different aspects of the scene, including brightness levels, contrast, color, and motion. Pertaining to the luminance levels found in the real world, direct sunlight at noon can be on the order of in excess of 10^7 cd/m^2, while a starlit night can be on the order of 10^{-1} cd/m^2. This corresponds to more than eight orders of magnitude. Thus, there is a large range of luminance present in different real-world scenes. The human eye also has the remarkable capability to perceive a large dynamic range (about 13 orders of magnitude), especially with sufficient adaptation time (Mather, 2006). An intuitive example of adaptation is when we arrive in a low-lit room on a sunny day. We cannot immediately perceive the visual data in the room and it takes a few minutes before we become accustomed (adapted) to the new luminance levels. In contrast, the instantaneous (ie, without adaptation) human vision range is smaller and the eyes are capable of dynamically adjusting so that a person can see about five orders of magnitude throughout the entire range. However, conventional display devices cover only up to three orders of magnitude. Consequently, the scenes viewed on typical low dynamic range (LDR) displays have lower contrast and a smaller color gamut than what the eyes can perceive. This leads to loss of visual details and in some cases can even lead to misrepresentation of the scene information. To overcome such limitations, high dynamic range (HDR) has recently gained popularity in both academia and industry. By representing the scene in terms of physical luminance information, HDR can achieve very high contrasts and a wider color gamut, in effect matching the human instantaneous vision range. Because it allows more scene information to be represented, HDR helps to capture very fine details which are otherwise difficult to retain with traditional photography. This leads to better visual experience for viewers, and this is particularly relevant in the context of the recent paradigm shift toward quality of experience (QoE)-based multimedia signal processing.

Such QoE-driven multimedia systems have increasingly come into focus in recent years, from both research and industry perspectives. The aim is to capture the end users' aesthetic expectations rather than simply to deliver content based on a technology-centric approach. As discussed, given the specific characteristics of HDR, it is one of the exciting fields toward providing end users with a more immersive and realistic viewing experience. As a result, HDR imaging is widely seen as an attractive alternative to improve the viewing experience of users in terms of more realistic content delivery. In light of this, it is important to define and quantify the user QoE in a more holistic fashion by our taking into account the relevant factors. This chapter focuses on discussing some of them and highlighting their role in HDR QoE.

QoE is a term that is especially relevant in the context of immersive communication technologies such as HDR imaging. According to a White Paper (Callet et al., 2013), QoE is defined as "the degree of delight or annoyance of the user of an application or service. It results from the fulfillment of his or her expectations with respect to the utility and/or enjoyment of the application or service in the light of the user's personality and current state." This definition includes several complex factors that contribute to the QoE in a specific application scenario. In a similar way, HDR QoE can include several dimensions. Thus, the HDR viewing experience should be viewed in the light of a broader framework rather than just signal fidelity. More specifically, perceptual fidelity has often been used as a proxy for QoE in many LDR video processing applications. Fidelity is defined as the measure of "closeness" of a processed image/video to its source. By definition, this imposes limits on the expectation and/or enjoyment of the end user based only on the characteristics of original source content. Such a narrow approach can fail in many cases; for instance, a processed content can possibly appear better despite its low fidelity with source content (eg, in the case of contrast enhancement, color correction/enhancement). Clearly, replacing QoE with just fidelity is an oversimplification because the user experience is much more complex and depends on several factors, including emotions, aesthetics, the surroundings, and the application scenario. With next-generation video technologies, the inadequacies of a fidelity-based QoE will be more apparent. Thus, fidelity should ideally not be substituted for QoE; rather, it is one of the several components/dimensions. Other factors include immersiveness, glare, naturalness, color, context of visual details (Narwaria et al., 2014a), and visual attention. In the next section, we elaborate on these dimensions.

16.2 DIMENSIONS IN HDR QoE

As pointed out in the introduction, defining and measuring HDR QoE is a multidimensional problem. In this section, we discuss some of the most prominent aspects that HDR QoE comprises.

- Immersiveness: The first and probably the most attractive feature of HDR is the immersiveness that it can offer. This is enabled because HDR imaging aims to overcome the inadequacies of the LDR capture and display technologies via better video signal capture, representation, and display so that the dynamic range of the video can better match the instantaneous range of the eye. In particular, the major distinguishing factor of HDR imaging (in comparison with traditional LDR imaging) is its focus on capturing and displaying scenes as natively (ie, how they appear in the real world) as possible by considering the physical luminance of the scene in question. This allows the capture and display of intricate scene details and a wider color gamut, which can enhance the realism of the captured scene, and allows the user to be more involved (immersed) in video viewing. Obviously, the concept of depth (as in three-dimensional video) and higher resolutions (eg, ultra high definition) could possibly be added to HDR imaging. Therefore, the degree or extent to which the user gets involved (or immersiveness) is probably the most important dimension in HDR QoE.
- Perceptual fidelity: As pointed out previously, fidelity is another component of HDR QoE. In the light of HDR video representation, the concept of perceptual fidelity is based on (relative) luminance values, and will thus involve models that can account for the relatively higher luminance conditions. Specifically, perceptual fidelity in HDR imaging should be based on visual details, color, and naturalness among other aspects. Thus, fidelity in HDR imaging by itself is expected to be a more complex.
- Visual attention: One of the key thrusts of HDR imaging is to preserve artistic intention. Artistic intention in simple terms refers to the message that the artist intends to convey through the photograph/video. This can involve the creation of focal points or regions of interest where the viewer should ideally look when viewing the content. Because HDR imaging provides a bigger scope in terms of providing more accurate visual details and colors, it would be useful in conveying more precise scene information to the viewer. Visual attention, which is the ability of humans to focus on certain regions of the scene, is therefore a useful tool toward capturing eye movements and thereby providing valuable information on the visually important (salient) regions in the content. Obviously, the visual attention behavior ideally should be the same for both processed and original content.
- Glare: Because HDR viewing involves much higher luminance in comparison with LDR viewing, the phenomenon of glare will be more apparent and needs consideration. Glare is a sensation when the emitted luminance exceeds the value at which the eyes are adapted at that point in time. It follows that glare will depend not only on the magnitude of emitted luminance but also on the

angle between the light source and eye level, in addition to the eye adaptation level. The glare effect also strongly depends on the content. Thus, measurement of the glare and analysis of its impact on the viewing experience will be necessary to quantify the HDR QoE more accurately.

- Naturalness: Naturalness is a typical feature of visual content. Its genesis lies in the premise that our visual system is adapted and used to processing a sensory input from real-world scenes. Thus, content that would appear *similar* to the adapted model will be natural. It is, of course, nontrivial to quantify naturalness on the basis of such an abstraction, and it is a highly subjective phenomenon. As HDR imaging aims to replicate the visual scene information as natively as possible, original or unprocessed HDR content is expected to be highly natural. The subsequent processing may, however, affect this and is therefore an important component in quantification of the HDR QoE.

The above-mentioned dimensions are by no means exhaustive but are probably the most prominent ones with typical use cases and applications. Also, these dimensions are not necessarily independent. Thus, loss of fidelity could decrease naturalness or a change in visual attention behavior need not directly affect the naturalness. This, in turn, implies that careful considerations are needed in order to accurately quantify these dimensions with both subjective and objective approaches.

Processing of HDR content during different stages of a typical HDR delivery chain is likely to alter one or more of the mentioned dimensions and hence affect the QoE. Thus, each HDR content processing module should be developed in a more QoE-aware fashion so that the artistic intent is rightfully conveyed to the end user. Because most existing technologies (including processing and display) were designed for LDR content, there is a need to extend them for use in the HDR domain, and at the time of writing work on this is ongoing. For instance, there have been recent efforts within the Moving Picture Experts Group (MPEG) to extend High Efficiency Video Coding (HEVC) to HDR. Likewise, the JPEG has announced extensions that will feature the original JPEG standard with support for HDR image compression. Such efforts typically seek to use legacy coders by range reduction (formally referred to as tone mapping) of the HDR video and then use additional transformations to obtain a compressed HDR signal. Further, tone mapping is needed to render and visualize HDR content on LDR displays. Even the existing HDR displays cannot accommodate the entire luminance range/contrast ratios found in the real world. Thus, it is more accurate to mention that both HDR and LDR displays require tone mapping although the nature and extent of such mapping can be different in the two cases. Thus, tone mapping is one of the key operations that will be required both to extend existing video processing tools to HDR and for visualization. Apart from this, processing-related artifacts (eg, codec induced distortions) will affect HDR QoE.

The next important issue is that of measuring HDR QoE. To that end, subjective viewing tests remain the most reliable and accurate given appropriate laboratory conditions and a sufficiently large subject panel. Although an objective approach (use of a computational model) could also be used, it may not always mimic the subjective opinion. Therefore, in the next section, we discuss a few important aspects in subjective assessment of HDR QoE.

16.3 MEASURING HDR QoE: A FEW CONSIDERATIONS

Subjective test results are considered gold standards when human factors are involved in evaluating and testing algorithms. HDR subjective tests serve the same purpose as traditional LDR subjective tests but there are some differences worth highlighting.

16.3.1 EFFECT OF DISPLAY

While HDR values are related to the actual scene luminance, they are typically not equal to it. Thus, unless there is a camera calibration, the HDR values represent the real-world luminance up to an unknown scale. Because the maximum luminance can vary for each scene, there is no fixed white point in this case. So HDR values must be interpreted on the basis of the display used to view the HDR video (this is in contrast to the case of LDR, where the white point is fixed; eg, for an eight-bit representation, the maximum value is fixed at 255 irrespective of the actual maximum luminance). This also introduces the need to carefully preprocess the HDR stimuli before they are displayed on the HDR display. HDR displays typically have a front liquid crystal panel and a backlight array of LEDs, so controlling the signals sent to these provides an effective way to control the output on the HDR display. Thus, with HDR there can possibly be a higher impact of the way the HDR stimuli are ultimately displayed. In contrast, there are no (or minimal) such considerations in the LDR case for most applications.

16.3.2 HIGH LUMINANCE CONDITIONS AND VISUAL DISCOMFORT

The most important distinction between HDR and LDR is with respect to the luminance range (which in turn leads to HDR). Traditional LDR has a white point which depends on the maximum displayable luminance and contrast ratio (both are usually insufficient to accurately render real-world scenes considering typical LDR displays). Moreover with LDR, the pixel values are typically gamma encoded and perceptually uniform. As a result, a change in the pixel values can be directly related to the change in visual perception. However, with HDR there is more flexibility and one can accurately represent the real luminance (generally up to an unknown scale factor). Consequently, there is no fixed white point in HDR that can correspond to the maximum luminance (as it can vary from scene to scene). There is only brighter (or darker) scene intensity. Of course, in practice, we still need to define a white point for rendering content on HDR displays (because of hardware limitations) but it is typically much higher than in the LDR case. Therefore, HDR viewing will involve higher levels of brightness, in general. Because human vision is sensitive to the luminance ratio (rather than absolute luminance), changes in the luminance may not necessarily lead to the same change in visual perception of HDR. High luminance can also be a source of visual discomfort for observers and should be carefully tackled. Because the HDR display luminance is relatively much higher (eg, the SIM2 Solar HDR display[1] has a maximum displayable luminance of 4000 cd/m^2) than that of conventional display devices, much higher background illumination is needed to reduce the visual discomfort of observers. Improper settings can result in glare, leading to maladapted viewing conditions. For HDR, there are currently no standard illumination settings. Recommendation ITU-R BT.500-13 recommends the illumination should be approximately 15% of the peak display luminance. With an HDR display, this means approximately 600 cd/m^2. But given the reduced sensitivity of the human eye at high luminance, values typically in the range of 150–200 cd/m^2 were found to be adequate in our tests (Narwaria et al., 2014d).

16.3.3 OBSERVERS

The general conditions of the LDR domain apply (eg, the observers should not have been involved in setting up the experiment, the observer panel is sufficiently large). Because most observers may not

---

[1]http://www.sim2.com/HDR/.

be familiar with how an HDR video appears, specific instructions should be provided to the observers especially when the task is to compare HDR and LDR stimuli. A training session for observers can therefore assume more importance in some cases for HDR subjective tests.

16.3.4 VIEWING CONDITIONS

The display resolution should ideally be full high definition or greater so that the immersive experience offered by HDR is not compromised by screen resolution. The viewing distance and angle should follow similar guidelines as for the LDR domain (eg, viewing distance of three times the active screen height in case of high definition).

16.3.5 SOURCE CONTENT SELECTION

While the general considerations of LDR (such as spatial and temporal information) apply to HDR, there is an additional factor of dynamic range. For example, scenes with different dynamic range must be selected to challenge the algorithm under consideration. The classical definition of dynamic range (ratio of maximum and minimum luminance) can be used but it is important to remember that it may suffer from drawbacks such as susceptibility to outliers and can be misleading in some situations (eg, a tiny patch of very dark and bright pixels can increase the dynamic range). More sophisticated and recently proposed solutions (eg, the solution by Narwaria et al., 2014b) could also be used as an alternative index for source content selection based on certain perceptual considerations. Thus, HDR source video selection is a three-dimensional problem (with spatial, temporal, and dynamic range as factors) although the spatial information may possibly be combined with dynamic range.

16.3.6 PAIRED COMPARISON TESTS IN AN HDR SETTING

Care must be taken that the HDR stimuli to be compared are at same (or similar) luminance levels. Thus, comparing two stimuli from different source content via paired comparison can be tricky especially for videos, where the luminance could vary greatly over time. Studying the impact of tone mapping is an interesting use case where observers watch both HDR and LDR stimuli simultaneously. Because the peak brightness of the displays can be very different, arriving at a comfortable illumination level is not easy. An alternative is to use higher illumination around the HDR display, while the diffused light can act as the illumination source for the LDR display. Paired comparison is generally assumed to be more effective as the observers have an easier task. Given the luminance considerations in HDR imaging, this calls for greater care in setting up a subjective experiment.

16.4 IMPACT OF TONE MAPPING OPERATORS ON QoE DIMENSIONS

Tone mapping operators (TMOs) play an important role in HDR content processing as well as visualization. Hence, evaluation of their impact on different dimensions such as visual attention, fidelity, naturalness, and color is necessary.

TMOs can be broadly classified into two categories — namely, local operators and global operators. As the name implies, local operators use a spatially varying mapping which depends on the local image content. In contrast, global operators use the same mapping function for the whole image. Chiu et al. (1993) introduced one of the first local TMOs by using a local intensity function based on a low-pass filter to scale the local pixel values. The method proposed by Fattal et al. (2002) is based on compression of the magnitudes of large gradients and solves the Poisson equation on the modified gradient field to obtain tone-mapped images. Durand and Dorsey (2002) presented a TMO based on the assumption that an HDR image can be decomposed into a base image and a detail image. The contrast of the base layer is reduced by an edge-preserving filter (known as the bilateral filter). The tone-mapped image is obtained as a result of multiplication of the contrast-reduced base layer with the detail image. Drago et al. (2003b) adopted logarithmic compression of the luminance values for dynamic range reduction in HDR images. They used adaptively varying logarithmic bases so as to preserve local details and contrast. The TMO proposed by Ashikhmin (2002) first estimates the local adaptation luminance at each point, and this is then compressed with a simple mapping function. In the second stage, the details lost in the first stage are reintroduced to obtain the final tone-mapped image. Reinhard et al. (2002) applied the dodging and burning technique (traditionally used in photography) for dynamic range compression. A TMO known as iCAM06 (Kuang et al., 2007a) has also been developed. It is based on the sophisticated image color appearance model (iCAM) and incorporates the spatial processing models in the human visual system for contrast enhancement, photoreceptor light adaptation functions that enhance local details in highlights and shadows. With regard to global TMOs, the simplest one is the linear operation in which the maximum input luminance is mapped to the maximum output value (ie, maximum luminance mapping) or the average input luminance is mapped to the average output value (ie, average luminance mapping). Another global TMO was proposed by Greg (1994) and focuses on the preservation of perceived contrast. In this method, the scaling factor is derived from a psychophysical contrast sensitivity model. Tumblin et al. (1999) have reported a TMO based on the assumption that a real-world observer should be the same as a display observer. The list of TMOs mentioned here is by no means exhaustive and the keen reader may refer to existing work for more details on the topic.

16.4.1 EFFECT ON PERCEPTUAL FIDELITY

It is worth pointing out that for assessment of the impact of TMOs, one can choose different test settings. These include the following:

- Evaluation with a real-world reference: This typically involves the use of a real scene as a reference. This is probably ideal because the source content in this case is absolutely unprocessed. However, such a setup is hardly practical because luminance conditions can change dramatically, especially in the case of outdoor scenes. This coupled with the fact that there will be incoherent motion of objects within the scene, potentially reducing its utility. It can, nevertheless, be used in controlled laboratory conditions with static images.
- Evaluation with a reference on an HDR display: This is probably the more scientific approach in terms of comparing tone-mapped content with an HDR reference. However, it should always be kept in mind that even HDR monitors cannot display the entire range of real-world luminance and require range reduction (tone mapping). So, the reference is not entirely pristine but is generally of higher visual quality than an eight-bit tone-mapped version.

- Evaluation without a reference: This is the most straightforward method. The advantage lies in its simplicity and it not requiring an HDR display. Such a method will, however, typically measure the preference on the basis of only the tone-mapped content. This may lead to a larger subjectivity in the results obtained.

Thus, depending on the requirements, one of the above methods can be used to evaluate the impact of TMOs. The studies in the literature used one of the mentioned settings to measure the fidelity of the tone-mapped HDR content subjectively. We first briefly describe some of the existing studies related to subjective evaluation of TMOs.

The psychophysical experiments performed by Drago et al. (2003a) aimed to evaluate six TMOs with regard to similarity and preference. Three perceptual attributes — namely, apparent image contrast, apparent level of detail (visibility of scene features), and apparent naturalness (the degree to which the image resembled a realistic scene) — were investigated. It was found that naturalness and details are important attributes for perceptual evaluation of TMOs. In the study by Kuang et al. (2007b) three experiments were performed. The first one aimed to test the performance of TMOs with regard to image preference. For this experiment, 12 HDR images were tone-mapped with six different TMOs and evaluation was done by the paired comparison method. The second experiment dealt with the criteria (or attributes) observers used to scale image preference. The attributes that were investigated included highlight details, shadow details, overall contrast, sharpness, colorfulness, and the appearance of artifacts. The subsequent regression analysis showed that the rating scale of a single image appearance attribute is often capable of predicting the overall preference. The third experiment was designed to evaluate HDR rendering algorithms for their perceptual accuracy of reproducing the appearance of real-world scenes. To that end, a direct comparison between three HDR real-world scenes and their corresponding rendered images displayed on an LDR LCD monitor was used. Yoshida et al. (2005) conducted psychophysical experiments which involved the comparison between two real-world scenes and their corresponding tone-mapped images (obtained by the application of seven different TMOs to the HDR images of those scenes). Similarly to other studies, this study also aimed at assessing the differences in how tone-mapped images are perceived by human observers and was based on four attributes: image naturalness, overall contrast, overall brightness, and detail reproduction in dark and bright image regions. In the experiments conducted by Ledda et al. (2005), the subjects were presented with three images at a time: the reference HDR image displayed on an HDR display and two tone-mapped images viewed on LCD monitors. They had to choose the image closest to the reference. Because an HDR display was used, factors such as screen resolution, dimensions, color calibration, viewing distance, and ambient lighting could be controlled. This is in contrast to the use of a real-world scene as a reference, which might introduce uncontrolled variables. Ledda et al. (2005) also reported the statistical analysis of the subjective data with respect to the overall quality and to the reproduction of features and details. Differently from the previously mentioned studies, Cadik et al. (2008) adopted both a direct rating (with reference) comparison of the tone-mapped images with the real scenes and a subjective ranking of tone-mapped images without a real reference. They further derived an overall image quality estimate by defining a relationship (based on multivariate linear regression) between the attributes: reproduction of brightness, color, contrast, detail, and visibility of artifacts. The analysis further revealed that contrast, color, and artifacts are the major contributing factors in the overall judgment of the perceptual quality. However, it was also argued that the effect of attributes such as brightness is indirectly incorporated through other attributes. Another conclusion from this study was that there was agreement between the ranking (of two tone-mapped images) and rating (with respect to a real scene) experiments. In contrast to this last observation, Ashikhmin and Goyal (2006) found that there were significant differences in subjective opinions depending on whether a real scene is used as a reference or not. Narwaria et al. (2014c) studied the impact of TMOs in HDR image compression and found that the visual quality of compressed HDR content can be significantly modified by TMOs. A recent survey evaluating TMOs for HDR video can be found in Eilertsen et al. (2013).

Most of these studies either ranked the TMOs on the basis of the performance in the respective subjective experiments or outlined the factors affecting visual quality of the tone-mapped content. However, it might be misleading to generalize the results from these studies because the number of HDR stimuli was limited. Nevertheless, all of them establish beyond doubt that tone mapping (of both still images and videos) not only tends to reduce the visual quality but also affects the naturalness of the processed HDR content (in addition for video stimuli there could be visible temporal artifacts). Because the underlying philosophy of TMOs is concerned with reducing the range, they inevitably saturate visual information, leading to loss of details.

16.4.2 VISUAL ATTENTION MODIFICATION

It is well known that human eyes tend to focus more on certain areas in an image/video than others, and therefore possess a remarkable ability to find and focus on relevant information (within a given context) quickly and efficiently (Rensink, 2006). To study this phenomenon, a saliency map is generally obtained as a two-dimensional representation to reveal the locations where an average human observer tends to look when viewing the scene (this is done with an eye tracker). In the context of QoE, eye tracking is a free-viewing task — that is, the observers are instructed to watch the scene as they would in a natural setting without their having to do any other task (this encourages attention tracking mainly based on bottom-up behavior).

To begin the analysis for visual attention, we recall that TMOs tend to destroy visual details by damaging contrast (local and global). This means that the regions that may have been salient (attracted eye attention) in the HDR content may become nonsalient. Conversely, a TMO might introduce a false edge or structure and may create salient areas in the tone-mapped content which were absent in the original HDR content. Hence, the artistic intention can be modified by TMOs (Narwaria et al., 2014d). To visually exemplify this, we consider Fig. 16.1. The images in Fig. 16.1A–D are the tone-mapped versions of the HDR image processed by the TMO of Drago et al. (2003b) and iCAM06 (Kuang et al., 2007a) and the corresponding human priority maps (visual attention maps), respectively. It can be seen that the two "red mats" (highlighted by red boxes; dark gray in print versions) are more clearly visible in the image processed by iCAM06 because there high contrast is preserved in and around that region. Consequently, one can see from the corresponding visual attention map that these indeed are salient regions for human observers. On the other hand, in the image processed by the TMO of Drago et al. (2003b) there is much lower contrast in the said regions. As a result, these attract much less eye attention as seen in the corresponding visual attention map. A second set of examples is shown in Fig. 16.1E–J. Fig. 16.1E–G shows three tone-mapped versions of an HDR image (tone mapped by the TMO of Tumblin et al. (1999), iCAM06, and a linear TMO), while the corresponding visual attention maps are shown below them in Fig. 16.1H–J. Again, one finds that the "orange spot" (highlighted by the red box; dark gray in print versions) is a salient region only in the case of the linear TMO (see the visual attention map in Fig. 16.1J) because this TMO destroys contrast in other regions, which makes

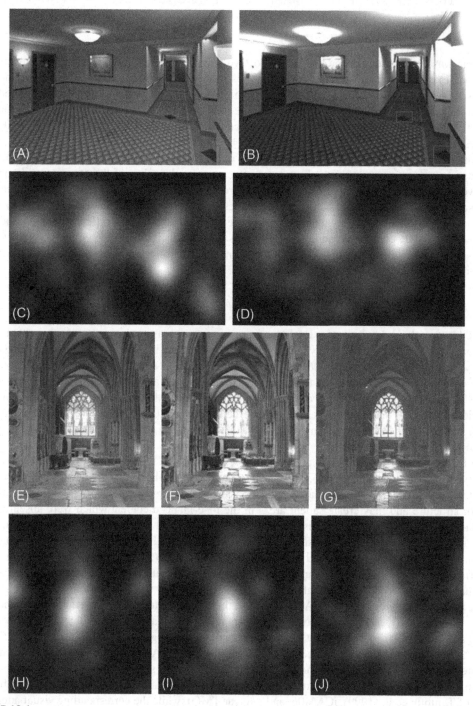

FIGURE 16.1

Effect of TMOs on visual attention: (A) image processed by iCAM06; (B) image processed by the TMO of Drago et al. (2003b); (C) visual attention map for (A), and (D) visual attention map for (B); (E)–(G) images processed by the TMO of Tumblin et al. (1999), iCAM06, and a linear TMO, respectively; and (I)–(J) visual attention maps for the images in (E)–(G), respectively. The red boxes (dark gray in print versions) highlight the area(s) in the images which become salient or nonsalient depending on the overall impact of the TMO.

 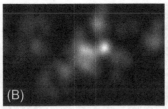 

(A) (B) (C)

FIGURE 16.2

Effect of TMOs on visual attention: (A) tone-mapped version of "moto" (TMO of Reinhard et al., 2002); (B) visual attention map of the image in (A) obtained from eye-tracking; and (C) visual attention map of the "moto" HDR image obtained from eye-tracking.

the "orange spot" stand out and thus eye-catching. As opposed to this, the TMO of Tumblin et al. (1999) and iCAM06 provide much better contrast in other parts of the image as well. So the "orange spot" is nearly nonsalient in these two images as the observers' attention is attracted to other parts.

Tone mapping can also be viewed in terms of reduced signal contrast due to tone mapping. With the reduced contrast, regions that may have attracted the observers' attention in the HDR content might be reduced. As a direct consequence, the number of salient regions in tone-mapped HDR content tend to decrease. To visually exemplify this, consider Fig. 16.2, where Fig. 16.2A shows a tone-mapped version (obtained with the TMO of Reinhard et al. (2002)) of an HDR image. In this image, we can easily identify the foreground (mainly comprising the headlight and front wheel of the motorbike) and the background (bicycles and the door). Fig. 16.2B shows the visual attention map of the image in Fig. 16.2A, while the HDR visual attention map is shown in Fig. 16.2C. Observe how the visual attention map in Fig. 16.2B indicates very few salient points in the background. As opposed to this, the HDR visual attention map shows that the background also had regions which attracted eye attention. The reason is obvious: tone mapping in this case destroys details mainly in the background but the foreground is fairly well preserved in terms of contrast. As a result, the number of salient points in the background reduce drastically. The last visual example is shown in Fig. 16.3. Fig. 16.3D–F shows the visual attention maps corresponding to the images shown in Fig. 16.3A–C. Because the HDR image does not display properly, we have shown the image processed by iCAM06 (instead of the original HDR image) for the sake of explanation. Also note that we have used a green box to highlight the area(s) that attracted maximum human attention and a blue box for area(s) with relatively lower attention.

Considering the HDR visual attention map in Fig. 16.3D, we see there are four main regions which are salient according to human observers. These have been highlighted by rectangular boxes at the corresponding image locations in Fig. 16.3A. Notice that the "outside area seen through the door" attracts more attention that the other three identified regions. Now we observe the effect of tone mapping on these four identified regions. We find that the image processed by the TMO of Ashikhmin (2002) (Fig. 16.3B) shows that now there are two regions (highlighted by the two middle rectangular boxes) which attract the maximum attention (see the corresponding visual attention map in Fig. 16.3E). Thus, tone mapping modified the visual signal in such a manner that a region which was less salient in the original HDR content has become more salient. Likewise, looking at the visual attention map in Fig. 16.3F, we find that there is only one region that now attracts maximum attention (this is marked by the rightmost rectangular box in the image in Fig. 16.3C), while the attention for other regions

FIGURE 16.3

Effect of TMOs on visual attention: (A) "dani_belgium" HDR image (for sake of better visualization, the tone-mapped image processed by iCAM06 is shown); (B) tone-mapped version of "dani_belgium" processed by the TMO of Ashikhmin (2002); (C) tone-mapped version of "dani_belgium" processed by the TMO of Drago et al. (2003a); (D) visual attention map of the HDR image; and (E) and (F) corresponding visual attention maps of the images in (B) and (C).

reduced considerably. Thus, tone mapping can change attentional regions in addition to increasing or decreasing the magnitude of attention. The eventual result is that a nonattentional region in the HDR image becomes an attentional one in the tone-mapped version. The opposite case is that in which structural information is destroyed by tone mapping. In such cases, an attentional region in the HDR image becomes less important (or less eye-catching) in the tone-mapped image. For example, a contrast that was visible in the HDR image becomes invisible in the processed image (loss of visible contrast).

Video signals differ from images because of the addition of a temporal dimension in addition to the spatial one. Given this, it is interesting to analyze the changes in visual attention behavior caused by tone mapping of HDR video sequences. The analysis of the visual attention maps from different video stimuli leads to similar conclusions as for still images. That is, TMOs have a large impact on the visual attention behavior as compared with the HDR video. We present an example in Fig. 16.4. This is from the video sequence[2] "Tunnel1," in which a car is shown to enter a tunnel (with normal traffic). Inside the tunnel, there is relatively lower illumination and so as the car enters it, there is a large change in scene illumination. Fig. 16.4A shows the car inside the tunnel and another car just behind it which also enters the tunnel. Before this time, we found that car was the main region of the subjects' attention from the start of the video. However, when the other car enters the frame from behind, it attracts the

---

[2]This video sequence was shot as part of the NEVEx project FUI11, related to HDR video chain study.

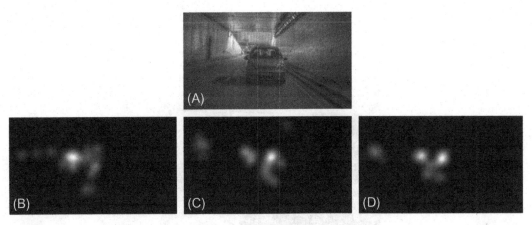

FIGURE 16.4

Change in visual attention behavior in videos: (A) video frame; (B) HDR visual attention map; (C) visual attention map of the frame tone-mapped by the TMO of Tumblin et al. (1999); and (D) visual attention map of the frame tone-mapped by the TMO of Durand and Dorsey (2002).

subjects' attention. This is expected because the entry of the new car in the frame is a "new" or a "rare" event (up to this point the subjects' attention is focused on the first car only). That directs the attention to the second car. This is what was observed when the HDR video was viewed on an HDR screen. The corresponding HDR visual attention map is shown in Fig. 16.4B, where one can see that the "second" car is the main region of attention. However, when tone-mapped video was shown to the subjects, we observed different behavior. In this case, it was found that the first car still remains the main focus of attention. This can be clearly observed in the visual attention map corresponding to the TMO of Tumblin et al. (1999) shown in Fig. 16.4C. That is, despite the occurrence of a new event (the entry of the second car), attention behavior did not change. We can attribute this to the fact that the TMO of Tumblin et al. (1999) could not maintain proper contrast at the tunnel entrance, where there is a large change in intensity (dark inside the tunnel and bright outside it). Because of this, the subjects' attention was not fully diverted toward the second car. A different observation was, however, made for in case of the TMO of Durand and Dorsey (2002). This TMO could maintain relatively better contrast at the tunnel entrance. Because of this, we have a situation where both the first car and the second car became the regions of attention. This can be seen from the visual attention map shown in Fig. 16.4D. Thus, depending on the TMO, we have different visual attention behavior for the same scene in the video. This suggests that similarly to the case of still images, tone mapping changes visual attention behavior over time.

16.4.3 IMPACT ON OTHER ASPECTS

Apart from fidelity and visual attention behavior, TMOs may also alter naturalness and modify color/luminance rendering (Narwaria et al., 2014a). This can lead to a quite different appearance of the tone-mapped content in comparison with the unprocessed reference HDR content. The most

FIGURE 16.5

Tone-mapped content preference in the presence of an HDR reference.

reliable approach to find the impact of TMOs on these dimensions is also via subjective studies where supplementary feedback can be taken from the observers. Also, in this case, it is more beneficial to provide the observers with an HDR reference for comparison. This is because in many cases the tone-mapped content would be preferred even though it may not have the same naturalness or color as the reference content. However, as noted before, such a setup will require two luminance conditions. This is shown in Fig. 16.5, where the reference content is displayed on an HDR display and the LDR content is displayed on either side of it.

Because there are two types of displays, the room illumination should be adjusted. In particular, with the HDR display (brighter) in the center, the illumination at the center (just above the HDR monitor) should be higher. Such a setup ensures a more suitable illumination setting for observers. To obtain observer preference, the next step is to select an appropriate method. Because these tests are not directly related to artifact visibility, the paired comparison method is typically more appropriate. The observers need to be instructed clearly. This can, for instance, be "Please choose the image (left or right) that is more similar to the reference image (center)." The observers can also be asked their reason for discarding the nonselected image. The question can be "Why did you discard this image?" To answer this, certain choices need to be provided, which can depend on the characteristic that is being tested. For example, the observers can be given three choices: low fidelity of colors/luminance, loss of details, and lack of naturalness. Because all the observers are generally naive for the purposes of the study (not experts in image or video processing), the physical meaning of each of these choices should be described in detail on a separate sheet during the experiment.

In one of our studies (Narwaria et al., 2014a), we found that the context of visual details is crucial. In other words, merely preserving details is not enough, and their location in the scene can play a role. To illustrate this point, an example is given in Fig. 16.6; the local sharpness maps of two tone-mapped LDR stimuli are shown in Fig. 16.6C and D. Both these images were almost equally preferred by the observers. In the local sharpness maps, brighter color implies more sharpness. In these images, we have

FIGURE 16.6

Illustration of context of details: (A) and (B) LDR content; (C) and (D) sharpness maps.

also highlighted (by a red box; dark gray in print versions) a local area where more details have been preserved by iCAM06 in comparison with the single-exposure content. One can see that the highlighted area belongs mainly to the background. As a result, the presence of more details was less or not noticed by an average observer. Therefore, the local context of where details appear is more important. Apart from that, the appearance of the details themselves (eg, unrealistic or overenhanced details) can affect user preference.

Another dimension that can be affected by TMOs is naturalness. The goal of a TMO is not only to preserve details but also to maintain the overall contrast and appearance. However, most TMOs tend to trade off one at the expense of the other. In other words, preserving local details aggressively can make the content appear more unnatural, while overall appearance preservation will come at the expense of details. An example is shown in Fig. 16.7, where the image in Fig. 16.7A appears more natural despite the loss of details in the background. In contrast, the image in Fig. 16.7B has more details preserved but the trees in the foreground appear overcontrasted, thus contributing to an unnatural appearance.

TMOs may also alter HDR scenes in a way that the information provided may change. For instance, depending on the contrast and color, an evening scene may not appear that way to the observer because of the illumination reproduced by the TMO. In such cases, an HDR reference probably helps more because otherwise the tone-mapped versions may be preferred. We present an example in Fig. 16.8. The image in Fig. 16.8A was preferred if the HDR reference was provided. However, with the absence of a reference scene, the image in Fig. 16.8B was preferred, most probably for its colorfulness and details.

FIGURE 16.7

Natural and unnatural images with different levels of details.

FIGURE 16.8

The scene appearance can change depending on color, details, and whether an HDR reference was shown to the observers.

Before we conclude this section, it is worth reiterating that TMOs can affect HDR QoE by impacting one or more constituent dimensions. In particular, in this section we have we presented an analysis and discussion of the impact of TMOs on perceptual fidelity, visual attention behavior, naturalness, context of visual details, etc. Hence, TMO-based HDR content processing should consider these aspects so that the HDR QoE is not severely affected. Another operation that is sometimes used in HDR content rendering is that of inverse tone mapping, and its impact on QoE is discussed by a case study in the next section.

16.5 CASE STUDY: QUALITY ASSESSMENT OF DYNAMIC RANGE EXPANSION OF VIDEO SEQUENCES

In this section we consider an example of how to set up a subjective study for HDR content, with a specific application scenario: the range expansion of video sequences for display on HDR reproduction devices.

HDR displays (Seetzen et al., 2004) are about to make their debut on the television market, and will probably become the most widespread television technology in the coming years, as they can reproduce far higher luminance and contrast levels than their LCD or CRT predecessors (see Chapter 14 for a description of HDR display technology). However, most of the existing image and video content is LDR and needs to be converted to HDR to be viewed on these displays. Dynamic range expansion, also known as inverse tone mapping, is the operation that enables this conversion (Rempel et al., 2007). Various *expansion operators* have been proposed (Banterle et al., 2006, 2008; Meylan et al., 2006, 2007; Rempel et al., 2007; Wang et al., 2007; Didyk et al., 2008; Huo et al., 2014; Kuo et al., 2012). Evaluating these techniques plays a key role in the design of new expansion operators and helps to improve the characteristics of HDR displays.

A few studies in the literature have compared the performance of expansion operators (Akyüz et al., 2007; Banterle et al., 2009; Masia et al., 2009). Assessing the quality of expanded HDR content is not simply a matter of *fidelity* to the original LDR, but entails complex and challenging *aesthetic* considerations, which are difficult to model in an objective quality metric. As a consequence, these studies are mainly based on subjective visual quality assessment experiments. While these studies have considered expansion of still images only, it has been conjectured that the very same expansion operators that produce visually pleasing results in the case of still images could perform poorly when applied to video, creating artifacts such as flickering (Banterle et al., 2008) or unnatural illumination behaviors on scene changes (Kuo et al., 2012). However, these conjectures have not been validated by any subjective studies designed specifically for video.

In the following, we describe a subjective study aimed at understanding whether existing expansion operators are adequate to support video, and, among existing algorithms, which of them yield the best visual results. We target the problem of real-time displaying of high-quality professional content, which is representative of a typical broadcasting scenario. Thus, expansion operators that cannot be run in real time, because of heavy processing or time-consuming off-line training, have been excluded from the comparison. The study starts with the selection of appropriate source LDR sequences, and representative expansion operators from the literature. Next, the actual study is designed and implemented, in this case as a pairwise comparison of videos, in order to rank four expansion operators of different algorithmic complexity, representative of the main approaches proposed so far. A complete description of this study can be found in De Simone et al. (2014).

16.5.1 SOURCE LDR CONTENT AND EXPANSION OPERATORS CONSIDERED

Several LDR video sequences, downloaded from YouTube, have been analyzed as potential test material to be included in the experiment. All the sequences considered had 4K spatial resolution, but we reduced them to high definition (1920 × 1080 pixels) to meet our display's resolution. We focused on high-quality video footage: no visible compression artifacts are present, the content is well exposed, and has been filmed and postprocessed professionally (as in a typical broadcast scenario). Furthermore, no video presents visible compression artifacts. Differently from previous studies (Akyüz et al., 2007; Banterle et al., 2009), the selected sequences were natively LDR and no HDR ground truth was available. For each content, we computed the spatial and temporal indexes (ITU-T, 2008), which describe the spatial and temporal complexity of the content, as well as two luminance and contrast-related features: the image key (Akyüz and Reinhard, 2006), which gives a measure of the brightness of the image; and the Michelson contrast (Michelson, 1927), computed locally with a 32 × 32 pixel

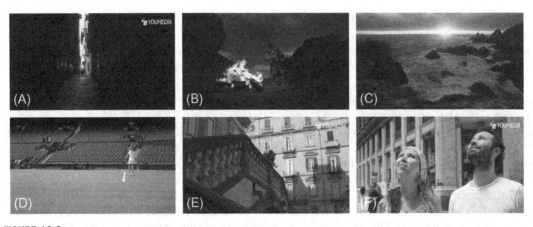

FIGURE 16.9

Sample frames from the six test sequences used for the test: (A) "vespa"; (B) "firedance"; (C) "sunset"; (D) "tennis"; (E) "stairs"; and (F) "naples".

sliding window. On the basis of the semantic interest of each content and on the diversity of the considered characteristics, we selected the six video sequences shown in Fig. 16.9. Three additional video sequences, not reported here, were used to train the subjects. All videos have a duration of between 4.5 and 10 s, and except for "naples" and "tennis," which have two scene changes; the rest of the sequences feature the same visual scene. We excluded longer sequences with several scene changes for two reasons. First, in a pilot test with expert viewers, we observed that scene changes did not influence the performance of the expansion operators considered. Second, the performance of expansion operators changes with the characteristics of the scene (eg, key, contrast), as discussed in Section 16.5.4. As a result, a video segment with multiple scenes would have time-varying quality, which complicates the subjective quality assessment task.

To produce the HDR videos, we considered four expansion operators, representative of the different approaches proposed in the literature. The algorithms include (uppercase letters are used in the rest of this chapter to refer to each algorithm) simple linear scaling[3] of LDR values to meet the HDR display luminance range (A) (Akyüz et al., 2007); the highlight enhancement method of Meylan et al. (2007) (with $\rho = 0.67$) (M); the method based on the expand map of Rempel et al. (2007) (R); and the perceptual algorithm of Huo et al. (2014) (H).

16.5.2 DISPLAY, TEST ENVIRONMENT, AND VIEWING CONDITIONS

Expanded HDR videos were displayed on a SIM2 HDR47 display,[4] which has HD1080 resolution with declared contrast ratio higher than 4×10^6. Using a light probe, we verified the linear response of the

---

[3]To be robust to obvious artifacts due to frame-by-frame oscillations of the minimum and maximum input LDR values, we clipped 1% of the darkest and brightest pixels before the expansion.
[4]http://www.sim2.com/HDR/.

monitor and measured a peak luminance of 4250 cd/m^2 when 60% of the screen surface is white (the maximum luminance is lower when all the surface area is lit because of the power constraints of the device; see Chapter 14). $L_{max} = 4250$ cd/m^2 was taken as the reference maximum display luminance for the four expansion operators. Display-referred HDR values for our screen are obtained by division of physical candela per square meter values by the luminance efficacy at equal-energy white (ie, by the constant factor 179).

We set up a test space with a midgray nonreflective background, isolated from external sources of light, as recommended in ITU-R (2012a,b). Differently from the conclusions reported by Rempel et al. (2009), we assessed during a pilot test that viewing sessions longer than a few minutes in a completely dark environment might cause visual fatigue. Therefore, we placed a lamp at a color temperature of 6500 K behind the HDR screen to ensure ambient illumination while avoiding the presence of any direct light source (apart from the HDR display) in the field of view of the user. The resulting ambient light measured in front of the screen, when this is off, was approximatively 10 cd/m^2. One viewer at a time participated in our viewing sessions, sitting at a distance of 1.6 times the diagonal of the display (2 m), which corresponds to a 30-degree viewing angle.

16.5.3 TEST METHOD

A mixed, quantitative and qualitative, test approach was used in our experiment. The viewer was presented twice with a pair of video sequences, a and b, played sequentially, with a 2-s midgray screen showing letter a or b before the rendering of the corresponding video; a and b depict the same content, expanded with two different expansion operators. After the presentation, the viewer was asked to answer the following questions: (1) *Did you prefer a, b or did you have no preference?* (ie, three-forced-choice paired comparison method; David 1969); (2) *If you had a preference, why did you prefer a or b?* (ie, qualitative evaluation to motivate the paired comparison choice); (3) *Would you say the quality of the preferred stimulus (or both if you did not have a preference) is acceptable if you would get this video on your TV?* (ie, acceptability rating); and (4) *How would you rate the overall quality of the preferred stimulus (or both if you did not have a preference)?* (ie, overall quality absolute categorical rating (ACR) using a five-level discrete quality scale; ITU-T 2008). Each viewer was left free to take as much time as needed to answer these questions, by directly handling the interface (ie, pausing and resuming the playout once the "vote" message on the screen had been reached).

A complete paired comparison method was used and each pair of stimuli was presented in both orders: thus, with n being the number of expansion operators compared, 12 pairs ($n \times (n - 1)$) were considered for each content, and in total 72 stimuli (12 × 6 contents) were presented to each viewer. To avoid viewing sessions longer than 20 min, the entire test was split into three sessions: each viewer performed a session on a different day. The list of test stimuli included in each session and their order were randomized for each viewer, with the constraint that no consecutive presentations of the same content would occur. Two dummy presentations were included at the beginning of each session to stabilize viewers' scores.

Before the first session, each viewer was asked to fill in a form to characterize his/her familiarity with HDR technology and multimedia user's habits. Each viewer was also screened for correct visual and color acuity. A training session, including three video scenes (different from the test material) and the same expansion operators as those considered in the test, was held before the first session to familiarize the viewer with the interface as well as with the paper rating sheet.

16.5.4 RESULTS AND DISCUSSION

A panel of 13 persons, sex-balanced and with an average age of 28.8 years (minimum 22 years, maximum 38 years), took part in the test. Most of them were not familiar with HDR imaging and declared they did to have a high-definition television at home. To detect and remove possible outliers from the panel, we leveraged the transitivity property of the ordering induced by paired comparison tests, as proposed by Lee et al. (2011). We found that there was no outlier in the paired comparison tests.

Table 16.1 shows the normalized winning frequencies (w_{ij}), averaged across the six video contents, of having algorithm i (on the rows) preferred to algorithm j (on the columns), where $i, j = 1, 2, \ldots, n$ and n is the number of different expansion operators. These are computed as $w_{ij} = p_{ij} + t_{ij}/2$, where p_{ij} is the normalized frequency of stimulus i being preferred to stimulus j (*I prefer a (b)*) and t_{ij} is the normalized frequency of the tie (*I have no preference*). The resulting ranking of the expansion operators, global and per content, is shown in Table 16.2. In terms of overall ranking and winning frequencies, it can seen that, in most cases, the simple linear expansion (algorithm A) ranks first, while algorithm R ranks last. Considering the results of the qualitative evaluation, we can explain this by the fact that in general the expanded HDR video obtained with use of algorithm R was judged as "too dark" and "not colorful," giving the impression of being "unnatural" to viewers.

Considering the acceptability scores, the expanded videos were almost always judged acceptable in terms of visual quality (only 8% of the video stimuli was judged unacceptable by at least one-third of the panel). From analysis of the results of the ACR quality assessment (Fig. 16.10), it can be seen that, in most cases, the best performing algorithms produce good or even excellent results. This is true for all the scenes apart from "vespa" and "firedance": in these cases, the ACR scores are between fair and

Table 16.1 Overall Winning Frequencies (w_{ij}), Averaged Over the Six Videos

	A	H	M	R
A	0	0.75	0.65	0.87
H	0.25	0	0.47	0.82
M	0.35	0.53	0	0.75
R	0.13	0.18	0.25	0

Table 16.2 Algorithms Ranking Across Test Videos

	1st	2nd	3rd	4th
Vespa	A	M	H	R
Firedance	H	M	A	R
Sunset	A	H	R	M
Tennis	A	H	M	R
Stairs	A	M	H	R
Naples	A	H	M	R
Average	A	H	M	R

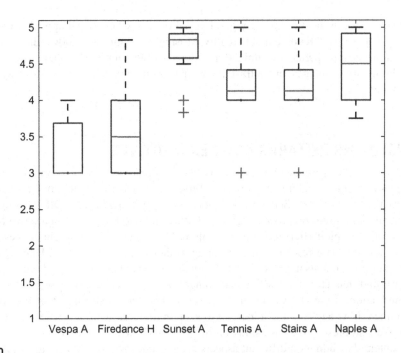

FIGURE 16.10

Box plot of the ACR results (1 = "bad," 2 = "poor," 3 = "fair," 4 = "good," 5 = "excellent") across the panel of viewers for the best ranking algorithm for each content (indicated by tick labels on the x axis).

good, indicating that there is still room for improvement of expansion operator performance. From an analysis of users' qualitative feedback, as well as the content characteristics, this can be explained by the fact that when the scenes are dark and have low local contrast, the resulting expanded videos are perceived as "noisy" (vespa A) or "too dark" (firedance H). Interestingly, even in the cases when the quality of the stimulus was considered to be not acceptable, viewers did not signal any specific temporal artifacts such as illumination changes or flickering: "dark" and "noisy" are the most frequent adjectives used to describe the worst stimuli.

Qualitative comments also reveal that the preferred stimulus is usually the one which is deemed as "brighter" and "with more details." This somehow confirms the findings of Akyüz et al. (2007), which showed that viewers prefer the brightest stimulus, as far as it displays a good contrast. This explains why the simple linear expansion algorithm (A) ranked first in the test: because the input LDR material was of high quality and well exposed, the linear expansion enables one to increase the dynamic range while enhancing details and giving the same overall impression of contrast as in the original LDR. On the other hand, if the input LDR is too dark, linear expansion amplifies noise, and thus, in some cases, algorithms that enhance contrast but do not uniformly increase brightness (such as algorithm H) could be preferred.

In conclusion, this subjective study confirms some of the results previously found in the literature for the expansion of still images — that is, simple algorithms such as linear range expansion can provide visually acceptable and, in most cases, very good visual quality when visual appeal is judged, rather than fidelity to the original HDR content or to reality. Most interestingly, the study indicates that none of the frame-based expansion operators examined introduce visible temporal artifacts, such as flickering due to global illumination changes, in contrast to what has been conjectured in previous work on range expansion for video.

16.6 CONCLUDING REMARKS AND PERSPECTIVES

HDR imaging is an emerging area within the realm of visual signal processing. It brings to the table two major advantages over traditional imaging systems. First, it can provide a more immersive and realistic viewing experience for the users. Second, the higher bit depth required in HDR imaging will allow more signal manipulation (eg, preprocessing toward efficient encoding) as compared with traditional imaging. However, to exploit HDR technology to its fullest potential, several challenges remain and this chapter has focused on a few of them pertaining to their impact the overall HDR QoE. We first identified a few key components in HDR QoE and then discussed the impact of TMOs on a few of them. In particular, we find that TMOs can affect perceptual fidelity, visual attention behavior, naturalness, and color reproduction of the HDR content. Because of this, the artistic intent of the HDR content may be lost or altered. As a result, it is important to keep in mind these effects when one is processing HDR content. We also presented a case study on inverse TMOs which sheds light on user preference of expanded content depending on different aspects such as naturalness and brightness. A take-home message from this chapter is that HDR QoE is a complex, multidimensional problem. Hence, we should address it by considering different dimensions and not by a simple fidelity-based approach.

REFERENCES

Akyüz, A.O., Reinhard, E., 2006. Color appearance in high-dynamic-range imaging. J. Electron. Imaging 15 (3), 033001.

Akyüz, A.O., Fleming, R., Riecke, B.E., Reinhard, E., Bülthoff, H.H., 2007. Do HDR displays support LDR content? A psychophysical evaluation. In: ACM SIGGRAPH 2007 Papers, SIGGRAPH '07. ACM, New York, NY.

Ashikhmin, M., 2002. A tone mapping algorithm for high contrast images. In: Proceedings of the 13th Eurographics Workshop on Rendering, EGRW '02. Eurographics Association, Aire-la-Ville, Switzerland, pp. 145–156.

Ashikhmin, M., Goyal, J., 2006. A reality check for tone-mapping operators. ACM Trans. Appl. Percept. 3 (4), 399–411.

Banterle, F., Ledda, P., Debattista, K., Chalmers, A., 2006. Inverse tone mapping. In: Proceeding of the 4th International Conference on Computer Graphics and Interactive Techniques in Australasia and Southeast Asia, GRAPHITE '06. ACM, New York, NY, pp. 349–356.

Banterle, F., Ledda, P., Debattista, K., Chalmers, A., 2008. Expanding low dynamic range videos for high dynamic range applications. In: Proc. of the 24th Spring Conference on Computer Graphics. ACM, New York, NY, pp. 33–41.

Banterle, F., Ledda, P., Debattista, K., Bloj, M., Artusi, A., Chalmers, A., 2009. A psychophysical evaluation of inverse tone mapping techniques. Comput. Graph. Forum 28 (1), 13–25.

Cadik, M., Wimmer, M., Neumann, L., Artusi, A., 2008. Evaluation of HDR tone mapping methods using essential perceptual attributes. Comput. Graph. 32 (3), 330–349.

Callet, P.L., Moller, S., Perkis, A., 2013. Qualinet White Paper on Definitions of Quality of Experience (2012). European Network on Quality of Experience in Multimedia Systems and Services (COST Action IC 1003).

Chiu, K., Herf, M., Shirley, P., Swamy, S., Wang, C., Zimmerman, K., 1993. Spatially nonuniform scaling functions for high contrast images. In: Proceedings of Graphics Interface 93, pp. 245–253.

David, H.A., 1969. The Method of Paired Comparisons. Charles Griffin, London.

De Simone, F., Valenzise, G., Lauga, P., Dufaux, F., Banterle, F., 2014. Dynamic range expansion of video sequences: a subjective quality assessment study. In: Proc. IEEE Global Conference on Signal and Information Processing (GlobalSIP), Atlanta, GA, pp. 1063–1067.

Didyk, P., Mantiuk, R., Hein, M., Seidel, H., 2008. Enhancement of bright video features for HDR displays. Comput. Graph. Forum 27 (4), 1265–1274.

Drago, F., Martens, W.L., Myszkowski, K., Seidel, H.P., 2003a. Perceptual evaluation of tone mapping operators. In: ACM SIGGRAPH 2003 Sketches & Applications, SIGGRAPH '03. ACM, New York, NY, USA, p. 1.

Drago, F., Myszkowski, K., Annen, T., Chiba, N., 2003b. Adaptive logarithmic mapping for displaying high contrast scenes. Comput. Graph. Forum 22 (3), 419–426.

Durand, F., Dorsey, J., 2002. Fast bilateral filtering for the display of high-dynamic-range images. In: Proceedings of the 29th Annual Conference on Computer Graphics and Interactive Techniques, SIGGRAPH '02. ACM, New York, NY, USA, pp. 257–266.

Eilertsen, G., Wanat, R., Mantiuk, R.K., Unger, J., 2013. Evaluation of tone mapping operators for HDR-video. Comput. Graph. Forum 32 (7), 275–284.

Fattal, R., Lischinski, D., Werman, M., 2002. Gradient domain high dynamic range compression. ACM Trans. Graph. 21 (3), 249–256.

Greg, W., 1994. A contrast-based scalefactor for luminance display. In: Graphic Gems IV. Academic Press, Boston, pp. 415–421.

Huo, Y., Yang, F., Dong, L., Brost, V., 2014. Physiological inverse tone mapping based on retina response. Vis. Comput. 30 (5), 507–517.

ITU-R, 2012a. General viewing conditions for subjective assessment of quality of SDTV and HDTV television pictures on flat panel displays. ITU-R Recommendation BT.2022.

ITU-R, 2012b. Methodology for the subjective assessment of the quality of television pictures. ITU-R Recommendation BT.500.

ITU-T, 2008. Subjective video quality assessment methods for multimedia applications. ITU-T Recommendation P. 910.

Kuang, J., Johnson, G.M., Fairchild, M.D., 2007a. iCAM06: a refined image appearance model for HDR image rendering. J. Vis. Commun. Image Represent. 18 (5), 406–414. Special issue on High Dynamic Range Imaging.

Kuang, J., Yamaguchi, H., Liu, C., Johnson, G.M., Fairchild, M.D., 2007b. Evaluating HDR rendering algorithms. ACM Trans. Appl. Percept. 4 (2), Article 9.

Kuo, P., Tang, C., Chien, S., 2012. Content-adaptive inverse tone mapping. In: Proc. IEEE Visual Communications and Image Processing, pp. 1–6.

Ledda, P., Chalmers, A., Troscianko, T., Seetzen, H., 2005. Evaluation of tone mapping operators using a high dynamic range display. ACM Trans. Graph. 24 (3), 640–648.

Lee, J., Simone, F.D., Ebrahimi, T., 2011. Subjective quality evaluation via paired comparison: application to scalable video coding. IEEE Trans. Multimedia 13 (5), 882–893.

Masia, B., Agustin, S., Fleming, R.W., Sorkine, O., Gutierrez, D., 2009. Evaluation of reverse tone mapping through varying exposure conditions. ACM Trans. Graph. 28 (5), Article No. 160.

Mather, G., 2006. Foundations of Perception. Psychology Press, Hove, East Sussex.

Meylan, L., Daly, S., Susstrunk, S., 2006. The reproduction of specular highlights on high dynamic range displays. In: Proc. of the 14th Color Imagining Conference.

Meylan, L., Daly, S., Susstrunk, S., 2007. Tone mapping for high dynamic range displays. In: Proc. IS&T/SPIE Electronic Imaging: Human Vision and Electronic Imaging XII, vol. 6492.

Michelson, A., 1927. Studies in Optics. Chicago Press, Chicago.

Narwaria, M., Da Silva, M.P., Le Callet, P., Pepion, R., 2014a. Single exposure vs tone mapped high dynamic range images: a study based on quality of experience. In: 2014 Proceedings of the 22nd European Signal Processing Conference (EUSIPCO), pp. 2140–2144.

Narwaria, M., Mantel, C., Da Silva, M.P., Le Callet, P., Forchhammer, S., 2014b. An objective method for high dynamic range source content selection. In: 2014 Sixth International Workshop on Quality of Multimedia Experience (QoMEX), pp. 13–18.

Narwaria, M., Da Silva, M.P., Le Callet, P., Pepion, R., 2014c. Impact of tone mapping in high dynamic range image compression. In: Proc. Video Processing and Quality Metrics (VPQM), pp. 1–6.

Narwaria, M., Da Silva, M.P., Le Callet, P., Pepion, R., 2014d. Tone mapping based HDR compression: does it affect visual experience? Signal Process. Image Commun. 29 (2), 257–273.

Reinhard, E., Stark, M., Shirley, P., Ferwerda, J., 2002. Photographic tone reproduction for digital images. ACM Trans. Graph. 21 (3), 267–276.

Rempel, A.G., Trentacoste, M., Seetzen, H., Young, H.D., Heidrich, W., Whitehead, L., Ward, G., 2007. LDR2HDR: on-the-fly reverse tone mapping of legacy video and photographs. In: ACM SIGGRAPH 2007 Papers, SIGGRAPH '07. ACM, New York, NY.

Rempel, A.G., Heidrich, W., Li, H., Mantiuk, R., 2009. Video viewing preferences for HDR displays under varying ambient illumination. In: Proc. of the 6th Symposium on Applied Perception in Graphics and Visualization, APGV '09. ACM, New York, NY, pp. 45–52.

Rensink, R.A., 2006. Visual Attention. John Wiley & Sons, Ltd, New York.

Seetzen, H., Heidrich, W., Stuerzlinger, W., Ward, G., Whitehead, L., Trentacoste, M., Ghosh, A., Vorozcovs, A., 2004. High dynamic range display systems. In: ACM SIGGRAPH 2004 Papers, SIGGRAPH '04. ACM, New York, NY, pp. 760–768.

Tumblin, J., Hodgins, J.K., Guenter, B.K., 1999. Two methods for display of high contrast images. ACM Trans. Graph. 18 (1), 56–94.

Wang, L., Wei, L., Zhou, K., Guo, B., Shum, H., 2007. High dynamic range image hallucination. In: Proc. 18th Eurographics Conference on Rendering Techniques, pp. 321–326.

Yoshida, A., Blanz, V., Myszkowski, K., Seidel, H.P., 2005. Perceptual evaluation of tone mapping operators with real-world scenes. In: Proc. of SPIE, vol. 5666, pp. 192–203.

HDR IMAGE AND VIDEO QUALITY PREDICTION 17

M. Narwaria*, P. Le Callet*, G. Valenzise†, F. De Simone†, F. Dufaux†, R.K. Mantiuk‡

University of Nantes, Nantes, France
Telecom ParisTech, CNRS LTCI, Paris, France†
Bangor University, Bangor, United Kingdom‡

CHAPTER OUTLINE

17.1 INTRODUCTION

In Chapter 16, the concept of quality of experience (QoE) was discussed and it was highlighted how high dynamic range (HDR) video processing can impact different aspects of HDR QoE. This chapter is devoted to the analysis of state-of-the-art objective methods for measuring the quality of HDR image and video, as part of evaluating one of the components in the overall QoE. First, we mention a few additional considerations in HDR video quality measurement. These include the following:

- HDR video compression may suffer from new distortions in addition to the low dynamic range (LDR) ones (eg, loss of details due to saturation from an inverse tone mapping operator in addition to blockiness). This calls for investigation of methods that can deal with the new situation.

High Dynamic Range Video. http://dx.doi.org/10.1016/B978-0-08-100412-8.00017-6

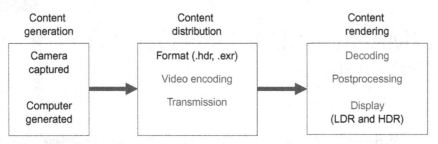

FIGURE 17.1

Simplified diagram illustrating the components of a typical HDR video delivery chain.

- Traditional LDR rendering usually does not consider the display used because the format and specifications are more standardized. By contrast, HDR rendering will need to take into account the display used (ie, LDR or HDR display). Both displays would need specific preprocessing, which can introduce artifacts. As a result, there is a need to analyze and quantify the possible effects of display toward HDR content rendering.
- The loss of details due to distortions (arising from preprocessing, compression, transmission errors, and so on) is an issue in LDR imaging. This will remain in HDR imaging but with the additional consideration of the artistic intent of HDR content, which can be altered by different processing. This problem is potentially more prominent in HDR imaging as it focuses on incorporating very high contrasts.

We can elaborate on and explain the above points further by considering a simplified diagram of a typical HDR video delivery system as shown in Fig. 17.1.

As shown, there are three main blocks — namely, content generation, distribution, and rendering. The first block pertains to HDR content generation. This can be realized if the scene is captured with a camera (single shot or multiexposure fusion based) or content can be computer generated. The second and third blocks concern content distribution and rendering, respectively, and are the main focus of this chapter.

The goal of HDR video encoding is to reduce the number of stored or transmitted bits. A video encoder exploits both temporal and spatial redundancies in the video signal to compress it. However, this inevitably leads to loss of visual quality and the encoder always needs to deal with a trade-off between data reduction and visual quality. Similarly, video quality measurement is required toward the optimal design of transmission system parameters so that the end user receives at least acceptable-quality content.

For content rendering, postprocessing is also usually applied to maintain or enhance visual quality. Finally, one needs to measure the impact of display-specific processing on the rendered HDR video. In our context, we can consider either HDR or LDR displays. Even an HDR display needs to preprocess (tone map) video data so that the maximum displayable luminance[1] is not exceeded. On the other hand,

---

[1] Hardware and power consumption constraints will limit the peak display luminance.

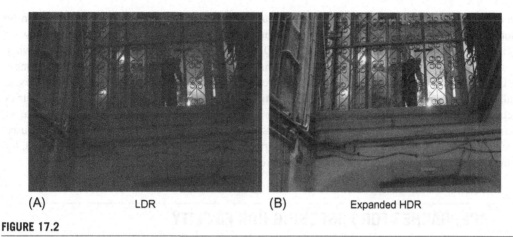

(A)   LDR   (B)   Expanded HDR

FIGURE 17.2

Amplification of noise from LDR to HDR.

LDR displays as such cannot provide high luminance conditions and thus require tone-mapped content for rendering. However, tone mapping is known to be a nontransparent operation which can reduce the levels of details as well as modify artistic intent (Narwaria et al., 2014b). Thus, quality measurement is an important aspect all along the HDR delivery chain, and we have highlighted a few scenarios for it.

The HDR processing chain entails that distortion might appear in new forms, even when it was not present or visible in analog LDR circumstances. For instance, consider the LDR image in Fig. 17.2A. The luminance of the scene was rather low; thus, the picture contains noise in the dark regions. However, when the image is displayed on a standard dynamic range display, this noise is hardly visible in dark areas at the bottom of the picture and the quality of the image is high. Fig. 17.2B shows the result of expanding the dynamic range of this image to match that of an HDR display (the result has been tone-mapped for this illustration). The inverse tone mapping algorithm is the linear expansion described in Akyüz et al. (2007). Although an HDR reproduction device is necessary to appreciate the content of this picture, the tone-mapped version in Fig. 17.2B clearly shows that the noise in the dark regions is boosted to a visible level. This example illustrates that perceived distortion can be significantly affected by the HDR processing chain.

Other examples of HDR-specific distortion are banding, due to a limited resolution of HDR encodings, and color changes due to the much wider gamut color provided by HDR. Finally, oversmoothing may happen in tone mapping-based HDR video compression (see Chapter 10 for a comprehensive treatment of this topic): areas where the original HDR frames contain details that are tone-mapped to very close LDR values might then be smoothed out by transform coding. As a result, these details will be lost after inverse tone mapping at the decoder.

As discussed in Chapter 16, the different aspects of HDR QoE (including video quality) can be measured by subjective and objective methods. The former involves the use of human subjects to judge and rate the quality of the test stimuli. With appropriate laboratory conditions and a sufficiently large subject panel, it remains the most accurate method. The latter quality assessment method uses a computational (mathematical) model to provide estimates of the subjective video quality. While such objective models may not mimic subjective opinions accurately in a general scenario, they can be reasonably effective in specific conditions/applications. Hence, they can be an important tool toward automating the testing and standardization of HDR video processing algorithms, especially when subjective tests may not be feasible. Therefore, this chapter deals with objective quality assessment of HDR content and elaborates on the issues and challenges that arise. We also discuss and present details of the existing efforts on the topic. Particularly, the focus is on full-reference HDR metrics which take as input two HDR signals (one of them is always assumed to be the reference). Hence, in the context of this chapter, the term "quality" can also be interpreted as "fidelity," and both can be used interchangeably. Another use case is that of comparing HDR and LDR signals, and this is needed, for instance, when HDR content is tone-mapped to be rendered on an LDR display.

17.2 APPROACHES FOR ASSESSING HDR FIDELITY

LDR pixel values are typically gamma encoded. As a consequence, LDR video encodes information that is nonlinearly (the nonlinearity arising from the gamma curve which can approximate the response of human eye to luminance) related to the scene luminance. This implies that the changes in LDR pixel values can be approximately linearly related to the actual change perceived by the human visual system (HVS). However, HDR pixel values are related (proportional) to the physical luminance. Hence, direct pixel-based differences (between reference and distorted HDR signals) may not be meaningful. This issue can tackled by two approaches. The first one is to explicitly model certain mechanisms of the HVS. The second approach is to transform the luminance values to a perceptually relevant space, and then use an LDR method for quality assessment (ie, adaptation of LDR metrics). The two approaches are briefly discussed next.

17.2.1 HVS-BASED MODELS FOR HDR QUALITY MEASUREMENT

The first strategy for assessing HDR fidelity is to model at least some of the relevant properties of the HVS explicitly. In theory, such an approach is more intuitive and probably more accurate. However, it may lead to practical difficulties because accurate theoretical modeling of the HVS is neither possible nor perhaps desirable in order to develop tractable solutions. HDR-VDP-2 (HDR visual difference predictor 2) proposed by Mantiuk et al. (2011) follows such an approach for predicting HDR quality. HDR-VDP-2 uses an approximate model of the HVS derived from new contrast sensitivity measurements. Specifically, a customized contrast sensitivity function was used to cover a large luminance range as compared with conventional contrast sensitivity functions. HDR-VDP-2 is essentially a visibility prediction metric. That is, it provides a 2D map with probabilities of detection at each pixel point, and this is obviously related to the perceived quality because a higher detection probability implies a higher distortion level at the specific point.

However, in several applications the global quality score is more desirable in order to quantify the overall annoyance level due to artifacts. Thus, the perceptual errors in different frequency bands must be appropriately pooled (combined) to compute a global quality score. This is not merely related to error (distortion) detection but requires the computation of pooling weights that will indicate the importance of local errors in the overall quality. One such study has been reported by Narwaria et al. (2015a), and we discuss it in Section 17.3.

17.2.2 ADAPTATION OF LDR METRICS FOR MEASURING HDR QUALITY

Many popular fidelity metrics used for LDR content are based on the assumption that pixel values are somehow linearly related to the perceived brightness. Thus, comparing pixel values (or functions of them) is possible without the need for an accurate model of the HVS, and simple metrics, such as the peak signal to noise ratio (PSNR) and the structural similarity (SSIM) index, are perceptually meaningful. In the case of HDR content, this is no longer true, and well-known objective quality metrics used for assessment of LDR fidelity cannot be directly applied to HDR images and video. This is mainly due to two reasons.

First, HDR values describe the physical luminance of the scene, and thus could span a range that could extend, in the brightest pixels, to 10^8 cd/m^2. Clearly, reproducing such luminance values is neither possible — HDR display technology nowadays can radiate up to 10^4 cd/m^2 — nor desirable (for obvious reasons). Therefore, taking into account the actual capabilities of an HDR display is fundamental to obtain perceptually meaningful results.

Second, while HDR images pixel values are proportional to the physical luminance, the HVS is sensible to luminance ratios, as expressed by the Weber-Fechner law. Thus, pixel operations should take into account and compensate for this effect. Notice that this is implicitly done for LDR content, because pixel values are *gamma corrected* in the sRGB color space (Anderson et al., 1996), which not only compensates for the nonlinear luminance response of legacy CRT displays, but also accounts for the nonlinear response of the HVS. In other words, the nonlinearity of the sRGB color space provides a pixel encoding which is approximately linear with respect to perception. However, the sRGB correction postulates a maximum display luminance of 80 cd/m^2; for brighter screens the sRGB gamma is not accurate enough to provide perceptual uniformity (Mantiuk et al., 2015, Chapter 2). In Section 17.4 we discuss the performance of adapting LDR metrics toward measuring HDR content quality.

17.3 FROM SPATIAL FREQUENCY ERRORS TO GLOBAL QUALITY MEASURE OF HDR CONTENT: IMPROVEMENT OF THE HVS-BASED MODEL

In Section 17.2.1 we briefly described the HDR-VDP-2 method, and outlined the need to obtain the pooling weights for errors in different frequency bands. In the original implementation of HDR-VDP-2, the pooling weights were determined by optimization on an existing LDR dataset. There are, however, three limitations of that approach, especially in the context of dealing with LDR and HDR conditions. First, Mantiuk et al. (2011) used only an LDR image quality dataset, which did not include any HDR images. Second, the optimization was done on a relatively small number of images. Finally, because the optimization was unconstrained, it lead to negative pooling weights that may not be easily interpretable. These limitations were addressed by Narwaria et al. (2015a) in HDR-VDP-2.2, which was built upon HDR-VDP-2.

In HDR-VDP-2, the following expression is used to predict the quality score Q_{HDRVDP} for a distorted image with respect to its reference:

$$Q_{\text{HDRVDP}} = \frac{1}{F \cdot O} \sum_{f=1}^{F} \sum_{o=1}^{O} \mathbf{w}_f \quad \log \left(\frac{1}{I} \sum_{i=1}^{I} \mathbf{D}_{\text{p}}^2[f, o](i) + \varepsilon \right), \tag{17.1}$$

where i is the pixel index, $\mathbf{D}_p$ denotes the noise-normalized difference between the fth spatial frequency ($f = 1$ to F) band and the oth orientation ($o = 1$ to O) of the steerable pyramid for the reference and test images, $\varepsilon = 10^{-5}$ is a constant to avoid singularities when $\mathbf{D}_p$ is close to 0, and I is the total number of pixels. In the above expression, $\mathbf{w}_f$ is the vector of per-band pooling weights, which one can determine by maximizing correlations with subjective opinion scores. However, unconstrained optimization in this case may lead to some negative $\mathbf{w}_f$. Because $\mathbf{w}_f$ determines the weight (importance) of each frequency band, a negative $\mathbf{w}_f$ is implausible and may indicate overfitting. Therefore, a constraint is introduced on $\mathbf{w}_f$ during optimization.

Let $\mathbf{Q}_{\text{HDRVDP}}$ and $\mathbf{S}$, respectively, denote the vector of objective quality scores from HDR-VDP-2 and the vector of subjective scores for a given set of N images. Then the aim is to maximize the Spearman rank-order correlation between the two vectors, with $\mathbf{w}_f$ being the optimized variables. To that end, we first rank the values in $\mathbf{Q}_{\text{HDRVDP}}$ and $\mathbf{S}$ from 1 to N and obtain new vectors $\mathbf{R}_{\text{HDRVDP}}$ and $\mathbf{R}_{\text{subjective}}$, which consist of the respective ranks. Further, we define $\mathbf{E} = \mathbf{R}_{\text{HDRVDP}} - \mathbf{R}_{\text{subjective}}$ as the rank difference vector. Then, the optimization problem can be formulated as

$$\underset{\mathbf{w}_f}{\text{Maximize}} \left(1 - \frac{6 \sum_{i=1}^{N} \mathbf{E}_i^2}{N(N^2 - 1)} \right), \text{ subject to } \mathbf{w}_f \geq \mathbf{0}. \tag{17.2}$$

Also note that in this case the optimization can be solved by the Nelder-Mead method, which does not require the computation of gradients. This is because the objective function is not continuous and differentiable as Spearman rank-order correlation is used. Because the aim was to calibrate the metric so that it can handle both HDR and LDR conditions, Narwaria et al. (2015a) optimized $\mathbf{w}_f$ on the basis of a set of subjectively rated LDR and HDR images. The details of the HDR datasets can be found in Narwaria et al. (2013, 2014a), while the LDR datasets (TID2008 and CSIQ) were developed by Ponomarenko et al. (2008) and Larson and Chandler (2010).

17.3.1 CROSS-VALIDATION RESULTS AND ANALYSIS

As reported by Narwaria et al. (2015a), there were 65 source content in total and 2932 distorted content (obtained by the application of different distortion types and levels to the source content). For the cross-validation studies, all the distorted images from 45 source contents was selected as the training set to find the optimal $\mathbf{w}_f$ vector, and the remaining images from 20 source content were used as a test set. To enable a more robust estimate of the prediction performance, the division into training and test sets was randomly repeated over 1000 iterations, and it was ensured that the two sets were different in terms of the source content. Hence, in each of the 1000 iterations, the prediction performance was assessed only for untrained content, thus providing a reasonably robust approach toward content-independent verification. With this data partition (45 source contents as the training set and remaining source contents as the test set), there were an average of 2032 and 900 images, respectively, in the training and test sets during each iteration.

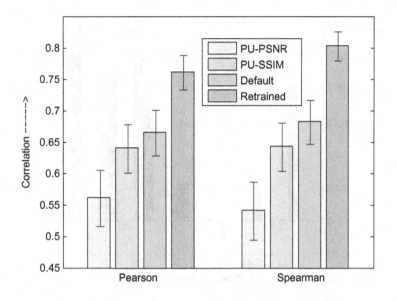

FIGURE 17.3

Cross-validation test results, 900 test images in each iteration. Error bars indicate 95% confidence intervals.

The experimental results are shown in Fig. 17.3 (the bars from left to right correspond to the key entries from first to last), where the performance is measured in terms of mean (over 1000 iterations) values of Pearson and Spearman correlation values (a higher value implies better performance for these measures). One can see that the prediction performance with the weights obtained from the training set is better than that with the default weights as well as that of the two modified LDR methods. The 95% confidence intervals are also shown in Fig. 17.3 (denoted by error bars) to provide an indication of uncertainty in the measured values. As can be seen, the confidence intervals do not overlap, indicating a better performance with the trained weights from statistical considerations.

Finally, the retrained and default weights are compared via the frequency versus weight plot shown in Fig. 17.4. The frequency is expressed in cycles per degree and the left and right bars at each special frequency indicate default and retrained weights, respectively. The retrained weights reduce the importance of low-frequency bands. However, they are not necessarily related to the contrast sensitivity function because the goal of pooling is to quantify quality (or the annoyance level), which might not always be at the level of visibility thresholds. Also note that the negative weights found in the original HDR-VDP-2 could cause an increase of quality with a higher amount of distortion. This situation is valid only in very specific cases such as denoising and contrast enhancement (where visual quality may be enhanced). However, because this condition is not included in any of the datasets that we used, the retrained weights result in better physical interpretability (because all of them are positive, quality will decrease with increased level of distortion).

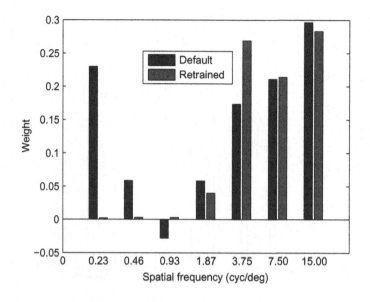

FIGURE 17.4

Plot of retrained and default weights.

17.4 ADAPTED LDR METRICS FOR MEASURING HDR IMAGE QUALITY IN THE CONTEXT OF COMPRESSION

This section provides a comparative analysis of the performance of adapted LDR metrics on compressed HDR images. The study was conducted by Valenzise et al. (2014), and the dataset used by them is publicly available for downloading.[2]

17.4.1 METRICS AND ADAPTATION

As discussed in Section 17.2.2, the adaptation of LDR metrics entails a preprocessing of the native HDR pixel values. In that context, Valenzise et al. (2014) considered logarithmic and perceptually uniform (PU) encoding-based transformation. Two existing LDR metrics — namely, PSNR and SSIM — were selected, and these were computed on either a logarithmic mapping (log-PSNR and log-SSIM) or a PU encoding.[3] They also compared the performance of the two versions of HDR-VDP-2 (HDR-VDP-2.1 and HDR-VDP-2.2), which compute quality directly on the basis of luminance values. Because pixel values in the HDR images considered were display referred, before the logarithmic mapping or the PU encoding was applied, they were converted to actual luminance values by their multiplication by the luminance efficacy at equal-energy white (ie, by the constant factor 179). Then, the response of the HDR display, which is approximately linear within its black level (0.03 cd/m^2) and maximum

---

[2]http://perso.telecom-paristech.fr/~gvalenzi/sw/HDR-compressionDB.v1.zip.
[3]A publicly available implementation can be found at http://resources.mpi-inf.mpg.de/hdr/fulldr_extension/.

luminance (4250 cd/m^2) and saturates over this value, was simulated. The resulting luminance values were used as input for both versions of the HDR-VDP-2 metric.

17.4.2 DATASET DESCRIPTION

Five HDR images were used, and these are shown in Fig. 17.5. These were taken from the HDR Photographic Survey dataset (Fairchild, 2007) in such a way to span a wide range of diverse content-specific characteristics, including spatial information, dynamic range, and overall brightness. All the selected images were adapted to the display resolution of 1920 × 1080 pixels. The test material was produced by compression of the source (reference) images with different codecs and coding conditions. In particular, three codecs were considered for test material generation: (a) JPEG; (b) JPEG 2000; (c) JPEG XT (Richter, 2013), which is the new standardization initiative (ISO/IEC 18477) of JPEG for backward-compatible encoding of HDR images. The HDR images were displayed on a SIM2 HDR47 display (SIM2, 2014), which has HD1080 resolution with a declared contrast ratio higher than 4×10^6:1 and a maximum luminance of approximately 4250 cd/m^2. The subjective quality evaluation was performed following the double stimulus impairment scale method (ITU-R, 2012). Particularly, pairs of images (ie, the original image and the compressed image) were sequentially presented to the user, who was told that the first image was the reference and asked to rate the level of annoyance of the visual defects that she/he may observe in the second stimulus using a continuous quality scale ranging from 0 to 100, associated with five distinct descriptions ("very annoying," "annoying," "slightly annoying," "perceptible," and "imperceptible"). The pairs of stimuli were presented in random order, different for each viewer, with the constraint that no consecutive pairs concerning the same content will occur. More details on the experimental results can be found in Valenzise et al. (2014).

(A)   "AirBellowsGap"   (B)   "LasVegasStore"   (C)   "MasonLake(1)"

(D)   "RedwoodSunset"   (E)   "UpheavalDome"

FIGURE 17.5

HDR images used for the test (tone-mapped version): (A) "AirBellowsGap"; (B) "LasVegasStore"; (C) "MasonLake(1)"; (D) "RedwoodSunset"; (E) "UpheavalDome."

17.4.3 RESULTS AND ANALYSIS

The subjective data (collected from 15 observers) were processed to compute the mean opinion score and the 95% confidence interval, assuming that the scores are following a Student's t distribution.

The performance of the metrics was evaluated in terms of the Spearman rank-order correlation coefficient, computed on the entire set of mean opinion scores. The use of a nonparametric correlation coefficient avoids the need for nonlinear fitting to linearize the values of the objective metrics, which may be questionable because of the relatively small size of the subjective ground truth dataset. The values of the modulus of the correlation coefficient for each metric, per content and per codec, are reported in Table 17.1. Fig. 17.6 summarizes the performance of the metrics and shows also the 95% confidence interval of the Spearman correlation coefficient (modulus). As can be seen, overall the best performing LDR metric is PU-SSIM, followed by log SSIM. We also report results for HDR-VDP-2.1 and HDR-VDP-2.2, with the latter improving significantly over the older version of the metric, as discussed in the previous section. Interestingly, PU-SSIM has performance almost equivalent to that of HDR-VDP-2.2, albeit at lower computational cost.

In terms of content dependency and distortion dependency, these results confirm widely known observations concerning the scope of validity of most objective metrics (Huynh-Thu and Ghanbari, 2008). On one hand, the codec-dependent results show that all the metrics suffer to some extent from content dependency in their prediction capability. On the other hand, the content-dependent results clearly indicate that even perfect (ranking) prediction is reachable for some contents (ie, the results of PU-SSIM for content "LasVegasStore" and "RedwoodSunset"). Of course, one must interpret these results by taking into account the limited set of distortions which characterize the test database. Finally, it is interesting to notice that the range of PSNRs obtained in this work is significantly greater than that commonly encountered in the case of LDR image compression, as reported, for example, by De Simone et al. (2011).

Table 17.1 Spearman Correlation Coefficients (Modulus) Calculated for Each Content and Codec, With Maximum Correlation Values for Each Column Highlighted in Bold

	Overall	"Air Bellows Gap"	"Las Vegas Store"	"Mason Lake(1)"	"Redwood Sunset"	"Upheaval Dome"	JPEG	JPEG XT	JPEG 2000
PU-PSNR	0.794	**0.976**	0.963	**0.976**	0.952	0.987	0.591	0.797	0.835
PU-SSIM	0.923	0.952	**1**	0.952	**1**	0.975	**0.942**	**0.944**	0.887
log PSNR	0.866	**0.976**	0.963	**0.976**	0.976	0.987	0.753	0.832	0.887
log SSIM	0.904	0.952	0.975	0.928	0.952	0.975	0.907	0.881	0.872
HDR-VDP-2.1	0.889	0.952	0.987	**0.976**	0.952	0.987	0.802	0.909	**0.924**
HDR-VDP-2.2	**0.946**	0.929	0.985	0.952	0.970	**1**	**0.942**	0.927	0.958

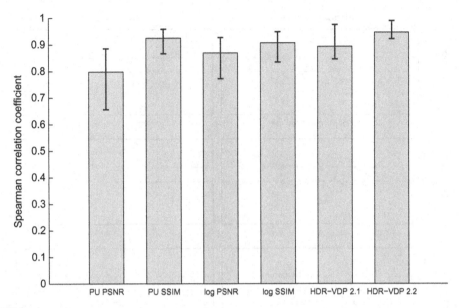

FIGURE 17.6

Spearman correlation coefficients for LDR metrics. For comparison, results for HDR-VDP-2.1 and HDR-VDP-2.2 are also shown. Error bars indicate 95% confidence intervals.

17.5 **TONE MAPPING AND DYNAMIC RANGE-INDEPENDENT METRICS**

In the previous sections, we discussed objective quality prediction of HDR content. Recall that in this approach it is assumed that both reference and distorted images have a similar dynamic range (hence, they can be either scene referred or display referred). However, there are many situations in which HDR video needs to be tone-mapped for it to be displayed on a typical LDR display, and this is where dynamic range-independent metrics can be useful.

We begin with the fact that tone mapping inherently produces images that are different from the original HDR reference. To fit the resulting image within the available color gamut and dynamic range of a display, tone mapping often needs to compress contrast and adjust brightness. A tone-mapped image may lose some quality as compared with the original seen on an HDR display, yet the images often look very similar and the degradation of quality is poorly predicted by most quality metrics. Smith et al. (2006) proposed the first metric intended for prediction of loss of quality due to local and global contrast distortion introduced by tone mapping. However, the metric was used only in the context of controlling a countershading algorithm and was not validated against experimental data. Aydın et al. (2008) proposed a metric for comparing HDR and tone-mapped images that is robust to contrast changes. The metric was later extended to video (Aydın et al., 2010). Both metrics are invariant to the change of contrast magnitude as long as that change does not distort contrast (reverse its polarity) or affect its visibility. The metric classifies distortions into three types: loss of visible contrast,

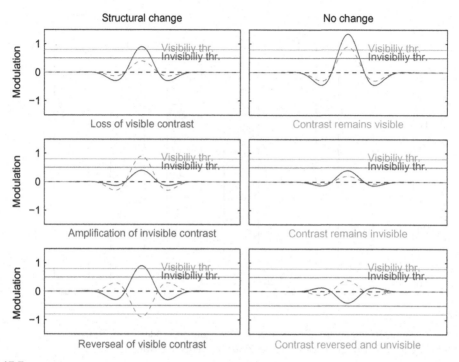

FIGURE 17.7

The dynamic range-independent metric distinguished by the change of contrast that does and does not result in structural change. The continuous blue line shows a reference signal (from a band-pass pyramid) and the dashed magenta line shows the test signal. When contrast remains visible or invisible after tone mapping, no distortion is signalized (top and middle right). However, when the change of contrast alters the visibility of details, making visible details invisible (top-left), it is signalized as a distortion. (For interpretation of the references to color in this figure legend, the reader is referred to the web version of this article.)

Source: Reproduced with permission from Aydın et al. (2008), © 2008 Association for Computing Machinery.

amplification of invisible contrast, and contrast reversal. All three cases are illustrated in Fig. 17.7 for the example of a simple 2D Gabor patch. These three cases are believed to affect the quality of tone-mapped images. Fig. 17.8 shows the metric predictions for three tone-mapped images. The main weakness of this metric is that the distortion maps produced are suitable mostly for visual inspection and qualitative evaluation. The metric does not produce a single-valued quality estimate and its correlation with subjective quality assessment has not been verified. The metric can be conveniently executed from a Web-based service available at http://drim.mpi-sb.mpg.de/.

Yeganeh and Wang (2013) proposed a metric for tone mapping which was designed to predict the overall quality of a tone-mapped image with respect to an HDR reference. The first component of the metric is the modification of the SSIM index proposed by Wang et al. (2004), which includes the contrast and structure components, but does not include the luminance component. The contrast component is further modified to detect only the cases in which invisible contrast becomes visible and visible contrast becomes invisible, in a spirit similar to that in the dynamic range-independent

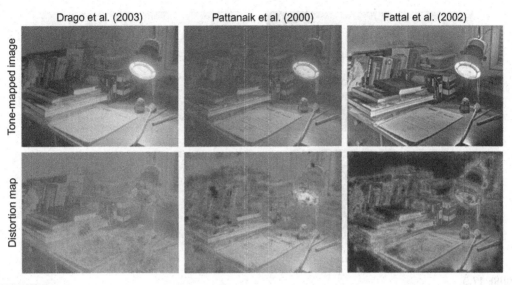

FIGURE 17.8

Prediction of the dynamic range-independent metric (Aydın et al., 2008) (top) for tone-mapped images (bottom). Green denotes the loss of visible contrast, blue denotes the amplification of invisible contrast, and red denotes contrast reversal; see Fig. 17.7. (For interpretation of the references to color in this figure legend, the reader is referred to the web version of this article.)

Source: Reproduced with permission from Aydın et al. (2008), © 2008 Association for Computing Machinery.

metric of Aydın et al. (2008) described above. This is achieved by the mapping of local standard deviation values used in the contrast component into detection probabilities by a visual model, which consists of a psychometric function and a contrast sensitivity function. The second component of the metric describes "naturalness." The naturalness is captured by the measure of similarity between the histogram of a tone-mapped image and the distribution of histograms from the database of 3000 LDR images. The histogram is approximated by the Gaussian distribution. Then, its mean and standard deviation are compared against the database of histograms. When both values are likely to be found in the database, the image is considered natural and is assigned a higher quality. The metric was tested and cross-validated with three databases, including one from Čadík et al. (2008) and the authors' own measurements. The Spearman rank-order correlation coefficient between the metric predictions and the subjective data was approximately 0.8. Such a value is close to the value for a random observer, which is estimated as the correlation between the mean and random observer's quality assessment.

An objective method for finding the optimal parameter space toward HDR tone mapping was proposed by Krasula et al. (2015). The method was aimed at addressing the nontrivial issue of proper tone mapping operator parameter selection because tone mapping operators usually need content-specific parameters in order to produce sharp and perceptually appealing pictures. It works by computing the percentage of area covered in a tone-mapped image. In the context of this method, the percentage of area covered can be used as an indicator of saturated and/or underexposed pixels. Hence, the method attempts to find a parameter space where the percentage is minimized. The method can be potentially applied in HDR content rendering and also security-related applications as suggested

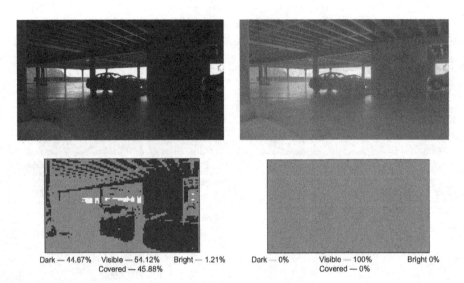

FIGURE 17.9

Example to illustrate the concept of percentage of area covered proposed by Krasula et al. (2015).

by Krasula et al. (2015). Fig. 17.9 provides an example to illustrate the concept of percentage of area covered. The two images in the top row were obtained with two different settings of a simple linear tone mapping operator, leading to images that will have different levels of details (it can be seen that the image on the right in the top row has more visible details). The maps (obtained on the basis of gradient and subsequent thresholding) shown below the corresponding images indicate areas where loss of details occurs because of very low (dark, shown in black) or very high (bright, shown in white) luminance, and the corresponding percentages of such areas (these are computed on the basis of the number of such overexposed/underexposed pixels and the total number of pixels in the image) are also indicated. The percentage of area covered will be equal to the sum of the percentages of the bright and dark areas. Obviously, a higher percentage of area covered will lead to loss of more details, and hence the concept can be used to tune tone mapping operator parameters to maximize the visibility of details in tone-mapped HDR content.

17.6 EXTENSIONS TO VIDEO

The previous sections discussed approaches for predicting quality in HDR images, and these can obviously be applied in the case of HDR video by applying them on frame-by-frame basis. However, this ignores the effect of temporal factors, which are also important for video quality measurement. Also note that application of HDR-VDP-2.2 for video can be computationally expensive because of the design of the metric. Therefore, this section discusses a method for objective measurement of HDR video quality proposed by Narwaria et al. (2015b), and has lower computational complexity. It is referred to as HDR-VQM and is based on PU encoding (recall this is one way of adapting a metric as discussed in Section 17.4). However, unlike image-based methods, HDR-VQM computes local quality considering spatiotemporal regions, and applies short-term and long-term pooling to derive the global quality score. Hence, it adopts true temporal pooling in contrast to image-based methods such as HDR-VDP-2.2.

17.6.1 BRIEF DESCRIPTION OF HDR-VQM

A block diagram outlining the major steps in HDR-VQM is shown in Fig. 17.10. HDR-VQM takes as input the source and the distorted HDR video sequences. As shown in Fig. 17.10, the first two steps are meant to convert the native input luminance to perceived luminance. These can therefore be seen as preprocessing steps, and the first step is performed if the user provides emitted luminance values. Next, the impact of distortions is analyzed by comparison of the different frequencies and orientation subbands in the source (reference video sequence) and the hypothetical reference circuit (distorted video sequence). The last step is that of error pooling, which is achieved via spatiotemporal processing of the subband errors. This comprises short-term temporal pooling, spatial pooling, and finally, a long-term pooling. A separate block diagram explaining the error pooling in HDR-VQM is shown in Fig. 17.11.

As seen from Fig. 17.10, the first step is to convert native HDR values to display-referred luminance. Such a step is required because distortion visibility can be affected by the processing adopted to display HDR video. The second step is to convert the resulting luminance to perceived values (this approach was discussed in Section 17.2.2). Next, frequency domain filtering is used to obtain the subband errors. The subband error signal is then further processed via error pooling, which is elaborated in Fig. 17.11. It basically comprises short-term temporal pooling, which aims to pool or fuse the data in local spatiotemporal neighborhoods. This is followed by spatial pooling, which results in short-term quality scores, which can be seen as indicators of temporal quality. The final step uses a long-term temporal pooling in order to pool the short-term quality measures into a global video quality score.

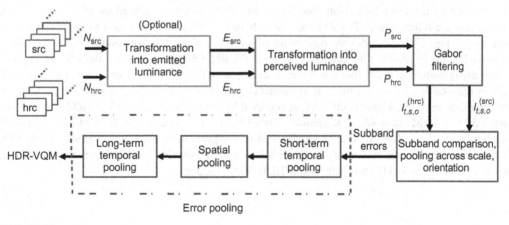

FIGURE 17.10

Block diagram of HDR-VQM. *src*, source; *hrc*, hypothetical reference circuit.

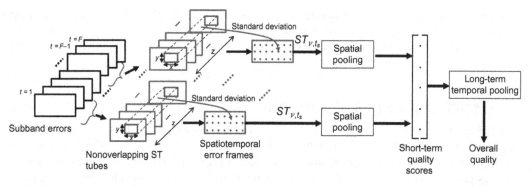

FIGURE 17.11

Error pooling in HDR-VQM.

The reader is referred to Narwaria et al. (2015b) for more detailed discussions on the HDR-VQM formulation. We now compare the performance of HDR-VQM with that of other objective methods, including the adapted metrics and HDR-VDP-2.2.

17.6.2 PREDICTION PERFORMANCE COMPARISON

The performance of HDR-VQM was analyzed on an HDR video dataset with a total of 90 HDR videos. These were obtained by compression of 10 source HDR sequences with a backward-compatible HDR video compression scheme. More details on dataset development can be found in Narwaria et al. (2015b).

A comparison of Pearson and Spearman correlation coefficients of different methods is shown in Fig. 17.12 (the bars from left to right correspond to the key entries from first to last), from which one can see that HDR-VQM performs better. The 95% confidence intervals for these correlation values are indicated by the error bars. Note that SSIM and PSNR were computed on PU-encoded values (hence these are referred to as P-SSIM and P-PSNR in Fig. 17.12). With regard to the computational complexity,[4] Fig. 17.13 provides the percentage of outliers and the relative computational complexity (expressed as the execution time relative to relative PSNR). Obviously, lower values for an objective method along both axes imply that it is better. It can be seen that the relative execution time for HDR-VQM is reasonable considering the improvements (ie, smallest percentage of outliers) in performance over the other methods. More specifically, as reported by Narwaria et al. (2015b), HDR-VQM (Linux cluster, 32 GB RAM) took 432 s (ie, about 7.2 min) for a 10-s sequence (with 250 frames). With the same hardware setup, HDR-VDP-2.2 took about 24 min to process the same video. Hence, HDR-VQM offers a less complex and reasonably accurate solution for HDR video quality prediction although it does not explicitly model HVS functions like HDR-VDP-2.2 does.

---

[4]All the methods were compared on the basis of their MATLAB implementations.

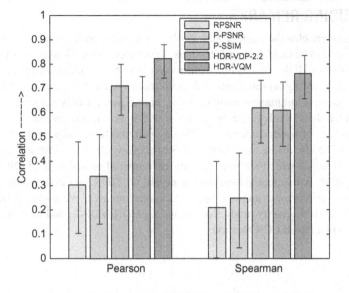

FIGURE 17.12

Comparison of Pearson and Spearman correlation coefficients of different methods on an HDR video dataset.

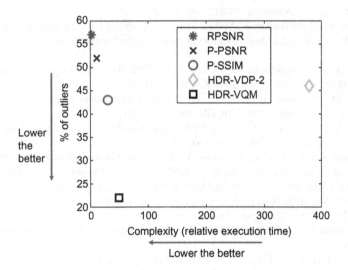

FIGURE 17.13

Percentage of outliers and the relative computational complexity (expressed as relative execution time with respect to RPSNR) for different methods.

17.7 CONCLUDING REMARKS

While HDR imaging enables the capture and display of higher-contrast videos, it brings with it new challenges which need to be tackled in order to make consumer applications a reality. In this regard, an important issue is that of HDR video quality estimation, and it is challenging primarily because HDR videos encode much more signal information than do traditional videos. Hence, in HDR imaging, the information is stored in a luminance-related format, unlike perceptually scaled pixel values in LDR signals. This chapter therefore focused on issues in HDR visual quality measurement. One of the important issues is that of handling scene- or display-referred luminance, and we highlighted a few methods to tackle it. We also discussed how the HDR processing chain can introduce additional artifacts (eg, amplification of noise) which are not typically encountered in traditional LDR imaging systems. Because tone mapping is inherently involved in nearly all HDR processing systems, we provided insights into existing efforts on dynamic range-independent metrics. We also outlined state-of-the-art objective methods for HDR quality prediction, including HDR-VDP-2 (and its extension HDR-VDP-2.2), HDR-VQM, and adapted LDR methods.

REFERENCES

Akyüz, A.O., Fleming, R., Riecke, B.E., Reinhard, E., Bülthoff, H.H., 2007. Do HDR displays support LDR content? A psychophysical evaluation. ACM Trans. Graph. 26 (3).

Anderson, M., Motta, R., Chandrasekar, S., Stokes, M., 1996. Proposal for a standard default color space for the internet — sRGB. In: Color and Imaging Conference, Society for Imaging Science and Technology, vol. 1996, pp. 238–245.

Aydın, T.O., Mantiuk, R., Myszkowski, K., Seidel, H.P., 2008. Dynamic range independent image quality assessment. ACM Trans. Graph. (Proc. SIGGRAPH) 27 (3), 69.

Aydın, T.O., Čadík, M., Myszkowski, K., Seidel, H.P., 2010. Video quality assessment for computer graphics applications. ACM Trans. Graph. (Proc. SIGGRAPH Asia) 29 (6), 1.

Cadík, M., Wimmer, M., Neumann, L., Artusi, A., 2008. Evaluation of HDR tone mapping methods using essential perceptual attributes. Comput. Graph. 32 (3), 330–349.

De Simone, F., Goldmann, L., Lee, J., Ebrahimi, T., 2011. Performance analysis of VP8 image and video compression based on subjective evaluations. In: SPIE Optics and Photonics, Applications of Digital Image Processing XXXIV, International Society for Optics and Photonics, pp. 81350M–81350M-11.

Fairchild, M.D., 2007. The HDR photographic survey. In: Color and Imaging Conference, Society for Imaging Science and Technology, vol. 2007, pp. 233–238.

Huynh-Thu, Q., Ghanbari, M., 2008. Scope of validity of PSNR in image/video quality assessment. Electron. Lett. 44 (13), 800–801.

ITU-R, 2012. Methodology for the Subjective Assessment of the Quality of Television Pictures. ITU-R Recommendation BT.500-13.

Krasula, L., Narwaria, M., Fliegel, K., Le Callet, P., 2015. Rendering of HDR content on LDR displays: an objective approach. In: Proc. SPIE.

Larson, E.C., Chandler, D.M., 2010. Most apparent distortion: full-reference image quality assessment and the role of strategy. J. Electron. Imaging 19 (1), 011006.

Mantiuk, R., Kim, K., Rempel, A., Heidrich, W., 2011. HDR-VDP-2: a calibrated visual metric for visibility and quality predictions in all luminance conditions. ACM Trans. Graph. 30 (4), 40:1–40:14.

Mantiuk, R.K., Myszkowski, K., Seidel, H.P., 2015. High dynamic range imaging. In: Wiley Encyclopedia of Electrical and Electronics Engineering. Wiley, New York, pp. 1–81.

Narwaria, M., Da Silva, M.P., Le Callet, P., Pepion, R., 2013. Tone mapping-based high-dynamic-range image compression: study of optimization criterion and perceptual quality. Opt. Eng. 52 (10), 102008.

Narwaria, M., Da Silva, M.P., Le Callet, P., Pepion, R., 2014a. Impact of tone mapping in high dynamic range image compression. In: International Workshop on Video Processing and Quality Metrics, pp. 1–6.

Narwaria, M., Da Silva, M.P., Le Callet, P., Pepion, R., 2014b. Tone mapping based HDR compression: does it affect visual experience? Signal Process. Image Commun. 29 (2), 257–273. Special Issue on Advances in High Dynamic Range Video Research.

Narwaria, M., Mantiuk, R.K., Da Silva, M.P., Le Callet, P., 2015a. HDR-VDP-2.2: a calibrated method for objective quality prediction of high-dynamic range and standard images. J. Electron. Imaging 24 (1), 010501.

Narwaria, M., Da Silva, M.P., Le Callet, P., 2015b. HDR-VQM: an objective quality measure for high dynamic range video. Signal Process. Image Commun. 35, 46–60.

Ponomarenko, N., Lukin, V., Egiazarian, K., Astola, J., Carli, M., Battisti, F., 2008. Color image database for evaluation of image quality metrics. In: IEEE 10th Workshop on Multimedia Signal Processing, pp. 403–408.

Richter, T., 2013. On the standardization of the JPEG XT image compression. In: Picture Coding Symposium (PCS), pp. 37–40.

SIM2, 2014. http://www.sim2.com/hdr/.

Smith, K., Krawczyk, G., Myszkowski, K., 2006. Beyond tone mapping: enhanced depiction of tone mapped HDR images. Comput. Graph. Forum 25 (3), 427–438.

Valenzise, G., Simone, F., Lauga, P., Dufaux, F., 2014. Performance evaluation of objective quality metrics for HDR image compression. In: Proc. SPIE, vol. 9217, pp. 92170C–92170C-10.

Wang, Z., Bovik, A., Sheikh, H., Simoncelli, E., 2004. Image quality assessment: from error visibility to structural similarity. IEEE Trans. Image Process. 13 (4), 600–612.

Yeganeh, H., Wang, Z., 2013. Objective quality assessment of tone-mapped images. IEEE Trans. Image Process. 22 (2), 657–667.

APPLICATIONS

PART

IV

APPLICATIONS

HDR IMAGING IN AUTOMOTIVE APPLICATIONS

18

U. Seger

Robert Bosch GmbH, Leonberg, Germany

CHAPTER OUTLINE

High Dynamic Range Video. http://dx.doi.org/10.1016/B978-0-08-100412-8.00018-8

18.1 HISTORY AND MOTIVATION FOR HIGH DYNAMIC RANGE SENSORS AND CAMERAS

Investigations of the root causes of traffic accidents by authorities such as the German Bundesanstalt für Straßenwesen (BAST), which is responsible for research into improved safety and efficiency of street traffic, and their systematic analysis have been the motivation to start massive efforts to develop technical driver assistance systems (DAS) to avoid or mitigate the yearly death toll paid for the increasing mobility possibilities given by affordable motor vehicles. Enke (1979) pointed out that 50% of collision accidents could be avoided by a reaction that started 1 s earlier. In 1979 Enke doubted that technical systems will be able to provide sufficient detection and recognition capabilities.

The European EUREKA Programme for a European Traffic System with Highest Efficiency and Unprecedented Safety (PROMETHEUS), which started in 1986, has been a European answer to that challenge and the approach to fight the increased risk of dying in street traffic.

European car companies such as Audi, BMW, Daimler, Jaguar, PSA Peugeot Citroën, and Volkswagen as well as their suppliers and subsuppliers teamed up with research institutes active in the area of remote sensing, fast signal computing, and actuation in PROMETHEUS. The target of the research activities was the development of systems and infrastructures to prevent or at least mitigate accidents. Common to all the approaches was the need to sense the environment so as to recognize potential risky traffic situations. Numerous remote sensing technologies from radar sensors to ultrasound sensors, ultraviolet, visible, near-infrared (NIR), and far-infrared radiation sensors, and lidar sensors have been proposed in all kinds of accident mitigation or prevention systems.

Pro-Chip, a subprogram of PROMETHEUS, focused on silicon sensors and neuronal processing chips in VLSI technology. Pro-Chip covered different remote sensing and signal processing areas, and one of these was the high dynamic range (HDR) sensor development. Focusing on optical sensing, it defined the goal to create an eyelike camera as an input for a DAS.

After the goal had been defined and initial calculations and tests had been performed, it became obvious that there was a significant gap between the seeing capabilities of a human eye and the technical performance of state-of-the-art image sensors for electronic cameras at that time. At that time mostly CCD cameras were used for machine vision. The signal-to-noise ratio in low-light conditions below $11\times$ and the dynamic range were by far not usable for the intended applications in uncontrolled light situations as required in standard road traffic situations.

To assist the driver in recognizing traffic situations especially in critical situations, the camera system should have performance capabilities similar to those of the driver's eyes. Further requirements arose from the intention to recognize what are also most important for drivers: traffic signs and road markings. The human observer as a reference model had been implied in the Vienna Convention on Road Signs and Signals, which had become effective by 1978. The visibility of signs for the driver is explicitly referenced on many occasions.

A most challenging goal was defined: the development of a video system with visual abilities similar to those of a human.

A lot of necessary spadework had already been done at this time by scientists in a different domain. In "Exploration of the primary visual cortex," Hubel (1982) had developed a good understanding of the visual cortex and the function of the retinal structure. Shortly after PROMETHEUS started, Dowling (1987) published his comprehensive study on the retina and Mead and Mahowald (1988) had started to transfer the biological model into the electronics VSLI domain. Chamberlain and Lee (1984) had already begun to expand the dynamic range of the already mature CCDs, and Mendis et al. (1994) were to begin the development of what were later called "active pixel structures" in CMOS technology around the same time.

Shortly after PROMETHEUS had begun, the Defense Advanced Research Projects Agency and the National Science Foundation sponsored a comparable research program, the MIT Vision Chip Project, with the target "to design and build prototype analog early vision systems that are remarkably low power, small, and fast" as Wyatt et al. (1991) described.

All the ingredients for a silicon seeing machine where known or under development, but it took some years before a combination of advanced CMOS technology, a suitable design approach, and a software implementation led to a successful demonstration of an HDR CMOS camera at the Institute for Microelectronics Stuttgart (Seger et al., 1993) within the PROMETHEUS framework.

In the automotive world the term "high dynamic range CMOS imager" has been established for an imager with an intrascene dynamic range similar to that of the human eye (eg, > 120 dB).

The picture of a fully illuminated 100-W light bulb (see Fig. 18.1) with good recognizable details of the filament, the printing on the glass bulb, and the bulb fitting had become a comprehensive symbol for the large dynamic range of the HDR imaging devices although the scene is not as challenging as an outdoor street scene in situations with both the sun and shadowed regions in the field of view.

At the same time all the different research activities mentioned above generated numerous seeing chips, vision chips, artificial retinas, etc. Some research spin-outs reached a technical level which allowed industrialization by different companies. Although some approaches stayed close to the biological archetype, with a close-to-ideal logarithmic conversion characteristic, providing a constant contrast resolution over six to eight decades of received intensities (Seger et al., 1999), others tried to approximate the logarithmic transfer function by means of a piecewise linear approximation (Yang et al., 1999).

The viability proof for all the different concepts, however, was the harsh environment of the automotive application. Temperature stability from −40 to 120°C, intense solar radiation focused on the chip for several hundred hours, electrostatic discharge, etc., brought many weaknesses of the research approaches to the surface and these were resolved in many small steps.

Fourteen years after PROMETHEUS started the first real car application development started in 2000 and a vision of a complete electronic safety cocoon for vehicles woven by various sensors (see Fig. 18.2) was presented to the public (Seger et al., 2000).

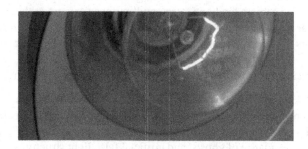

FIGURE 18.1

A 100-W light bulb depicted by a high dynamic range sensor.

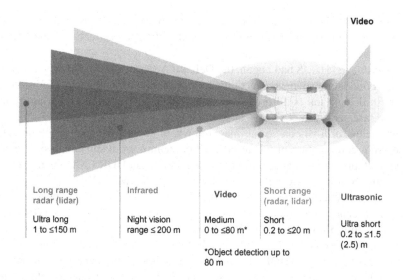

FIGURE 18.2

Safety cocoon for vehicles.

Although this vision was considered as being too progressive at that time, things turned out differently. When Daimler presented its 2005 Mercedes-Benz S-Class model (model 220), one of the innovative highlights was a night vision camera using an HDR CMOS-based sensor with enhanced NIR sensitivity in combination with an active NIR headlamp. This was both the first implementation of a CMOS night vision camera in a series automobile and the first appearance of an HDR sensor in an automotive application. Because of their HDR capability, HDR CMOS sensors outperformed CCD sensors at that time. Around the same year, the first time in history, more CMOS sensors than CCD sensors were produced worldwide. Eight years later it was possible to buy a Mercedes-Benz S-Class automobile (model 222) equipped with:

- two near-range radar sensors, one long-range radar sensor, one night vision NIR camera, one night vision far-infrared camera, and one stereo multipurpose camera at the front;
- two near-range radar sensors and one multimode radar sensor at the back;
- 12 ultrasonic sensors distributed around the car (4 at the front, 4 at the back, and 4 at the sides);
- four surround-view cameras (distributed in the grill, the side mirrors, and the trunk).

One of key technologies enabling this vision was HDR sensors with "eyelike" performance. Within 15 years of the development of automotive cameras it has been possible, in cooperation with sensor and lens suppliers, to resolve most of the optical problems such as optical cross talk, unwanted reflections from all optical surfaces, avoidance of ghosts, and artificial false light structures, caused by inappropriate materials or structures used in the optical path. HDR imaging became a success story and today we see a high percentage of HDR sensors in all developed automotive video sensing solutions. Some viewing applications still rely on linear sensors but will convert to HDR sensors as the benefits of HDR sensing are recognized. The reason for this success is explained in Section 18.2.

18.2 REQUIREMENTS FOR AUTOMOTIVE CAMERA SENSORS

The requirements in automotive video applications are as diverse as the applications themselves. From night vision to pedestrian detection, road sign recognition, lane detection, oncoming headlamp recognition, cross traffic alert, time-to-collision calculation, surround and rear vision, driver monitoring, etc., the applications are widespread. However, there is one common factor and one common goal behind all those applications: they have to deal with natural and technical illuminations and with an unpredictable variance in object appearance and have to work reliably and safely over a large temperature and voltage span.

18.2.1 THE ILLUMINATION CONDITION

"Natural illumination condition" means that unlike industrial image capture and processing systems and unlike any photographer, an automotive video system cannot avoid particular bad situations but has to deal with uncontrolled illumination situations:

- Fast changing illumination situations in tree alleys or galleries. When one drives past windows in arcades (see Fig. 18.3), the light situation changes rapidly from bright sunshine to darkness, which is difficult even for human eyes to accommodate. The same effects can be observed when one leaves or enters tunnels.
- Relevant objects with only little contrast to the background. The bicyclist hidden in the fog (see Fig. 18.4) might decide to cross the road and should be detected even in situations worse than this.
- As long as the information on the low contrast between the bicyclist and the background is preserved, postprocessing of the image will help to detect details (see Fig. 18.5). The human eye especially preserves contrasts and allows the detection of even small differences in reflectance.
- Challenging situations — for example, the sun from various directions. The surfaces of objects change their appearance with a in the light conditions. The reflectance of a material describes how

FIGURE 18.3

Challenging light conditions in covered arcades at Col du Verdon (France).

FIGURE 18.4

Low contrast in a foggy scene.

FIGURE 18.5

Enhanced contrast in a foggy scene.

much light is absorbed and how much is reflected in which direction. The appearance changes with the relative position of the camera to the object and the illuminant of the object. An example can be demonstrated by the simulation of the appearance under different illumination conditions (see Figs. 18.6 and 18.7).

- Nearly no illumination on relevant objects on an overcast dark night. Automotive systems are expected to operate during the day and at night. Although the illumination condition differs largely within the opening angle of the headlamp and aside, an automotive video system is expected to see contrast in both dark and bright regions (Fig. 18.8).

FIGURE 18.6

Different light situations, here noon, sun from above.

FIGURE 18.7

Different light situations, here early afternoon sun.

FIGURE 18.8

Simulation of an overcast night with an unlit vehicle.

FIGURE 18.9

Strong intensity variations as the sun is in the field of view.

- Challenging light situations also appear at night, when oncoming traffic leads to headlamp beams directly in the viewing angle, or in daytime, when the sun appears in the field of view. Every photographer always respects the advice never to take a photograph with the sun in the viewing angle, unless you use specific filters or you are trying to take special effect photographs. The strong variance in illumination typically overstrains linear sensors, resulting in either saturation or missing information due to missing gray value resolution (Fig. 18.9).

A further challenge for automotive sensors are motion effects in the image, which are of special importance as the automotive systems are themselves moving, which leads to shifted image contents with every acceleration or attenuation. All facilitating means which would normally be used in image capturing (flashlights to allow short integration times so as to reduce motion smear, long integration times with static objects, large or cooled sensors for high-sensitivity imaging in low-light situations) are not at hand for an automotive camera designer. Active illumination is not always possible, flashing is not allowed, integration time prolongation is not a particularly good idea if the camera is mounted in a moving car, and the super sensitive imagers with large pixels are good but far too expensive for a mass market product. The same holds for cooled sensors. Any kind of image postprocessing such as denoising, filtering, or tone mapping has to be done with limited resources in video real time, which means within 33–40 ms, corresponding to 25–30 frames/s.

In the context of a *viewing application*, video real time is whatever a human observer sees as a continuous video image (33–40 ms, corresponding to 25–30 frames/s). In the context of *safety-relevant applications*, real time is equal to or faster than the system reaction time, which for actuation of warning, restraint, braking, or steering systems is typically less than 150 ms.

The possibility to improve the illumination situation by a car-borne illumination system is very limited, unless variable active projection systems known as "matrix beams" or "multibeam headlamps" (see Section 18.4.4) are used. For cost reasons, however, the application of these illumination systems is still limited to top-range car models.

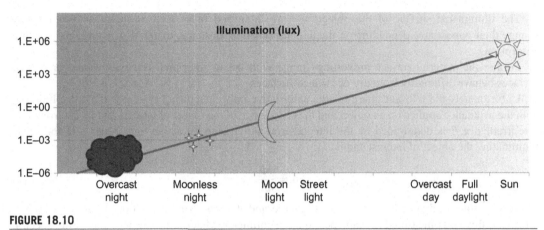

FIGURE 18.10

Luminance dynamic range.

Further, the supporting effect of the illumination system is limited to a certain illumination field of view and is limited in range, because for a luminous flux of more than 2000 lm, United Nations Economic Commission for Europe Regulation 48 restricts the maximum luminance to 430,000 cd for all illumination components involved. Finally, outside the illuminated area there is darkness.

The light sources we encounter in road traffic are global sources such as sunlight, sky illumination ranging from clear blue sky to nightglow on a moonless night, the same but with an overcast sky, street lamps of various types, our vehicle's own headlamps and those from other cars, different kinds of technical light sources such as building illumination flashing signals, and recently also LED and laser illumination systems. The dynamic range caused by the different intensity of light sources easily exceeds nine orders of magnitude; however, for automotive cameras a range of "only" around six to seven orders of magnitude is relevant (see Fig. 18.10).

The radiant flux Φ (W) of the light source is given by

$$\Phi = \frac{dQ}{dt},$$

(18.1)

where Q (J) is the energy radiated from a source over optical wavelengths from the ultraviolet to the infrared (wavelengths from 0.01 to 1000 μm) within a unit time (s).

18.2.2 THE OBJECT REFLECTANCE

The variance in surface properties of interesting objects is of relevance similar to that of the variance in illumination. Surfaces can be

- specularly reflecting (eg, flashing garnish molding, windscreens, or the cladding of modern buildings),
- diffuse reflecting (eg, matt-finish paintings or roughened surfaces),
- a close-to-ideal absorber such as black fur, which in the worst case is worn by an pedestrian on a dark winter night.

The illuminated surface of the objects we are interested in as well as their background may change their appearance depending on the angular position, distance, material structure, and spectral absorption.

Further the material surface properties change with the weather and usage, as wetted or frosted surfaces behave differently from dry or cleaned surfaces.

Unfortunately the relevant figure for the automotive image capturing system is the convolution of both the irradiance applied on an object and the reflectance of the object of interest and its environment. The irradiance E is determined by the integral irradiation from all active light sources in the scene illuminating the object of interest and is

$$E = \frac{d\Phi}{dA},\qquad(18.2)$$

where Φ is the irradiance flux of a radiant source and A is the part of the object surface which is hit by the flux. E is measured in watts per square centimeter. Material properties such as the spectral absorption coefficient and the surface orientation define the portion of radiant flux which will be sent in the direction of the observer, which in our case is the camera for the DAS.

Figs. 18.6–18.8 show the same scenery with different illumination conditions. Figs. 18.6 and 18.7 show daylight conditions with the sun in different locations and Fig. 18.8 shows an overcast night situation with only the vehicle's headlamps as the light source. Fig. 18.9 shows a hard shadow and objects with similar reflectance (eg, pavement) on the two sides of the shadow line.

18.2.3 THE MEDIUM BETWEEN THE OBJECT AND THE IMAGE

The medium between the camera and the object and the light source is in the best case transparent but sometimes behave like a distortion filter (eg, a windscreen) but in the worst case behaves as a low-pass filter (fog) or an image disturbing noise source (reflected light from rain or snow). Fig. 18.4 shows the impact of fog on driver assistant systems. The luminance contrast of objects in the relevant distance for decision making (eg, braking or not braking) is significantly lowered and can reach values as low as 1%. For comparison, under optimal illumination conditions the human eye can distinguish between objects which appear to have a difference in brightness of 1% with respect to their neighboring objects. The insert in Fig. 18.5 shows the potential of image postprocessing to recover information from an HDR image as long as the image content is still above the noise level.

18.2.4 FUNCTIONAL PERFORMANCE

Functions bundled in complex DAS such as lane detection, pedestrian detection, headlamp detection, and traffic sign recognition lead to irradiation requirements that span several orders of magnitude . At the same time, applications that require a steadily increasing field of view lead to a high probability that very different appearing objects will be detected at the same (integration) time by a single sensor.

A solution to reduce data complexity right at the location of detection in the focal plane, as it occurs in the human retina, has been investigated from time to time and has resulted in many approaches, such as the massive lateral processing in the focal plane proposed by Tarr et al. (1996) or Schultz et al. (2014). These approaches aim to allow local, cell-based processing like local signal inhibition, local amplification, or even digitization so as to decrease the requirements for an "overall correct pixel setting." However, these efforts have not led to an automotive-applicable solution. Still the "off focal array" image processing requires image data as undistorted, unsaturated, and contrast preserving as possible. The envelope requirements are driven by various applications in different domains and result in time resolution better than 1/30 s, contrast resolution better than 1%, Delta E resolution close to 2, latency better than 40 ms, spatial resolution better than 1200×900 pixels, and intrascene dynamic range on the order of 1,000,000:1 between maximum and minimum irradiation. The combination is challenging enough to result in a nearly empty solution space, especially if contradicting requirements such as high resolution of object motion and at the same time suppression of beat frequency of modulated light sources are rated with similar priorities.

18.2.5 OPERATING RANGE

A challenge for automotive systems is the requirement of an operational temperature range.

The automotive system has to work within a temperature range of more than 120°C, or to be exact has to withstand temperatures from −40 to 125°C and has to perform fully within an operational range from −40 to 85°C ambient temperature. Self-heating reduces life expectancy for current imagers and even may lead to damages if the junction temperature T_j exceeds 125° for some 100 h, and self-heating induces leakage currents, doubling for every 7–9°C difference in temperature. With use of image sensor process technology with a feature size on the order of 50 nm and back-side illumination (BSI-technology) available for recent image sensors, self-heating might easily result in an artificial signal which is not distinguishable from real light.

18.2.6 RELIABILITY

No drift effects are expected to happen during the lifetime of the sensor (which might be 20 years or more in car applications) and the sensor cannot be protected against significant solar radiation, thermal cycling, humidity, and mechanical shock, which means it must be resilient to those conditions.

18.2.7 POWER CONSIDERATIONS

Operational temperature and reliability as described in earlier lead to an additional indirect requirement for camera sensors. As self-heating leads to a reduced operational range, the power for an HDR image becomes a relevant figure. Intrinsic HDR sensors which generate an HDR image in a single frame might outmatch reconstructive approaches which need storage for multiple integration steps and processing elements to reconstruct the image.

As the energy needed for each HDR image is a further key performance indicator, it is obvious that a target of 3 mJ per HDR image is a quite challenging target to reach if you need a considerable amount of memory and processing logic to reconstruct an HDR image from multiple exposures instead of deriving the HDR image by controlled charge skimming during integration.

18.2.8 COST OR THE DEMOCRATIZATION OF A NEW TECHNOLOGY

In 2005 a night vision system providing a brilliant glare-free image displayed in the cockpit could be ordered for €1900 as optional equipment for a Mercedes-Benz S-Class vehicle. In 2018 a back-over avoidance systems will become mandatory in the United States for every new passenger car to reduce the yearly more than 200 fatalities and more than 15,000 injuries. It is assumed that equipment such as a back-over avoidance system will be affordable for customers of all car classes (not only for buyers of luxury cars) so that it is effective in reducing the number of traffic accident victims reported in official reports — for example, the US National Highway Traffic Safety Administration (NHTSA). At the same time, economies of scale will help to produce cost-effective components as automotive video systems originally intended to be a niche market product.

18.2.9 SAFETY

The New Car Assessment Program (NCAP) initiated by the NHTSA and the corresponding other regional NCAP labels judge the safety of new cars brought onto the market. The mission of the NHTSA with the NCAP is "to save lives, prevent injuries, and reduce traffic-related health care and other economic costs associated with a motor vehicle use and highway travel" (NHT, 2007). Regional NCAP organizations — for example, in Europe, Japan, Australia, New Zealand, and Latin America, adopted the target but made adaptations according to the regional differences in accident statistics. The approach is to test new cars independently from the car manufacturers and publish test results so as to inform the customer about the safety aspects of new cars and change their buying habit. The first standardized tests were released in May 1979 and the first results of crash tests were published in October 1979. Whereas in the beginning the tests focused on safety for passengers and rated the probability of being injured in collisions, the newest ratings are oriented more and more toward the avoidance of accidents and the protection of individuals involved in potential accidents such as pedestrians as well. Although improvements in the past were achieved by the implementation of safety belts, air bags, and impact protection structures, the prediction and anticipative working systems came into the focus in recent years. Systems such as a lane departure warning system, a front collision avoidance system, and automated emergency braking need remote sensing techniques as provided by radar, lidar, and video systems.

In the United States, the NHTSA considers new technology concerning the potential impact on fatalities and decides on changes to the rating system depending on the expected effect of new technologies. Each star of the five-star rating system corresponds to a certain protective feature of a new car. But the rating is adapted because of technical progress, which means that what is good for five stars today will not be sufficient in the future. Thus the NCAP keeps driving the safety standard and equipment development as the NCAP stars are good marketing arguments. Among the different local NCAP ratings, the European New Car Assessment Programme (Euro NCAP) tests are seen to be some of the most demanding tests. For currently requested features and their rating in the domains adult occupant protection, child occupant protection, pedestrian protection, and safety assist, see Euro NCAP Strategy Working Group (2015) and Euro NCAP Ratings Group (2015). The requirements are subject to ongoing adaptation. In the road map for 2020, especially in the sections on pedestrian protection and safety assist, new cars can receive high scores only if remote sensing systems are onboard. In most cases, signals from radar-, lidar-, and video-based systems are fused to allow highly reliable detection and the implementation of staged escalation systems starting with acoustic or tactile warnings for the driver, preparing reactions such as preparing the brake booster for fast interaction, and finally take-over action if the driver did not react in time to prevent an accident. To judge the situation video systems working reliably even under harsh environmental conditions are a key factor. Although HDR sensors are doing a great job in managing the large variety of lighting conditions and fast changing situations, further nonfunctional requirements exist.

The ISO 26262 standard (International Organization for Standardization Technical Committee ISO/TC 22/SC3, 2012) is another important driver for the development of automotive safety equipment and includes a video-based DAS. It was the reason to rethink the in situ testability of image sensors and marked the start of Automotive Safety Integrity Level (ASIL) support features on imagers. ISO 26262-10:2012 came into effect in July 2012. The standard is applicable to all safety-related electronic equipment for road vehicles of up to 3.5 t. In simple words, ISO 26262 requires that all electronic equipment developed and brought onto the market has to prove that it can fulfill the safety targets and is still functional within the system reaction time. Once the safety goals for a new car have been defined by the manufacturer, a safety plan has been established, and a decomposition of safety requirements has been done, it is up to the engineers to judge the impact of failures of electronic components on the safety goal (eg, "do not drive over a pedestrian entering the driving lane as long as it is physically possible to brake before reaching the pedestrian"). How can an image sensor contribute to that goal you might ask.

After a hazard and risk analysis has been performed and the ASIL-classification of the system and its components has been derived, a failure budget is given to each component of the system, which should allow the safety goal to be reached. Largely depending on the architecture and implementation of the system, the expected numbers of failures in time are in the range from 1000 ($<10^{-6}$/h) in an ASIL A case to 10 ($<10^{-8}$/h) in an ASIL D case. Note that this is for the complete system! Whereas a warning function might be classified as ASIL A, an automatic braking or steering system might be classified as ASIL C or D.

For each single component it has to be decided if redundancies are needed, special control measures are to be implemented, etc.

The image sensor as a key component and a potential source of failures which might lead to wrong decisions (eg, do not brake if a pedestrian is in the driving lane as the pedestrian has not been recognized because of saturation in the image) came into focus. A good understanding of which failures are relevant to the system is developed by the system designer and the component designer and leads to key failures per application as well as to strategies to correct failures or detect a critical system status. In the above-mentioned situation, automatic braking for pedestrians, a single-pixel defect will not play a major role unless the rate of defects becomes too high. Failures in the register setting for the integration time, however, might have a fatal impact. ASIL support features have been added to automotive imager designs to allow a functional test of analog and digital circuitries within the system reaction time. For automatic braking and steering the system reaction time is typically less than 150 ms, which means that for practical reasons the functionality must be checked within each individual image frame!

Automotive imagers currently on the market allow one to read out test lines or columns which provide information on the correct function of basic circuitries such as column gain amplifiers and analog-to-digital converters but also to check if transferred register settings have been implemented correctly.

Automotive HDR imagers with their ability to cover the complete range of physically possible input intensities in the application contribute significantly to the safety aspect as a functional test over the complete potential intensity range is possible within a single frame time.

18.3 HDR IMPLEMENTATIONS

Currently, several implementations of HDR sensors are available on the sensor market. We distinguish between *intrinsic HDR* concepts and *reconstructed HDR* concepts. Whereas intrinsic HDR concepts allow in situ generation of an HDR image by controlled charge skimming or reset, the reconstructed HDR concepts need a considerable amount of buffer memory to hold several captured images and rebuild an HDR image from several subsequently recorded images.

The main difference in the principles can be demonstrated in situations with a high illumination dynamic range, in combination with moving objects or modulated light sources. Sensors of the intrinsic HDR type generate no or at least fewer motion artifacts than sensors that use a reconstruction from images which have been integrated at different time slots.

18.3.1 INTRINSIC HDR SOLUTIONS

Among the intrinsic HDR solutions, a differentiation can be made according to the similarity of the transfer function to an ideal logarithmic behavior. An imager which allows one to produce an analog pixel output which is nearly an ideal $V_{out} = \log \Phi$ response or sensors which achieve a nonlinear conversion of generated cell output voltages which achieve a similar quasi-logarithmic behavior with sufficient resolution exhibit significantly better contrast resolution than their linear-approximated counterparts with only a few knee points in their transfer characteristic.

Fig. 18.11 shows the transfer characteristics of three different sensor implementations with direct logarithmic conversion representing an infinite number of knee points (resulting in a smooth transition characteristics) compared with the quasi-logarithmic transfer characteristics with piecewise linear approximations with either six or two knee points (resulting in a segmented transfer characteristics).

For simplification the sensor signal is assumed to be noiseless and the quantization of the signal is assumed to be equidistant in voltage steps in a linear-converting analog-to-digital converter. Whereas the difference in the linear intensity space shows only slight differences, the effect of the differences becomes quite obvious in the logarithmic intensity space as shown in Fig. 18.12. The introduction of a high number of knee points allows a smaller deviation from the logarithmic transfer characteristic (resulting in a constant increase of gray values when plotted vs a log illuminance scale) compared to a festoon like curve which shows alternating regions of gray value resolution versus a log scale.

Equidistant voltage steps represent, depending on the conversion function $V_{out} = F(\Phi)$, different contrast steps. In a logarithmic case, each of the 2048 quantization steps of an assumed 11-bit analog-to-digital converter represents an identical contrast step representing a 1.0068-fold intensity which allows one to distinguish objects with a difference in reflectance of 0.68%. The calculation is too optimistic by far as the noise contribution will consume a considerable part of the separation capability.

The difference in contrast which is necessary to reach the next least significant bit level with a characteristic deviating from the ideal logarithmic transfer characteristic is depicted in Fig. 18.13, which shows the minimum resolvable contrast across the complete dynamic range. Sensors with a log transfer characteristic show a constant contrast resolution while piecewise approximated transfer characteristic results in peaks in the minimum resolved contrast at the knee points of the transfer function. Standard natural scenes (unless it is foggy) often have an intensity difference in the range from 1:30,000 to 1:80,000 and in an extreme situation it reaches 1:500,000, and it is obvious that a

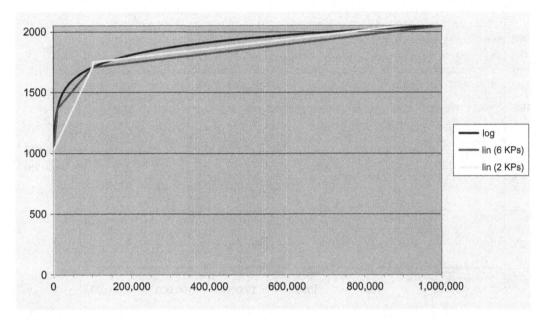

FIGURE 18.11

Transfer characteristic of logarithmic and piecewise linear sensors in the linear intensity domain.

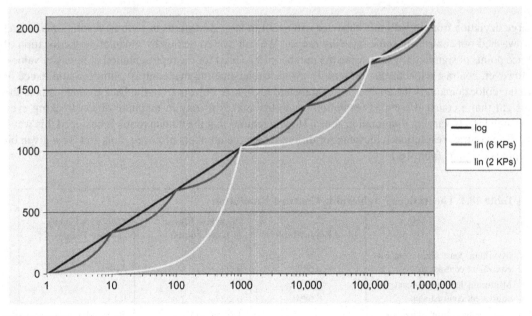

FIGURE 18.12

Transfer characteristic of logarithmic and piecewise linear sensors in the logarithmic intensity domain.

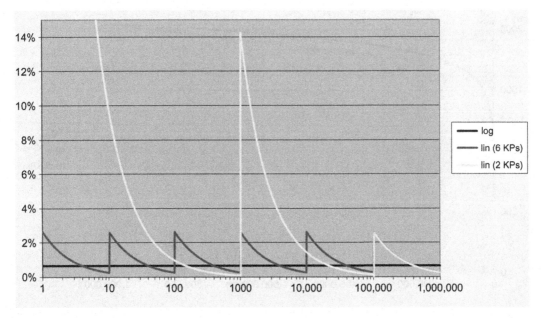

FIGURE 18.13

Contrast separation achieved by different sensor transfer characteristics.

large deviation from logarithmic behavior will result in large deviations in contrast resolution and lead to washed-out contrast in some intensity ranges. We can overcome this by modulating the location of knee points or significantly increasing the number of bits used for the representation of intensity values. However, contrast wipeouts are difficult to avoid. Understanding that contrast, either in luminance or in the color domain, is the enabler to separate and recognize objects from the background leads to the insight that a constant contrast resolution capability makes it easy to maintain object tracking even through largely varying illumination zones. How ingenious that the human retina is designed this way!

The contrast resolution achievable for different discrete numbers of knee points and for a given bit depth is given in Table 18.1.

Table 18.1 Theoretically Achievable Contrast Resolution			
	Ideal Logarithmic	**Piecewise Linear (6 Knee Points)**	**Piecewise Linear (2 Knee Points)**
Maximum least significant bit-equivalent contrast step	0.68%	2.6%	14%
Minimum least significant bit-equivalent contrast step	0.68%	0.26%	0.14%
Contrast resolution gap (1%)	–	15% of digits	30% of digits

Table 18.1 shows that under the assumption that the output signal is converted with 11-bit accuracy, each of the 2048 digitization levels of a piecewise linear-converting sensor represents different contrast levels. Constant contrast resolution can be achieved only with a purely logarithmic transfer characteristic. That means that with a purely logarithmic characteristic each digitization level corresponds to the same factor in contrast change to reach the next digit in the output signal.

The maximum least significant bit-equivalent contrast step indicates what contrast increase is necessary for two object areas to appear different to the sensor used. Here the contrast step is defined as the ratio of irradiances received from the objects.

The contrast resolution gap describes the probability of not being able to distinguish between surfaces with 1% different brightness as the converted gray value is in an inappropriate region of the transfer curve.

18.3.2 RECONSTRUCTED HDR SOLUTIONS

Whereas the intrinsic HDR concept (with a sufficient number of knee points or number of bits used) allows close to constant contrast resolution in moving or light-modulated scenes, it is restricted in the minimum contrast resolution by the position of the knee point because around the knee point the resolution is degraded. Superfine contrast on the order of less than 0.5% cannot be resolved with considerable effort with use of this technique. Reconstructed HDR images from multiples exposures with a single or multiple slope linear sensing devices can achieve sophisticated contrast resolution by combining the image values of multiple overlapping exposures. However, it is required that the illumination is controlled during the taking of multiple exposures and the objects should not move significantly nor should the light be modulated during the different exposures unless a high computational effort is made to compensate for the object motion and light modulation artifacts. Similar techniques are of interest for still image photography as well, but are not a preferred image recording technique for automotive use.

18.4 HDR VIDEO-BASED DRIVER ASSISTANCE APPLICATIONS

A good understanding of the value of HDR sensors and cameras in the automotive field can be achieved by a description of the major performance drivers for requirements by application.

18.4.1 NIGHT VISION

Night vision was one of the first applications on the European market for different reasons. One reason was the transparency of the function to the end user, which allowed the end user to see in operation a new type of sensor entering the automotive field and share his/her impression with others; it is totally different from other systems (eg, the air bag trigger system, which nobody was keen to work on). Another major reason was the possibility for customers to differentiate their vehicles from those of their competitors and demonstrate a technological lead. At the beginning of the millennium, a camera in a car was still an eye-catcher.

The market penetration of this first vision product was not overwhelming on a global scale but served as a showcase for video-based applications and allowed a lot of experience to be gained with a very demanding application and confidence was generated in its operation as the night vision system was able to show images of high quality even under situations challenging for drivers' eyes. The quality of the displayed image was the key performance indicator, and only an always saturation-free image with good preserved contrast as well as a temperature-stable and low-noise image could convince the end user that a night vision system served as a valuable support for driving at night.

To achieve a good contrasting image with low noise under precarious illumination conditions, at least from the perspective of a camera developer, meant dealing with differences in irradiation of up to six orders of magnitude while being able to distinguish contrast in luminance on the order of around 2%.

18.4.2 FROM LANE DETECTION TO LANE KEEPING SUPPORT

The change from monochrome displaying functions such as night vision or the colorful images of a rear-view camera to computer vision systems was accompanied by the introduction of imagers with sparse color filter arrays. Color separation was required for only one purpose in lane detection systems, to distinguish mainly regular white lane markings from yellow lanes used in construction zones as well as to recognize blue lines used in some countries.

No full-color Bayer color filter array was necessary, rather a coarse discrimination of lane markings in color categories was required, but at the same time high resolution was needed to maintain the lane recognition at greater distance providing a safe look-ahead range.

For lane recognition systems the HDR concept is essentially helpful in situations at night in which retroreflective fillers in the lane marking material cause very large gradients in reflectance that depend on the position of the lane marking relative to the headlamp. The resulting difference in irradiated intensity shows large variance; however the contrast between the lane marking and surrounding pavement is close to constant so the HDR sensor output is very similar over a large range of distances.

Although simple systems provided a warning once the calculated time to lane crossing became too small for a reasonable reaction of the driver before a lane was crossed, lane-keeping support became the more elaborate function. This function provides a carefully controlled momentum which is superposed on the steering column to maintain the car's trajectory between the lane markings when they are present. It is still a comfort function, because the driver is always responsible and can easily overrule the system. This system is a precursor for autonomous driving as knowledge of the location and moving direction relative to the car's own lane is a prerequisite to navigate a car.

18.4.3 TRAFFIC SIGN RECOGNITION TO SIGN READING

Very early on the detection and recognition of traffic signs also became a popular and easy-to-experience function of video-based DAS. Made to be detected visually by the driver, traffic signs are predestinated for automated visual detection systems. Because traffic signs are largely standardized by the Vienna Convention on Road Signs and Signals, are limited in the number of categories, and have expressive symbols; their recognition seemed to be an easy task for an optical detection system. But again it was the uncontrolled illumination and unpredictable surroundings which presented some challenges to the developers. Traffic signs that appear near the sun during sunset or diluted contrast in foggy conditions resulted in situations in which a standard linear camera-based system needs several attempts to obtain the information needed, resulting in a significant additional processor load, which is not necessary if only the sensor's dynamic range has enough margin to resolve details of several decades of intensities. Future systems will need higher resolution to expand the optical system from a symbolic pattern matching machine to a real character reading and syntax understanding system. Still, resolution and viewing angle requirements as well as the availability of high-resolution HDR sensors is limiting the application in this field, but with upcoming technology changes we will see three- to five-megapixel HDR sensors which will be suitable for tasks of this kind. Another development brought some challenges for the traffic sign recognition community. The appearance of change signs equipped with LED illuminators but with nearly no standardization in the frequency domain caused camera makers to think about techniques to cope with millisecond pulses, which are not easy to detect with a standard camera sensor.

18.4.4 HEADLAMP CONTROL

Also a comfort system but one already with some impact on safety is headlamp control. Oncoming vehicles are detected several hundred meters before they pass. The headlamps of oncoming vehicles are distinguished from those of leaving vehicles and from other random light sources in the vicinity.

Again, an HDR of more than six decades helps to distinguish the different headlamps as a pure detection of the presence of light sources is even easier to achieve with a saturated linear sensor.

A higher level of sophistication both in detection and in processing is the application of matrix beams, grid light, intelligent headlamps, or multibeam headlamps as they are sometimes called. Those headlamps allow separately controlled single beams to mask oncoming or leaving vehicles while dynamically forming a variable shaped beam illuminating the surroundings as known from a conventional high beam. Even beam shaping to follow a curved road has become possible. A precondition for the application of this advantageous lighting system is a safe and absolutely reliable detection of other road users as oncoming or leaving traffic. Again it is the HDR sensor which is superior to detect the 10,000 times brighter contour of an oncoming car's headlamp and at the same time the light contour of the car's own shaped beam as well as the contour of nonilluminated objects beside the headlamp corridor.

Beam shaping technology is being developed especially to increase safety at night for both the driver and the accompanying traffic as unintended blinding by headlamps is avoided and the seeing conditions are improved. This also shows a new dimension in the functionality of DAS. Whereas all systems mentioned before increase the comfort or the safety of the driver and the fellow passengers, headlamp control also takes care of the safety of others.

18.4.5 FROM REAR VIEW TO AUTONOMOUS PARKING

Early rear-view cameras were NTSC cameras adapted for automotive use and in the very beginning even used CCD sensors but they became one of the fastest developing areas and developed from a camera to look behind the car to a surround-view system allowing even low-speed maneuvering with and without guidance from the driver. A new chapter in autonomous driving was opened by this kind of system.

Three-dimensional reconstruction of images from different cameras simulating a bird's view camera floating above the car virtually controllable by a joystick from the driver's seat showing the image on a high-definition monitor in the cockpit requires high uniformity in sensor performance and because of the huge field of view an extremely high level of mastering various different illumination conditions. Somewhere in the 360-degree image there is always the sun, a bright headlamp, a strong reflective building, or an especially dark corner, and the problems they pose can be overcome only with the use of cameras with extreme dynamic range capabilities. Illumination conditions with different color temperatures in every direction from the car (eg, caused by a sunset and artificial light in opposite directions) create challenging situations which require not a single saturated color channel if the absence of color noise or disturbing appearances is expected. Again this is the domain of HDR sensors.

18.4.6 FROM OBJECT DETECTION TO PREEMERGENCY BRAKING

The first pedestrian-detecting camera was introduced in 2007 when night vision systems were upgraded to warn the driver of pedestrians close to the driving corridor or intruding on the driving corridor.

Again this development was the precursor for a significant step toward greater safety in road traffic. The system has been used to gather experience in the field of pedestrian detection without the high risk of an automated reaction. For the warning systems a false positive is annoying but not critical, but for an automatic braking system, which came onto the market in 2009, a much higher level of detection has been achieved.

18.4.7 AUTONOMOUS DRIVING

The new hype in the development of cars and in the public discussion since the first Google car autonomously navigated through the streets of North America is autonomous driving.

The first automatic parking assist systems were introduced on the market in 2003 by Toyota, and today parking assist systems are available for a large variety of cars and demonstrate that autonomous driving at low speed is both affordable and safe. The use of autonomous driving (or at least largely assisted driving) can help to avoid accidents caused by inattentive drivers or can enable drivers with handicaps to still be flexible in their choice to use a car. Google announced at the end of 2014 that it would start road tests of its autonomous driving prototypes by the beginning of 2015, and autonomous driving was one of most discussed topics at the Consumer Electronics Show (CES) in Las Vegas in 2015. So DAS are moving from convenient companions to pilots and driving becomes traveling.

As a logical consequence, automotive safety systems have not only to monitor the environment, measure distances to other road user, estimate trajectories, and plan the driving speed and path but also have to monitor the driver and prepare for situations when handing the responsibility back to the driver is foreseen. Driver-monitoring cameras will be needed as well as surround-view cameras and other remote sensors.

The requirements for HDR sensing imagers will increase as will the pressure to find cost-effective ways to implement the systems to democratize the technology so as to get a benefit from many cars taking care for their passengers.

18.4.8 WHAT IS THE DRAWBACK OF HDR IMAGING?

Not every automotive camera is an HDR camera today (but the proportion is steadily reaching 100%). But what is the drawback? The drawback is that 60 years of linear thinking in imaging has produced a significant amount of literature dealing with the shortcoming of video systems — blooming, saturation, color aberration, etc. — which is confusing to newcomers to this area as well as to customers and users. This book might help to create a profound overview and a better understanding of the advantages of nonlinear imaging and the genius archetype, the human eye. It is a great honor for me to contribute a short chapter which might help to obtain an even better understanding of what HDR sensors can contribute to safe driving.

ACKNOWLEDGMENTS

Without the inspiring working conditions and colleagues and later the excellent working groups I have had the chance to work with throughout the past 25 years of my professional life both in research (with Institute of Microelectronics Stuttgart) and in product development (with Robert Bosch GmbH) and without the patience of our promoters in the research phase and our customers in the product phase, all the experiences with HDR sensor design and camera development would not have been possible. Driven by the goal to develop systems which help to avoid accidents or at least to mitigate the impact on the individuals involved, we can be proud of the current status although there is still some need for improvement to reach the goal of "injury-free driving" with cars equipped with DAS or with autonomous driving cars. It has always been a motivation for me to come closer to the perfection of the best camera I ever experienced, the human eye, with its connection to an unrivaled effective control unit, the visual cortex. Finally, I thank my family, who patiently accompanied me and sometimes suffered from my enthusiasm for my job that sometimes required long business hours and some weekends to get things done with the necessary degree of perfection.

REFERENCES

Chamberlain, S., Lee, J., 1984. A novel wide dynamic range silicon photodetector and linear imaging array. IEEE J. Solid State Circuits 19 (1), 41–48.

Dowling, J., 1987. The Retina. An Approachable Part of the Brain. Harvard University Press, Cambridge, MA.

Enke, K., 1979. Possibilities for improving safety within the driver vehicle environment loop. In: 7th International Technical Conference on Experimental Safety, Paris.

Euro NCAP Ratings Group, 2015. Euro NCAP Rating Review 2015.
http://euroncap.blob.core.windows.net/media/16470/ratings-group-report-2015-version-10-with-appendix.pdf.

Euro NCAP Strategy Working Group, 2015. Roadmap 2020. European New Car Assessment Programme, March 2015. http://euroncap.blob.core.windows.net/media/16472/euro-ncap-2020-roadmap-rev1-march-2015.pdf.

Hubel, D., 1982. Exploration of the primary visual cortex, 1955–78. Nature 299, 515–524.

International Organization for Standardization Technical Committee ISO/TC 22/SC3, 2012. ISO 26262-10:2012(en).

Mead, C., Mahowald, M., 1988. A silicon model of early visual processing. Neurol Netw. 1, 91–97.

Mendis, S., Kemeny, S.E., Fossum, E.R., 1994. CMOS active pixel image sensor. IEEE Trans. Electron Devices 41 (3), 452–453.

National Highway Traffic Safety Administration, 2007. NHTSA DOT HS 810 698 (Department of Transportation documentation system).

Schultz, K.I., Kelly, M.W., Baker, J.J., Blackwell, M.H., Brown, M.G., Colonero, C.B, David, C.L., Tyrrell, B.M., Wey, J.R., 2014. Digital-pixel focal plane array technology. MIT Lincoln Lab. J. 20 (2), 36–51.

Seger, U., Graf, H.-G., Landgraf, M.E., 1993. Vision assistance in scenes with extreme contrast. IEEE Micro 13 (1), 50–56.

Seger, U., Apel, U., Höfflinger, B., 1999. HDRC-imager for natural visual perception. In: Handbook of Computer Vision and Applications, vol. 1, Sensors and Imaging. Academic Press, New York, pp. 223–235.

Seger, U., Knoll, P.M., Stiller, C., 2000. Sensor vision and collision warning systems. In: Convergence Conference, Detroit.

Tarr, G.L., Carreras, R.A., DeHainaut, C.R., Clastres, X., Freyss, L., et al., 1996. Intelligent sensors research using pulse-coupled neural networks for focal plane image processing. In: Proc. SPIE 2760, Applications and Science of Artificial Neural Networks II, p. 534. http://dx.doi.org/10.1117/12.235942.

Yang, D., El Gamal, A., Fowler, B., Tian, H., 1999. A 640×512 CMOS image sensor with ultrawide dynamic range floating-point pixel-level ADC. IEEE J. Solid State Circuits 34 (12), 1821–1834.

Wyatt Jr., J.L., Standley, D.L., Yang, W., 1991. The MIT Vision Chip Project: analog VLSI systems for fast image acquisition and early vision processing. In: Proceedings of the IEEE International Conference on Robotics and Automation Sacramento, California, April 1991.

AN APPLICATION OF HDR IN MEDICAL IMAGING

19

G. Ramponi*, A. Badano†, S. Bonfiglio‡, L. Albani‡, G. Guarnieri*

University of Trieste, Trieste, Italy*
US Food and Drug Administration, Silver Spring, MD, United States†
Barco FIMI, Saronno (VA), Italy‡

CHAPTER OUTLINE

19.1 INTRODUCTION

The "slow revolution" that brought the field of diagnostic radiology from analog film to digital imaging paved the way to important benefits that were progressively achieved in the last few decades: above all, the possibility for medical personnel to rapidly share huge amounts of patient-related information.

High Dynamic Range Video. http://dx.doi.org/10.1016/B978-0-08-100412-8.00019-X

However, some of the consequences of the analog-to-digital shift were problematic. One issue, whose importance is often underestimated, was the visual quality of the images presented to the physician. As will be described in detail herein, the requirements of the different specialties with regard to medical images span a wide range; the weakest link in the processing chain, which determines the overall quality, is frequently the display equipment.

In particular, the dynamic range of the luminance that the device can emit and its relation to the typical illumination of the environment in which the device operates should be considered with care. Old analog films placed on a lightbox could yield peaks of 4000 cd/m^2, and could span a luminance range of several log units: this made details well visible even in a highly illuminated room. The best medical CRT displays at the time of the digital transition, instead, had a maximum luminance in the several hundred candelas per square meter (cd/m^2) range. Only later did LCDs reach 1000 cd/m^2, at the price of an undesired higher luminance for "black" pixels.

In this chapter we will describe a solution aimed at overcoming dynamic range problems in the visualization of medical images. We will first analyze the specific requirements for a high dynamic range (HDR) display that should operate in the medical (diagnostic) context, and the procedures that are followed in a laboratory of the US Food and Drug Administration to evaluate such a display. We will then describe in some detail an LCD composed of two stacked liquid crystal panels, a prototype of which is presently being developed at Barco.

19.2 REQUIREMENTS OF HDR VISUALIZATION IN THE MEDICAL FIELD
19.2.1 GENERAL REQUIREMENTS FOR MEDICAL IMAGING DISPLAYS

In critical tasks such as diagnostic procedures, medical images should provide sufficient information to allow clinicians to detect diseases and to make medical decisions with the highest possible degree of accuracy (Martin et al., 1999). Here an important role is played by the visualization part of the medical imaging chain; a suitable display permits one to achieve the optimal compromise between productivity (short image interpretation time) and accuracy (low false positive and false negative rates) of the diagnosis. According to Reiner et al. (2003) the frequency of false positive readings is on the order of 2–15%, while false negative readings are much more frequent — that is, on the order of 20–30%. Approximately 50% of the total errors are due to perceptive issues. A good display also has other advantages; for example, it can permit a reduction of the radiation dosage (Kubo et al., 2014).

Characteristics such as image accuracy, spatial resolution, grayscale resolution (bit depth), color discrimination, black level and dynamic range, viewing angle, video response time, and absence of artifacts (eg, noise, Moiré effect) have different impacts on the outcome of the medical task depending on the specific application.

Another important requirement for medical applications is the consistency of the image rendering across different displays and therefore the need for a standardization. For grayscale radiographic images used for primary diagnosis, the Digital Imaging and Communications in Medicine (DICOM) grayscale standard display function (GSDF) has been universally accepted by regulatory authorities throughout the world as a standard (ACR and NEMA).

Moreover, much attention has to be devoted to perceptual issues (eg, visual acuity, contrast sensitivity), without forgetting that they show a significant amount of interobserver and intraobserver variability: different observers may have quite different performance because of different vision capabilities, age, and experience level; the same observer can show a lack of consistency in the assessment of the same image ("reader jitter").

Finally — considering the intensive use of images in some clinical procedures — ergonomic requirements such as viewing comfort and absence of aftereffects of the vision experience should be taken into account.

19.2.2 THE ADDED VALUE OF HDR DISPLAYS IN SPECIFIC MEDICAL APPLICATIONS

The display's weaknesses in terms of dynamic range and maximum luminance may be unacceptable in medical applications where a large number of luminance levels and fine details with very small luminance differences need to be discriminated. The display of a high-quality diagnostic image depends on at least three factors: (i) the capability of the detector to provide a large number of levels (bit depth of the acquired data set), (ii) a sophisticated mapping that converts source data into suitable driving levels, and (iii) the visualization of each of them as a distinctly perceived luminance value on the display screen.

A joint and harmonized advance in the whole image chain (acquisition, ie, sensors, image processing/mapping, and image visualization, ie, display) is needed. With regard to the visualization part, even if source data are represented by a large number of bits, the detectability of distinct luminance levels associated with each grayscale step depends on the luminance range of the device. Indeed, if the grayscale steps are too closely spaced, some may fall below the threshold of human perception.

In consumer applications, dynamic range compression or tone mapping techniques allow HDR images to be visualized on conventional displays (Ashikhmin, 2002; Meylan and Suesstrunk, 2006; Mantiuk et al., 2006), but their application on medical images poses serious concerns because the photometric distortion that is intrinsically introduced by the processing can cause the loss of clinically relevant details (Guarnieri et al., 2008b). In the medical field *window-and-level* adjustment techniques are used, at the price of a long analysis time and with the risk of missing details in the search phase (Yeganeh et al., 2012). Alternatively, an approach was proposed still based on the standard eight-bit gray level resolution but supported by eye-tracking techniques that dynamically process the display image by optimizing the luminance and contrast of the inspected area (Cheng and Badano, 2010).

Recent advances in the display industry have contributed to the development of HDR LCDs with extended luminance range, capable of effectively generating perceivable scales that extend to beyond 14 bits and of reproducing the original image with no theoretical distortion and no loss of information with respect to grayscale values. They will allow radiologists and physicians to perceive subtle and medically significant details in images where there is a large variation of the dynamic range in the data. Various technologies for HDR displays have been proposed (see Part IV of this book and Section 19.4). Regardless of the technological option adopted for the display, much attention has to be devoted to the implementation of the solution because an increased luminance range may come at a price with respect to other image quality parameters, such as increased veiling glare, optical crosstalk, and visual adaptation.

19.2.3 GRAY LEVEL MAPPING IN HDR MEDICAL IMAGES

Medical images evidence and translate into luminance levels some properties of the patient's tissues (eg, density in the case of radiography and computed tomography, or water concentration in the case of magnetic resonance imaging). Some sort of mapping is needed at the display stage to convert the source data into luminance values. One desirable property the mapping should have is that equal changes in the input values should produce equal changes in the perceived luminance regardless of the background to ensure that parts with the same density have the same visibility on any background. Because of the intrinsically nonlinear behavior of the human visual system, a nonlinear mapping that takes into account the properties of the human visual system and possibly adapts to the luminance range of the individual display device is needed (Albani et al., 2013). As mentioned earlier, the solution of choice for standard medical displays is the DICOM GSDF (ACR and NEMA). The function is based on the model of the contrast sensitivity of the human eye developed by Barten (1999) and uses the concept of just noticeable difference (JND), which represents the minimum variation of luminance that an average observer is able to detect in specified conditions. Unfortunately, the DICOM GSDF was developed for regular dynamic range displays (currently found in clinical use) and it is inadequate for the newly developed HDR equipment featuring a black level below 0.01 cd/m^2 — a value that lies outside the interval in which the DICOM GSDF is defined (0.05–4000 cd/m^2). In particular, if the DICOM GSDF is used, the details in the dark portions of the image are less visible than the details in the bright portions. This suggests that Barten's model overestimates the sensitivity of the eye at very low luminance levels.

It is expedient to propose an extension of the DICOM GSDF. To do so, perceptual issues should be reconsidered; the DICOM curve is indeed based on a theoretical approach that is often not applicable to the case of radiological images having as a background a mixture of white and dark areas that produce a changed perception (McCann and Rizzi, 2011).

The extension of the DICOM GSDF is still an open problem; a possible solution will be presented in Section 19.4.3.

19.3 EVALUATION OF MEDICAL HDR DISPLAYS

Tisdall et al. (2008) compared signal detectability in an HDR display and a standard medical LCD using a low-resolution eight-bit rear panel for the backlight and a high-resolution eight-bit front panel to obtain a 16-bit grayscale. They quantified the performance for a detection task with the display at both settings (standard LCD and HDR display). The results showed that subjects performed similarly in a detection task on the standard LCD and on an HDR display. Tisdall et al. concluded that further experiments were needed to verify this claim and that careful characterization of HDR devices is required to understand the benefits and drawbacks of HDR display technology.

In this section, we summarize the current understanding of how to evaluate medical HDR displays. Areas of image quality that are of particular relevance to HDR technologies for medical displays include the luminance and color response and uniformity, spatial noise and resolution, and veiling glare. For a more complete treatment, the reader is encouraged to read the literature on medical display characterization (Samei et al., 2005; Saha et al., 2006; Vogel et al., 2007; VESA, 2003).

19.3.1 LUMINANCE

Luminance response is typically measured with a luminance meter connected to a computer that automatically sweeps the range of digital driving levels and record the output. Although 18 measurement points have been recommended in the past, it is advisable to record the response to 256 levels to accommodate nonlinear intrinsic behavior — for instance, in the case of LCDs. In HDR displays, the complete set of available grayscale combinations can include 65,536 measurements, making the entire set difficult to acquire. In this situation, a subset of graylevels can be measured and then analyzed to develop a full model of the output luminance. One area that needs further research is the appropriate mapping of the luminance output in the range below 0.1 cd/m^2. Even if the viewing is performed in a dark room, the reflected luminance will increase the minimum luminance of the model applied. Thus, it is uncertain at the moment how to calibrate HDR displays. In addition, as already mentioned, the well-known and widely adopted GSDF (ACR and NEMA) is not defined for very low luminance, and even the underlying Barten model that constitutes the foundation of the GSDF model was not developed for that range.

19.3.2 UNIFORMITY

Luminance uniformity is another area of concern and is measured according to American Association of Physicists in Medicine (AAPM) Task Group 18 (TG18) (Samei et al., 2005) using a high-gain photodiode with the tip of the probe 1 mm from the center of each display. To determine changes with respect to viewing angle, measurements are taken at 0°, 15°, 30°, and 45°. The luminance nonuniformity metric is calculated as $100(L_{max} - L_{min})/L_{max}$, where L_{max} and L_{min} are the maximum and minimum luminance, respectively.

Color display systems are becoming commoner in medical imaging. Color monitors are replacing grayscale monitors to accommodate more imaging modalities as well as to increase the functionality of the visualization. Color coordinates (u', v') are measured following the 1976 CIE color scheme. The color nonuniformity is defined as the maximum distance between the color coordinates (u', v') for the various points on the display screen:

$$\Delta(u', v') = \sqrt{(u'_1 - u'_2)^2 + (v'_1 - v'_2)^2}. \tag{19.1}$$

The maximum change in (u', v') is the reported color nonuniformity parameter, which should be less than 0.01 for an acceptable clinical workstation according to AAPM TG18. Measurements of luminance and color coordinates are taken at five screen locations (four corners and the center), for uniform backgrounds at gray levels of 10, 50, 128, 200, and 255 on an eight-bit scale. Recently, a number of reports have highlighted the relevance of performing colorimetric measurements on medical displays (Fan et al., 2007; Krupinski et al., 2011; Roehrig et al., 2010; Silverstein et al., 2012; Cheng and Badano, 2011).

19.3.3 GRAYSCALE TRACKING

A complementary metric often used in the display industry is grayscale tracking, which is defined as the variation in color coordinates of the grayscale gradation.

Grayscale tracking can be characterized by measurement of color coordinates (u', v') with a color probe at all graylevels. A recent approach being considered by AAPM Task Group 196 to determine the variation in color of a grayscale gradation is summarized in the following. Assuming a display is calibrated to the GSDF, use TG18-LN test patterns (TG18-LNi , $i = 1, 2, \ldots, 18$) and record the luminance and color coordinates (u', v'). For the analysis, it is expedient to consider only measurements from patterns with a recorded luminance higher than 5 cd/m^2 or 1% of the luminance measured for TG18-LN18 (full white). Denote the remaining measurements as N. One possible definition of a quantity of interest ($\mathcal{T}$) is the average distance in the u', v' plane between the measurement of full white (TG18-LN18) and the nondiscarded measurements:

$$\mathcal{T} = \frac{1}{N-1} \sum_{j=18-N+1}^{17} \Delta j, \tag{19.2}$$

where

$$\Delta j = \sqrt{(u'_{xj} - u'_{x18})^2 + (v'_{xj} - v'_{x18})^2}.$$

This metric $\mathcal{T}$ quantifies how close the grayscale chromaticity is to the chromaticity of full white, with a lower value representing an improved grayscale tracking performance.

19.3.4 SPATIAL NOISE AND RESOLUTION

The analyses of subpixel structures and noise in HDR displays involve the acquisition of uniform patterns with a CCD camera. The camera should be cooled to reduce dark counts caused by thermal noise. A macro lens is needed to get magnified images on the CCD chip. A single line set at gray level 128 is placed both vertically and horizontally (Badano et al., 2004). It is useful to test different conditions with HDR displays to see the effects of different gray levels on the pixel.

19.3.5 TEMPORAL RESPONSE

Of relevance for image browsing typical in cross-sectional medical imaging, the temporal response of the display device needs to be characterized. Temporal response measurements are typically performed with gray level transitions relying on a measurement system composed of a display driver for pattern generation and a photomultiplier tube for light detection (Liang and Badano, 2006, 2007). Gray level transition time is measured by the capturing of the optical pulse corresponding to a luminance change. The rise time is usually defined as the time required for the luminance to change from 10% to 90% of the luminance difference, while the fall time is from 90% to 10%. One must take care when measuring all the components associated with the time response of an HDR display particularly when different components of the device might have different time constants (eg, modulated backlight and primary shutter display component). To date, this topic has not been investigated in the literature.

19.3.6 VEILING GLARE

HDR technologies with a large luminance range and a deep bit depth allow more information to be offered to a reader. However, a major limitation of HDR is the long-range light scattering from bright areas in the image that can reduce perceived contrast. This phenomenon is known as veiling glare. Operating over too wide a luminance range may cause readers to miss contrast in dark regions due to adaptation to bright areas and veiling glare effects in the human eye, or, alternatively, miss edges in bright regions due to short-range scattering or blurring (Tisdall et al., 2008).

Glare in human vision is caused by scattering of light in the cornea, lens, and retina and diffraction in the coherent cell structures on the outer radial areas of the lens. These effects result in "bloom" and "flare lines" perceived by readers in the vicinity of bright objects. Many attempts have been made to empirically model the veiling glare effect in the human visual system in terms of the equivalent veiling luminance (Vos, 1984; Stiles, 1929; Holladay, 1929). Investigations into a subject's visual threshold under different conditions of glare found it to be dependent on the angular separation between the source and the object and to increase with increased illumination of the glare source. Other studies, summarized in Vos (1984), give different values for thresholds depending on the validity domain. One cause of this variability was reported to be the age of the subjects in each experiment. Vos (2003a,b) introduced age into the CIE general disability glare equation:

$$\left(\frac{L_{eq}}{E_{gl}}\right)_{age\ adjusted} = \frac{10[1 + (Age/70)^4]}{\theta^2}, \quad 1° < \theta < 30°. \tag{19.3}$$

This equation does not include individual variations due to other factors, such as disease and ocular pigmentation, and was used to analyze disability glare particularly in traffic. Veiling glare is prominent in other components of the imaging chain, including acquisition (Raskar et al., 2008) and display (Flynn and Badano, 1999). Hardware glare can significantly degrade the quality of an image on different displays as characterized by the veiling glare response function by Flynn and Badano (1999). They developed an experimental technique to measure the degradation in image quality due to veiling glare in display devices such as radiographic film, monochrome monitors, and color monitors with antireflective surface coating. In HDR medical displays, the combined effect of veiling glare in the display and in the human visual system can negatively influence the detectability of lesions, as represented in Fig. 19.1.

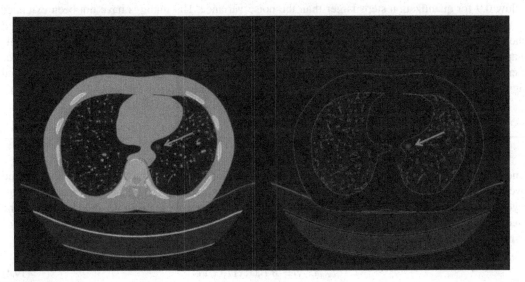

FIGURE 19.1

Left: A chest computed tomography image with a simulated nodule in the middle right of the image.
Right: After clipping of the bright structures in the computed tomography image, the simulated tumor nodule becomes more apparent.

19.3.7 QUANTIZATION

Among the many differences between reading a digital mammogram on transilluminated film and utilizing electronic display devices is the quantization of the image data. Quantization is defined as a lossy data compression technique by which intervals of data are grouped or binned into a single value (or quantum). Quantization processes are used in image acquisition, transmission, archiving, and display. Most display systems are limited to eight-bit luminance scales, but several approaches have been implemented to present extended grayscales. The chief examples of these technologies are the temporal and subpixel (Flynn et al., 1995) modulation techniques common in modern medical imaging workstation displays. Such techniques have successfully expanded the grayscale from eight bits, which is the commonest resolution for current LCDs, to 10–11 bits. However, these techniques have drawbacks with regard to other display quality characteristics, including spatial and temporal resolution.

While medical images are acquired with grayscales of 12–16 bits, most imaging systems quantize the data to eight bits for display. It is assumed that this loss of information does not translate into a degradation of visual task performance. This assumption relies primarily on literature reports that state that the human visual system is not able to perceive more than a small number of distinct luminance levels (Barten, 1999). These claims are sometimes based on human visual system internal noise considerations or on the effects of a wide luminance range on optimal perceptual adaptation. However, these claims have not been verified for the case of medical image interpretation. Previous attempts to study similar questions in the area of image quantization and observer performance have not been conclusive (Krupinski et al., 2007; Reiner et al., 2001). Burgess (1985) studied the effect of image quantization in the presence of white (uncorrelated) noise and found that performance decreased below 0.9 for quantization steps larger than the noise variance. His findings have not been extended to correlated noise images. Krupinski et al. (2007) studied the effect of 8-bit and 11-bit display of chest radiographs, comparing a conventional LCD with a similar device but with extended grayscale mapping of 11 bits. Both devices were calibrated to the same luminance range, forcing the luminance increase between grayscale steps to be reduced in the 11-bit device. Krupinski et al. found no statistical significance between the two devices but reported decreased reading times for the 11-bit presentation.

19.4 THE DUAL-LAYER APPROACH

One possible method of increasing the dynamic range of an LCD is to stack a strong backlight and two liquid crystal panels, one in front of the other (Visser et al., 2005). The transmittance of this "dual-layer LCD" is equal to the pointwise product of the transmittances of the two individual layers, and the theoretical dynamic range is squared for equal liquid crystal panels. More precisely, in ideal conditions (which will later be discussed), the output luminance $L_{\text{out}}(x, y)$ at pixel location (x, y) is given by

$$L_{\text{out}}(x, y) = B\, T_{\text{b}}(x, y)\, T_{\text{f}}(x, y), \tag{19.4}$$

where B is the backlight intensity (cd/m^2) and $T_{\text{b}}(x, t)$ and $T_{\text{f}}(x, y)$ are the adimensional transmittances of the two panels. Dual-layer displays, calibrated for a white level of 500 cd/m^2, are able to achieve a measured black level of around 0.003 cd/m^2. Moreover, the bit depth also increases, because the light is modulated by two eight-bit values,[1] although the overall bit depth of the dual-layer display is less than the theoretical 16 bits because output levels corresponding to the 2^{16} possible inputs are partially overlapped and nonuniformly spaced.

A drawback of this approach is that the backlight intensity must be greater than that of a standard display, because a medical-grade grayscale liquid crystal panel transmits around 30% of the light when fully on, and this value is reduced to 10% for color panels because of the additional presence of the color filters. For this reason, only grayscale displays are currently practical.

19.4.1 IMAGE SPLITTING AND THE PROBLEM OF THE PARALLAX ERROR

A dual-layer LCD requires the use of dedicated "image splitting" algorithms to generate the two images which drive the back panel and the front panel (Guarnieri et al., 2008a). Designing a splitting algorithm is a complex task that involves an accurate study of the behavior of the individual panels and the human visual system, and requires the use of advanced mathematical techniques for an efficient implementation.

A splitting algorithm also has the additional objective of minimizing the parallax error that is caused by the unavoidable slight misalignment of the two panels and that is described in the following.

Fig. 19.2 shows the effect on spatial resolution from the use a dual-layer approach to achieve HDR. Fig. 19.2A–D depicts a horizontal line image on a uniform background as visualized in different implementations of the display. In the synchronized mode, both layers of the display present the same image data. The same data when shown only by the front (back) layer are presented in Fig. 19.2B and C. For comparison, Fig. 19.2D shows the same image on a single-layer display. These photographs of the display emissions directly describe how the spatial resolution properties of the HDR display change depending on the addressing modes. While these images should all contain similar representations of the data, it is clear that the spatial resolution achieved in the image in Fig. 19.2C is lower than that achieved in the other images. The same effect but for a vertical line can be seen in Fig. 19.2E–H. The data shown in Fig. 19.2 can also be analyzed by Fourier analysis to show the change of the modulation transfer function for different addressing modes. The effect of parallax is added to this effect if one looks at the line images from a nonperpendicular direction. Fig. 19.2I–N shows images of the same patterns as in Fig. 19.2A–H but taken at 5 degree in the orthogonal direction from the line. The same degradation effect described above is seen at an angle but compounded with the addition effect of parallax. The images in Fig. 19.2 have been scaled down from their actual dynamic range to allow for reproduction in this media. The analysis of these images with the methods described in Yamasaki et al. (2013) would provide a quantitative characterization of the importance of this effect in HDR displays that rely on dual emission layers.

Parallax effects on spatial resolution are problematic for medical image visualization for two reasons. Firstly, parallax affects the spatial resolution of the display device in a way that makes it variable across the screen. This is problematic because detail in medical images will appear with different content in spatial frequencies depending on the location of the features examined. In addition,

---

[1]The pixels of grayscale LCD panels are actually made of three independent subpixels driven by the three RGB input channels. This permits the use of *subpixel dithering* to improve the perceived bit depth of the device.

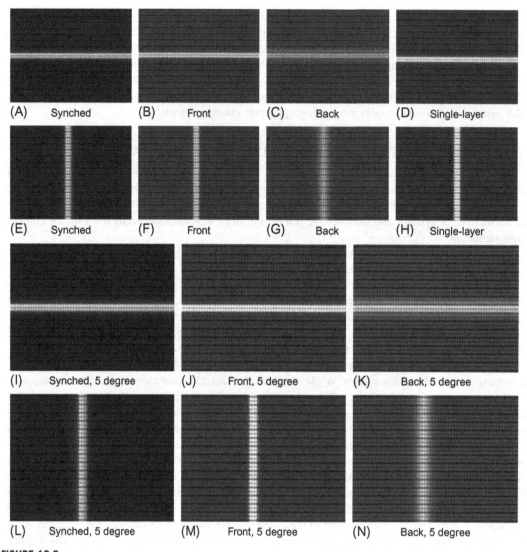

FIGURE 19.2

Horizontal and vertical lines displayed on the front and back panels of a dual-layer HDR prototype taken perpendicularly to the display surface (A)–(H) and at a 5 degree angle (I)–(N). The line is at a gray level of 255 on a background of 0.

this variability will make clinical decisions become inconsistent between readers looking at the same image on a device. Secondly, the effect of parallax also incorporates a dependency on the driving level used. This implies that the spatial frequency content of an image displayed on the HDR display will depend on the gray level, which is not an absolute, characteristic feature of the image data and can therefore lead to interpretation errors. An example of this effect can be appreciated in that the sharpness of a microcalcification cluster in a mammogram would be seen differently depending on the baseline gray level of the surrounding area. This characteristic is not desired in medical imaging visualization because it adds variability to the reading process that is difficult to control or compensate for through image processing.

We will now describe different splitting algorithms, point out their drawbacks, and use this knowledge to design a more advanced technique. Only the conceptual approach to these algorithms is described in this chapter; in Guarnieri et al. (2008a,b,c), the interested reader can find the whole mathematical formulation, together with the details of its numerical implementation and a feasible solution of the problem.

19.4.2 TECHNIQUES FOR IMAGE SPLITTING

The simplest possible technique is to perform the splitting on a pixel-by-pixel basis. More precisely, if we indicate with $L_{in}(x, y)$ the luminance of the input image at pixel location (x, y) and follow the notation used in Eq. (19.4), the splitting algorithm takes the form

$$T_b(x, y) = F(L_{in}(x, y)), \quad T_f(x, y) = \frac{L_{in}(x, y)}{B\,T_b(x, y)}. \tag{19.5}$$

In other words, we compute the transmittance of the back panel by mapping the input luminance with a suitable nonlinear function $F(\cdot)$, and we subsequently compute the transmittance of the front panel by division in order to guarantee that the product of the two images reproduces the input. The splitting is computed on linear data; the nonlinear encoding of the source image (if present) and the response of the liquid crystal panels are compensated appropriately by the mapping of the data before and after the processing. An intuitive choice for $F(\cdot)$ is a square root (Penz, 1982): in this way, each panel displays the same image.

If this simple technique is used on an actual dual-layer display, a problem immediately becomes visible. There is a small but not negligible distance between the liquid crystal cells of the two panels, due to the presence of glass sheets and polarizing filters in between. Therefore, if the observer looks at the display from an oblique angle, the two images appear misaligned. More precisely, instead of seeing the correct image $L_{out}(x, y) \triangleq B\,T_b(x, y)\,T_f(x, y)$, an observer focusing on position (x, y) on the display surface sees a distorted image $\tilde{L}_{out}(x, y) \triangleq B\,T_b(x + \Delta x, y + \Delta y)\,T_f(x, y)$, where the displacements Δx and Δy depend on the viewing angle. This form of distortion is the *parallax error* that was described in Section 19.4.1. It can be proved that in order to minimize the parallax error, a splitting algorithm should minimize the norm of the relative gradient of the back panel (ie, the back panel should be smooth).

However, the back-panel values cannot be generated by a simple low-pass filter, because of limitations in the dynamic range of the panels and the need to obtain a perfect or quasi-perfect reconstruction of the output image (Guarnieri et al., 2008a). One possible solution is to generate, by means of a constrained optimization algorithm, a back panel that minimizes the parallax error and from which a front panel with suitable black-to-white range can be generated. To do so, it is computationally advantageous to minimize the mean squared value of the parallax error, subject to constraints that can be easily derived by the panel specifications. If perfect reconstruction is required, the transmittance of the front panel can be computed by division, as in Eq. (19.5).

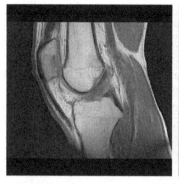 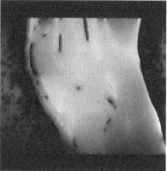 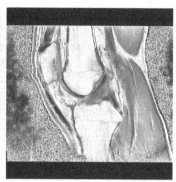

FIGURE 19.3

Example of splitting. From left to right, original image, back panel, and front panel.

Source: Reprinted with permission from Guarnieri, G., Albani, L., Ramponi, G., 2008. Image splitting techniques for a dual layer high dynamic range LCD display. J. Electron. Imaging 17 (4), 043009-1–9.

An example of splitting of an actual medical image is shown in Fig, 19.3.

Fig. 19.3 shows that the back panel still contains sharp edges. This happens when the input image contains edges which have a greater magnitude than the dynamic range of the panels. In this case, the front panel alone is not able to completely reproduce the edge, and a fraction of its magnitude must be transferred onto the back panel.

One possible solution is to allow some distortion in the visualized image. The properties of the visual system suggest that the details on the dark side of a high-contrast edge are less visible; this can be used to relax the requirement of perfect reconstruction in those portions. As a consequence, the back-panel transmittance $T_b(x, y)$ and the front-panel transmittance $T_f(x, y)$ are no longer linked by Eq. (19.5), and are both treated as unknowns in the optimization problem. The constraints are simple bounds, constant over the entire image:

$$T_{min} \leq T_b(x, y) \leq T_{max}$$
$$T_{min} \leq T_f(x, y) \leq T_{max} \quad \forall (x, y), \tag{19.6}$$

where T_{min} and T_{max} are the black and white levels of the panels.

The reconstruction error metric to be used in this case must satisfy two strongly conflicting requirements. It should be kept as simple as possible, because it is being used inside an optimization algorithm; this excludes advanced methods such as visible difference predictors (Daly, 1993; Mantiuk et al., 2004). At the same time, the expression should be spatially adaptive because, as mentioned above, the eye sensitivity depends on the context.

We can obtain a pointwise estimate of the visibility of the reconstruction error by taking the ratio between the luminance error and the JND. We obtain the JND by mapping the adaptation level L_{ad} with a *threshold versus intensity* (TVI) function (Ashikhmin, 2002); possible choices for the latter are discussed in Section 19.4.3. In turn, we compute L_{ad} by filtering the input image with a low-pass filter; in this sense the metric is spatially adaptive (in particular, it correctly predicts the lower visibility of the details near brighter portions), but at the same time the computational cost is low. The function to be minimized takes the following form:

$$E_{\text{total}} = \iint \left\{ k_{\text{p}} \left\| \nabla \log(T_{\text{b}}) \right\|^2 + \frac{L_{\text{in}}^2}{\text{TVI}(L_{\text{ad}})^2} \left[\log(T_{\text{b}}) + \log(T_{\text{f}}) - \log\left(\frac{L_{\text{in}}}{B}\right) \right]^2 \right\} dx\, dy, \qquad (19.7)$$

where k_{p} is a user-adjustable scalar parameter that balances the relative weight of the reconstruction and the parallax errors.

A full analysis of the variational problem (Eq. 19.7) is beyond the scope of this chapter; it can be found in Guarnieri et al. (2008c) together with a discrete formulation of the problem itself and its numerical solution.

19.4.3 DEVELOPMENT OF AN HDR DISPLAY FUNCTION

The DICOM GSDF described in Section 19.2.3 has been universally adopted to convert the original image data into luminance values, but has some drawbacks which limit its use on HDR displays. One issue is that the black level of HDR displays can be lower than 0.05 cd/m^2, which is the lower limit of the DICOM GSDF. It is possible to extrapolate the low-luminance values with the original equations of Barten's model, but this "extended" DICOM curve does not produce satisfactory results. Indeed, experiments performed on a dual-layer LCD showed that DICOM-mapped images appear too dark and that the details in the dark portions are less visible than those in the bright portions, therefore failing the requirement of perceptual uniformity. Examination showed that Barten's model overestimates the eye sensitivity at low luminance levels, because it predicts JNDs that tend to zero as the luminance decreases. This implies that the eye has infinite sensitivity and is able to detect arbitrarily small luminance differences. Other models of visual JNDs have been introduced — for instance, by Blackwell CIE (1981) and Ferwerda et al. (1996), and the predicted JNDs tend to a constant nonzero limit as the luminance decreases. Display functions computed with these models produce a more uniform visibility of the details but result in compatibility issues, because the experiments used to derive the models are different from Barten's: Barten measured the contrast sensitivity by means of static sinusoidal gratings, whereas Blackwell and Ferwerda et al. measured the temporal sensitivity by means of brief flashing dots. This motivated a series of psychophysical experiments, aiming to measure the visual JND for a wide range of luminance levels and using a test pattern as similar as possible to the one used for the DICOM GSDF.

The experiments were performed on volunteers at the University of Trieste. The recently proposed *two-alternative forced-choice method with denoising* (García-Pérez, 2010) was used to collect JND values at different background levels throughout the dynamic range of the dual-layer display. The experiments were conducted in a dark room, with use of a chin rest to ensure a constant viewing distance; although unrealistic, these conditions were essential to produce consistent results.

To compute a display function, it is necessary to interpolate the JNDs, measured at specific background levels, with a continuous TVI function. For this purpose, Blackwell's model was taken as a reference, because it is simple and provides good results in practice (the units are candelas per square meter):

$$\text{TVI}(L) = 0.0594\,(1.219 + L^{0.4})^{2.5}. \qquad (19.8)$$

The model of Ferwerda et al., instead, has a piecewise definition which makes it more difficult to handle. The analytical form was kept the same and the constants were adjusted by means of a maximum-likelihood fit to the experimental data. One important property is that the image obtained with a display function is not influenced by multiplicative scale factors in the display function itself or in the TVI function used to derive it, and this suggested a procedure to "average" the measurements of different observers: in the maximum-likelihood fit, a different leading scale factor was allowed for each observer, but the same value was enforced for all the other constants. The family of curves obtained with this procedure is as follows:

$$\text{TVI}(L) \propto (0.572 + L^{0.4})^{2.5}. \tag{19.9}$$

19.4.4 A DUAL-LAYER LCD PROTOTYPE

Barco demonstrated a prototype of a dual-layer HDR display (Albani et al., 2013); its schematic structure and a photograph are shown in Figs. 19.4 and 19.5.

The prototype is based on 19-inch SXGA (1280×1024) monochrome liquid crystal panels with an in-plane switching pixel design to ensure consistency of image quality over a wide viewing angle. As already mentioned, a dual-layer LCD requires a backlight solution with high brightness; to achieve this, a novel LED direct backlight with a high number of LEDs was designed. This is not common, for cost reasons, in conventional LCDs. The form factor of the LEDs and their matrix structure result in efficient heat sinking and a uniform temperature distribution, solving in this way the problem of the spatial nonuniformity of the output luminance evidenced in previous dual-layer LCDs using a backlight with multiple cold cathode fluorescent lamps.

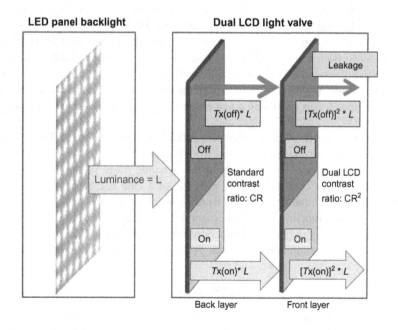

FIGURE 19.4

Block scheme of the prototype display.

FIGURE 19.5

Photograph of the prototype display (on the right).

Special care and dedicated tools were used to get a proper alignment of the two liquid crystal layers and reduce the parallax error; nevertheless, a small displacement between the two panels is unavoidable. The industrialization of the product will require minimization of the gap between the two liquid crystal layers by use of panels with thin glass. To avoid a visible change in luminance, it was necessary to blur the image that is displayed on the back panel, and to process the image on the front panel to compensate for this blurring (see Section 19.4.1).

Image processing in the dual-layer LCD requires a powerful computational platform to achieve real-time behavior; for this reason a graphics processing unit running on a proprietary graphics board was used (Szydzik et al., 2013). It creates the opportunity of using HDR displays in real-time medical procedures such as interventional X-ray.

In the prototype, a field-programmable gate array-based platform drives the hardware components of the dual-layer LCD (the two panels and the backlight controlling the color and light emitted by the LED matrix), while the image processing algorithm was implemented with OpenCL for multitask processing on a multicore architecture. The choice of the OpenCL implementation ensures a high level of hardware and platform independency, which provides flexibility toward future hardware platforms.

The performances of the dual-layer display were analyzed with the procedures described in Section 19.3, and satisfactory results were obtained: the black level was 0.007 cd/m^2, the maximum luminance was 1000 cd/m^2, and the static contrast was $140,000{:}1$.

The prototype permitted the initiation of clinical validations of HDR visualization; a first observers-based study was performed and related to the comparison of the performance of low dynamic range and HDR displays in the detection of nodules in X-ray chest images. The study evidenced a shorter time for the execution of the diagnostic task when the HDR display was used (20.52 s per case on average for the HDR display, 23.32 s per case on average for the low dynamic range display) and a statistically significant difference (2.65%) in favor of the HDR display in the accuracy of the detection of nodules mainly with difficult or very difficult degree of detection (small size, very low contrast).

19.5 CONCLUSIONS

An HDR display is beneficial in all medical applications requiring the visualization of images with a wide range of gray levels. Examples are as follows:

- Analysis of chest radiographs: Chest radiography shows low-density details, such as nodules, blood vessels, or lesions in the soft tissues, which are located over a background that can contain very wide density variations because of the presence of the bones of the rib cage.
- Mammography-based diagnosis: This task involves the detection of masses and microcalcifications of very small size and low density; mammographic images are typically rich in details, with very fine luminance differences. Furthermore, the improved visibility of HDR displays in the darker areas of the image permits better delineation of the anatomical structures.
- Ophthalmologic applications: Tests performed in Trieste (Italy) evidenced how an HDR display allows better recognition of different retinal layers.
- Endoscopic and minimally invasive applications: Color and moving images are involved here; endoscopes currently available on the market already have HDR. In addition, often the light source of the endoscope generates a dark area around the light spot, and an HDR display will allow the detection of details there.

In all the above-mentioned applications the benefits of HDR displays will consist in improved detection accuracy and/or faster execution of search and decision tasks. Some first evidence was already gathered in studies reported in the literature and comparing low dynamic range and HDR displays (Badano et al., 2009).

Other critical issues need to be considered in the design of an HDR imaging chain for medical applications.

In the area of luminance mapping, future developments may involve the personalization of the display function for the individual observer, taking into account in particular the loss of visual acuity, especially in the dark portions, which occurs with aging. One difficulty comes from the length and complexity of the psychophysical experiments that are needed to build an individual vision model, which may render the observers unwilling to take the test. For this reason, faster but still accurate tests should be developed. A second issue concerns the influence of ambient light. It is well known that the ambient light reflected off the screen surface reduces the contrast of the visualized image, and recommendations exist which relate the display black level to the maximum allowable ambient light (American Association of Physicists in Medicine, 2005). However, for an HDR display, this would require a very dark reading room, which is impractical. Moreover, a bright display visualized in a dark room is fatiguing for the eye, and indeed some experiments showed an improved visibility (ie, a lower JND) with the lights *on*. For this reason, it is advisable to design reading rooms with an ambient light that illuminates the visual field around the display, in order to produce comfortable visual surroundings, without illuminating the surface of the displays (Xthona, 2003).

An additional problem is represented by the presence of very bright and very dark regions in the same image; this can introduce a further loss of detail visibility because of various causes:

- Veiling glare, which was described in Section 19.3.6 (Choi et al., 2012, 2014).
- The underadaptation condition: The DICOM GSDF and its alternatives are derived from JND models that measure the eye sensitivity by means of low-contrast test patterns under the assumption that the observer's eye is adapted to a uniform luminance level; this kind of model does not describe accurately the perception of an image that contains wide luminance variations (Pattanaik et al., 2000).
- Different backgrounds generate different perceptions of the same gray value: areas shown on a black background are perceived to be lighter than the same observed areas on a bright background and a drastic reduction of the perceived luminance range is produced by the context (Rizzi and McCann, 2009).

Performance characterization merits more work too. First, further resolution and noise measurements need to be performed to determine the effect of different gray levels on the magnitude of display interpixel and intrapixel noise. In the same area, more work is needed to determine a metric for the influence of the parallax effect on resolution and noise in the context of a medical image where a set of edges can be superimposed on a slowly varying anatomical background field.

Finally, for a further large and long-term performance improvement, a technological discontinuity is needed: future HDR medical displays will probably use more efficient and chromatically richer quantum dot backlights, or they will switch from LCD technology to organic LED technology.

ACKNOWLEDGMENTS

This work was supported by the CHIRON European project, co-funded by the EU-ARTEMIS Joint Undertaking program under grant agreement no. 100228. Badano acknowledges partial support from Cooperative Research and Development Agreement between FIMI/Barco and Center for Devices and Radiological Health.

The mention of commercial products, their sources, or their use in connection with material reported herein is not to be construed as either an actual or implied endorsement of such products by the Department of Health and Human Services.

REFERENCES

ACR and NEMA. Digital Imaging and Communications in Medicine (DICOM), Part 3.14, Grayscale Standard Display Function. Technical report. American College of Radiology and National Electrical Manufacturers Assoc.

Albani, L., De Paepe, L., Bonfiglio, S., Guarnieri, G., Ramponi, G., 2013. HDR medical display based on dual layer LCD. In: International Symposium on Image and Signal Processing and Analysis, ISPA, pp. 511–515.

American Association of Physicists in Medicines, 2005. Assessment of Display Performance for Medical Imaging Systems. American Association of Physicists in Medicine, Task Group 18.

Ashikhmin, M., 2002. A tone mapping algorithm for high contrast images. In: 13th Eurographics Workshop on Rendering, pp. 1–11.

Badano, A., Gagne, R.M., Jennings, R.J., Drilling, S.E., Imhoff, B.R., Muka, E., 2004. Noise in flat-panel displays with subpixel structure. Med. Phys. 31 (4), 715–723.

Badano, A., Guarnieri, G., Ramponi, G., Albani, L., 2009. Quantization in medical imaging displays: initial observer results for a high-luminance-range, dual layer LCD. In: SID Symposium Digest of Technical Papers, 40 (1), 923–926.

Barten, P.G.J., 1999. Contrast Sensitivity of the Human Eye and Its Effects in Image Quality. SPIE Press, Bellingham, WA.

Burgess, A., 1985. Effect of quantization noise on visual signal detection in noisy images. J. Opt. Soc. Am. A 2 (9), 1424–1428.

Cheng, W.C., Badano, A., 2010. A gaze-contingent high-dynamic range display for medical imaging applications. In: Progress in Biomedical Optics and Imaging — Proceedings of SPIE.

Cheng, W.C., Badano, A., 2011. Virtual display: a platform for evaluating display color calibration kits. In: SID Symposium Digest of Technical Papers, pp. 1030–1033.

Choi, M., Sharma, D., Zafar, F., Cheng, W.C., Albani, L., Badano, A., 2012. Does veiling glare in the human eye hinder detection in high-dynamic-range displays? J. Display Technol. 8 (5), 273–282.

Choi, M., Wang, J., Cheng, W.C., Ramponi, G., Albani, L., Badano, A., 2014. Effect of veiling glare on detectability in high-dynamic-range medical images. J. Display Technol. 10 (5), 420–428.

CIE, 1981. An analytic model for describing the influence of lighting parameters upon visual performance — vol. 1. Technical foundations.

Daly, S., 1993. The Visible Differences Predictor: An Algorithm for the Assessment of Image Fidelity. MIT Press, Cambridge, MA, pp. 179–206.

Fan, J., Roehrig, H., Krupinski, E., 2007. Characterization of color-related properties of displays for medical applications. In: Medical Imaging. International Society for Optics and Photonics, p. 65160X.

Ferwerda, J.A., Pattaniak, S.N., Shirley, P., Greenberg, D.P., 1996. A model of visual adaptation for realistic image synthesis. In: Computer Graphics, pp. 249–258.

Flynn, M., Badano, A., 1999. Image quality degradation by light scattering in display devices. J. Digital Imaging 12, 50–59.

Flynn, M.J., McDonald, T., DiBello, E., et al., 1995. Flat panel display technology for high performance radiographic imaging. Proc. SPIE 2431, 360–371.

García-Pérez, M.A., 2010. Denoising forced-choice detection data. Br. J. Math. Stat. Psychol. 63 (1), 75–100.

Guarnieri, G., Albani, L., Ramponi, G., 2008a, Image-splitting techniques for a dual-layer high dynamic range LCD display. J. Electron. Imaging 17 (4), 043009-1–9.

Guarnieri, G., Albani, L., Ramponi, G., 2008b. Minimum-error splitting algorithm for a dual layer LCD display — part I: background and theory. J. Display Technol. 4 (4), 383–390.

Guarnieri, G., Albani, L., Ramponi, G., 2008c. Minimum-error splitting algorithm for a dual layer LCD display — part II: implementation and results. J. Display Technol. 4 (4), 391–397.

Holladay, L., 1929. The fundamentals of glare and visibility. J. Opt. Soc. Am. 12, 271–319.

Krupinski, E.A., Siddiqui, K., Siegel, E., Shrestha, R., Grant, E., Roehrig, H., Fan, J., 2007. Influence of 8-bit vs 11-bit digital displays on observer performance and visual search: a multi-center evaluation. J. Soc. Inform. Disp. 15, 385–390.

Krupinski, E.A., Roehrig, H., Matsui, T., 2011. High luminance monochrome vs. color displays: impact on performance and search. In: SPIE Medical Imaging. International Society for Optics and Photonics, p. 79661R.

Kubo, T., Ohno, Y., Kauczor, H., Hatabu, H., 2014. Reduction dose reduction in chest CT — review of available options. Eur. J. Radiol. 83 (10), 1953–1961.

Liang, H., Badano, A., 2006. Precision of gray level response time measurements of medical liquid crystal display. Rev. Sci. Instrum. 77, 065104.

Liang, H., Badano, A., 2007. Temporal response of medical liquid crystal displays. Med. Phys. 34 (2), 639–646.

Mantiuk, R., Myszkowski, K., Seidel, H.P., 2004. Visible difference predicator for high dynamic range images. In: Proceedings of IEEE International Conference on Systems, Man and Cybernetics, pp. 2763–2769.

Mantiuk, R., Myszkowski, K., Seidel, H.P., 2006. A perceptual framework for contrast processing of high dynamic range images. ACM Trans. Appl. Percept. 3 (3), 286–308.

Martin, C., Sharp, P., Sutton, D., 1999. Measurement of image quality in diagnostic radiology. Appl. Radiat. Isotopes 50 (1), 21–38.

McCann, J., Rizzi, A., 2011. The Art and Science of HDR Imaging. John Wiley & Sons, Ltd, New York.

Meylan, L., Suesstrunk, S., 2006. High dynamic range image rendering with a Retinex-based adaptive filter. IEEE Trans. Image Process. 15 (9), 2820–2830.

Pattanaik, S.N., Tumblin, J., Yee, H., Greenberg, D.P., 2000. Time-dependent visual adaptation for fast realistic image display. In: Proceedings of the 27th Annual Conference on Computer Graphics and Interactive Techniques, SIGGRAPH '00. ACM Press/Addison-Wesley Publishing Co., New York, NY, pp. 47–54, http://dx.doi.org/10.1145/344779.344810.

Penz, P.A., 1982. Stacked electro-optic display. U.S. Patent No. 4,364,039.

Raskar, R., Agrawal, A., Wilson, C., Veeraraghavan, A., 2008. Glare aware photography: 4D ray sampling for reducing glare effects of camera lenses. ACM Trans. Graph. 27.

Reiner, B.I., Siegel, E.L., Hooper, F.J., Pomerantz, S., Dahlke, A., Rallis, D., 2001. Radiologists productivity in the interpretation of CT scans: a comparison of PACS with conventional film. Am. J. Roentgenol. 176, 861–864.

Reiner, B., Siegel, E., Krupinski, E., 2003. Digital radiographic image presentation and display. In: Advances in Digital Radiography. RSNA Press, pp. 79–89.

Rizzi, A., McCann, J.J., 2009. Glare-limited appearances in HDR images. J. Soc. Inform. Display 17.

Roehrig, H., Rehm, K., Silverstein, L.D., Dallas, W.J., Fan, J., Krupinski, E.A., 2010. Color calibration and color-managed medical displays: does the calibration method matter? In: SPIE Medical Imaging, p. 76270K.

Saha, A., Liang, H., Badano, A., 2006. Color measurement methods for medical displays. J. Soc. Inform. Disp. 14 (11), 979–985. http://link.aip.org/link/?JSI/14/979/1.

Samei, E., Badano, A., Chakraborty, D., Compton, K., Cornelius, C., Corrigan, K., Flynn, M.J., Hemminger, B., Hangiandreou, N., Johnson, J., Moxley-Stevens, D.M., Pavlicek, W., Roehrig, H., Rutz, L., Shepard, J., Uzenoff, R.A., Wang, J., Willis, C.E., 2005. Assessment of display performance for medical imaging systems: executive summary of AAPM tg18 report. Med. Phys. 32 (4), 1205–1225.

Silverstein, L.D., Hashmi, S.F., Lang, K., Krupinski, E.A., 2012. Paradigm for achieving color-reproduction accuracy in LCDs for medical imaging. J. Soc. Inform. Disp. 20 (1), 53–62.

Stiles, W., 1929. The scattering theory of the effect of glare on the brightness difference threshold. Proc. R. Soc. Lond. 105, 131–145.

Szydzik, T., Nunez, A., De Paepe, L., Albani, L., 2013. Contributions to visualization algorithm enabling GPU-accelerated image displaying for dual panel high dynamic range LCD display. In: International Symposium on Image and Signal Processing and Analysis, ISPA.

Tisdall, M., Damberg, G., Wighton, P., Nguyen, N., Tan, Y., Atkins, M., Li, H., Seetzen, H., 2008. Comparing signal detection between novel high-luminance HDR and standard medical LCD displays. J. Display Technol. 4 (4), 398–409.

VESA, 2003. Flat panel display measurements standard, version 2.0. Technical report, VESA.

Visser, H.M., Rosink, J.J.W.M., Raman, N., Rajae-Joordens, R., 2005. Invited paper: tuning LCD displays to medical applications. In: EuroDisplay 2005, pp. 74–77.

Vogel, R., Saha, A., Chakrabarti, K., Badano, A., 2007. Evaluation of high-resolution and mobile display systems for digital radiology in dark and bright environments using human and computational observers. J. Soc. Inform. Disp. 15 (6), 357–365. http://link.aip.org/link/?JSI/15/357/1.

Vos, J., 1984. Disability glare — a state of the art report. CIE-J. 3, 39–53.

Vos, J., 2003a. On the cause of disability glare and its dependence on glare angle, age and ocular pigmentation. Clin. Exp. Optom. 86, 363–370.

Vos, J., 2003b. Reflections on glare. Lighting Res. Tech. 35, 163–176.

Xthona, A., 2003. White Paper: Designing the Perfect Reading Room for Digital Mammography. Barco.

Yamasaki, A., Liu, P., Cheng, W.C., Badano, A., 2013. Spatial resolution and noise in organic light-emitting diode displays for medical imaging applications. Opt. Exp. 21 (23), 28111–28133.

Yeganeh, H., Wang, Z., Vrscay, E., 2012. Adaptive windowing for optimal visualization of medical images based on a structural fidelity measure. Lect. Notes Comput. Sci. (Including Subseries Lect. Notes Artif. Intell. and Lect. Notes Bioinform. 7325 (Part 2), 321–330.

HIGH DYNAMIC RANGE DIGITAL IMAGING OF SPACECRAFT

20

B. Karr*,†, A. Chalmers†, K. Debattista†

Kennedy Space Center, FL, United States*
University of Warwick, Coventry, United Kingdom†

CHAPTER OUTLINE

20.1 INTRODUCTION

A transition from film cameras to digital imagers is anticipated in the post–Space Shuttle era. Analysis of digital imaging systems and comparison with the film baseline is important for determining the transitional road map. Resolution, dynamic range, uniformity, color response, sharpness, timing, etc., are examples of imaging metrics for consideration. Traditional imaging methods are unable to capture the wide range of luminance present in a natural scene. High dynamic range (HDR) imaging is an

High Dynamic Range Video. http://dx.doi.org/10.1016/B978-0-08-100412-8.00020-6

exception, enabling a much larger range of light in a scene to be captured (Banterle et al., 2011; Reinhard et al., 2010). The focus of this chapter is the development and testing of digital imaging systems with respect to the HDR imaging of spacecraft. A review of the dynamic range of film, the current engineering imaging baseline, is presented along with relevant characteristics. Past work related to the imaging of the Space Shuttle is included. Field experiments are then described, with the inclusion of sample image frames. Finally, current work is discussed, including laboratory capture and processing of imager dynamic range, and characterization of display device output luminance.

20.2 BACKGROUND

The NASA requirement for engineering-quality imagery is defined as being visually lossless — that is, no loss of perceived quality once compression is applied, as compared with the original captured file. The ability to capture engineering-quality imagery with a wide degree of dynamic range during rocket launches is critical for postlaunch processing and analysis (US Columbia Accident Investigation Board and Gehman, 2003; United States Congress, House Committee on Science and Technology, 1986). Rocket launches often present an extreme range of lighting when viewed by image analysts during both day and night launches. During the Space Shuttle program, engineering imagery was collected at Kennedy Space Center primarily through the use of 16 and 35 mm Photosonics high-speed film cameras. The film cameras include automatic exposure control, shifting the available dynamic range in response to scene luminance.

HDR images may be captured through a single sensor with use of multiple exposure settings, dual gain sensors, or multiple sensors in combination with beam splitters and neutral density filters (Mann and Picard, 1995; ARRI, 2013b). In addition, new classes of imaging sensors are being developed that are capable of directly capturing a larger dynamic range — for example, those proposed by Fuji and Panasonic (Cade, 2013). Workflows in typically lossless HDR formats such as Radiance RGBE (Ward, 1991), allowing half-float representations, as well as OpenEXR (Bogart et al., 2003) and TIFF, allowing half-float and full-float representations, are required for HDR data (Ward, 1991). Lossy compression formats such as goHDR and HDR MPEG (Rafa et al., 2006) may also be used for bandwidth reduction and smaller file size. The ability to display HDR imagery directly on HDR displays is also evolving rapidly with the advent of LED-modulated liquid crystal HDR displays. These HDR capture, processing, and display workflows aim to reproduce the simultaneous contrast ratio capability of the human visual system, and indeed even more dynamic range than the human eye can see (Seetzen et al., 2004).

The entire capture, processing, and display workflow must be considered when one is evaluating dynamic range, as each subsystem can potentially enhance or degrade the overall system capability. Benefits have been realized by use of complete system capability data in the planning and initial design of a next-generation imaging system.

20.2.1 AN ENGINEERING IMAGERY EXAMPLE

One of the issues facing the switch to digital technology can be illustrated by night launches. Night launches present a twofold problem, capturing detail of the vehicle and scene that is masked by

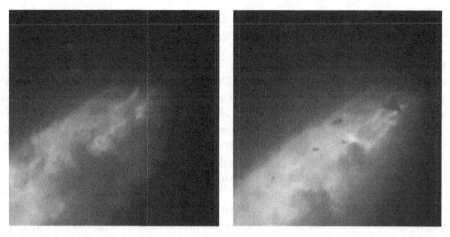

FIGURE 20.1

Engine debris during a night mission.

darkness, while also capturing detail in the engine plume. Fig. 20.1 contains two frames of a debris event as captured by traditional non-HDR imaging. The images originated from an eight-bit, 4:2:2, Rec. 709 color space data file during a spacecraft night mission. Fig. 20.1 (left) illustrates debris to the left and above the vehicle plume in the night sky. The camera exposure was set to capture detail in the flame, while the limitation in dynamic range results in little to no observable detail in the vehicle area.

Fig. 20.1 (right) includes debris in front of the engine plume several frames later. While the debris can be observed in and near the plume area, there is a loss of useful visual information as to the origination point from the vehicle, how the debris was generated, and the remaining state of the vehicle. Detail of the debris itself is also reduced as a result of the limited dynamic range, among other factors such as image resolution and optical system sharpness.

20.3 FILM BASELINE

Sixteen and 35 mm films were used as the engineering imagery baseline during the Space Shuttle program. Investigation is now ongoing to consider replacing or at least complementing film with advanced digital technology.

20.3.1 DYNAMIC RANGE OF FILM

The dynamic range of "film" from a system perspective should include the negative, processing, and printing/display combination. Modern motion picture original color negative stocks such as Kodak Vision3 capture images with red, green, and blue records representing up to a 13-stop scene (Eastman Kodak, 2010). Kodak estimates that the usable digital scan range is 10–13 stops depending on the scanning and encoding scheme (Eastman Kodak, 2009). One method for retaining the full density range is through the use of adjusted density mapping.

Film negatives, however, capture more than can be reproduced on the print, and follow a characteristic "film" curve for relative log exposure. The classical approach to densitometry (scientific analysis of exposure) was devised by Hurter and Driffield and is called the H&D curve (Davis and Walters, 1922). The H&D curve (also called the $D \log E$ curve, $\log E$ curve, or relative log exposure) plots the amount of exposure E in logarithmic units along the horizontal axis, and the amount of density change D along the vertical axis. A sample $D \log E$ curve is shown in Fig. 20.2 (Eastman Kodak, 2006).

In practice the exposure curve indicates that film response is not "linear" throughout the exposure range. A theoretical perfectly linear response would have a 45-degree slope. The slope of a relative exposure response in practice is a measure of its contrast, while the slope of the straight-line portion of the curve (essentially ignoring the shoulder and toe) is referred to as the gamma. The upper and lower end of the film emulation response differs than that of the gamma section. At the low end, the film does not initially respond as it "sees" radiant energy and then responds sluggishly for initial increases. At the upper end, the response is compressed as the emulsion becomes overloaded. These areas are indicated by the toe and shoulder regions in the H&D curve. The logarithmic data could be linearized with a nominal film transform such as a linearized printing display. Use of a linear scene-referred representation would require detailed characterization of each film emulsion, which film manufacturers often consider confidential. Additionally, the inverse linearization of the film H&D curve would produce very high gain in the nonlinear toe and shoulder regions, potentially producing high-gain artifacts (Kennel, 2014).

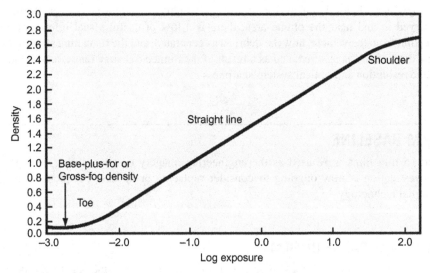

FIGURE 20.2

Characteristic curve for a negative emulsion.

From: Kodak, E., 2006. Basic Sensitometry and Characteristics of Film. Eastman Kodak Company, Rochester, NY.

The film laboratory "timer" sets the exposure range of the negative that the print will be exposed to in a development process using reducing chemicals. The process is stopped either by washing or use of a stop bath. This is a "relative" process where the exposure is determined by the timer. After exposure, printing, and digital scanning have been performed, the process will need to be repeated if an adjustment in overall exposure is required. This process is both time-intensive and costly. For Space Shuttle engineering film processing, the film was flown from Kennedy Space Center, Florida, to a processing facility near Atlanta, Georgia, processed and scanned, with the goal to return the digital transfers within 24 h. As film processing laboratories have reduced in number, processing cost continues to increase.

20.3.2 AUTOEXPOSURE AND CONTRAST RATIO

Film cameras implemented autoexposure systems during Space Shuttle launch imaging. At times there was some confusion with regard to the available dynamic range with the use of autoexposure.

Contrast ratio, the ratio of luminance of the lightest and darkest elements of a scene or image, can be used to describe different aspects of autoexposure. When autoexposure is implemented, the lighting information captured often changes with time. Therefore two measures of contrast ratio may be used (Poynton, 2003). Simultaneous contrast ratio (static) is the contrast ratio at one instant of time. Sequential contrast ratio (dynamic) is the contrast ratio separated in time (ie, light and dark elements over a series of images).

Human vision may have a simultaneous contrast ratio of close to four orders of magnitude (3.7 density, 12.3 stops) (Kunkel and Reinhard, 2010), while human sequential contrast ratio via adaption can be as great as 10 orders of magnitude, or approximately 33 stops (Ferwerda, 2001). Film and digital imagers can both be defined in terms of simultaneous and sequential contrast ratio.

20.3.3 LINEAR AND NONLINEAR QUANTIZATION

In both scanned film and digital-based imaging of spacecraft, the final result is often a digital data file. In nontypical cases, the film print may be used directly in a projection system. In the film case, the processed negative is scanned within the capabilities and bit depth of the scanning system to a digital file generally referred to as a digital intermediate. In the digital imager case, the analog sensor voltages are converted to linear digital values within the capabilities and bit depth of the camera sampling system.

A primary constraint in both cases is the digital bit depth, or the number of discrete quantization levels (or codes) available to represent a given relative luminance range. There are two primary methods for bit depth encoding that are implemented by digitizer manufacturers: linear and nonlinear scales. Nonlinear scales are often referred to as log or log-c, even though the functions are often in practice based on a power law as it can be a better approximation to human visual response.

In general, linear scale quantization is used when the sampling resolution is to be kept constant over the entire exposure. Nonlinear scale quantization is used when the sample resolution follows the logarithmic response similar to human perception of luminance. When nonlinear quantization is implemented, fewer bits are required to theoretically match the viewing capability of human luminance response; the visual threshold. The trade-off is that a reduced subset of total exposure information is stored, limiting image analysis capability when detail within a subset of the contrast range is required.

A typical industry scanning metric is SMPTE 268M Digital Moving Picture Exchange (DPX), logarithmic with 10-bit integer calibrated over a 2+ density range (SMPTE, 2003). Density range is defined in the following equation:

$$\text{Density range} = \log_{10}\left(\frac{\text{Number of photons transmitted through the film}}{\text{Number of photons hitting the film}}\right).$$

The relationship to dynamic range in terms of stops is defined by the following equation:

$$\text{Stops} = \log_2 10^{\text{density range}}.$$

In practicality, a Cineon digital film scanner is calibrated for a 2.048 density range (Kennel and Snider, 1995). HDR imagery, on the other hand, is typically stored with use of float point values.

20.3.4 PLAYBACK AND DISPLAY

The final, yet equally important, piece of the dynamic range workflow for both film and digital technologies is the playback software, hardware, and display. Regardless of the capture and quantization methods, the end user will be able to view the simultaneous contrast ratio only within the limitations of the end playback software, hardware, and display system. An example film workflow using a digital intermediate and a typical 8-bit display is shown in Fig. 20.3.

In this example, the full dynamic range benefit is not realized as a result of playback and display limitations. Limitations in dynamic range were present in the review of Space Shuttle imagery, depending on the equipment available at each participating NASA center. One example display device, a 10-bit 4K projection system, is limited in dynamic range because of a peak luminance of only 60 cd/m^2 measured at the projection screen. For redundancy, image analysts were available at Johnson Space Center, Marshall Space Flight Center, and Kennedy Space Center, albeit each with different

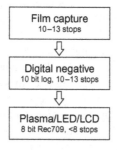

FIGURE 20.3

Film-to-digital transfer with display on an eight-bit monitor.

equipment and capabilities. Although the differences between centers were a result of several factors, including areas of expertise, program responsibilities, and budget, the differences do have a benefit of a more comprehensive and overlapping analysis approach.

Fig. 20.3 is dependent on the specifics of the DPX scan, as well as the setup and calibration of the monitor. For conversion from the full-range 10-bit log digital negative to an 8-bit Rec. 709 video representation, the preferred method is to limit the density range that is translated to that of a normal exposure with scene contrast range of 100:1, or six 2/3 stops. The resulting scene will have reasonable contrast when displayed on a standard graphics display monitor (Kennel and Snider, 1995) where the Dmin code value of 95 and the 90% white card code value of 685 are mapped to an output range of 0–255 in the eight-bit representation.

20.3.5 SUMMARY OF FILM USE

During the Space Shuttle program, 16 and 35 mm high-speed film cameras provided greater dynamic range capability than digital imagers available at the time. The film assets were only minimally utilized, on average for 4.5 Space Shuttle flights per year. With an ongoing maintenance program, they did not require regular replacement. The development of digital imaging technology, along with increases in dynamic range capability, occurred as the Space Shuttle program reached the end of its life. As the program ended, an upgrade of film assets to digital technology was cost-prohibitive in terms of equipment, infrastructure, training, and operation.

A transition to digital technology for future programs, however, can provide several benefits. Although the film emulsion can capture up to 13 stops, in most cases the film is transferred to a digital intermediate with reduced dynamic range, often displayed via a limited contrast ratio display. The process of developing film is time-consuming and costly, and is becoming more difficult to procure. Operation and maintenance of film systems requires specialized skills sets. After launch, the exposed films are collected, transported to a local airport, and flown to a development house often out of state. The film is developed and scanned, and the digital files are transported back to Kennedy Space Center after as much as a 24-h delay.

In the case of digital capture, files are available much faster, often with access to the RAW data quickly after launch. HDR data capture surpassing the dynamic range capability of film is becoming more prevalent, HDR file containers are available that support HDR data types, and HDR workflows allow the processing of HDR data. HDR displays are with expanded contrast ratio are becoming available. The entire HDR workflow aims to fulfill a primary requirement of image analysis, the ability to process, display, and analyze imagery with increased simultaneous contrast ratio.

20.4 HDR IMAGING OF SPACECRAFT FIELD EXPERIMENTS

Testing of HDR digital image capture began in 2009 by the Advanced Imaging Laboratory (AIL), in preparation for an eventual film-to-digital transition. Experiments were initially conducted with equipment currently on hand or available via loan, while preparations were made to obtain the latest-generation digital technology. As the Space Shuttle program was coming to a close, field experiments were often planned in advance of laboratory testing, in order to capture actual launch imagery while the launch opportunities were still available.

20.4.1 HIGH DEFINITION CAMERA TEST DURING STS-129

The first HDR experiment used two industrial eight-bit HD1100 2/3-inch sensor cameras with similar optics to capture similar fields of view, albeit with different exposures. The experiment was conducted during STS-129 (Clements et al., 2009) with the imaging systems mounted on a common tracking system. The A camera was mounted on the upper side of the tracker right wing and the B camera was mounted on the underside. At the time of the experiment, the HD1100 shutter speed was limited to 1/300 of a second. The difference in exposure from the A camera to the B camera was two 1/4 stops on the basis of available exposure times of 1/60 and 1/300 of a second. Both cameras were mounted on Meade eight-inch-aperture telescopes, focal length of 2032 mm. The 2/3-inch sensors (8.8 mm × 6.6 mm) have an equivalent 35 mm crop factor of 3.9. The telescope and crop factor result in a tight field of view with the tracker distance of approximately 3 miles from the Space Shuttle launch pad.

Atmospheric effects of ground heating resulted in a reduction of overall image quality, as can be observed in Figs. 20.4 and 20.5. The effects are particularly pronounced with daytime launches during

FIGURE 20.4

HD1100 A camera (1/60 exposure).

FIGURE 20.5

HD1100 B camera (1/300 exposure).

the Florida summer months, where temperatures are commonly above 30°C. Limitations in shutter speed also contributed to the softness of the imagery during the actual launch.

Mechanical errors in the camera alignment stage are evident when one compares Figs. 20.4 and 20.5, complicating efforts to recombine the synchronized frames into single HDR frames. Independent vibration affected each of the two image sequences, primarily a result of the effect of the launch acoustics on the mechanical support structures. Recombination efforts did not result in acceptable video because of the misalignments and vibration effects.

20.4.2 RED ONE M CAMERA TEST DURING STS-131

As a result of misalignment issues noted with the two HD1100 imagers, market research was conducted to identify the latest single sensor imagers advertised with increased dynamic range capability. One contender identified was the RED ONE imager from RED. RED specifies the dynamic range as 11.3 stops (assumed to be an MX sensor). The AIL measured 10.4 stops with the M sensor using log gamma, and 11.5 stops with the MX sensor using logFilm gamma. Both measurements were made with respect to 0.5-stop root mean square (RMS) noise, as discussed in Section 20.5. M sensor versions of the cameras were available and configured for STS-131 (Clements et al., 2010) to acquire imagery of the launch from five camera sites around the Space Shuttle pad perimeter as illustrated in Fig. 20.6. Deployed in pairs to each site, the cameras provided imagery not only for HDR testing purposes but also to generate stereoscopic 3D image sets. A diagram of the Space Shuttle pad perimeter camera site locations is shown in Fig. 20.7.

The RED ONE M experiment was successful in illustrating the capability of a single-exposure digital imager at the time. Clearly the solid rocket booster (SRB) plume is saturated and outside the available dynamic range. Detail of the vehicle is present, primarily lit from the rocket plume, with the addition of some ground-based spotlights. The five camera locations had similar exposure settings, resulting in similar dynamic range detail for all image sets.

Fig. 20.8 shows a full frame from the RED ONE M at camera site 3, southeast of the launch pad, illustrating the available lightness detail of the full plume. Launch debris may be present throughout the field of view, ranging from the SRB bright plume to the dark shadow areas of the plume cloud.

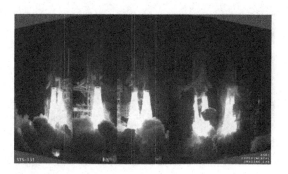

FIGURE 20.6

RED ONE composite of five camera sites.

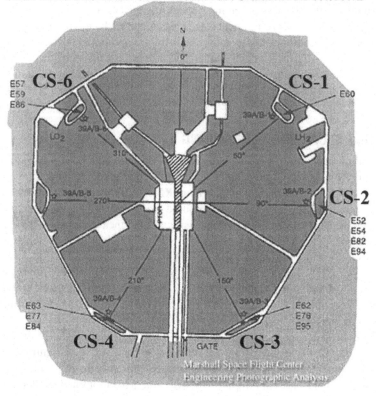

FIGURE 20.7

Space Shuttle pad perimeter camera site locations.

FIGURE 20.8

RED ONE camera site 3, single frame.

If the shadow areas are comparable to starlight, and the SRB bright plume is considered to be greater than direct overhead sunlight conditions, the total dynamic range of the launch environment is greater than eight orders of magnitude. Efforts will be made to quantify the launch environment total dynamic range.

20.4.3 HDR IMAGING DURING THE FINAL SPACE SHUTTLE FLIGHT STS-135

Market research was expanded for the final Space Shuttle flight of STS-135 to identify the growing number of available single-sensor imagers with increased dynamic range capability. New systems identified included the Vision Research Phantom HD Gold High Speed, RED EPIC with 5K resolution sensor, Cooke DiMax High Speed, Photron SA2 High Speed, and the ARRI Alexa (Lane et al., 2011). The RED ONE with M sensor was again used as a baseline for comparison.

20.4.3.1 RED ONE M test STS-135
A RED ONE camera with M sensor was deployed to the infield location near camera site 4. A 25-mm RED lens was used for the wide angle field of view. The resulting imagery was similar in dynamic range response to that in previous tests. A sample frame is shown in Fig. 20.9. Although the SRB plume is saturated, the overall imagery is of high quality, with detail observed throughout the field of view. An additional benefit is that the RED ONE imager is capable of 4K resolution.

20.4.3.2 ARRI Alexa test
Two ARRI Alexa cameras were placed within the Space Shuttle pad perimeter. The A camera was configured to capture a "film look" exposure with 180 degree shutter at 30 frames/s (16.67 ms exposure time) with a 50-mm lens. The B camera was configured to capture an "engineering" exposure with 22.5 degree shutter at 120 frames/s (0.52 ms exposure time) with a 40-mm lens. Whereas the 180 degree shutter creates the familiar film look motion blur, the engineering 22.5 degree shutter has the goal of reduced motion blur and sharp individual frames. Film look is a subjective term. It is commonly associated with the human visual experience when one is viewing a film-capture-based motion picture.

FIGURE 20.9

RED ONE M sample frame.

FIGURE 20.10

ARRI Alexa 120 frames/s B camera sample frame.

Typical characteristics of "film look" include a frame rate of 24 frames/s, 180 degree shutter angle, narrow depth of field, and film grain-type noise. An engineering exposure, alternatively, has the goal of producing as sharp an image as possible in each individual frame, with high frame rate.

The 180 degree A camera required the use of a neutral density filter because of the longer exposure time. For the B camera, this test was one of the first using ARRI's 120 frames/s feature. The ARRI Alexa is marketed as capable of 14 stops of dynamic range. The AIL laboratory test measured 13.9 stops at 0.5-stop RMS noise in log mode. Fig. 20.10 shows a sample frame from the "engineering" 120 frames/s B camera. Although the SRB plume is saturated, the overall imagery is of high quality. When digitally cropped on a subsection of the digital image that includes fast moving content, such as the orange SRB exhaust plume, the 180 degree shutter A camera image is less sharp than the 22.5 degree shutter B camera. The Alexa imagery workflow was considered one of the simplest while still producing high-quality imagery. The straightforward application of a 3D lookup table (LUT), generated from the online ARRI LUT Generator, to the log data files resulted in realistic looking color and lightness without further grading being required (ARRI, 2013a). The LUT, in the case of AIL processing, was generated for the DaVinci Resolve color grading system.

20.4.3.3 RED EPIC test

The RED EPIC without HDR mode was laboratory tested by the AIL as 11.5 stops at 0.5-stop RMS noise, logFilm mode. REDlogFilm applies a log curve to the linear sensor data without the addition of a film-like contrast curve. LogFilm replicates the characteristic curve of Cineon film scans (RED, 2013). For the STS-135 test, HDR mode was available but had not yet been tested, so it was not used. The RED EPIC has the benefit of a 5K resolution imager, upgradable to the DRAGON 6K sensor. The RED EPIC was configured to capture at 96 frames/s at a resolution of 5210×2700, with a 100-mm lens and a 1-ms shutter period.

The imagery captured was similar in dynamic range to that captured by other imagers tested in that the SRB plume was saturated. Overall the RED EPIC footage was of high quality and the increased pixel count allowed higher-quality cropping within the imagery. A sample RED EPIC frame is shown in Fig. 20.11.

FIGURE 20.11

RED EPIC sample frame.

FIGURE 20.12

Photron SA2 sample frame with debris item inset.

20.4.3.4 Photron SA2 test

A Photron SA2 high-speed digital camera with 2K resolution with a 500-mm lens was deployed near camera site 3. The frame rate was set to 500 frames/s. Fig. 20.12 illustrates a frame from the Photron SA2 showing a debris item that was observed.

The field of view was selected specifically to image the Space Shuttle main engine hydrogen burn off "sparklers." Portions of the Space Shuttle main engine flame are saturated, although detail can be seen in other portions as evident in the identification of a debris item passing behind the plume.

20.4.3.5 Cooke DiMax test

A Cooke DiMax high-speed digital camera with an 85-mm lens was mounted on a tracker near camera site 2. The camera resolution was 2016 × 2016, capturing images at 250 frames/s. The camera exhibited

FIGURE 20.13

Cooke DiMax sample frame.

some banding immediately surrounding the saturated section of the plume as seen in Fig. 20.13, but otherwise operated as expected. On reviewing the imagery after launch, the manufacturer recommended a darker capture exposure setting.

20.4.3.6 Vision Research Phantom HD Gold test

A Vision Research Phantom HD Gold high-speed digital camera was located at the 275-ft level of the Space Shuttle fixed service structure. A 16-mm film camera was also located at this level, with similar fields of view configured for comparison. A sample frame from the Phantom HD Gold is shown in Fig. 20.14, while a sample frame from the 16-mm film camera is shown in Fig. 20.15. Both cameras were operated at 400 frames/s. The film was scanned via a film to high definition 10-bit log transfer. The Phantom HD Gold captured 14-bit linear data with gamma applied during the color grading process. Both frames have been cropped per their respective aspect ratios.

The darkened film image in Fig. 20.15 is a result of the automatic exposure control. More detail is observed in the plume area in the film image as compared with the digital image, at the expense of the darkened overall image. The Phantom HD Gold camera did not utilize autoexposure and was configured for proper exposure of the vehicle. The SRB plume is observed to saturate in the Phantom HD Gold image.

The changing exposure level of the film represents sequential contrast ratio, complicating comparison over the entire image sequence. From a single frame perspective as shown, simultaneous contrast ratio can be observed, albeit at different exposure levels. The dynamic range of the Phantom HD Gold imagery was better than expected as the SRB plume was saturated but did not bloom into surrounding pixels.

FIGURE 20.14

Vision Research Phantom HD Gold sample frame.

FIGURE 20.15

Sixteen-millimeter film/high definition transfer sample frame.

20.4.4 **RED EPIC HDR MODE DURING DELTA 4 WGS 5**

The RED EPIC capability to capture in HDR mode, allowing the selection of an additional one to six stops, was tested in the post–Space Shuttle period. The RED EPIC stores two different frames in HDR

FIGURE 20.16

RED EPIC HDR+ 6 mode illustrating dual exposure, A frame.

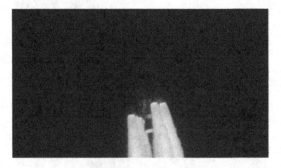

FIGURE 20.17

RED EPIC HDR+ 6 mode illustrating dual exposure, X frame.

mode, the A frame and the X frame. The RAW A frame contains the "standard exposure" portion of the capture, and the RAW X frame contains the "highlights" portion of the capture. "RAW" refers to the high-quality encoding format of the RED EPIC camera known as REDCODE, which is a proprietary RED un-debayered linear wavelet-based compression (RED, 2012).

An example RED EPIC HDR capture was made during the Delta 4 WGS 5 mission (Karr, 2013). This HDR+ 6 mode captured approximately 17 stops. To illustrate this range in limited print form, or via a low dynamic range monitor, the imagery must be shown as two separate frames. The A frame is shown in Fig. 20.16 and the X frame is shown in Fig. 20.17.

The capability to capture this range of brightness detail is unprecedented for a single imager-based system, and is an excellent example of the current progress in HDR image capture.

20.4.5 REVIEW OF HDR FIELD EXPERIMENTS

At the close of the Space Shuttle program, field testing of digital imagers provided insight into the current state of technology, the ever-increasing complexity of working with a multitude of digital formats, and the benefits and trade-offs of the different imaging systems.

20.4.5.1 State of technology

How captured digital data are provided to the user is manufacturer dependent. For example, ARRI uses a dual gain architecture with two paths of different amplification for each pixel (ARRI, 2013b). ARRI processes the dual path data internal to the camera, providing a single output file available as RAW (with external recorder), LogC, or Rec. 709. Generation and application of postprocessor-specific LUTs is straightforward, resulting in a particularly user friendly workflow.

RED provides its HDR digital data as two separate file sequences comprising A and X frames. Access to both exposure frames has the benefit of increased flexibility in postprocessing, at the cost of increased complexity in understanding the tone-mapping process. The RED software CineXPro includes two tone-mapping functions, Simple Blend and Magic Motion, assisting the user in the tone-mapping process. Overall, the RED HDR mode had the highest dynamic range measured, as will be discussed in Section 20.5.

The high-speed camera manufacturers such as Vision Research, Photron, and Cooke DiMax all provide the capability to export data from manufacturer formats to either linear or gamma-corrected image sequences. In most cases, 16-bit short integer linear TIFF sequences were used, as HDR-type formats using half floats such as LogLuv TIFF were not available (Larson, 1998). The accompanying software packages from the high-speed camera vendors generally include basic image processing tools, with the expectation that most of the image processing will be conducted by a third-party application. As the RED and ARRI cameras are considered cinematic cameras encompassing tools and workflows to aid the user in producing cinematic-like images, high-speed cameras have traditionally been used in industrial applications such as vehicle safety crash testing, spacecraft launch environments, and munitions testing.

20.4.5.2 Digital formats

A significant challenge in testing a large sample of digital imagers is developing data workflows. Just as there are many vendors, so are there many proprietary data formats, encoding algorithms, export formats, software packages, and postprocessing guidelines. Metadata differ and are nonstandard. Color range and exposure were observed to differ from camera to camera, even within manufacturers, potentially requiring custom LUTs per individual imager for precise characterization of color and luminance response as required.

In some cases, file wrappers, encoding schemes, gamma, and color space functions are vendor specific, such as with RED Redcode files, but have wide acceptance and are importable and manageable either through vendor-provided software or popular third-party applications. In other cases such as with ARRI, file formats typically use ARRI LUTs for gamma and color space, but are wrapped in standard ProRes422 or ProRes444 wrappers. Some high-speed cameras such as the Vision Research Phantom have file formats that are natively importable to popular third-party applications, while others are not and must be exported as an image sequences for import to third-party applications.

20.4.5.3 Benefits and trade-offs

Finally, although the primary focus of this chapter is dynamic range, there are many additional considerations for a spacecraft imaging system. Accurate frame timing is extremely important for the syncing of debris events across multiple imager fields of view. Trigger mechanisms for remote start, resetting of false triggers, exposure adjustments, and black shading are other example requirements. The capability to withstand and operate in high-shock and high-vibration environments is essential for many camera locations. Environmental concerns, including temperature and humidity limits, need to be considered. Nonvolatile memory is beneficial in the event of power loss. Resolution, color response, sharpness, and sensor uniformity are also testable parameters of interest.

20.5 CALIBRATED MEASUREMENT OF IMAGER DYNAMIC RANGE

A test procedure was developed to measure the dynamic range of imagers with the goal of an objective comparison. A DSC Labs Xyla 21 test chart was procured that includes 21 stops of dynamic range via a rear-lit voltage-regulated light source (Fig. 20.18). The Xyla 21 test chart features a stepped xylophone shape for minimizing flare interference from brighter steps. Additional techniques for reducing glare can be considered in future tests, such as imaging only one step at a time, although this will result in additional measurement and processing time. The Xyla 21 test chart is preferred to front-lit reflective charts as front-lit charts are more difficult to evenly relight over time, as test configurations change or the test environment is altered. Rear-lit grayscale stand-alone films require the use of an illuminator such as a light box, where special care is required to monitor the light source and voltage. The Xyla 21, being an all enclosed calibrated single unit, simplifies test setup and measurement.

The camera under test is mounted on a tripod and placed in a dark test room with no light leakage along with the Xyla 21 test chart. Camera lights and all reflective surfaces are masked to prevent reflections off the chart surface. A 25-mm Prime PL mount lens is attached to the camera. The camera distance to the chart is adjusted to maintain approximately 50 pixels per patch horizontal resolution, required by the Dynamic Range module of the analysis software program Imatest (Koren, 2014). For large-format sensors, a longer focal length lens may be required, such as an 85-mm lens for the RED DRAGON 6K.

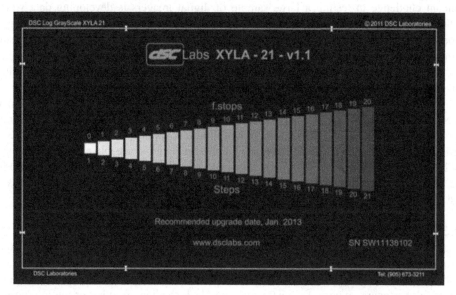

FIGURE 20.18

DSC Labs Xyla 21 test chart.

Source: Courtesy DSC Laboratories.

A one-foot straight edge is placed across the front plane of the lens. Measurements are taken at each end of the straight edge to the face of the test chart to ensure the two planes are parallel to approximately 3 mm. The lens aperture is set to fully open, and the focus is set by the temporary placement of a Siemens star, as seen on an ISO 15775 chart, on the face of the test chart. Once the focus has been verified, the camera is set to the manufacturer-recommended native ISO, and data collection begins. Before the ambient lights in the room are turned off, a few seconds of video is captured of the brightly lit chart, to be used for alignment and autoprocessing in Imatest. The ambient room lights are then turned off and the door to the test room is closed so that only the light from the test chart reaches the camera under test. The lens remains at the lowest f-stop (wide open) to ensure the luminance of at least the first two or three chips will saturate the image sensor at the slowest exposure time (generally 1/24 or 1/30 of a second). A few seconds of video is collected, and then the exposure time is reduced by half (one stop). Again a few seconds of video is collected, and the process is repeated for a total of at least five measurements. The sample data are collected and processed with the appropriate vendor software, DaVinci Resolve, or a combination of proprietary vendor and commercial applications as summarized in Table 20.1.

Table 20.1 Dynamic Range Postprocessing Workflow Summary

Imager	Capture Format	Processing Steps
ARRI Alexa	12-bit LogC, ProRes444, 1920 × 1080	LogC — export 16-bit TIFF from DaVinci Resolve Rec. 709 — apply 3D LUT, export 16-bit TIFF from DaVinci Resolve
RED ONE M/MX	12-bit linear, Redcode36, 4096 × 2034	Log/LogFilm Gamma — apply with CineXPro, export 16-bit TIFF Gamma3/Color3 — apply with CineXPro, export 16-bit TIFF
RED EPIC/DRAGON	12-bit linear, Redcode 8:1, EPIC 4K 16:9, DRAGON 6K 16:9	LogFilm Gamma — apply with CineXPro, export 16-bit TIFF Apply Simple Blend, Magic Motion tone-mapping as required.
Toshiba HD Hitachi DK	Uncompressed YUV, Blackmagic Hyperdeck Pro Recorder, .MOV wrapper	Rec. 709 gamma applied by camera. Export 16-bit TIFF from DaVinci Resolve
Blackmagic	Film mode, Apple ProRes 422HQ, .MOV wrapper, 3840 × 2160	Film gamma applied by camera. Export 16-bit TIFF from DaVinci Resolve
Canon 5DM3 Magic Lantern RAW Still	ML RAW, Full frame	Convert ML RAW with RAW2DNG. Export 16-bit TIFF DaVinci Resolve
Canon 5DM3 Magic Lantern HDR Movie	ML H.264, 1920 × 1080	Convert H.264 HDR file with AVISynth to HDR JPEG

A sample frame is then processed through the Dynamic Range module of the test software program Imatest. Imatest has a number of Dynamic Range settings that may be configured in the software. Two primary metrics of interest are as follows:

1. Average density of the grayscale patches in terms of "normalized" exposure pixel value (normalized to 0–255)
2. Dynamic range (stops for several maximum RMS noise levels; high to low image quality: 0.1, 0.25, 0.5, and 1.0 f-stop RMS noise)

The measurement of dynamic range in terms of stops corresponds to the relative luminance differences the human eye perceives. By quantifying the range of tones in terms of a maximum specified RMS noise, we can make objective comparisons between imaging systems. Quite often dynamic range data are provided by manufacturers without their defining if the measurement includes all noisy shadow areas or a noise limit.

The normalized exposure is first plotted against the step value of the Xyla 21 chart for each exposure time captured during the measurement. Sample normalized exposure data for the ARRI Alexa are shown in Fig. 20.19. The goal of processing the one-stop-separated normalized exposures is to ensure the proper exposure, resulting in the greatest dynamic range, is in the dataset. To determine which exposure resulted in the greatest dynamic range, the dynamic range is plotted against the RMS noise for each exposure time captured. Dynamic range data for the ARRI Alexa are shown in Fig. 20.20.

In general, as is the case for the ARRI Alexa, the exposure setting that most often resulted in the peak dynamic range was the one where the brightest chip was just saturated. In the case of the ARRI

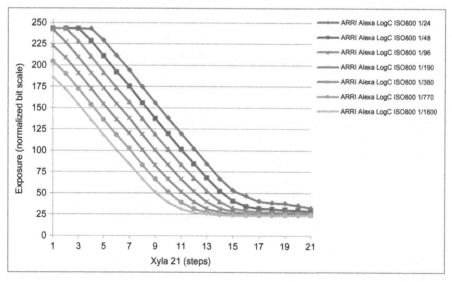

FIGURE 20.19

Normalized exposure data for the ARRI Alexa.

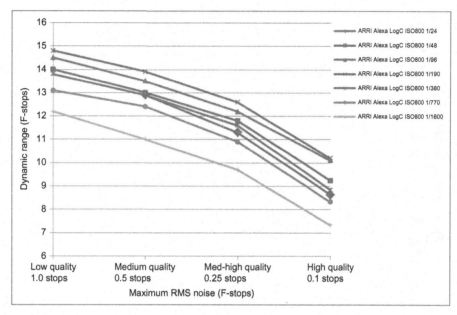

FIGURE 20.20

Dynamic range data for the ARRI Alexa.

Alexa data, this was the exposure time setting of 1/190 of a second. This value is taken as the peak dynamic range.

The RED EPIC and DRAGON HDR modes are treated similarly; however, an estimate of the total dynamic range is made on the basis of a combination of both A frame and the X frame, by equating similar scene luminance values from both frames. HDR exposure data for the RED EPIC are shown in Fig. 20.21, including a "shifted" X frame normalized to the sixth A frame step.

Results for imagers tested to date are shown in Table 20.2, including the native ISO values used. The dynamic range measurements are stated in terms of "medium quality," corresponding to 0.5 stops of RMS noise. This level of noise was selected on the basis of measurements from a number of different camera manufacturers and comparisons with the manufacturers' published dynamic range. The RED EPIC and DRAGON total measurements are approximate as they include the combination of the two measured A and X frames. Data for the RED tone-mapping functions Simple Blend and Magic Motion have been included.

In some cases, as with the ARRI Alexa data, the vendor specification corresponds closely with the measured data. In this case the difference is 0.1 stops, well within the one-third stated repeatability error of the measurement (IMATEST, 2014). In other cases the measured dynamic range is less than the vendor-specified range. This could be a result of the vendor-specified data determined using a different maximum RMS noise, or perhaps including all noisy areas as part of the definition of total dynamic range.

Although film was not yet exposed with the Xyla 21 chart, with the digital transfer data processed through Imatest, the stated dynamic range of film at approximately 10–13 stops can still be compared with the measured data. Imagers such as the ARRI Alexa and RED EPIC/DRAGON appear to, at a

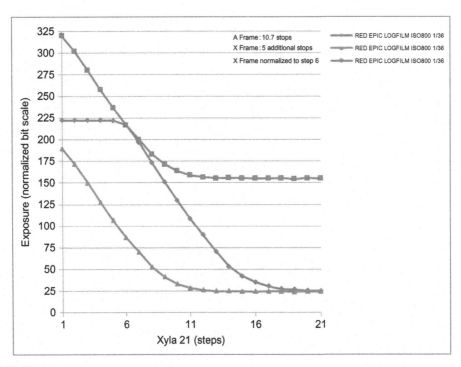

FIGURE 20.21

Normalized exposure data for the RED EPIC A and X frames.

minimum, match the dynamic range capabilities of film, and at best have exceeded it. Further work comparing film transfers at a similar maximum RMS noise would be of interest.

The stepped xylophone shape of the Xyla 21 chart helps to minimize, not eliminate, flare interference. Flare, or veiling glare, is stray light in lenses and optical systems caused by reflections between lens elements and the inside barrel of the lens.

20.6 HDR WORKFLOW AND DISPLAY DEVICE LUMINANCE

Evaluation of display devices is equally important in the HDR workflow. Ideally, the entire dynamic range that is captured and processed in the HDR workflow would also be displayed in the full range (Ward, 2008; Wanat et al., 2012). Current-generation HDR monitors such as the SIM2 HDR47ES4MB and the Dolby PRM are tested along with sample commercial monitors, including a Toshiba LED monitor and a Samsung plasma monitor.

A 15-stop HDR test card digital file was created in Adobe Photoshop in 32-bit floating point format and stored as an EXR file. A representation of the test card is shown in Fig. 20.22. Light values were measured via a Sekonic L-758Cine light meter in luminance spot meter mode. The built-in tube was used to minimize glare.

Table 20.2 Dynamic Range Measurement of Digital Imagers

Imager — Mode, Gamma	Lens/Aperture	Stops at 0.5 RMS Noise	Native ISO	Vendor Specification
RED DRAGON — HDR+ 6, LogFilm, Total	RED 85 mm PL, T1.8	11.8+ ~ 3	250	16.5+
RED EPIC — HDR+ 6, LogFilm, Total	RED 85 mm PL, T1.8	10.7+ ~ 5	800	<18
RED EPIC — HDR+ 6, LogFilm, Simple Blend	RED 85 mm PL, T1.8	15.0	800	NA
ARRI Alexa — Log	RED 25 mm PL, T1.8	13.9	800	14.0
RED EPIC — HDR+ 6, LogFilm, Magic Motion	RED 85 mm PL, T1.8	13.1	800	NA
RED Dragon — HDR Off, LogFilm	RED 85 mm PL, T1.8	12.4	250	NA
ARRI Alexa — Rec. 709	RED 25 mm PL, T1.8	11.8	800	NA
RED EPIC — HDR Off, LogFilm	RED 85 mm PL, T1.8	11.5	800	13.5
RED ONE MX — LogFilm	RED 25 mm PL, T1.8	11.5	800	13.0+
RED ONE MX — Gamma 3	RED 25 mm PL, T1.8	11.4	800	13.0+
Canon 5DM3 — Magic Lantern H.264 (ISO 400/1600)	Canon 50 mm EF, F1.4	11.4	100	14.0
Toshiba IK-HR1S HD — Gamma 0	Goyo 16 mm C mount, F1.4	11.1	0 dB	NA
Canon 5DM3 — RAW Still	Canon 50 mm EF, F1.4	10.6	100	NA
Hitachi DK-H100 — Gamma On, G-3	Goyo 16 mm C mount, F1.4	10.6	0 dB	10.0+
RED ONE M — Log	RED 25 mm PL, T1.8	10.4	320	11.3
RED ONE M — Gamma 3	RED 25 mm PL, T1.8	9.9	320	NA
Hitachi DK-Z50 — Gamma On, G-3	Goyo 16mm C mount, F1.4	9.3	0 dB	9.6
Blackmagic Cinema 4K — film	Canon 50 mm EF, F1.4	9.0	400	12.0

NA, not available.

In the case of the SIM2 HDR47ES4MB monitor, the test card EXR file was gamma corrected and played on an HDR player created by goHDR via MacBook Thunderbolt to DualLink DVI (GOHDR, 2014). In the case of the Dolby, Samsung, and Toshiba monitors, DaVinci Resolve was used as the player via a Blackmagic video card. Linear-to-gamma conversion was accomplished in DaVinci Resolve via the addition of the appropriate 3D LUT. The Dolby monitor uses a 10-bit hardware and

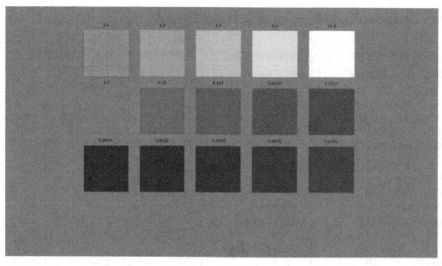

FIGURE 20.22

15 Stop HDR test card.

software workflow, and was measured for several color spaces, including sRGB, Rec. 709, and DCI. The Samsung and Toshiba monitors used an eight-bit workflow (Rec. 709 color space).

The raw measured simultaneous luminance data are included in Table 20.3. The bottom of the table includes rows for the maximum and minimum luminance, contrast ratio, and stops.

The measured contrast ratio data do not always correspond with manufacturer-stated data. This may be due to differences in the test methods, as well as differences in the test conditions. Contrast is typically measured by two methods, on/off and American National Standards Institute (ANSI) (International Committee for Display Metrology, 2012). On/off measures first a full black screen, followed by a full white screen, and therefore is a sequential, full range, type contrast measurement. ANSI uses a checkerboard pattern displaying black and white at the same time, and is therefore a simultaneous-type contrast measurement. The ANSI measurement is also dependent on the configuration of the test room, and the amount of light that is reflected back onto the screen, potentially reducing overall contrast. For the 15-stop HDR test card, the measurement data presented in Table 20.3 are similar to the ANSI simultaneous-type measurement data.

As an example, SIM2 states an ANSI contrast of greater than 20,000:1 (14.3 stops) as compared with the contrast for the measured data of 7407:1 (12.9 stops) (SIM2, 2014). The measured data in Table 20.3 were based on the created 15-stop HDR test card as opposed to an ANSI test card, and the exact SIM2 test room conditions were unknown. This may account for some of the 1.43 f-stop difference between the measured contrast ratio and manufacturer-stated contrast ratio. Further testing is required to verify the differences.

A graphical representation of the luminance data is shown in Fig. 20.23. From the graph the additional dynamic range of the SIM2 LCD with LED backlight monitor as compared with devices

Table 20.3 Simultaneous Display Device Output Luminance (cd/m^2)							
RGB Value	Stop	SIM2 HDR47ES4MB	Dolby PRM sRGB	Dolby PRM Rec. 709	Dolby PRM DCI	Toshiba 4K LED	Samsung Plasma
16	4	4000	480	480	450	240	40
8	3	2000	480	480	450	240	40
4	2	960	480	480	450	240	40
2	1	400	480	480	450	240	40
1	0	180	480	480	450	240	40
0.5	−1	80	220	220	220	170	15
0.25	−2	45	110	120	130	90	7
0.125	−3	24	50	56	70	50	3
0.0625	−4	12	23	32	40	23	1.6
0.0313	−5	7	9	15	20	7.5	1
0.0156	−6	2.8	4	7.5	10	2.6	0.54
0.0780	−7	1.2	1.7	4	6	0.93	0.35
0.0039	−8	0.6	0.76	2.3	3.3	0.47	0.25
0.0020	−9	0.6	0.44	1.3	2	0.38	0.25
0.0010	−10	0.54	0.44	0.9	1.3	0.35	0.25
Maximum luminance		4000	480	480	450	240	40
Minimum luminance		0.54	0.44	0.9	1.3	0.35	0.25
Contrast ratio		7407:1	1091:1	533:1	346:1	686:1	160:1
Stops		12.9	10.1	9.1	8.4	9.4	7.3

is observed in the region from 450 cd/m^2 to a peak luminance of 4000 cd/m^2. The total dynamic range of the SIM2 HDR47ES4MB monitor was measured as 12.9 stops.

The Dolby PRM monitor, as configured, had a peak luminance of 450 cd/m^2 in the sRGB color space. The total dynamic range of the Dolby PRM monitor in this case was 10.1 stops. The Toshiba 4K LED monitor had a peak luminance of 240 cd/m^2. Care had to be taken to ensure the dynamic range autoadjust function was disabled, as measurements were taken in a dark environment, in which the autoadjust feature reduces the output luminance. The total dynamic range of the Toshiba 4K LED monitor was 9.4 stops. The Samsung plasma monitor had a peak luminance of 40 cd/m^2. The total dynamic range of the Samsung plasma monitor was 7.3 stops.

The addition of an HDR workflow and a display device provides the capability to increase the simultaneous contrast ratio, and ultimately the observable luminance dynamic range. The workflow must include data structures such as 16-bit floats that are capable of storing the data, and display devices with significant contrast ratio to display it. Ultimately, efficient HDR encoding algorithms will be needed to encode and allow real-time playback at various quality standards. End user hardware and software will be required that is compatible with the HDR data, including integration with standard postprocessing vendor applications.

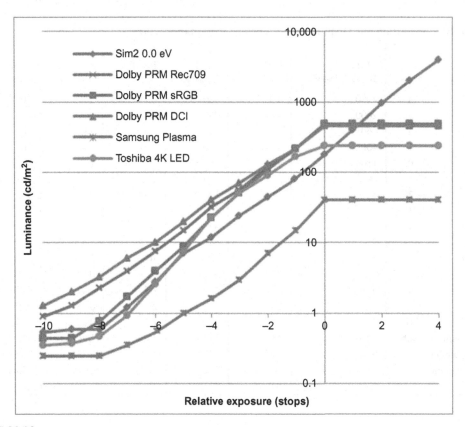

FIGURE 20.23

Display device output luminance.

20.7 **CONCLUSIONS**

HDR is a critical capability of imaging systems for use as engineering documentation of rocket launches. Digital imaging systems have now surpassed the dynamic range capability of film cameras used during the Space Shuttle program. The capability and use of digital imagers requires testing and evaluation, both in a controlled laboratory environment and during actual launch events. Field testing provided real-world lessons and experience with the current state of digital image systems. This experience transferred to the laboratory in terms of qualification testing of dynamic range capture, development of data workflows, and measurement of display device output luminance capability.

It is clear that future dynamic range workflows will require an examination of the entire digital process from capture to display. Handling of linear and nonlinear data formats, application of appropriate gamma response curves, conversion to low dynamic range and HDR, tone mapping, metadata, and display hardware and software must be considered. The digital imager marketplace is rapidly expanding and experiencing an ever-increasing number of new product releases. More access to the raw data is commonplace, creating opportunities for analysis and increasing complexity.

Appropriate HDR test methods will need to continually evolve in order to match the advancement of technology.

The ultimate goal of implementation of HDR technology with digital imaging of spacecraft is to increase both the observable simultaneous and the overall sequential contrast ratio. This will allow image analysts the ability to "see" more of what is happening in the launch environment in a single playback, without adjustment of the available dynamic range, rescanning of film, or multiple iterations of playback.

With such an array of imagers available at a wide variety of price points, a better understanding of imager capabilities will aid in the making of more objective and cost-conscious purchase decisions regarding the implementation of new image systems. Future work will include the evaluation and prediction of the required dynamic range of specific scenes under variable lighting conditions. This work could include an evaluation of the required accuracy of the lighting model as well as the verification process. Finally, remote and automatic control of digital imagers adjusted on the basis of real-time lighting conditions will also be investigated further.

ACKNOWLEDGMENTS

Advanced Imaging Laboratory team members have included Cole Alcock, Sandra Clements, Amy Elliott, Charles Harnden, Philip Haun, Matt Malczyk, Brian Moravecky, Herbert Oleen, Jamie Peer, John Lane, Omar Rodriguez, and Jay Shropshire. NASA Space Shuttle mission sponsorship was provided by Douglas England and Robert Page. NASA sponsorship in the post–Space Shuttle era was provided by the Ground Systems Development and Operations Program, Prentice Washington, and Dave Connolly. Additional NASA support was provided by Rodney Grubbs, Walt Lindblom, and Darrell Gaddy of the Marshall Space Flight Center, and Larry Craig of the Launch Services Program. Test support was provided by RED, ARRI, PCO, Photron USA, Vision Research, United Launch Alliance, and Space Exploration Technologies. This project is also partially supported by EU COST Action IC1005.

REFERENCES

ARRI, 2013a. Alexa LUT Generator: Look Up Tables (Online). www.arri.com. http://www.arri.com/camera/alexa/tools/lut_generator/ (Accessed 8/10/2014).

ARRI, 2013b. ARRI Imaging Technology — Alexa's Sensor (Online). ARRI.com: ARRI Inc. http://www.arri.com/camera/alexa/technology/arri_imaging_technology/alexas_sensor/ (Accessed 26/08/2014).

Banterle, F., Artusi, A., Debattista, K., Chalmers, A., 2011. Advanced High Dynamic Range Imaging: Theory and Practice. AK Peters, Natick, MA.

Bogart, R., Kainz, F., Hess, D., 2003. OpenEXR image file format. In: ACM SIGGRAPH 2003, Sketches & Applications.

Cade, D.L., 2013. Fuji and Panasonic's New Organic Sensor Boasts Insane 29.2 Stop Dynamic Range (Online). Petapixel. http://petapixel.com/2013/06/11/fuji-and-panasonics-new-organic-sensor-boasts-insane-14-6-stops-dynamic-range/ (Accessed 8/10/2014).

Clements, S., Haun, P., Karr, B., Oleen, H., Peer, J., 2009. ASRC Imaging Development Lab STS-129 Experimental Imagery. NASA, KSC.

Clements, S., Haun, P., Karr, B., Oleen, H., 2010. ASRC Experimental Imaging Lab STS-131-Experimental Imagery. NASA, KSC.

Davis, R., Walters, F.M., 1922. Sensitometry of photographic emulsions and a survey of the characteristics of plates and films of American manufacture. Sci. Pap. Bur. Standards 18, 10.

Eastman Kodak, 2006. Basic Sensitometry and Characteristics of Film. Eastman Kodak Company, Rochester, NY.

Eastman Kodak, 2009. Scanning Recommendations for Extended Dynamic Range Camera Films. Eastman Kodak Company, Rochester, NY.

Eastman Kodak, 2010. Kodak VISION3 500T Color Negative Film 5219/7219/SO-219. Eastman Kodak Company, Rochester, NY.

Ferwerda, J.A., 2001. Elements of early vision for computer graphics. IEEE Comput. Graph. Appl. 21, 22–33.

GOHDR, 2014. goHDR Player Software. http://gohdr.com/website/gohdr-products/.

IMATEST, 2014. Step Chart Module Documentation (Online). www.imatest.com. http://www.imatest.com/support/modules/stepchart/ (Accessed 03/02/2015).

International Committee for Display Metrology, 2012. Information Display Measurement Standard. Society for Information Display.

Karr, B., 2013. Advanced Imaging Lab Vimeo Website, United Launch Alliance Delta 4 WGS 5 Mission. NASA Engineering Services Contract.

Kennel, G., 2014. Color and Mastering for Digital Cinema. Focal Press, Burlington, MA.

Kennel, G., Snider, D., 1995. Conversion of 10-Bit Log Film Data to 8-Bit Linear or Video Data Ver. 2.1. The Cineon Digital Film System. Eastman Kodak Company, Rochester, NY.

Koren, N., 2014. Imatest. 3.10 ed.

Kunkel, T., Reinhard, E., 2010. A reassessment of the simultaneous dynamic range of the human visual system. In: Proceedings of the 7th Symposium on Applied Perception in Graphics and Visualization. ACM, Los Angeles, CA.

Lane, J., Karr, B., Haun, P., Oleen, H., 2011. Experimental Imaging Lab Activities in Support of STS-135 Launch and Landing. NASA, KSC.

Larson, G.W., 1998. LogLuv encoding for full-gamut, high-dynamic range images. J. Graph. Tools 3, 15–31.

Mann, S., Picard, R., 1995. On Being 'undigital' With Digital Cameras: Extending Dynamic Range by Combining Differently Exposed Pictures. Perceptual Computing Section, Media Laboratory, Massachusetts Institute of Technology, Cambridge, MA.

Poynton, C.A., 2003. Digital Video and HDTV: Algorithms and Interfaces. Morgan Kaufmann, San Francisco, CA.

Rafa, M., Efremov, A., Myszkowski, K., Seidel, H.P., 2006. Backward compatible high dynamic range MPEG video compression. In: ACM SIGGRAPH 2006 Papers. ACM, Boston, MA.

RED, 2012. Overview of the Redcode File Format (Online). Red.com: RED. http://www.red.com/learn/red-101/redcode-file-format (Accessed 21/06/2014).

RED, 2013. Understanding Red LogFilm and RedGamma (Online). Red.com. http://www.red.com/learn/red-101/redlogfilm-redgamma (Accessed 25/08/2014).

Reinhard, E., Heidrich, W., Debevec, P., Pattanaik, S., Ward, G., Myszkowski, K., 2010. High Dynamic Range Imaging: Acquisition, Display, and Image-Based Lighting. Elsevier Science, Amsterdam.

Seetzen, H., Heidrich, W., Stuerzlinger, W., Ward, G., Whitehead, L., Trentacoste, M., Ghosh, A., Vorozcovs, A., 2004. High dynamic range display systems. ACM Trans. Graph. 23, 760–768.

SIM2, 2014. SIM2 HDR47 Series Monitor (Online). www.sim2.com. http://www.sim2.com/HDR/hdrdisplay/hdr47e_s_4k (Accessed 03/02/2015).

SMPTE, 2003. SMPTE Standard for File Format for Digital Moving: Picture Exchange (DPX) Ver 2.0. Society of Motion Picture and Television Engineers.

United States Congress, House Committee on Science and Technology, 1986. Investigation of the Challenger Accident: Hearings Before the Committee on Science and Technology, House of Representatives, Ninety-Ninth Congress, Second Session, U.S. G.P.O.

US Columbia Accident Investigation Board, Gehman, H.W., 2003. Columbia Accident Investigation Board Report. Columbia Accident Investigation Board.

Wanat, R., Petit, J., Mantiuk, R., 2012. Physical and perceptual limitations of a projector-based high dynamic range display. In: Theory and Practice of Computer Graphics, pp. 9–16.

Ward, G., 1991. Real pixels. In: Graphics Gems II. Academic Press, Boston.

Ward, G., 2008. 59.2: defining dynamic range. In: SID Symposium Digest of Technical Papers, 39, pp. 900–902.

THE DYNAMIC RANGE OF DRIVING SIMULATION

21

R. Brémond, N.-T. Dang, C. Villa

Paris Est University, Marne-la-Vallée, France

CHAPTER OUTLINE

21.1 INTRODUCTION

Driving simulation has become a quite common virtual reality tool, addressing various fields, such as video games, vehicle design, driving lessons, and behavioral studies (Fisher et al., 2011). As in other fields of virtual reality, driving simulator providers have developed a number of visual effects in order to render a variety of environmental situations. For instance, night driving, rain, fog, and glare can be simulated with the use of state-of-the-art techniques from the computer graphics (CG) literature to mimic either a physical phenomenon (eg, beam pattern in automotive lighting) or its effect on driver perception (eg, fog), and to minimize the perceptual gap between the displayed image and the computed image with tone mapping operators (TMOs) (Reinhard et al., 2005). According to Andersen (2011), some visual factors are critical for the external validity of driving simulations (ie, validity with respect to actual driving). He emphasizes luminance and contrast as the most important visual factors for the main critical driving situations: night driving, and driving in rain, fog, and complex urban environments.

In this context, high dynamic range (HDR) rendering and display may improve the realism of a number of visual cues relevant to driving. However, the implementation of CG algorithms is not straightforward, and a trade-off is needed among cost (financial and computational), performance, and market (or user) demand. Moreover, in many driving situations, the driver's visibility is good, and looking at the relevant visual cues is easy, both in real and in virtual environments. In these situations, the benefit of HDR images is often seen as too low, considering the associated costs and constraints.

High Dynamic Range Video. http://dx.doi.org/10.1016/B978-0-08-100412-8.00021-8

In this chapter, we discuss to what extent HDR rendering and display has improved, or can improve driving simulators, and why HDR has not invaded the field yet. We will focus on driving simulation as a tool for behavioral studies; automotive design is considered in another chapter of this book.

Section 21.2 provides some evidence that HDR issues are not considered in most current driving simulator studies, even in vision-demanding situations. Then, we argue (Section 21.3) that some low-level visual factors, relevant to driving, are photometry dependent, and thus should benefit from a real-time HDR imaging and display system. After a short discussion in Section 21.4 of HDR rendering and display issues in CG, Section 21.5 reviews the few existing driving simulation studies where photometric issues and HDR components have been considered, and discusses what kind of realism is now available, or will be soon, in terms of realism, whether physical, perceptual, or functional (Ferwerda and Pattanaik, 1997), and what for. This includes a small number of simulations with a true HDR display device. We conclude (Section 21.6) by discussing the reasons why HDR has not yet invaded the field of driving simulation: it is interesting, in our opinion, to better understand the obstacles in order to push the development of HDR rendering and display with better efficiency. We also present some technical and experimental perspectives, toward a more intensive development of HDR video for driving simulations and vision science issues relevant in driving situations.

21.2 NO NEED FOR HDR VIDEO IN DRIVING SIMULATIONS?

One cannot say that driving simulator developers do not care about rendering issues. For instance, some level of realism may be important in video games (they ought to be the state of the art with respect to the video game market), and improves the driver's sense of immersion. But perceptive realism is needed only when visual performance or visual appearance issues arise, such as at night, in low-visibility situations, or in complex situations where the visual saliency of objects in the scene may attract the driver's visual attention (Hughes and Cole, 1986), and needs to be carefully simulated if one wants the driver's visual behavior in the simulator to be similar to a real driver's behavior.

This is the main point: perceptual realism is not a key issue in mainstream applications. For instance, in a recent review of the driver's visual control of the vehicle's lateral position (Lappi, 2014), low-visibility conditions are not even mentioned. Only a few people around the world are concerned with visual performance or visual appearance in a car: first, because it helps in the vehicle design (see Chapter 19); second, because it is needed for behavioral realism in driving situations where visual perception is a complex task (Brémond et al., 2014).

Night driving is a good example of a driving situation where both high and low luminance levels are expected to occur, leading to a high luminance range. Automotive and road lighting sources may appear in the field of view (with luminance values up to 25×10^9 cd/m^2 with xenon automotive lighting), while dark areas at night are in the mesopic range (below 3 cd/m^2), and may be in the scotopic range in some areas (below 0.005 cd/m^2), where the human visual system behavior and performance are different from what happens in the daylight photopic range (CIE, 1951, 2010).

Evidence show that visual performance lowers for driving at night (Owens and Sivak, 1996; Wood et al., 2005), and some measures of that performance (labeled as *focal vision*) are more impaired than other (*ambient vision*) (Owens and Tyrrell, 1999). Moreover, the usual low dynamic range (LDR) display devices cannot display scotopic and low mesopic luminance values, nor glaring lights. Thus, one would expect nighttime driving simulation to carefully consider illumination issues and glare, and take advantage of HDR rendering and display. But this is not what happens. For instance, it is striking that in their review of perceptual issues in driving simulations, Kemeny and Panerai (2003) did not even mention driving at night as an issue.

Considering the number of driving simulation studies in the last 15 years, the number of published studies which include a night driving condition is unexpectedly small, and to the best of our knowledge, only around 30 studies have done so (see also Wood and Chaparro, 2011). Moreover, available nighttime simulations almost never provide any information about the technical settings or performance in nighttime conditions. For instance, the *Material and methods* section of articles might not even mention night driving (Panerai et al., 2001) or will offhandedly state that "the main part of the evaluation consisted of eight spells of driving, featuring different combinations of lighting condition (day/night)" (Alexander et al., 2002). Interestingly, most of these studies have been published in medical and biological science journals (Gillberg et al., 1996; Banks et al., 2004; Campagne et al., 2004; Contardi et al., 2004; Pizza et al., 2004; Åkerstedt et al., 2005; Silber et al., 2005; Schallhorn et al., 2009), and address hypovigilance and drug use issues. In a few articles, the lack of information about night driving settings is mitigated by a figure showing the visual appearance of the night driving simulation (Konstantopoulos et al., 2010; Schallhorn et al., 2009); this somehow enforces the feeling that the experimenters have a low level of control over illumination issues. This also appears in a series of driving simulation experiments at night, where the ambient luminance is controlled by neutral density filters (Alferdink, 2006; Bullough and Rea, 2000) or goggles (Brooks et al., 2005) in daylight simulated scenes (the "day for night" cinematographic technique) at the cost of unrealistic visual environments.

This lack of reported technical or photometric details also occurs with fog. For instance, in an important driving simulation study by Snowden (1998), showing that speed perception is altered in fog, little information was provided about the simulated fog. Indeed, in most articles reporting driving simulator studies in fog (Saffarian et al., 2012), no information is given about the fog density; no technical information is provided either, and one is left to guess that the simulator used OpenGL fog — that is, a contrast attenuation associated with the object's distance. This means that a minimal model of fog is deemed acceptable, as we have seen for night simulation; it is possible with OpenGL to fit the physical law of contrast attenuation in fog (Koschmieder's law; see Middleton, 1952); however, no information is given in this respect in the articles cited above. For instance, Broughton et al. (2007) compared the driver's behavior in three visibility conditions: two fog densities are compared with a no-fog condition. The fog conditions are described in terms of a "visibility limit," which probably means that the authors used a nonphysical tuning of the OpenGL fog. Moreover, simulation of artificial lighting (automotive lighting, road lighting, etc.) with OpenGL is rather complex (Lecocq et al., 2001).

21.3 **VISUAL FACTORS WHICH IMPACT DRIVING BEHAVIOR**

It is common knowledge that vision is the main sensory channel to collect information during driving (Allen et al., 1971; Sivak, 1996), and in a driving simulator, CG images are supposed to provide the driver's visual information.

The link between the displayed images and driving behavior is not direct; it is mediated by notions from vision science, such as luminance and contrast (Andersen, 2011), the visibility level (Adrian, 1989; Brémond et al., 2010b), the adaptation luminance (Adrian, 1987a; Ferwerda et al., 1996), motion, distance, and speed perception (Snowden, 1998; Cavallo et al., 2002; Caro et al., 2009), scotopic and mesopic vision (Gegenfurtner et al., 1999), glare (Spencer et al., 1995), and the visual saliency of the relevant/irrelevant objects in a scene (Brémond et al., 2010a), such as road markings (Horberry et al., 2006) and advertising. These visual factors first impact the visual performance and then the driving behavior. These are photometry-based concepts, and were not controlled for in the above-cited driving simulator studies.

For all these issues, photometric control of the images is mandatory. In some cases, an HDR display may be needed, or alternatively, TMOs may help to minimize the gap between ideal and displayed visual information. For instance, road lighting design needs some criterion, and the visibility level has been proposed as the visibility for the driver of a reference target on the road (Adrian, 1987b). The American standard includes this concept in the small target visibility assessment of road lighting (IESNA, 2000), and the French standard also includes this visibility level index (AFE, 2002).

To assess an operator's quality, one needs a quality criterion. This is not so easy, and for instance, the correlation is weak among visual appearance, visual performance, and visual saliency in an image (Brémond et al., 2010b), so a choice is needed. In previous evaluations, visual appearance was considered first in most benchmarks (Eilertsen et al., 2013).

While visibility is considered by practitioners as a key perceptual issue in night driving, is it possible to preserve the visibility level of objects with a TMO? Some authors have proposed operators in order to control some kind of visual performance (Ward, 1994; Pattanaik et al., 1998), and Grave and Brémond (2008) proposed an operator focusing on preserving the visibility level. For dynamic situations, Petit and Brémond (2010) proposed a TMO preserving visibility, based on the work of Irawan et al. (2005) and Pattanaik et al. (2000). So, some efforts have been made to design TMOs shaped by visual performance constraints. On the other hand, the main effort in TMO design has been devoted so far to appearance criteria, such as color appearance and lightness perception, rather than visual performance criteria.

21.4 HDR RENDERING

HDR issues have a specific flavor in CG. The split between image computation and image display is also relevant, but the problems are not the same. First, use of HDR virtual sensors for HDR image computation is now possible, because graphics processing units can manage *float* values. The main constraint is to run in real time, rather than sensor design or noise issues. Indeed, it is possible with pixel shaders to allocate some sensitivity to the virtual sensors (ie, compute the CG images in float units), even if the image computation does not simulate light propagation in the virtual scene in physical units.

The situation is quite different for HDR image display. HDR display devices are now available (Seetzen et al. (2004) demonstrated a prototype of an HDR display device at SIGGRAPH in 2004; see Part IV of this book for an update on HDR display). Commercial HDR display devices based on Seetzen's ideas are now available (first from Brightside, now from Dolby). But this technology is still very expensive compared with conventional displays, and as a result, HDR display devices are very rare in human factors laboratories.

An important issue for driving simulators is that a large field of view is often needed, which is almost impossible to address with existing HDR display devices. For instance, most low-cost driving simulators use three displays, and in many driving situations, a field of view of 150° is needed (eg, if you have to cross an intersection). Virtual reality helmets can be viewed as an alternative as far as the field of view is concerned, but HDR displays are not available for these devices at the moment.

So the eight-bit frontier is still hard to cross for the driving simulator's display, and the CG pipeline, which is expected to link the rendering part to the display part of the loop, tends to use TMOs in order to overcome the lack of HDR displays. As we have mentioned, in the case of visual performance preservation, this led to a number of TMOs (see Reinhard et al., 2005), followed by some concerns about the evaluation of these operators.

What would be a good design criterion for a TMO dedicated to driving simulation? Real time is mandatory. Second, temporal fluidity is needed, in order to avoid rapid changes and oscillations in the visual adaptation level, which may be due to a light source appearing in (or leaving) the field of view (Petit et al., 2013). This can be done by simulation of the time course of visual adaptation (Pattanaik et al., 2000). Third, as we have emphasized already, fidelity in terms of visual performance (rather than visual appearance) is relevant in most driving simulation applications, because the main goal of driving simulation experiments is the study of driver behavior, which in turn depends on the visual cues the driver finds in his or her environment.

21.5 PHOTOMETRIC CONTROL OF CG IMAGES IN DRIVING SIMULATIONS

Maybe the reader has so far found this chapter a bit pessimistic about the success of an HDR approach in driving simulation. The picture should be mitigated, however, and it is worth mentioning some articles where the photometric tuning of night driving simulations is taken seriously. The first one, to our knowledge, is from Mortimer (1963), with a very special driving simulator; however: at that time, in 1963, there were no personal computers available, and the simulator was purely electromechanical. In the computer era, Featherstone et al. (1999) conducted a field study, collecting reference data about car's headlights at night, and tuned the rendering of the simulator in terms of contrast, color, and luminance.[1] Kemeny and his team (Dubrovin et al., 2000; Panerai et al., 2006), at Renault Virtual Reality Centre, conducted several studies focusing on the simulation of automotive lighting, based on a simulation of light propagation, projecting lightmaps from the headlamps to the road surface (see also Weber and Plattfaut, 2001). Horberry et al. (2006) and Brémond et al. (2013) attempted to control the luminance map of the rendered images of night driving, which makes sense because their articles address road marking and road hazard visibility, respectively.

At night, glare is a key issue, as the usual displays cannot produce a glare sensation. To overcome this problem, Spencer et al. (1995) proposed a biologically inspired algorithm which simulates the effects of glare on vision (bloom, flare lines, lenticular halo) in CG images. Some technical solutions have also been proposed to simulate fog with a control on the luminance map, with a display calibration and a physical OpenGL tuning (Cavallo et al., 2002; Espié et al., 2002), and the simulation of halos around light sources (Lecocq et al., 2002; Dumont et al., 2004). The main issue with fog is contrast attenuation, rather than luminance values.

---

[1]Considering the available display devices at that time, it is unlikely that they could tune the luminance range to glaring situations, as suggest in the article.

In addition, many driving simulator developments have not been published, because they are conducted by industrial firms which do not want to make their internal development public. They open discuss, however, HDR issues, and HDR rendering is mentioned in the technical documents of some driving simulation software. For example, SCANeR HEADLIGHT Interactive Simulation (OKtal, 2015) supports HDR rendering for realistic night driving experiments. Note that the OKtal software is widely used by automobile companies in France (eg, Renault, Valeo, PSA). In Germany, VIRES also supports HDR rendering (VIRES Virtual Test Drive software) (Vires, 2015). HDR rendering is also mentioned among OpenDS features (Math et al., 2013; OpenDS, 2015), a recently developed open-source driving simulator, which originated from the European Union Seventh Framework Programme project GetHomeSafe (GetHomeSafe, 2015). Some details of these technical developments are sometimes published, as in the case of Pro-SiVICTM, software developed by CIVITEC, where HDR textures are used in the sensor simulation for prototyping of advanced driver assistance systems (Gruyer et al., 2012). Optis is also active with regard to HDR issues, with SPEOS and Virtual Reality Lab (Optis, 2015).

The direct use of an HDR display device in driving simulations is still in its infancy. Shahar and Brémond (2014) were the first to use a true HDR display device (47-inch Dolby/SIM2 Solar), under photometric control, to conduct a driving simulation where night driving behaviors with and without LED road studs were compared. The automotive lighting, road lighting, and LED road stud beam pattern were tuned to realistic values, with use of direct photometric measurements of the road surface, the road markings, and the LED themselves on the screen.

The main issues was to run in real time, with three screens (1920×1080 pixels). The geometric configuration of the simulator was chosen in such a way that when the road studs were switched on, they were very likely to appear in the central screen, so it was decided to run the simulation with one HDR display device in front of the driver, and an LDR display device on each side of the driver. The main purpose of these lateral screens was to give the driver some sense of his/her own speed.

Several technical challenges needed to be addressed, among them the number of light sources and the lighting simulation itself (Dang et al., 2014). The IFSTTAR visual loop, developed under Open Scene Graph, supports two HDR renderings: one originated from a TMO proposed by Petit and Brémond (2010) and the other was adapted from a TMO proposed by Reinhard et al. (2002). Since thousands of road studs were required in this study, a particular way of controlling the light sources was adopted to guarantee a high frame rate, a key issue in real-time simulations. Each light source was simulated on the basis of the photometric characteristics of real LED road studs, as measured in the IFSTTAR photometry laboratory; their intensity was made dynamically controllable during the simulation. The LED road studs were divided into groups, each of them being controlled by a virtual group manager. This organization was particularly useful in this study, because the road studs were turned on/off automatically by these group managers depending on the vehicle's position. Another challenge was the simulation of realistic night driving conditions with high luminance range, with bright areas due to the studs, the road lighting, and the headlamps of incoming vehicles, while very dark areas were also needed in the nighttime countryside landscape.

This experiment used an HDR display device, but with eight-bit (LDR) input images from the driving simulation visual loop. Thus, the benefit of the HDR display was the luminance dynamics, not the luminance sensitivity. The next challenge will be to develop a full HDR driving simulator visual loop, and feed an HDR display device with HDR videos.

21.6 CONCLUSION

This rapid overview of the potential benefits of HDR rendering and display for driving simulations leads to a balanced conclusion. On the one hand, dynamic range is marginally addressed in current driving simulation studies. The main reason is that the expected benefits of HDR are associated with low-level vision issues. Although they are known to have an impact on the driver's behavior, this impact is limited to some specific situations, such as night and fog driving, or drivers with poor vision.

Whereas a smart tuning allows some qualitative control of the visual appearance in virtual environments, in quantitative behavioral studies there is a need for some photometric control of the displayed images. This includes physical, optical, and photometric data on light sources, participating media, and surfaces.

Of course, another reason for the limited interest in HDR in the driving simulation community is the cost. A photometric description is important for HDR imaging, but at some cost: photometric description of the virtual environments, need for photometric data for surfaces and light sources, real-time issues, etc. Further, HDR display devices are still very expensive, difficult to use with current video output formats, and seldom required, even in nighttime driving simulations to be published in peer-review journals. Also, most of the tone mapping literature (both on TMOs and on TMO evaluation) focuses on appearance criteria (eg, subjective fidelity) not on performance criteria (visibility, reaction time, etc.), which would be needed for driving simulations in low-visibility conditions.

But there is another side to this story, and we have tried to show that in some important cases, the control of image luminance map allows us to expect a much better fidelity between virtual reality and actual driving in terms of "Where do we look?" and "What do we see?" This, in turn, is known to impact driving behavior. Thus, HDR imaging, rendering, and display open the way to new driving simulation applications, to situations where the external validity was poor in previous studies: photometric parameters were known to impact the behavior, but they were not controlled.

This is why we plan to conduct experiments soon to assess the influence of an HDR display device on psychovisual and driving tasks. To that purpose, the impact on specific perceptive mechanisms underlying the driving behavior on a driving simulator will be compared, first for an LDR display (ie, HDR imaging followed by a TMO) then for an HDR display.

For instance, the contribution of an HDR display device on speed perception will be assessed in a driving context. We can do this by estimating the "time to collision," using moving stimuli on both kinds of display devices. As the perceived speed is expected to depend on the luminance contrast (Stone and Thompson, 1992), more specifically, at low contrast or at night, the perceived speed of an object is underestimated (Blakemore and Snowden, 1999; Gegenfurtner et al., 1999). Thus, it can be assumed that this bias is closer to the real one with an HDR display device compared with an LDR display device. Therefore, use of an HDR display device may allow the investigation of reduced-visibility situations such as nighttime driving or glare. More broadly, a benefit is expected when driving simulation studies are conducted on an HDR display device when speed perception is a key figure (either the driver's own speed or the speed of other vehicles), especially in reduced-visibility situations.

New lighting systems (signaling, lighting, headlamp beams, stop lights) may also benefit from HDR driving simulations. This was done, for instance, to study the impact of a motorcycle's lighting design on its perceived speed (Cavallo et al., 2013), as well as for the driver's behavior when the driver was facing a new concept of dynamic signaling with LED road studs (Shahar and Brémond, 2014).

REFERENCES

Adrian, W., 1987a. Adaptation luminance when approaching a tunnel in daytime. Light. Res. Technol. 19 (3), 73–79.

Adrian, W., 1987b. Visibility level under night-time driving conditions. J. Illum. Eng. Soc. 16 (2), 3–12.

Adrian, W., 1989. Visibility of targets: model for calculation. Light. Res. Technol. 21 (4), 181–188.

AFE, 2002. Recommandations relatives à l'éclairage des voies publiques. Association Française de l'Eclairage, Paris.

Åkerstedt, T., Peters, B., Anund, A., Kecklund, G., 2005. Impaired alertness and performance driving home from the night shift: a driving simulator study. J. Sleep Res. 14 (1), 17–20.

Alexander, J., Barham, P., Black, I., 2002. Factors influencing the probability of an incident at a junction: results from an interactive driving simulator. Acc. Anal. Prevent. 34, 779–792.

Alferdink, J.W., 2006. Target detection and driving behavior measurements in a driving simulator at mesopic light levels. Ophtalmic Physiol. Opt. 26, 264–280.

Allen, T.M., Lunenfeld, H., Alexander, G.J., 1971. Driver information needs. Highway Res. Board 36, 102–115.

Andersen, G.J., 2011. Sensory and perceptual factors in the design of driving simulation displays. In: Fisher, D.L., Rizzo, M., Caird, J.K., Lee, J.D. (Eds.), Driving Simulation for Engineering, Medicine and Psychology. CRC Press, Boca Raton, FL, pp. 1–11.

Banks, S., Catcheside, P., Lack, L., Grunstein, R., McEvoy, D., 2004. Low levels of alcohol impair driving simulator performance and reduce perception of crash risk in partially sleep deprived subjects. Sleep 6 (27), 1064–1067.

Blakemore, M.R., Snowden, R.J., 1999. The effect of contrast upon perceived speed: a general phenomenon? Perception 28 (1), 33–48.

Brémond, R., Petit, J., Tarel, J.P., 2010a. Saliency maps for high dynamic range images. In: Media Retargetting Workshop in Proc. ECCV 2010, Crete, Greece.

Brémond, R., Tarel, J.P., Dumont, E., Hautière, N., 2010b. Vision models for image quality assessment: one is not enough. J. Electron. Imaging 19 (4), 04304.

Brémond, R., Bodard, V., Dumont, E., Nouailles-Mayeur, A., 2013. Target visibility level and detection distance on a driving simulator. Light. Res. Technol. 45, 76–89.

Brémond, R., Auberlet, J.M., Cavallo, V., Désiré, L., Faure, V., Lemonnier, S., Lobjois, R., Tarel, J.P., 2014. Where we look when we drive: a multidisciplinary approach. In: Proc. Transportation Research Arena, Paris, France.

Brooks, J.O., Tyrrell, R.A., Frack, T.A., 2005. The effect of severe visual challenges on steering performance in visually healthy young drivers. Optom. Vis. Sci. 82 (8), 689–697.

Broughton, K.L., Switzer, F., Scott, D., 2007. Car following decisions under three visibility conditions and two speeds tested with a driving simulator. Acc. Anal. Prevent. 39 (1), 106–116.

Bullough, J.D., Rea, M.R., 2000. Simulated driving performance and peripheral detection at mesopic and low photopic light levels. Light. Res. Technol. 32 (4), 194–198.

Campagne, A., Pebayle, T., Muzet, A., 2004. Correlation between driving errors and vigilance level: influence of the driver's age. Physiol. Behav. 80, 515–524.

Caro, S., Cavallo, V., Marendaz, C., Boer, E., Vienne, F., 2009. Can headway reduction in fog be explained by impaired perception of relative motion? Hum. Factors 51 (3), 378–392.

Cavallo, V., Dumont, E., Galleé, G., 2002. Experimental validation of extended fog simulation techniques. In: Proc. Driving Simulation Conference, Paris, France, pp. 329–340.

Cavallo, V., Ranchet, M., Pinto, M., Espié, S., Vienne, F., Dang, N.T., 2013. Improving car drivers' perception of motorcyclists through innovative headlight configurations. In: Proc. 10th International Symposium on Automotive Lighting (ISAL), TU Darmstadt, Germany.

CIE, 1951. The standard scotopic sensitivity function. In: Proc. of the Commission Internationale de l'Eclairage, Paris, vol. 1(4), p. 37.

CIE, 2010. Recommended system for mesopic photometry based on visual performance. Technical report. Commission Internationale de l'Eclairage.

Contardi, S., Pizza, F., Sancisi, E., Mondini, S., Cirignotta, F., 2004. Reliability of a driving simulation task for evaluation of sleepiness. Brain Res. Bull. 63 (5), 427–431.

Dang, N.T., Vienne, F., Brémond, R., 2014. HDR simulation of intelligent LED road studs. In: Proc. Driving Simulation Conference, Paris (France), pp. 41.1–41-2.

Dubrovin, A., Lelevé, J., Prevost, A., Canry, M., Cherfan, S., Lecocq, P., Kelada, J.M., Kemeny, A., 2000. Application of real-time lighting simulation for intelligent front-lighting studies. In: Proc. Driving Simulation Conference, Paris (France).

Dumont, E., Paulmier, G., Lecocq, P., Kemeny, A., 2004. Computational and experimental assessment of real-time front-lighting simulation in night-time fog. In: Proc. Driving Simulation Conference, Paris (France), pp. 197–208.

Eilertsen, G., Wanat, R., Mantiuk, R.K., Unger, J., 2013. Evaluation of tone mapping operators for HDR-video. Comput. Graph. Forum 32 (7), 275–284.

Espié, S., Moessinger, M., Vienne, F., Pébayle, T., Follin, E., Gallée, G., 2002. Real-time rendering of traffic reduced visibility situations: development and qualitative evaluation. In: Proc. Driving Simulation Conference, Paris (France), pp. 89–98.

Featherstone, K., Bloomfield, J., Lang, A., Miller-Meeks, M., Woodworth, G., Steinert, R., 1999. Driving simulation study: bilateral array multifocal versus bilateral AMO monofocal intraocular lenses. J. Cataract Refract. Surg. 25 (9), 1254–1262.

Ferwerda, J.A., Pattanaik, S.N., 1997. A model of contrast masking for computer graphics. In: Proceedings of ACM SIGGRAPH, pp. 143–152.

Ferwerda, J.A., Pattanaik, S.N., Shirley, P., Greenberg, D.P., 1996. A model of visual adaptation for realistic image synthesis. In: Proceedings of the 23rd Annual Conference on Computer Graphics and Interactive Techniques, SIGGRAPH '96. ACM, New York, NY, USA, pp. 249–258.

Fisher, D.L., Rizzo, M., Caird, J.K., Lee, J.D. (Eds.) 2011. Driving Simulation for Engineering, Medicine and Psychology. CRC Press, Boca Raton, FL.

Gegenfurtner, K.R., Mayser, H., Sharpe, L.T., 1999. Seeing movement in the dark. Nature 398 (6727), 475–476.

GetHomeSafe, 2015. http://www.gethomesafe-fp7.eu.

Gillberg, M., Kecklund, G., Åkerstedt, T., 1996. Sleepiness and performance of professional drivers in a truck simulator — comparisons between day and night driving. J. Sleep Res. 5 (1), 12–15.

Grave, J., Brémond, R., 2008. A tone mapping algorithm for road visibility experiments. ACM Trans. Appl. Percept. 5 (2), Article 12.

Gruyer, D., Grapinet, M., De Souza, P., 2012. Modeling and validation of a new generic virtual optical sensor for ADAS prototyping. In: Intelligent Vehicles Symposium. IEEE, New York, pp. 969–974.

Horberry, T., Anderson, J., Regan, M., 2006. The possible safety benefits of enhanced road markings: a driving simulator evaluation. Transp. Res. Part F Traffic Psychol. Behav. 9 (1), 77–87.

Hughes, P.K., Cole, B.L., 1986. What attracts attention when driving? Ergonomics 29 (3), 377–391.

IESNA, 2000. American National Standard Practice for Roadway Lighting. RP-8-00. IESNA, New York.

Irawan, P., Ferwerda, J.A., Marschner, S.R., 2005. Perceptually based tone mapping of high dynamic range image streams. In: Proceedings of Eurographics Symposium on Rendering, pp. 231–242.

Kemeny, A., Panerai, F., 2003. Evaluating perception in driving simulation experiments. Trends Cogn. Sci. 7 (1), 31–37.

Konstantopoulos, P., Chapman, P., Crundall, D., 2010. Driver's visual attention as a function of driving experience and visibility. Using a driving simulator to explore drivers' eye movements in day, night and rain driving. Acc. Anal. Prevent. 42, 827–834.

Lappi, O., 2014. Future path and tangent point models in the visual control of locomotion in curve driving. J. Vis. 14 (12), 1–22.

Lecocq, P., Michelin, S., Arquès, D., Kemeny, A., 2001. Simulation temps-réel d'éclairage en présence de brouillard. Revue de CFAO et d'Informatique Graphique 16 (1), 51–66.

Lecocq, P., Michelin, S., Kemeny, A., Arquès, D., 2002. Lighting simulation with the presence of fog: a real time rendering solution for driving simulators. In: Proc. Driving Simulation Conference, pp. 65–74.

Math, R., Mahr, A., Moniri, M.M., Muller, C., 2013. OpenDS: a new open-source driving simulator for research. Technical report. GMM-Fachbericht-AmE

Middleton, W., 1952. Vision Through the Atmosphere. University of Toronto Press, Toronto.

Mortimer, R., 1963. Effect of low blood-alcohol concentrations in simulated day and night driving. Percept. Motor Skills 17, 399–408.

OKtal, 2015. http://www.scanersimulation.com/software/software-research/headlight-simulation.html.

OpenDS, 2015. http://opends.de.

Optis, 2015. http://portal.optis-world.com.

Owens, D.A., Sivak, M., 1996. Differentiation of visibility and alcohol as contributors of twilight road fatalities. Hum. Factors 38 (4), 680–689.

Owens, D.A., Tyrrell, R.A., 1999. Effects of luminance, blur and age on night-time visual guidance: a test of the selective degradation hypothesis. J. Exp. Psychol. Appl. 5 (2), 115–128.

Panerai, F., Droulez, J., Kelada, J.M., Kemeny, A., Balligand, E., Favre, B., 2001. Speed and safety distance control in truck driving: comparison of simulation and real-world environment. In: Proc. Driving Simulation Conference, Sophia Antipolis (France).

Panerai, F., Toffin, D., Paillé, D., Kemeny, A., Fadel, K., 2006. Eye movement patterns in nighttime driving simulation: conventional and swivelling headlights. In: Proc. VISION Conference.

Pattanaik, S.N., Ferwerda, J.A., Fairchild, M.D., Greenberg, D.P., 1998. A multiscale model of adaptation and spatial vision for realistic image display. In: Proceedings of SIGGRAPH. ACM, New York, NY, USA, pp. 287–298.

Pattanaik, S.N., Tumblin, J., Yee, H., Greenberg, D.P., 2000. Time-dependent visual adaptation for fast realistic image display. In: Proceedings of SIGGRAPH. ACM Press, New York, NY, USA, pp. 47–54.

Petit, J., Brémond, R., 2010. A high dynamic range rendering pipeline for interactive applications: in search for perceptual realism. Vis. Comput. 28 (6–8), 533–542.

Petit, J., Brémond, R., Tom, A., 2013. Evaluation of tone mapping operators in night-time virtual worlds. Virtual Reality 17, 253–262.

Pizza, F., Contardi, S., Mostacci, B., Mondini, S., Cirignotta, F., 2004. A driving simulation task: correlations with multiple sleep latency test. Brain Res. Bull. 63, 423–426.

Reinhard, E., Stark, M., Shirley, P., Ferwerda, J., 2002. Photographic tone reproduction for digital images. In: Proceedings of SIGGRAPH.

Reinhard, E., Ward, G., Pattanaik, S.N., Debevec, P., 2005. High Dynamic Range Imaging: Acquisition, Display, and Image-Based Lighting. Morgan Kaufmann, San Francisco, CA.

Saffarian, M., Happee, R., de Winter, J., 2012. Why do drivers maintain short headways in fog? A driving-simulator study evaluating feeling of risk and lateral control during automated and manual car following. Ergonomics 55 (9), 971–985.

Schallhorn, S., Tanzer, D., Kaupp, S., Brown, M., Malady, S., 2009. Comparison of night driving performance after wavefront-guided and conventional LASIK for moderate myopia. Ophthalmology 116 (4), 702–709.

Seetzen, H., Heidrich, W., Stuerzlinger, W., Ward, G., Whitehead, L., Trentacoste, M., Ghosh, A., Vorozcovs, A., 2004. High dynamic range display systems. ACM Trans. Graph. 23 (3), 760–768.

Shahar, A., Brémond, R., 2014. Toward smart active road studs for lane delineation. In: Proc. Transportation Research Arena, Paris la Défense (France).

Silber, B., Papafotiou, K., Croft, R., Ogden, E., Swann, P., Stough, C., 2005. The effects of dexamphetamine on simulated driving performance. Psychopharmacology 179, 536–543.

Sivak, M., 1996. The information that drivers use: is it indeed 90% visual? Perception 25, 1081–1089.

Snowden, R.J., 1998. Speed perception fogs up as visibility drops out. Nature 392, 450.

Spencer, G., Shirley, P., Zimmerman, K., Greenberg, D.P., 1995. Physically based glare effect for digital images. In: Proceedings of ACM SIGGRAPH, pp. 325–334.

Stone, L.S., Thompson, P., 1992. Human speed perception is contrast dependent. Vis. Res. 32 (8), 1535–1549.

Vires, 2015. http://www.vires.com.

Ward, G., 1994. A contrast based scale-factor for image display. Academic Press Professional, San Diego, CA, pp., 415–421.

Weber, T., Plattfaut, C., 2001. Virtual night drive. In: Proceedings of the 17th International Technical Conference on the Enhanced Safety of Vehicles, Amsterdam, The Netherlands, pp. 1–5.

Wood, J., Chaparro, A., 2011. Night-driving: how low illumination affects driving and the challenge of simulation. In: Fisher, D.L., Rizzo, M., Caird, J.K., Lee, J.D. (Eds.), Handbook of Driving Simulation for Engineering, Medicine and Psychology. CRC Press, Boca Raton, FL, pp. 1–12.

Wood, J.M., Tyrrell, M.A., Carberry, T.P., 2005. Limitations in drivers' ability to recognize pedestrians at night. Hum. Factors 47 (3), 644–654.

HDR IMAGE WATERMARKING

22

F. Guerrini*, M. Okuda†, N. Adami*, R. Leonardi*

University of Brescia, Brescia, Italy*
University of Kitakyushu, Kitakyushu, Japan†

CHAPTER OUTLINE

22.1 A BRIEF INTRODUCTION TO DIGITAL WATERMARKING

This section introduces the basic notions of digital watermarking, to give the reader a self-contained, succinct yet quite complete coverage of digital watermarking basics while presenting the watermarking system in a way congenial to the rest of this chapter.

Digital watermarking (Barni and Bartolini, 2004; Cox et al., 2008) belongs to the *data hiding* field. Data (or information) hiding is a field as old as history itself (Cox et al., 2008, Chapter 1). One can hide an object, as a piece of information, for many possible reasons to obtain more disparate intended results. The commonest reason for hiding information is to protect it from inappropriate or prohibited or sometimes even perfectly licit use by people who have not the authority to use it. A different but somewhat correlated reason is secrecy: one may want to hide information to keep its very existence unknown (secrecy is correlated with protection because the former can be a means for protection on its own and because secrecy and protection are often simultaneously present, and sometimes confused, in real-world applications). Hidden information may also be used in surveillance systems to trigger some actions in response to people, unaware of their presence, performing (potentially illicit) operations, this way constituting a repressive rather than a preventative way of protection. Finally, information could

High Dynamic Range Video. http://dx.doi.org/10.1016/B978-0-08-100412-8.00022-X

be hidden because, in spite of its perhaps indispensable presence, should it be in plain sight in some cases it would degrade the perceptual value of the object it corresponds to. An excellent introduction to data hiding history in its totality can be found in Petitcolas et al. (1999). For a mathematical analysis of the problem of secret communication, see the classic article by Shannon (1949).

Digital data hiding is the direct extension of these concepts to the digital world. Since digital content is only another representation of the same information the human senses naturally perceive, it is quite natural that needs the same as those previously mentioned arise in the digital world as well. Hence, digital information hiding applications attempt to hide some kind of digital information in digital documents.

Digital information hiding can be thought of as a combination of three main techniques: cryptography, steganography, and digital watermarking. The best known branch of information hiding is probably *cryptography* (Menezes et al., 1996), which is the art of hiding the content of a transmission between two subjects from a potentially malicious eavesdropper by making it unintelligible to all except the intended recipients. Another very old information hiding technique, although less well known than cryptography, is *steganography* (Provos and Honeyman, 2003), sometimes also called covert communication. In steganography, the very existence of the communication is hidden; the information is conveyed by proper, imperceptible modifications of an innocent-looking object. The most recent information hiding technique, *digital watermarking*, is the subject of this chapter, applied in the high dynamic range (HDR) imaging context.

Loosely speaking, digital watermarking tries to introduce information in a certain domain (this process is called *watermark embedding*, Barni and Bartolini, 2004, Chapter 4) inside a digital object while preserving its perceptual content (like steganography) and while being in a "hostile" environment (like cryptography) — that is, populated by intelligent attackers (embodying the so-called *watermark channel*, Barni and Bartolini, 2004, Chapter 7) interested in disrupting the watermarking system operativeness while preserving the perceptual quality of the host object. This information will then be retrieved by another entity, or perhaps the same one that performed the embedding process, thus performing the so-called *watermark recovery* (Barni and Bartolini, 2004, Chapter 6).

The need for digital watermarking first arose as an answer to the inherent deficiencies of the existing information hiding techniques to counter the digital multimedia piracy problem. In this case, a digital content owner wants to protect it (eg, against unauthorized copying). Steganography is useless in this case because the potential pirate is well aware of the owner's intentions and methods, thus failing the main hypothesis of the steganographic model; and cryptography can only ensure the encrypted content is not subject to eavesdropping during its distribution but can do nothing more when the content is finally decrypted for consumption by its intended recipient. Digital watermarking, then, could offer a solution to this problem, at least in principle, because it is contained within the content itself.

Later, it became clear that digital watermarking could also be used in totally different application contexts (Cox et al., 2008, Chapter 2; Barni and Bartolini, 2004, Chapter 2). For example, broadcast monitoring refers to the ability by a broadcaster to have precise reports of which shows are aired and how often they are aired, for reasons ranging from collection of royalties to marketing studies. It includes as applicative environments digital TV broadcasting and Internet TV services and has recently surfaced as a critical problem because of the proliferation of available channels. The main target of these recently released products is the identification of copyrighted content made available on the Internet by online viewers or peer-to-peer networks, at the very least to ask for its removal.

Other possible watermarking applications do not deal with intellectual property rights protection at all. As the watermark represents a side channel to convey information that is attached to the content, it could also carry useful metadata instead. For example, metadata watermarks could simplify content-based retrieval of multimedia objects. A related application which is experiencing growing interest is the embedding of linking information to enable a range of e-commerce services. The distinct advantage of the use of watermarking to attach metadata to a piece of content is its robustness to digital-to-analog and analog-to-digital conversions, processes that usually result in loss of any other type of metadata such as header-based metadata. The ability to insert as much information as possible into the host object is the main requirement, along with issues related to implementation such as complexity and speed of execution.

A rather ambitious proposed application of digital watermarking is enhanced coding of digital content, both source coding and channel coding. It is not clear from a theoretical point of view whether it could boost coding performance, especially in practical scenarios; nevertheless, some authors argue that some advantage is to be expected by data hiding techniques are applied for content coding. For the source coding part, it has been proposed that watermarking could help the compression process, achieving better compression by replacing (a lossless operation) the imperceptible part of the content with information data on the content that therefore no longer has to be stored in the bitstream. Digital watermarking could also be used for channel coding (ie, to counter transmission errors that are especially harmful to compressed content). It could be a valuable alternative to error concealment on the decoder side (to approximate lost information by means of some sort of filtering) and/or redundant coding on the encoder side (via error correcting codes), which usually suffer from backward-compatibility issues. In the case of the use of a digital watermark to embed redundant information, compatibility is automatically achieved because it can be safely ignored by the decoder. It is not clear at this point if the peak signal-to-noise ratio (PSNR) distortion (the traditional way to evaluate the goodness of a coding algorithm) introduced by the watermark is better than that achievable by means of other techniques.

The first domain considered by the digital watermarking community, and undoubtedly the most studied even today, is the one pertaining to still digital low dynamic range (LDR) images. Digital watermarking has been applied to other domains as well, especially audio, but also video (often by use of the methods designed for still images) and more exotic domains such as text and three-dimensional meshes. As we will see, the most recent entry in this list is still digital HDR images.

22.1.1 DIGITAL WATERMARKING REQUIREMENTS

The requirements of a digital watermarking system, whose simultaneous presence distinguishes it from the other data hiding techniques, are *capacity*, *robustness/security*, and *imperceptibility*. All of these requirements are discussed in the following.

As is often the case, the system designer must handle an application-dependent trade-off between the requirements. To picture their conflicting nature, they are often drawn on a so-called trade-off triangle, as in Fig. 22.1, to stress the fact that trying to favor one of these requirements always damages to some extent one or both of the other two. Note that robustness and security have been drawn at the same vertex; this can be acceptable in general, given that the boundary between these two requirements is very subtle (and even not considered by some authors). However, to be specific, security and robustness

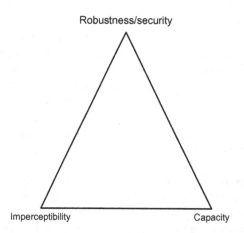

FIGURE 22.1

Digital watermarking trade-off triangle depicting how the requirements at the vertices conflict.

are sometimes themselves conflicting requirements, when treated as separate objectives (and we argue below that this should indeed be the case), so the trade-off triangle could actually be a trade-off tetrahedron.

The capacity (Barni and Bartolini, 2004, Chapter 3) is the quantity of information (usually measured in bits) that the watermark is able to convey. It is generally dependent on the size of the host object; thus, sometimes the capacity is expressed as a relative entity (eg, in the case of images the unit is a bit of information per pixel, or bpp). This way of expressing capacity is particularly common in steganography, where capacity is often the primary concern. Vice versa, the capacity of a watermark more strongly depends on its application; for example, an image watermark capacity could range from a single bit in the case of detectable watermarking (see later) to some thousands of bits for a single host image, while in steganography it is usually some fraction of a bit of information per pixel or more (which, given the number of pixels in an ordinary image, is several order of magnitude higher). Hence, the watermark capacity must be tailored for the aim of the intended application.

Once the watermark has been embedded into a host object, sooner or later it must be retrieved. How this can be achieved even after the host watermarked object has been possibly processed is referred to under the terms watermark security and watermark robustness. There is some confusion in the early literature about the definition of these two requirements. The philosophy adopted here — that is, to clearly distinguish security and robustness as separate and conflicting requirements — is probably the most acceptable in our opinion.

Robustness (Cox et al., 2008, Chapter 9) refers to the ability of the watermark to survive any nonmalicious data processing the host object happens to undergo. Such processing is not intended to remove the watermark but is applied to the host object for some other purpose. The set of acceptable data processing must be decided before the watermarking system is designed and obviously depends on the nature of the host object. For classic LDR images, lossy compression, noise addition, digital-to-analog and analog-to-digital conversion, geometric transformations (such as zooming, rotation, and cropping), linear and nonlinear filtering for image enhancement, and histogram modifications are all examples of possibly nonmalicious processing that can occur.

Security (Cox et al., 2008, Chapter 10) is related to the inability by a hostile entity (referred to as the attacker) to remove the watermark or disable in some way its recovery. The most basic instance of security, which is usually present even in watermarking systems which do not explicitly address security, is that only authorized entities are allowed to embed watermarks in or recover watermarks from host objects. This, in turn, implies that both the embedding process and the recovery process must in some way depend on knowledge of some *secret key*. In a well-designed system, security must obey Kerckhoffs's principle (Kerchoffs, 1883; Menezes et al., 1996) much like in cryptography — that is, the security of the system must be assured considering the attacker is aware of all the system details; the only thing he/she ignores is the secret key.

Some authors refer to security in a way related to cryptanalysis, which means it should be impossible for an attacker to guess the secret key. However, we prefer to use our previous definition of security because the former is only a (particularly harmful) instance of the latter, because an attacker could also be interested in simply disabling or removing any watermark present without knowing the secret key.

Finally, it is worth noting that repeated or strong use of a certain processing tool initially considered as nonmalicious (maybe the one against which the system is the most vulnerable) could be effectively considered a security attack, making the boundary between security and robustness very fuzzy. Therefore, it can be tempting to put both these requirements at the same vertex of the trade-off triangle, but we believe that neglecting the more sophisticated intelligence of a determined attacker can hurt the deployability of a watermarking system in a security-critical application. In our view, robustness and security are conflicting requirements in the sense that robustness tends to exploit any possible perceptual niche of the host object to survive processing, while security tries to make the watermark characteristics and location as unpredictable as possible.

Last, the watermark has to be imperceptible (Cox et al., 2008, Chapter 8) — that is, it should not degrade the perceptual content of the host object. Starting from this definition, it is obvious that perceptibility is a subjective matter, so it is impossible to give a universal measure of perceptibility. The best way to handle perceptibility is to study how humans perceive the environment in which they live means of models that approximate the mechanisms underlying perception. These models, primarily the human auditory system and the human visual system (HVS), have been extensively studied in the field of digital compression. Multimedia data compression tries to remove perceptually irrelevant parts of the original data to decrease the amount of information that must be conveyed to reproduce an acceptable "quality" of the data, so it is very important to include perceptual cues. In a sense, digital watermarking and data compression can be thought of as dual problems: the former, in fact, because it has to be imperceptible, must reside in the field of imperceptible data, the data that compression tries to eliminate from the original data. Unfortunately (or luckily depending on the point of view) perfect perceptual compression is not achievable, and therefore watermarks could be accommodated in imperceptible "niches" left by compression. Hence, to achieve imperceptibility, digital watermarking could exploit a whole mass of knowledge borrowed from multimedia compression.

Many systems do not explicitly use a perceptual model (which is usually difficult to implement), but instead rely on more classic approaches based on standard metrics (eg, PSNR) to minimize perceptual impairments; however, care has to be taken when one is comparing human perception with these kinds of absolute distortion measures. It is also very common to guarantee imperceptibility by the mere selection of an appropriate watermark domain.

An exception to the general rule of imperceptibility is *visible watermarking*, where the watermark is rendered perceptible to assess its presence (maybe for informative purposes, much like a logo) while retaining all the other watermark characteristics. In any case, even if the watermark is visible, there is a certain amount of distortion that the embedding process cannot exceed on the host object, so even in this case it is possible to define the imperceptibility requirement with some slight modification.

22.1.2 WATERMARKING SYSTEM EXAMPLES

To help convey the basics of watermarking systems as described above, we first provide a brief explanation of a couple of classic algorithms before introducing the watermarking system structure in general and abstract terms in the following sections.

Cox et al. (1997) introduced a widely used watermarking paradigm known as spread-spectrum watermarking. The simplified flowcharts of the watermark embedding and recovery processes are depicted in Fig. 22.2. On constructs the watermark by seeding a pseudorandom number generator

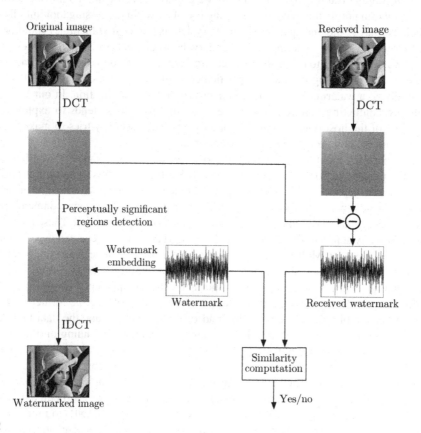

FIGURE 22.2

Example of a spread-spectrum watermarking process, the one presented in Cox et al. (1997). *IDCT*, inverse DCT.

with the secret key and then extracting a Gaussian random sequence w. It is argued that to render the watermark robust against common processing such as compression, the best way is to insert the watermark in the most perceptually significant portion of the host object. For the still image case, a full-frame two-dimensional discrete cosine transform (DCT) is computed and represents the watermark domain. Next, a perceptual mask is computed to identify the most perceptually significant coefficients, referred as the vector v. The watermark is then introduced to obtain the watermarked coefficients v' by application of one of the following formulas:

$$v'_i = v_i + \alpha_i w_i, \tag{22.1}$$

$$v'_i = v_i(1 + \alpha_i w_i), \tag{22.2}$$

for $i = 1, \ldots, n$, where n is the watermark length. The scaling factors α_i should be selected so as to ensure imperceptibility and thus can be dependent on the value v_i. The scheme's target is only to assess if a particular watermark is present or not (see the discussion of detectable watermarking later). The above formulas are sometimes referred to as additive spread-spectrum and multiplicative spread-spectrum watermarking, respectively.

The watermark recovery is nonblind (see later) — that is, the original, unwatermarked object is required by the entity that performs the recovery. To determine if the watermark is present, the received watermarked object (which is possibly further processed) undergoes the same two-dimensional DCT and the coefficients obtained are then subtracted from the original ones to obtain the recovered coefficients w^* and then a normalized correlation (or similarity measure) is computed as

$$\text{Sim}(w, w^*) = \frac{w \cdot w^*}{\|w\| \cdot \|w^*\|}. \tag{22.3}$$

Use of Gaussian distributed watermark coefficients increases security because it counters so-called collusion attacks, in which the attacker averages several watermarked images in the hope of obtaining an unwatermarked object: in this case all the watermarks are still present simultaneously. However, it should be noted that an attacker can apply the same perceptual mask to identify which coefficients carry the watermark and then perform subtler attacks in the DCT domain. Choosing just a random portion of all coefficients would increase security in this sense, but would surely hurt robustness.

The second watermarking system example we discuss is quantization index modulation (QIM), the precursor of many techniques that have appeared and are still appearing in the literature. It was proposed by Chen and Wornell (2001). QIM consists in a quantization of the host object features using a particular codebook Q associated with a given watermark. To be more specific, suppose we have a set U of $2^{|\mathbf{b}|}$ different quantizers, each identified by a specific string associated with a binary string $\mathbf{b}$ of length $|\mathbf{b}|$. If we wish to embed the watermark code $\bar{\mathbf{b}}$ in the host object, the latter's features must be quantized with use of the correspondent quantizer from the set U, giving the quantized (watermarked) features. When the watermark is to be retrieved, the received object features are requantized with use of the entire codebook set U (since the retriever does not know in advance which quantizer was used in the first place) and then one identifies to which particular codebook the quantized value belongs by taking the one in the entire set U with a quantized value at the minimum distance, thus retrieving $\bar{\mathbf{b}}$.

Scalar QIM (SQIM), also referred as dither modulation watermarking, is a very common subset of QIM algorithms. In these systems the reconstruction values associated with all the codebooks forming U are arranged in a regular, rectangular lattice; in this way, one can perform scalar feature quantization,

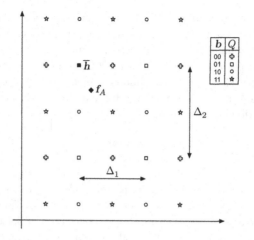

FIGURE 22.3

QIM example, specifically a two-bit scalar QIM is depicted.

one feature at time, thus avoiding vector quantization processes. A rearrangement of QIM to make it adhere to the SQIM paradigm is illustrated in Fig. 22.3, where two features are represented on each axis. As observable, now it is possible to perform feature quantization separately along the two axis, with in general two different quantization steps, as every bit of **b** is embedded in a single feature. In this case, the watermark code **b** is only two bits long, so U consists of four distinct quantizers, each represented by a different symbol. The key on the right describes the relation between each watermark code and its correspondent quantizer. The host feature point $\mathbf{f}_A$ is represented by the filled diamond. Assuming that we wish to embed the code $\overline{\mathbf{b}} = 01$, only the quantizer individuated by the squares is used for the embedding and so the filled square is selected as the quantized feature $\overline{\mathbf{h}}$ (with the watermark embedded *because* it is a square). Note that the quantized feature point $\overline{\mathbf{h}}$ is by no means the closest to $\mathbf{f}_A$ for the entire codebook U, but is so only for a suitable quantizer Q. When, because of the watermark channel, this feature point is moved (hopefully in a close neighborhood), its requantization using all the symbols performed by the decoder will output which symbol was used during the embedding process, hence allowing us to identify $\overline{\mathbf{b}}$. Note that as long as the codebook is known during the recovery phase, there is no need to use the original unwatermarked object (blind recovery, see later).

The trade-off triangle is easily observable in QIM systems. Robustness, heuristically, depends on the mutual distance between the reconstruction values belonging to different quantizers; imperceptibility depends on the quantization step Δ because it is related to how much the host feature point is moved; and capacity refers to the number of different codebooks available. It is obvious they are conflicting requirements, because increasing robustness means moving away reconstruction values of different codebooks, but this in turn affects imperceptibility; and increasing capacity increases the density of symbols, worsening robustness, whereas lowering the density increases the quantization step, harming imperceptibility. As a final note, one usually achieves security in QIM systems by shifting the codebook U by a random quantity, extracted from a uniform distribution to increase the uncertainty on its value, and depending on the secret key, so that an attacker cannot tell which symbol has been used in the

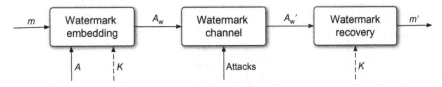

FIGURE 22.4

Elementary structure of a digital watermarking system. A watermark message **m** is embedded into a host object A, optionally with use of a secret key K. The watermarked object A_w goes through the channel in which it possibly undergoes some attacks. The recovery is then applied on the resulting $A_{w'}$. The output is the estimated watermark message **m'**.

quantization process. Because addition of this shift, which can be shared thanks to knowledge of the secret key, has no effect on the other requirements and is easily implemented, this solution is very commonly adopted.

22.1.3 STRUCTURE OF A WATERMARKING SYSTEM

In the literature the watermarking system structure has been proposed in many ways; here we adopt the view of the watermarking game as a digital communication one. The basic flowchart of a digital watermarking system is illustrated in Fig. 22.4. In this high-level flowchart, only mandatory variables (ie, variables always present in every watermarking system) are depicted (an exception is the dashed optional input, which is represented because it is almost always used).

The message (which can be represented as a bit string **m** without loss of generality) is the main input to the system; the objective of the watermarking game is to guarantee that it is correctly received at the end of the transmission chain. The watermark embedder introduces the message into the host object A following the suitable mix of watermarking requirements a priori selected by the system designer, producing a watermarked object A_w. As previously mentioned, this process is very often driven by a secret key K to ensure a certain amount of security: for example, the seed of a pseudorandom generator of the watermark as in spread-spectrum watermarking. After the embedding stage, the watermarked object A_w possibly undergoes some processing (both malicious and nonmalicious attacks, or no attacks at all) which is overall modeled by the so-called watermark channel. Finally, the watermark recovery is performed on the "received" object A'_w, aided by the secret key K which was shared with the embedder (if used), giving as output a bit string **m'** which represents an estimate of the original message.

22.1.3.1 Watermark embedding

The embedder block objective is to produce the watermarked object A_w; this can be summarized by the following formula:

$$A_w = \mathcal{E}(\mathbf{m}, A[, K]), \tag{22.4}$$

where $\mathcal{E}(\cdot)$ is referred to as the *embedding function*: Eq. (22.1) constitutes an example of an embedding function. In Eq. (22.4) the notation, $[K]$ indicates that the secret key K is an optional parameter whose presence depends on the considered watermarking paradigm. This also applies to formulas of the chapter. Notice how we consider the secret key K as an optional variable (enclosed in square

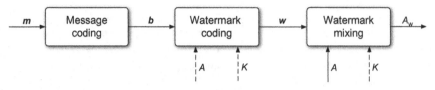

FIGURE 22.5

Waveform-based watermark embedding steps.

brackets), to be coherent with Fig. 22.4. Depending on how the task of Eq. (22.4) is implemented, we can distinguish between two different types of watermark embedding (and, by extension, watermarking systems): *waveform-based watermarking* and *direct embedding watermarking*.

Fig. 22.5 depicts the typical steps of a waveform-based embedding stage. Additive and multiplicative spread-spectrum watermarking can be classified as based on a waveform-based embedding process. First the message code **m** is coded into a bit string **b** (the watermark code or simply the *watermark*) with use of a code C; this operation is not always present, and in this latter case **m** = **b**.

After the preliminary message coding step, the embedding function is applied. Usually, the watermark domain is different from the host object domain — that is, the watermark embedding is accompanied by a feature extraction process $\mathcal{F}(\cdot)$ which transforms the host object A into a set of original host features $\mathbf{f}_A$ (the feature space is the *watermark domain*). For example, the full-frame two-dimensional DCT coefficients represent the watermark domain in our example spread-spectrum system (Cox et al., 1997). Analogously to $\mathcal{F}(\cdot)$, a *watermark coding* $\mathcal{W}$ must take place beforehand, and transforms the watermark **b** into a suitable *watermark signal* **w** (alternatively called a *watermark waveform*; see Eqs. 22.1 and 22.2 for examples), expressed in the feature domain, which is well suited to be embedded in the description of the host object A carried by its features. The embedding function $\mathcal{E}(\cdot)$ can then be thought of in this case as a *mixing* $\oplus$ of some kind of watermark signal **w** with the host features $\mathbf{f}_A$ to obtain the watermarked features $\mathbf{f}_{A_w}$. Note that not all the host features $\mathbf{f}_A$ need to be mixed with the watermark signal **w** (ie, the watermark signal dimensionality need not to be the same as that of the host features). The watermarked object A_w is finally obtained by a reverse mapping function $\mathcal{F}^{-1}(\cdot)$ from the watermarked features $\mathbf{f}_{A_w}$ to the host object domain (the inverse DCT in our spread-spectrum example). When the watermark domain coincides with the host object domain (eg, image watermarking in the pixel domain which works directly on pixel values), the feature extraction and its inverse revert to an identity function. This whole procedure can be illustrated as follows (optional arguments are again enclosed in square brackets):

$$\mathbf{b} = C(\mathbf{m}), \tag{22.5}$$

$$\mathbf{w} = \mathcal{W}(\mathbf{b}, [A, K]), \tag{22.6}$$

$$\mathbf{f}_A = \mathcal{F}(A[, K]), \tag{22.7a}$$

$$\mathbf{f}_{A_w} = \mathbf{f}_A \oplus_{[K]} \mathbf{w}, \tag{22.7b}$$

$$A_w = \mathcal{F}^{-1}(\mathbf{f}_{A_w}[, K]). \tag{22.7c}$$

Eqs. (22.5) and (22.6) pertain to the message coding and watermark coding blocks depicted in Fig. 22.5, respectively, while the operations referred to as Eq. (22.7) are performed by the mixing block.

The secret key K could drive both the watermark coding process and the watermark mixing process. For example, the watermark signal $\mathbf{w}$ could be randomly selected from a set of possible waveforms according to K; the secret key could also randomly select which features the watermark signal has to be mixed with and which it should not mixed with or randomize the feature extraction process itself out of a predetermined set.

On the other hand, in direct embedding techniques there is no watermarking signal defined before the manipulation of the host features. They are described in Fig. 22.6, where with respect to Fig. 22.5 the watermark coding step is missing. Hence, the bit string $\mathbf{b}$ is embedded directly into the host object A by modification of, in a controlled way, the host features $\mathbf{f}_A$. The QIM paradigm falls into this category. The set of equations describing the direct embedding paradigm is as follows:

$$\mathbf{b} = \mathcal{C}(\mathbf{m}), \tag{22.8}$$

$$\mathbf{f}_A = \mathcal{F}(A[, K]), \tag{22.9a}$$

$$\mathbf{f}_{A_w} = \mathcal{E}'(\mathbf{b}, \mathbf{f}_A[, K]), \tag{22.9b}$$

$$A_w = \mathcal{F}^{-1}(\mathbf{f}_{A_w}[, K]). \tag{22.9c}$$

Message coding of Eq. (22.8) is the same and serves the same purpose as Eq. (22.5). In this case, therefore, what is really different with respect to the waveform-based approach is the absence of any operation similar to Eq. (22.7b). Instead of mixing with a predefined signal $\mathbf{w}$, there is a function $\mathcal{E}'(\mathbf{b}, \mathbf{f}_A)$, described by Eq. (22.9b), which moves the host features to the watermarked features $\mathbf{f}_{A_w}$ in a way depending both on the initial position $\mathbf{f}_A$ and on $\mathbf{b}$. Referring to Fig. 22.3, the host feature point $\mathbf{f}_A$ must be moved to the new watermarked feature point $\overline{\mathbf{h}}$ without our explicitly defining a watermark signal $\mathbf{w}$. This process usually involves minimizing a cost function tied to the introduced perceptual distortion, possibly using an iterative algorithm (for SQIM, searching for the "nearest" quantized value).

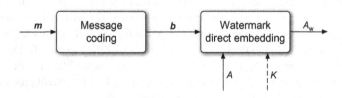

FIGURE 22.6

Direct watermark embedding steps.

22.1.3.2 Watermark channel

The watermark channel represents all the transformations applied to the watermarked object A_w before the watermark recovery is performed, so the latter is actually done on an attacked object A'_w. As previously mentioned, the attacks applied on a watermarked object could be either malicious (targeting watermarking security) or nonmalicious (targeting robustness instead), although this separation is not always simple. Nevertheless, if we keep this overlapping of meanings in mind, it is possible to define nonmalicious attacks as *robustness attacks* and malicious attacks as *security attacks*. In turn, malicious attacks can be further classified as *blind security attacks*, in which the attacker does not exploit any knowledge of the watermarking system but instead attacks the watermarked object with some operation which is not considered usual, hoping from the attacker's point of view that the system designer did not take it into account, and *nonblind security attacks*, in which the attacker knows all of the watermarking system, except the secret key used, and exploits this knowledge to attack the system at its weak spots — for example, by mounting special attacks allowed by the particular system implementation (such as the availability of watermark recovery tools and the ability to apply them to arbitrary objects) and/or exploiting knowledge of watermark localization or spectral properties to alter synchronization between the embedder and the recovery block or to filter out the watermark. These latter attacks are the most dangerous and at the same time they cannot be ignored by the system designer if the already mentioned Kerckhoffs's principle is to be respected; however, in some applications, it is reasonable to safely ignore such attacks simply because the intentional watermark removal or disabling is not in the attacker's own interest.

Robustness attacks are defined as all transformations which are applied on the host object without the aim of explicitly disabling or removing the watermark; consequently, they comprise all processing which belongs to the "normal" course of life of the host object. Which attacks one should consider as nonmalicious is a matter strongly dependent on the nature of the host object A; moreover, also the application intended for the digital watermarking system at hand plays an important role because many robustness attacks are important than others in certain contexts. There are so many types of robustness attacks that usually the system designer takes into account only a certain number of them (if any) and then hopes for the best for all the others.

Common signal processing, usually aimed at digital objects' perceptual content enhancement, is another form of robustness attack. It could be some kind of simple manipulation of features (eg, the so-called constant gain attack where the watermarked features are multiplied by a certain factor) or some more complex processing (eg, for digital images histogram modification is a classic type of image processing similar to a constant gain attack); it could also be represented by filtering, either linear (eg, low-pass filters) or nonlinear (eg, noise-suppressing filters).

Geometric manipulations are another very important class of robustness attacks, where geometry refers to any coordinate of the host object (spatial coordinates for images, temporal coordinates for audio, and both spatial and temporal coordinates for video): for this reason they are also called synchronization attacks. These attacks can be very tricky as they tend to break the consistency between the coordinates of the embedder and those used during the recovery process. To undo synchronization attacks, the system could either try to be as invariant as possible to modification of coordinates (losing robustness with respect to other types of attacks) or leave the task of recovering the original coordinate configuration to the recovery block (usually by exhaustive search).

Finally, object editing processes, such as cropping for images, can be considered robustness attacks; they usually intrinsically contain some amount of geometric manipulations when they are applied.

Among blind security attacks, the exhaustive use of some robustness attacks is the first thing that comes into mind, as the attacker may want to break the system by going beyond the robustness the watermark may tolerate.

Another way of attacking the system without delving into its details is to treat the watermark as noise and hence to use some noise-suppressing process to completely remove it in the simplest cases (in particular when the watermark is well modeled by an additive noise) or at times at least roughly estimate the watermark itself to remove it or to illicitly embed it into other objects (another example of how security and robustness concepts overlap). Other, more sophisticated threats should also be taken into account (eg, the already mentioned collusion attack where many watermarked copies of the same host object or many objects with the same watermark embedded are compared to estimate the watermark).

With regard to nonblind security attacks, the first observation that has to be made is that the secret key K is the most sensitive parameter of the system, because its unwanted leakage to an attacker aware of the system design could signify the nullification of the intended task of the watermark. This is effectively a very tough problem in some applications where the recovery process is to be performed by any user willing to do so (and this means the user has to use the secret key). This prompted the rise of asymmetric key schemes which use different keys during the embedding and recovery stages, although they introduce other problems (most notably a robustness decrease tendency).

Furthermore, when the attacker can repetitively perform the recovery process on any object, the so-called sensitivity attack can be adopted. In this attack, the attacker modifies little by little the watermarked object and then performs the watermark recovery; this will give a rather precise estimation of the detection/decoding boundaries, allowing the attacker to choose the most convenient unwatermarked object (eg, the one that minimizes distortion), and maybe to learn some information about the secret key used. Using complex detection boundaries or making the attack construction unfeasible thanks to computational complexity issues are the most common countermeasures to the sensitivity attack.

Obviously, a nonblind attacker could also use one of the attacks previously classified as blind if it is identified as a weak point of the system after an appropriate analysis of the system operations. As a last note, to better counter security attacks on a critical application, it is arguably mandatory to couple watermarking and cryptography technologies on a protocol level. Several examples are found in security-oriented applications, where cryptography is usually used to secure content distribution and to limit as much as possible unauthorized access to the object at hand and watermarking is used to tie security information with the content perception. A good survey of security issues in digital watermarking can be found in Cayre et al. (2005a,b).

22.1.3.3 Watermark recovery

The recovery stage is responsible for the extraction from the possibly attacked object A'_w of an estimate $\mathbf{m}'$ (the bit string representing m') of the original message string $\mathbf{m}$. Hence, the most general aspect of the recovery process is described by the following equation:

$$\mathbf{m}' = \mathcal{D}(A'_w[, K]). \tag{22.10}$$

A *recovery function* $\mathcal{D}(\cdot)$ is applied on A'_w to obtain the recovered message $\mathbf{m}'$. The form of the recovery function $\mathcal{D}(\cdot)$ very much depends on the nature of the watermarking algorithm — that is, whether the recovery consists in assessing that a certain watermark is present or in choosing which

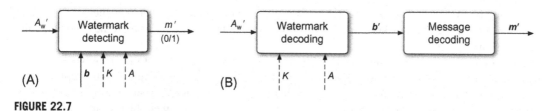

FIGURE 22.7

Watermark recovery process forms: (A) watermark detector; (B) watermark decoder.

watermark among those possible has been embedded. Therefore, Eq. (22.10) can be specialized into two different forms belonging, respectively, to *detectable* and *decodable* watermarking, which are separately depicted in Fig. 22.7.

In detectable watermarking, the watermark recovery process (now called the detector block) can be schematized as in Fig. 22.7A. Here we are interested only in assessing the presence or the absence of the watermark, so the message m can be reduced to a binary variable (thus, $\mathbf{m} = m$ as the string has unitary length); the fact that we are embedding a watermark inside a host object A means that $m = $"1." Consequently, the coding $\mathcal{C}(\cdot)$ of Eq. (22.5) is better described by a codeword selection (such as bit repetition rather than a channel code), as the watermark $\mathbf{b}$ is embedded into A only to mean that A_w is watermarked, regardless of the meaning of $\mathbf{b}$. The detector, then, looks for the watermark $\mathbf{b}$ in the attacked object A'_w and takes a decision about its presence or absence, thus outputting an estimated message m' which is 1 if the detector believes the watermark $\mathbf{b}$ has been embedded into A'_w and 0 otherwise. Eq. (22.3) exemplifies this process: a threshold selected on the similarity measure between the supposedly embedded watermark and the received one embodies the detection decisor. Therefore, the watermark detection can be expressed as

$$\mathcal{D}(\mathbf{b}, A'_w[, A, K]) = m' \in \{1, 0\}. \tag{22.11}$$

Notice how in this case the detector must know $\mathbf{b}$ in advance so it can achieve its task. In some literature, it is stated that since the message m conveys only one bit of information (the watermark presence or its absence), detectable schemes are also defined as one-bit watermarking. This could be confusing: it would be more correct to say that the *detector* obtains one bit of information, but it has to be kept in mind that there is only one possible message m.

Decodable watermarking is depicted in Fig. 22.7B. In this case, the recovery block, conveniently called the decoder, does not know in advance the watermark $\mathbf{b}$, so it has to read it from the attacked object A'_w (for this reason, this scheme is also called readable watermarking) to form an estimated watermark $\mathbf{b}'$. Here the original message m is meaningfully represented by the string $\mathbf{m}$, and the message coding $\mathbf{b} = \mathcal{C}(\mathbf{m})$, if present, could be, for example, a channel coding of the message $\mathbf{m}$ into the watermark $\mathbf{b}$. The first operation of the decoder is to decode an estimated watermark $\mathbf{b}'$ from the attacked object A'_w (using a function which is referred as $\mathcal{D}'(\cdot)$); then, given that the first stage of the embedding process is the message coding, the last stage of the decoding process is obviously the message decoding of the recovered string $\mathbf{m}' = \mathcal{C}^{-1}(\mathbf{b})$. As is easily imagined, decodable watermarking is also called multibit watermarking. Fig. 22.3 shows an example of multibit watermarking. Depending on the symbol over which the received feature point is quantized, a different pair of bits is decoded

Table 22.1 Watermark Recovery Function Forms		
	Blind Recovery	**Nonblind Recovery**
Detectable watermarking	$\mathcal{D}(\mathbf{b}, A'_w[, K]) = 1/0$	$\mathcal{D}(\mathbf{b}, A'_w, A[, K]) = 1/0$
Decodable watermarking	$\mathcal{D}(A'_w[, K]) = \mathbf{m}'$	$\mathcal{D}(A'_w, A[, K]) = \mathbf{m}'$

(hopefully, if we embedded the "square" as depicted by $\overline{h}$, the received feature point will still be nearer to it than to the other symbol to ensure correct decoding). The whole process is illustrated as follows:

$$\mathcal{D}'(A'_w[, A, K]) = \mathbf{b}', \tag{22.12a}$$

$$\mathbf{m}' = \mathcal{C}^{-1}(\mathbf{b}). \tag{22.12b}$$

Looking at Fig. 22.7, we see that the host object A plays the role of an optional input to the recovery process. In some applications the recovery process has access to the original host object A (perhaps because the embedding and the recovery are performed by the same entity), so in Fig. 22.7 we may add the original host object A as an additional input to the detector/decoder block. If this is the case, the watermarking scheme (or equivalently the recovery process) is said to be *nonblind* (and the optional input A in Eqs. (22.11) and (22.12) has to be considered); otherwise, if the recovery block does not have any knowledge of the host object A, it is called *blind*. The spread-spectrum watermarking technique described above and in Cox et al. (1997) is an example of nonblind watermarking, because the original unwatermarked image is needed during the recovery phase. On the other hand, the QIM paradigm as illustrated in Chen and Wornell (2001) is based on a blind watermark recovery stage.

Using realistic assumptions on the system, nonblind systems surely achieve better robustness over their blind counterparts, but one must keep in mind that such a framework is not always applicable in real-world applications; moreover, the performance gap is not as large as one would intuitively expect.

To summarize, the recovery function can assume one of the forms listed in Table 22.1.

22.1.4 WATERMARKING SYSTEM EVALUATION

Once a watermarking system has been implemented, its performance should be evaluated as objectively as possible (Cox et al., 2008, Chapter 7). Although additional parameters can play a role, such as complexity and memory usage, here we will focus on the satisfaction of the main requirements that we discussed earlier.

Imperceptibility of the watermarked image refers to its indistinguishability from the original, unwatermarked image, and it can be evaluated either subjectively or objectively. The former usually involves user tests performed in controlled environments, but they are seldom used in watermarking contexts. The latter relies on the computation of objective metrics which may or may not be driven by HVS models. In almost all the studies that are the subject of this chapter, imperceptibility was evaluated through either the PSNR, which, although providing a rough estimate of the embedding distortion, is not very well correlated with actual human perception, or more sophisticated metrics for HDR data such as HDR-VDP (Mantiuk et al., 2005) and its successor HDR-VDP-2 (Mantiuk et al., 2011) (see Chapter 17).

Evaluating security is a much a more challenging issue and is totally ignored in many studies, even in security-critical applications such as copyright protection. In most cases, the secret key is used only as the seed of a pseudorandom number generator responsible for constructing the watermark sequence. In this sense, only recipients with the correct secret keys can read the watermark; however, all the other security aspects are not covered by this approach. For example, an intelligent attacker can be interested in disabling the watermark recovery by simply trying to delete the watermark (eg, by embedding a spurious watermark, which is always possible if one assumes that he/she knows the details of the watermarking system and the location of the watermark in the embedding domain is invariant). Although theoretical analysis can done when the problem is simplified, mostly evaluating the security of a watermarking system is done empirically.

Finally, evaluating robustness is related to how the watermark is recovered. In every system, in some part of it a threshold (or more than one) needs to be chosen to differentiate between watermarked and unwatermarked images, or to decode bits of the embedded message. The threshold should be selected by minimization of a loss average, where the loss is a function describing the damage caused if an error occurs in the decision process. In particular, for decodable watermarking, the bit error rate (BER) of the recovered watermark is usually considered. In the case of detectable watermarking, if the system believes that the watermark is present even when this is not the case, we are in presence of a *false alarm*; conversely, if the system wrongly believes that the watermark is absent, a *miss* has occurred. The goal of the designer is to estimate the probability of a false alarm and the probability of a miss of the watermarking system given any host object and any other input (eg, any secret key). Watermarking systems are usually designed according to the Neyman-Pearson criterion (Neyman and Pearson, 1933), where the threshold is selected such that the false alarm probability is less than a given target figure and then the resulting miss probability is evaluated. Both probabilities are then depicted on a receiver operating characteristic (ROC), which one usually obtains by letting the false alarm probability vary, and then calculating the correspondent threshold (the order is inverted when experiments correspond to practical rather than theoretical evaluations) and finally observing the miss probability. Sometimes, the equal error rate, which is the point in the ROC where miss probability and false alarm probability are equal, is provided. An example of an ROC is given in the next section.

22.2 DIGITAL WATERMARKING FOR HDR IMAGES

As HDR images are gaining ground in day-to-day applications, early efforts with regard to more sophisticated requirements for applications have been pioneered. Digital watermarking has carved itself a niche in security-oriented applications but its deployment could represent a challenging problem, so it should not be surprising that at the time of this writing only a handful of studies have been published on the subject. In what follows, we will give a brief description of these studies, following an approximate temporal order, that includes their classification according to the criteria described earlier to give a better idea of how they relate to each other. Before that, we will discuss, in general terms, the requirements for HDR image watermarking.

22.2.1 REQUIREMENTS FOR HDR IMAGE WATERMARKING

In the case of HDR images, some peculiarities exist with respect to classic LDR images. These should be taken into account when one is designing a watermarking system. For this reason, it is generally impossible to directly port an established LDR image watermarking technique to the HDR domain because both imperceptibility and robustness would suffer greatly. It is obvious that a small modification in the LDR domain could become huge when ported to the HDR domain because the range of pixel values there is far more extended with respect to the usual eight-bit, [0, 255] range. Also, the pixel value itself may be redundant — that is, if one suitably modifies the exponent and mantissa of the floating-point representation, sometimes there is more than one way to express a given pixel value.

The pixels' value range difference in the two domains also has far-reaching implications for how humans perceive HDR images, so perceptual models should be corrected when we are dealing with them. That generally implies tuning the specific parameters of the usual perceptual masks, considering how sharper and richer the visual representation of HDR images is with respect to LDR images. For example, the contrast masking effect of the HVS is surely higher in the HDR domain given its richness of details.

One way to tackle these issues is to first transform the HDR image into an LDR one, watermark it by an LDR image watermarking process, and then revert to the HDR domain by some inverse transformation, as many systems do. In doing so, one should exercise particular caution to guarantee that the modifications in the LDR domain will not be perceptible in the HDR domain. An example of such reasoning can be found in the following when we describe Fig. 22.8.

Moreover, it is worth noting that the high visual value of HDR images is always put at a premium. Therefore, when one is considering what processing HDR images can (or should) undergo, only processing that does not severely alter their perceptual value should be studied. Most systems try to be robust against a single type of manipulation: tone-mapping operators. They are common nowadays to permit the rendering of HDR images on LDR displays, so naturally they are considered common signal processing in HDR imaging. Therefore, watermarking systems should be robust against tone mapping and possibly permit watermark recovery in both the original HDR image domain and the LDR of the tone-mapped versions, whatever specific operator has been used.

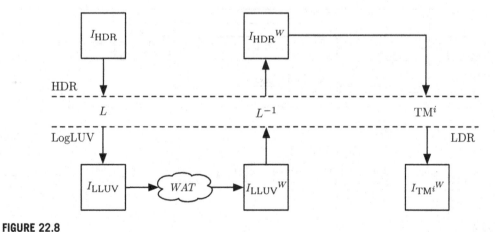

FIGURE 22.8

Conceptual representation of the framework in which the watermarking system of Guerrini et al. (2011) is expected to operate.

More comments on the applicability of digital watermarking in the HDR domain can be found in the last section, where we make some remarks based on the current state of the art, which we review next.

22.2.2 SURVEY OF THE CURRENT STATE OF THE ART

The work by Guerrini et al. (2008), further expanded in Guerrini et al. (2011), is to the best of our knowledge the first published work on data hiding for HDR images. It is a blind detectable watermarking method, so its capacity is a single bit. The capacity is sacrificed to favor the other requirements: robustness against tone mapping and noise addition, security, and imperceptibility.

The rationale behind this watermarking scheme is depicted in Fig. 22.8. Tone-mapping operators are well modeled through a logLUV process (L), so the HDR image is first transformed into the logLUV domain and then an LDR watermarking robust against nonlinear attacks and invariant to constant gain modifications is applied to the luminance component only. The HDR watermarked image is obtained by exponentiation L^{-1}. It is argued that each tone-mapping operator (TM^i) is only a mild nonlinear transformation away from the logLUV image, so in the end $I_{TM^i}^W$, the tone-mapped version of the HDR watermarked image, still retains the watermark.

The embedding method is quite complicated: the flowchart is shown in Fig. 22.9A. It is based on the QIM paradigm, which in this case encodes information into the shift of a nonuniform quantizer. The quantization is applied to the kurtosis of the approximation coefficients resulting from a wavelet decomposition, taken into randomly positioned, randomly shaped blocks. Imperceptibility is aided by use of an HVS-derived perceptual mask, computed with use of the detail subbands as well (that are left untouched by the watermarking process), taking into account brightness, neighborhood activity, and the presence of edges. Security is very high, because it relies on both the shift of the quantizer and the random position and shape of the blocks on which the kurtosis feature is computed, with the attacker

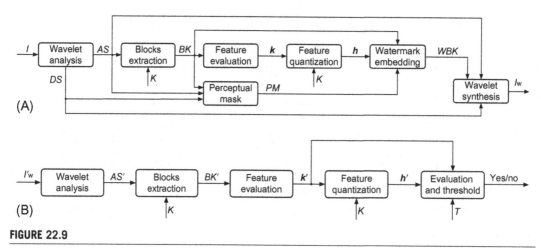

FIGURE 22.9

Watermark system flowchart for the system of Guerrini et al. (2011): (A) watermark embedder; (B) watermark detector.

needing to use an attack that is at least visually perceptible as the mask to disable the watermark. Watermark recovery, illustrated in Fig. 22.9B, is relatively straightforward. The secret key drives the extraction of blocks and the computed kurtosis feature in each of them is quantized by means of the same codebook used in the embedding phase. If the number of blocks correctly decoded is higher than a threshold T, the image is judged as watermarked.

The experiments were performed on 15 HDR images and with seven tone-mapping operators. The imperceptibility was measured with HDR-VDP, the average of which is below 0.5%, signifying that the watermark is almost imperceptible. If the false alarm probability is fixed at 10^{-5}, the miss probability can be as high as 10^{-2} for small images but goes well below 10^{-6} for larger images, which is the commonest case for HDR imagery. One of the ROCs obtained can be seen in Fig. 22.10, for an image watermarked with use of $N = 700$ blocks (which is reasonable for a medium-sized HDR image). As is usual for ROCs, the axis are drawn on a logarithmic scale and the curve in the cases of various robustness attacks (tone-mapping operators in this case) is obtained by the changing of the detection threshold. It is worth noting that no nonrealistic assumptions are made with regard to the distribution of errors; instead, a binomial distribution based on the actual block decoding error probability is assumed.

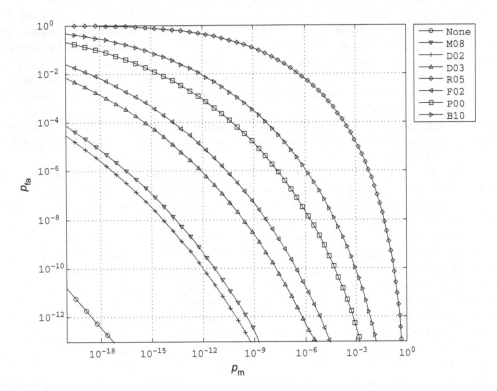

FIGURE 22.10

Example of an ROC. This is obtained with use of the method in Guerrini et al. (2011) setting the number of blocks $N = 700$. Each curve represents a different tone-mapping operator (see Guerrini et al., 2011 for further details).

Fig. 22.10 also suggests how hugely different can be the scale of introduced distortion for different tone-mapping processes.

The work in Guerrini et al. (2008) went mostly unnoticed until the appearance of the work in Guerrini et al. (2011). Meanwhile, other work appeared that was actually targeted toward steganography, so caution should be adopted when one is evaluating such work for watermarking purposes. Nevertheless, it is interesting to report here those efforts that, exploiting the characteristics of HDR images to hide data, inspired later work on HDR image watermarking.

The work in Cheng and Wang (2009) is the first that tackled the subject of steganography for HDR imagery. As is customary for steganography, the emphasis is put on capacity and imperceptibility at the expense of robustness. In this case, no manipulation is even expected between the message embedding and its subsequent recovery, so robustness has not been tested. The method is based on the so-called least significant bit (LSB) embedding, in which the LSB or LSBs carry the embedded message. LSB embedding can be considered a waveform-based, additive scheme dependent on the original image pixel values and is a popular choice in steganographic applications. Early watermarking techniques also proposed LSB embedding until robustness and security issues mostly barred further efforts in this direction. The embedded message recovery is blind, another common feature of steganography.

Imperceptibility is pursued through some heuristics: no explicit HVS model is used. First, 32-bit, RGBe-format pixels (Reinhard et al., 2005) are classified into flat and nonflat image classes and different embedding methods are used for each. This classification is performed by comparison of the exponent terms in neighboring pixels. Then, the number of LSBs that are modified to carry the message are computed in a way to be higher for dark and contrast image areas and smaller for bright and smooth areas. Also, RGB channels are weighted to reflect the higher sensitivity of the human eye to the red and green channels.

Security requirements are basic in this case. The (symmetric) secret key is used to encrypt and decrypt the plain text message, and a form of authentication through message digesting (eg, MD5) is also advised and implemented.

Given that this work is not concerned with robustness, the experiments focus on imperceptibility with PSNR as a metric and capacity. The tests were performed on seven standard HDR images and report values ranging from one to three bits per color channel per pixel (with a total capacity per image in the megabit range) and a PSNR above 30 dB.

The work in Li et al. (2011) is inspired by the work in Cheng and Wang (2009) and therefore embraces most of its premises. For instance, the suggested application is still steganography and, as such, in this work too, robustness is not taken into account. Security concerns are neglected as well, but one can assume that the same kind of basic requirements can be used for this scheme. Of course, if the same type of message digesting security as in Cheng and Wang (2009) is to be included, the net capacity is bound to decrease through the inclusion of security information in the embedded bits.

Again, no HVS model is considered. Instead, a simple assumption is made: to minimize the embedded information perceptibility, the total variation of each pixel's value should be minimized. The HDR format considered is logLUV TIFF (Reinhard et al., 2005), where the luminance value is floating point, and each image is first normalized to a given set of luminance exponent values before the embedding to cope with variable HDR ranges across different images.

The embedding strategy is once again based on the LSB paradigm. The information is embedded in the mantissa for luminance values and the exponent is then selected to minimize the difference between the new luminance value and the original one. The chrominance channels have a direct value representation; hence, classic LSB is used there. The best trade-off between capacity and imperceptibility is reported as 6-bit embedding for the luminance channel and 10-bit embedding for each chrominance channel.

The experiments once again consider only capacity and imperceptibility, using a testbed of 10 HDR images. Moreover, the PSNR is still chiefly used to measure imperceptibility. Li et al. (2011) reported an increase in capacity (more than doubled) and PSNR (up to 2–3 dB) with respect to the findings of Cheng and Wang (2009). However, the set of test images (10 in this case) is different, so these conclusions should be taken with caution. For a single image, HDR-VDP$_{75}$ and HDR-VDP$_{95}$ were also computed and both were under 1%.

The work in Yu et al. (2011) is again on HDR image steganography, but it considers a different set of requirements. Robustness is neglected in this work as well, and basic security is achieved by simple pseudorandom scrambling of the order of the pixels. The same assumptions as above about increasing security at the expense of net capacity still apply. Furthermore, the capacity could still be decreased when the presence of an intelligent attacker trying to detect the presence of a message is assumed because only a subset of the available pixels is used (see later). The embedded message recovery is blind.

With respect to the previously described methods, imperceptibility in this case is almost maximized, but the capacity is much lower. This method exploits the redundancy in the RGBe representation format for a given pixel, which is the fact that adding 1 to the exponent and halving the other channels (or subtracting 1 from the exponent and doubling the other channels), with some obvious extra assumptions on overflowing and underflowing and rounding, does not change the pixel value. Embedding information into pixels, therefore, consists in choosing one of these equivalent representations, once they have been sorted by means of the exponent value, using the embedded bits as index. Obviously, not all pixels admit equivalent representations, and the number of embeddable bits depends on the number of these representations.

The experiments are only concerned with capacity, as imperceptibility is all but guaranteed by this technique. Depending on the intended application, Yu et al. (2011) reported two different average capacities. In the less security critical environment of image annotation, it is in the 0.1 bpp range. When an intelligent attacker is assumed, the message is embedded only in a random, secret key driven subset of the available pixels. In particular, the selected pixels are such that altering them does not alter the statistics of the image with respect to the original image. In this case, the capacity is in the 0.001 bpp range.

The work in Wang et al. (2012) directly builds on that in Yu et al. (2011). The proposed variation is to group pixels for embedding purposes into so-called segments, instead of considering pixels one at a time, and it also uses a more sophisticated approach to scramble the pixel order to construct the segments. This mechanism makes a more effective use of the number of equivalent representations for groups of pixels and improves the capacity by just around 5%, so there is probably not much more room for improvement on the capacity side.

Following the above-mentioned articles, other watermarking methods started to appear in 2011. As already stated, the commonest requirement is robustness against tone-mapping operators, while the other requirements are variably addressed.

The work in Xue et al. (2011) discusses two approaches to implement a blind detectable (one-bit-capacity) watermarking scheme. Both techniques aim at robustness against tone mapping and do not consider security. Also, imperceptibility is not addressed explicitly but relies on the watermark embedding domain to ensure it is achieved. Both proposed techniques are based on the multiplicative spread-spectrum watermarking paradigm. Hence, correlation-based blind recovery is used in both, with a basic modicum of security provided by the secret key acting as the seed of the pseudorandom watermark.

The first technique approximates the tone-mapping process with a μ-law function applied on the HDR image, producing an LDR image and a ratio image (the original HDR image divided by its tone-mapped version). Then, the LDR image is wavelet decomposed and the watermark is embedded into the vertical and horizontal detail subbands. Last, the watermarked HDR image is obtained by multiplication of the wavelet-reconstructed watermarked LDR image by the ratio image.

The second technique applies bilateral filtering to the HDR image to obtain a large-scale part and subsequently a detail part by subtracting the large-scale part from the HDR pixel values, after having taken the logarithm of both. Then, the detail part undergoes the same processing described above: wavelet decomposition and reconstruction with the spread-spectrum watermark embedding in between. Finally, the watermarked detail part is summed to the large-scale part and, recalling we have applied the logarithm, the sum is exponentiated to obtain the watermarked HDR image.

The experiments conducted on five HDR images first address imperceptibility, giving PSNRs in excess of 50 dB for the first technique and lower values (32 dB in one image) for the second technique. Not surprisingly, the second technique is more robust against tone mapping according to a limited set of experiments, although some correlation values appear to be under the threshold, hence resulting in missed detection.

The work in Wu (2012) is a blind decodable watermarking technique that aims to embed a 4800-bit logo in HDR images. Aside from tone-mapping operators, robustness against noise addition, cropping, and blurring is sought, while security is completely neglected.

To be robust against tone mapping, a prototype tone-mapping operator is first applied to the HDR image and the watermark embedding is performed on the resulting LDR image. The HDR watermarked image is obtained again by the storing of a ratio image by which to multiply the watermarked LDR image. The LDR watermarking method is a variation of the classical spread-spectrum method in the DCT domain, where instead of direct embedding of the watermark with the additive rule of Eq. (22.1) on the DCT coefficients, the value difference between specific pairs of medium-frequency to low-frequency coefficients is modified. The watermark blind recovery is based on the computation of a correlation coefficient.

The experiments were performed on a single HDR image at a time. Imperceptibility was evaluated through the PSNR, which was reported around 70 dB. Robustness against attacks was given by the correlation coefficient, which, on average, ranges from 0.5 to 0.8 (with BERs approximately ranging from 12% to 23%), giving mostly human-readable recovered logos because the HVS tends to compensate for errors in logo images with those BERs.

The work in Solachidis et al. (2013a) is directly derived in the HDR domain. A just noticeable difference (JND) mask is obtained from the HDR image, and then it is used to embed 128 bits. Again, the target is robustness to tone-mapping operators, and a minimum of security is obtained by use of a secret key to scramble the data.

The imperceptibility is pursued by use of a mask based on a contrast sensitivity function, on bilateral filtering, and on the aforementioned JND. The embedding method is of the multiplicative spread-spectrum type, applied in the wavelet domain, and the blind recovery is based on a threshold. The mask is used to temper the embedded information depending on the mask value. Only the luminance channel is used to embed the watermark in this work as well.

The experiments were performed on three HDR images and used a set of seven tone-mapping operators. The watermark decoding can be performed on both HDR and LDR images obtained through tone mapping. The reported BER is, in the worst case, around 5%. Miss and false alarm probabilities were extrapolated under the assumption of Gaussian distributions for the bit errors, and almost negligible miss probabilities for false alarm probabilities equal to 10^{-10} are reported. No experimental evaluation on imperceptibility is reported.

The authors of Solachidis et al. (2013a) proposed two other techniques. The first, which is proposed in Solachidis et al. (2013b), is a blind detectable watermarking scheme (so the capacity is one bit) with, again, robustness against tone mapping as the main target.

Here, the original HDR image is decomposed into a set of LDR images, each representing a subset of the original dynamic range. On each LDR image, an LDR image watermarking already proposed in the literature is applied and the watermarked HDR image is then obtained by combination of the set of LDR watermarked images. The LDR watermarking scheme belongs to the additive spread-spectrum family, it works in the wavelet domain, and it uses an HVS-based mask to achieve imperceptibility. Security resides in the pseudorandom sequence to be added to the original wavelet coefficients, so the watermark cannot be read but it can be disabled by an intelligent attacker. Collusion attacks are also not considered. The recovery can be performed on both the HDR image directly and an LDR image obtained by scaling the dynamic of the HDR image.

The experiments were performed on six HDR images. The strength of the watermark was adjusted so as to have an HDR-VDP-2 under 5% for 90% of the image pixels, which is assumed as a good value for imperceptibility. The miss and false alarm probabilities are very small, when Gaussian distributions are assumed for the detection scores. The tests considered five different tone-mapping operators.

The second work (Maiorana et al., 2013) is a blind decodable watermarking method. In this case, the method used is completely different, although it also aims at being robust against tone-mapping operators.

Maiorana et al. (2013) propose considering one of the first-level wavelet subbands of the logarithm of the luminance ($\log L$) component of the HDR image, separating it into blocks, and then applying the Radon transform-DCT (Do and Vetterli, 2003), leaving the chrominance channels unmodified. Then, a QIM watermarking process is applied, hence classifying this method into the direct embedding family. The features to be quantized are the most energetic directions, so as to embed the information into the edges of the image to ensure maximum imperceptibility. The secret key controls the quantizer shift and is therefore the only security mechanism present. The capacity is the range of tens of kilobits and depends on the number of blocks present in the image. In this case as well, the blind watermark recovery can be done on both the watermarked HDR image and a tone-mapped LDR image.

Experiments were conducted seven images and five different tone-mapping operators were considered. The imperceptibility was best when the HH subband was used, where HDR-VDP-2 gives a 5% probability of detecting a modification on around 5% of the image. However, on average for the HH subband, the BER is as high as 22%.

The work in Autrusseau and Goudia (2013) also focuses on robustness against tone mapping. It is a detectable watermarking technique, so its capacity is one bit. Imperceptibility relies once again on the use of wavelet decomposition.

A nonlinear variant of the classical multiplicative spread-spectrum watermarking paradigm is applied on all the first-level subbands of a wavelet decomposition. Thus, security consists only of the seeding for pseudorandom noise, which is the watermark. The blind watermark recovery is, as usual for such watermarking methods, based on correlation between the extracted watermark and the original watermark.

With regard to robustness, considered for eight HDR images and six tone-mapping operators, no experimental values were given, but it was reported that false detection occurs more frequently than true detection for at least some combinations of the original HDR image and the tone-mapping operators. The imperceptibility was in this case also evaluated through HDR-VDP and was reported to be around 95% (except for a single image with 85%).

22.2.3 SUMMARY OF HDR IMAGE WATERMARKING SYSTEMS

In Table 22.2 we have reported a summary of the proposed HDR image watermarking systems described above. The point of the table is not to allow experimental comparisons (see the next section). Instead, here we just want to list the various approaches and point out a few characteristics that can be inferred by comparing them.

The first observation is how uneven the proposed embedding schemes are, including purely steganographic systems. In addition, some systems are of the detectable kind and others embed a variable quantity of information in the host image. This is not necessarily a bad thing, but indicates how a clear favorite application for HDR image watermarking (and a suitable requirements mix for it) has not emerged yet.

The second observation is that, besides a single work, security is mostly neglected and relies on simple strategies, such as the use of a secret key to construct the watermark sequence. While this prevents the unauthorized decoding or detection of the watermark, it does not guarantee that the watermark recovery cannot be impaired by more sophisticated attacks aimed directly at the recovery process. The latter fact, in general, forbids the deployment of the watermarking scheme in security-critical applications such as ones dealing with copyright protection.

Another couple of observations can be made with respect to how imperceptibility is handled. First, there is a discrepancy in the use of metrics to assess the perceptual distortion introduced by the watermark embedding process. Some of studies use the PSNR, which is probably unsuitable when applied to the high-quality content at hand: the breadth of work dedicated to the development of suitable perceptual metrics for HDR image quality is clear proof of this fact (see Chapter 17). Also, it is worth noting that no articles have even considered the possibility of implementing visible watermarking, but all aim to render the watermark imperceptible to retain the image content quality. In addition, most studies just rely on the watermark domain (eg, using the detail subbands of a wavelet decomposition) to achieve imperceptibility. Perceptual masks, possibly developed ad hoc for the HDR domain, can possibly provide a needed upgrade to guarantee that the watermarked image retains high visual quality.

There are some common points as well that arise from Table 22.2. For example, watermark recovery is always blind, a scenario justified for those applications in which it is assumed that the entities performing watermark embedding and recovery differ. Also, all the proposed techniques that aim to achieve robustness (ie, excluding the steganographic systems) recognize the importance of being robust against the tone-mapping operators, which constitute the most widely diffused form of processing that HDR images are currently expected to undergo.

Table 22.2 Summary of the Current State of the Art in HDR Image Watermarking

Reference	Embedding Domain and Algorithm	Recovery	Watermark Requirements			
			Capacity	Robustness	Security	Imperceptibility
Guerrini et al. (2008, 2011)	QIM (direct embedding) on the kurtosis of approximation subband (AS) coefficients in the logLUV domain (2-level wavelet decomposition)	Blind	1 bit (detectable)	7 TMOs, Gaussian noise (masked). Evaluated by ROC	Features computed in blocks of random shape and in random locations, random quantization shift	Use of a perceptual mask, based on brightness, activity, and edges. Experiments with HDR-VDP
Cheng and Wang (2009)	LSB embedding in HDR domain, RGBe format	Blind	3–9 bpp	None (steganographic)	In the clear, MD5 for message authentication	Experiments with PSNR
Li et al. (2011)	LSB embedding in HDR domain, logLUV TIFF format	Blind	26 bpp	None (steganographic)	In the clear	Experiments with HDR-VDP
Yu et al. (2011) and Wang et al. (2012)	Using redundant representation of HDR pixel values in RGBe format	Blind	0.001 bpp	None (steganographic)	Pixel scrambling	Completely imperceptible by design
Xue et al. (2011)	Bilateral filtering of HDR image, multiplicative spread spectrum in wavelet domain	Blind	1 bit (detectable)	4 TMOs. Evaluated by detection scores	Pseudorandom watermark	Experiments with PSNR
Wu (2012)	LDR domain (after TMO), variation of additive spread-spectrum watermarking applied to medium-frequency to low-frequency DCT coefficients	Blind	4800 bits	4 TMOs, noise, blurring, and cropping. Evaluated by BER	None	Experiments with PSNR

(Continued)

Table 22.2 Summary of the Current State of the Art in HDR Image Watermarking—Cont'd

| Reference | Embedding Domain and Algorithm | Recovery | Watermark Requirements | | | |
			Capacity	Robustness	Security	Imperceptibility
Solachidis et al. (2013a)	HDR domain, multiplicative spread spectrum applied to wavelet coefficients	Blind	128 bits	7 TMOs. Evaluated by BER	Pseudorandom watermark	Use of a perceptual mask, based on JND, contrast, and bilateral filtering. No assessment
Solachidis et al. (2013b)	LDR domain, obtained through bracket decomposition, additive spread-spectrum watermarking applied to wavelet coefficients	Blind	1 bit (detectable)	5 TMOs. Evaluated by the equal error rate obtained under the assumption of Gaussian error distributions	Pseudorandom watermark	Experiments with HDR-VDP-2
Maiorana et al. (2013)	QIM (direct embedding) of most energetic Radon-DCT directions, applied on the $\log L$ LDR domain	Blind	Tens of kilobits	5 TMOs. Evaluated by BER	Quantizer shift	Experiments with HDR-VDP-2
Autrusseau and Goudia (2013)	Wavelet transform of HDR pixel values, nonlinear variant of multiplicative spread spectrum	Blind	1 bit (detectable)	6 TMOs. Evaluated by BER	Pseudorandom watermark	Experiments with HDR-VDP

22.3 CONCLUDING REMARKS

In this chapter we briefly introduced the data hiding branch known as digital watermarking, with a particular focus on the still image case. Turning our attention to the HDR domain, we wished to provide a high-level overview of the current state of the art. To conclude our discussion, in this section we offer some remarks on the present status of digital watermarking applied to HDR images.

First, as can be inferred from our discussion in the preceding section, referring in particular to Table 22.2, we avoided comparing the HDR image watermarking techniques described. We did this purposefully: we wish to highlight here how difficult it is to properly conduct critical evaluations of the current state of the art for a number of reasons. First, the number of available studies is still too low to draw conclusions, suggesting that much more work is needed in the field. Second, there is a distinct lack of recognized "standard" HDR image databases, and the few that are readily available consist of a small number of images. In both these aspects, it is easy to conclude that HDR image watermarking is completely out of pace with respect to the huge amount of research conducted in the LDR image field. Instrumental to this detachment is also what we have previously pointed out: that it is, in general, not possible to directly transpose LDR image watermarking to the HDR domain.

These are not the only problems preventing a critical comparison between available work. More importantly, there is a feeling that no research group agrees on the requirements HDR image watermarking should satisfy. This is probably dictated by the complete absence of the deployment of a watermarking system in real-world scenarios. Once the need to explore watermarking in a real application will arise, it is likely that all these problems will be dealt with simultaneously and comparisons between proposed techniques will be made possible.

Speaking of watermarking requirements, looking again at Table 22.2, we find it somewhat surprising that all the proposed technique are based on the blind recovery paradigm. It is quite easy to imagine applications catering to the perceptual quality of HDR content that might make use of a nonblind recovery stage, considering the advantages in terms of robustness achievable by such a framework. For example, some company could sell its HDR images after having properly and imperceptibly watermarked them. Then, an interested party might want to verify the authenticity of the content of a tone-mapped version of one such image. To do that, it could upload that LDR image to a site controlled by the selling party, who would then extract the watermark by means of a nonblind watermark recovery system since it possesses the unwatermarked original image. The only security requirement in such an application would be the impossibility to completely remove the watermark to prevent misappropriation — simply disabling its recovery would damage the image quality and prevent its authentication. This application is perfectly viable, so we suggest that more effort should be put into nonblind watermarking. We also wish to point out that, with these assumptions in place, steganography could still arise as the killer application, and that only time will tell of this is the case.

In addition, as one can see from Table 22.2, imagining critical security-free applications such as the one described above also matches the poor attention that security has enjoyed until now in the literature. Techniques with tighter security can, of course, also proposed (eg, letting the buying party in the scenario above perform the authentication itself). Such applications would likely require additional security infrastructure besides the one provided by the watermarking system — for example, enlisting the aid of an asymmetric key cryptography framework: for example, a public key could be needed to recover the watermark, while a private key could be necessary to embed it.

As a brief note, robustness against tone-mapping operators seems to be the present focus of the proposed methods. However, it is easy to predict that HDR image watermarking will have to be robust against other types of processing as well: transcoding, as HDR image coding solidifies itself, springs to mind.

As a last note, a major absence from this chapter is the HDR video medium. In fact, to the best of our knowledge, no one has proposed an HDR video watermarking technique yet. However, given the infancy of even the still image field at the present time, that should be hardly surprising. We are quite certain that, soon after the still image case reaches an adequate level of maturity, work concerning HDR video watermarking will begin to appear.

REFERENCES

Autrusseau, F., Goudia, D., 2013. Nonlinear hybrid watermarking for high dynamic range images. In: Int. Conf. on Image Processing (ICIP), Melbourne, Australia, pp. 4527–4531.

Barni, M., Bartolini, F., 2004. Watermarking Systems Engineering: Enabling Digital Assets Security and Other Applications. CRC Press, Boca Raton, FL.

Cayre, F., Fontaine, C., Furon, T., 2005a. Watermarking security part one: theory. In: Proc. SPIE Security and Watermarking of Multimedia Contents VII, vol. 5681, pp. 746–757.

Cayre, F., Fontaine, C., Furon, T., 2005b. Watermarking security part two: practice. In: Proc. SPIE Security and Watermarking of Multimedia Contents VII, vol. 5681, pp. 758–768.

Chen, B., Wornell, G., 2001. Quantization index modulation: a class of provably good methods for digital watermarking and information embedding. IEEE Trans. Inform. Theory 47 (4), 1423–1443.

Cheng, Y.M., Wang, C.M., 2009. A novel approach to steganography in high dynamic range images. IEEE Multimedia 16 (3), 70–80.

Cox, I.J., Kilian, J., Leighton, F., Shamoon, T., 1997. Secure spread spectrum watermarking for multimedia. IEEE Trans. Image Process. 6 (12), 1673–1687.

Cox, I.J., Miller, M.L., Bloom, J.A., Fridrich, J., Kalker, T., 2008. Digital Watermarking and Steganography, second ed. Morgan Kaufmann Publishers Inc., San Francisco, CA.

Do, M.N., Vetterli, M., 2003. The finite ridgelet transform for image representation. IEEE Trans. Image Process. 12 (1), 16–28.

Guerrini, F., Okuda, M., Adami, N., Leonardi, R., 2008. High dynamic range image watermarking. In: Proc. of the Int. Conf. on Circuits/Systems, Computers and Communications, Shimonoseki, Japan, pp. 949–952.

Guerrini, F., Okuda, M., Adami, N., Leonardi, R., 2011. High dynamic range image watermarking robust against tone-mapping operators. IEEE Trans. Inf. Forensics Secur. 6 (2), 283–295.

Kerchoffs, A., 1883. La cryptographie militaire. J. Sci. Militaires IX, 5–38.

Li, M.T., Huang, N.C., Wang, C.M., 2011. A data hiding scheme for HDR images. Int. J. Innov. Comput. Inform. Control 7 (5A), 2021–2035.

Maiorana, E., Solachidis, V., Campisi, P., Lou, Y., 2013. Robust multi-bit watermarking for HDR images in the Radon-DCT domain. In: Proc. Int. Symposium on Image and Signal Processing and Analysis, Trieste, Italy, pp. 284–289.

Mantiuk, R., Daly, S., Myszkowski, M., Seidel, H.S., 2005. Predicting visible differences in high dynamic range images — model and its calibration. In: SPIE 17th Annual Symposium on Electronic Imaging, San Josè, CA, USA, vol. 5666, pp. 204–214.

Mantiuk, R., Kim, K.J., Rempel, A.G., Heidrich, W., 2011. HDR-VDP-2: a calibrated visual metric for visibility and quality predictions in all luminance conditions. ACM Trans. Graph. 30 (4), 1–14.

Menezes, A.J., van Oorschot, P.C., Vanstone, S.A., 1996. Handbook of Applied Cryptography. CRC Press, Boca Raton, FL.

Neyman, J., Pearson, E.S., 1933. On the problem of the most efficient tests of statistical hypotheses. Philos. Trans. R. Soc. Lond. 231, 289–337.

Petitcolas, F.A.P., Anderson, R.J., Kuhn, M.J., 1999. Information hiding: a survey. Proc. IEEE 87 (7), 1062–1078.

Provos, N., Honeyman, P., 2003. Hide and seek: an introduction to steganography. IEEE Secur. Priv. 1 (3), 32–44.

Reinhard, E., Ward, G., Pattanaik, S., Debevec, P., 2005. High Dynamic Range Imaging: Acquisition, Display, and Image-Based Lighting. Morgan Kaufmann Publishers Inc., San Francisco, CA.

Shannon, C.E., 1949. Communication theory of secrecy systems. Bell Syst. Tech. J. 28, 656–715.

Solachidis, V., Maiorana, E., Campisi, P., 2013a. HDR image multi-bit watermarking using bilateral-filtering-based masking. In: Proc. SPIE Image Processing: Algorithms and Systems XI, vol. 8655.

Solachidis, V., Maiorana, E., Campisi, P., Banterle, F., 2013b. HDR image watermarking based on bracketing decomposition. In: Proc. Int. Conf. on Digital Signal Processing, Santorini, Greece, pp. 1–6.

Wang, Z.H., Chang, C.C., Lin, T.Y., Lin, C.C., 2012. A novel distortion-free data hiding scheme for high dynamic range images. In: Proc. Int. Conf. on Digital Home, Guangzhou, China, pp. 33–38.

Wu, J.L., 2012. Robust watermarking framework for high dynamic range images against tone-mapping attacks. Watermarking 2, 229–242.

Xue, X., Jinno, T., Jin, X., Okuda, M., Goto, S., 2011. Watermarking for HDR image robust to tone mapping. IEICE Trans. Fund. Electron. Comm. Comput. Sci. E94-A (11), 2334–2341.

Yu, C.M., Wu, K.C., Wang, C.M., 2011. A distortion-free data hiding scheme for high dynamic range images. Displays 32 (5), 225–236.

Index

Note: Page numbers followed by *f* indicate figures and *t* indicate tables.

Printed in the United States
By Bookmasters